International Max Planck Research School (IMPRS)
for Maritime Affairs
at the University of Hamburg

Hamburg Studies on Maritime Affairs
Volume 32

More information about this series at
http://www.springer.com/series/6888

Sarah Fiona Gahlen

Civil Liability for Accidents at Sea

Springer

Sarah Fiona Gahlen
Walther Schücking Institute of International Law
University of Kiel
Kiel
Germany

Dissertation zur Erlangung der Doktorwürde an der Fakultät für Rechtswissenschaft der Universität Hamburg
vorgelegt von Sarah Fiona Gahlen
Erstgutachter: Prof. Dr. Rainer Lagoni, LL.M. (Colombia)
Zweitgutachter: Prof. Dr. Dr. h. c. Marian Paschke
Tag der mündlichen Prüfung: 28.07.2014

ISSN 1614-2462 ISSN 1867-9587 (electronic)
ISBN 978-3-662-45554-8 ISBN 978-3-662-45555-5 (eBook)
DOI 10.1007/978-3-662-45555-5
Springer Heidelberg New York Dordrecht London

Library of Congress Control Number: 2015932252

Printed on acid-free paper

Springer is part of Springer Science+Business Media (www.springer.com)

Acknowledgements

This book is the result of a doctoral thesis undertaken at the University of Hamburg and the Max Planck Institute for Comparative and International Private Law, supported by a scholarship from the International Max Planck Research School of Maritime Affairs. I therefore wish to express my gratitude to all persons within these institutions who made this project possible.

First and foremost, I am indebted to my supervisor, Professor Dr. Rainer Lagoni, whose constant advice on the topic and related fields of research has been of indefinite value to me. My sincere thanks also go to my friends and colleagues at the Research School and the Institute, namely to Solène Guggisberg, Lina Lumetzberger, Young-Kyung Yoon, Victor Chacón, Lief Bleyen and Tess Chemnitzer, who witnessed the progress of this work and made my time at the Institute so much more enjoyable. Barbara Schröder helped me with many practical aspects of this project, and Michael Friedman provided a lot of assistance in the final editing.

Parts of this study have greatly benefited from a stay at the Rotterdam Institute for Shipping and Transport Law at the Erasmus University Rotterdam. I therefore wish to thank Professor Frank Smeele, who initiated my visit, and everyone else at the Institute and the Erasmus School of Law for an insightful and very pleasant time in Rotterdam.

Last but not least, I want to thank my friends in Berlin, Hamburg and elsewhere for their support and encouragement. Mentioning everyone would render this paragraph a very long one—I trust that all of my friends know that they contributed, in many different ways, to the achievement of this work. Finally, I want to express my gratitude and my love to my parents, who have played a very unique role in supporting this venture.

Hamburg, Germany
September 2014

Sarah Fiona Gahlen

Abbreviations

BGBl.	Bundesgesetzblatt
BOPC	International Convention on Civil Liability for Bunker Oil Pollution Damage, 2001
CLC	International Convention on Civil Liability for Oil Pollution Damage, 1969
CMI	Comité Maritime International
ECJ	European Court of Justice
EEZ	Exclusive Economic Zone
EFTA	European Free Trade Association
EGBGB	Einführungsgesetz zum Bürgerlichen Gesetzbuch
EMSA	European Maritime Safety Agency
EU	European Union
FUND	International Convention on the Establishment of an International Fund for Compensation for Oil Pollution Damage, 1971, as amended
HNS	Hazardous and Noxious Substances
HNSC	International Convention on Liability and Compensation for Damage in Connection with the Carriage of Hazardous and Noxious Substances by Sea, 1996
ILM	International Legal Materials
ILO	International Labour Organization
IMCO	Inter-Governmental Maritime Consultative Organization (original name of IMO)
IMDGC	International Maritime Dangerous Goods Code
IMO	International Maritime Organization
IOPC Funds	International Oil Pollution Compensation Funds
ISM Code	International Safety Management Code
ITLOS	International Tribunal for the Law of the Sea

LLMC	International Convention on the Limitation of Liability for Maritime Claims, 1976
MARPOL	International Convention on the Prevention of Pollution from Ships 1973, as amended by the 1978 Protocol
OILPOL	International Convention for the Prevention of Pollution of the Sea by Oil, 1954
OMI	Organisation Maritime Internationale
PAL	Athens Convention relating to the Carriage of Passengers and their Luggage by Sea, 1974
RGBl.	Reichsgesetzblatt
s.	Section
SAR	International Convention on Maritime Search and Rescue, 1979
SDR	Special Drawing Right
SNDP	Substances nocives et potentiellement dangereuses
SOLAS	International Convention for the Safety of Life at Sea, 1974 as amended
UKTS	United Kingdom Treaty Series
UNCLOS	United Nations Convention of the Law of the Seas, 1982
UNTS	United Nations Treaty Series
US	United States
USC	United States Code
WRC	Nairobi International Convention on the Removal of Wrecks, 2007

Contents

Chapter 1
Introduction

Shipping and the maritime offshore industry are probably the most international businesses in the world. There are hardly any cases that would not affect parties in more than one country, and the operation of ships and facilities at sea can furthermore affect very different interests in a wide variety of relationships, such as economic rights, the interest in a safe and clean environment or the interest in life and health.

Shipping is, in absolute terms, a very safe business. Given the number of maritime transports effectuated every day, and given the number of vessels currently at sea, the absolute number of accidents is very small. Shipping is by far the most effective, and thus, the most environment-friendly form of transport. Still, the maritime enterprise is a large-scale business, and the sea does not have many borders that would confine adverse consequences to one region only. Unforeseen incidents at sea can cause disastrous damages, affecting the wide range of interests in a geographically large area.

This study deals with the civil liability for accidents at sea, in a broad sense and with a focus on the interests of parties that are not contractually participating in the maritime enterprise. It is obvious that parties directly involved in a shipment will have contractual relationships with each other, providing for means of compensation in the case of an accident. In general, it is rather easy to assess the value of the interests at stake in these contractual cases, according to the cost of the shipment, the value of a vessel or its cargo, and, at the end of the day, according to the price that the parties themselves have agreed on. Limitation of liability within transport agreements is of course also an important issue here, and there are many international conventions dealing with this matter. Still, liability and compensation in these contractual agreements are questions that arise between parties who are aware of their undertaking and of the risks involved.

Maritime accidents can, however, also have important consequences beyond the interests of parties contractually bound to the maritime enterprise, and these are the legal relationships that this study focuses on. The most important head of damage in this respect is of course the compensation for environmental pollution, for damage

S.F. Gahlen, *Civil Liability for Accidents at Sea*, Hamburg Studies on Maritime Affairs 32, DOI 10.1007/978-3-662-45555-5_1

done to the interests of States and coastal communities that are only affected by a maritime accident because of a rather fortuitous geographical link. Still, the damage caused to their interests, the costs of dealing with the pollution and the ensuing economic losses, can be many times higher than the value of a given shipment or the vessel as a whole. Similarly, even if passengers or crewmembers are contractually bound by a title of transport or a maritime labour contract, the contractual agreement is mainly concerned with the transport or the working relationship, and heads of damage such as death or personal injury are not the main subject matter of the contract, even where the agreement sets out rules as to them. In legal terms, this study deals first and foremost with tort relationships, even if contractual agreements between the parties or between other parties in the maritime enterprise may have an impact on them.

These relationships, which have been caused by an unforeseen event and thus link parties by virtue of an actual incident, are rather hard to administer and to adjudicate where they take place at sea. The oceans present a part of the planet where State powers are restricted in many ways. On the one hand, State jurisdiction may be reduced by virtue of public international law, the law of the sea setting out State sovereignty and States' sovereign rights according to a carefully drafted system of maritime zones. On the other hand, the exercise of State jurisdiction at sea, even if there are legitimate sovereign rights, may be difficult for practical reasons, as activities at sea are still hard to survey and State jurisdiction even harder to enforce. The quest for civil liability for accidents at sea thus presents a good example of the difficulties of practical ocean governance.

Accidents at sea are undesired events with adverse consequences and, in general, do not benefit anyone. States have become aware of the need to avoid accidents at an early point in time, through the actual occurrence of disasters at sea. It also became obvious that in order to have effective regulations that will prevent accidents for all ships and in all geographic regions, international rules will be necessary. It is interesting that within the various international maritime agreements, it is the avoidance of accidents and the avoidance of their undesirable consequences which is the topic most commonly addressed, albeit in many cases indirectly or only with regard to a given aspect.

The first part of the study (Chap. 2) will thus inquire into the international maritime regulation that may be applicable to accidents at sea, trying to analyse the way in which this international problem is defined and tackled. This will entail an overview over the applicable preventive regulation, even if the focus of this study is the civil liability for accidents and, hence, questions of responsibility and compensation.

The second part (Chaps. 3–5) deals with questions of civil liability that have been addressed by international regulation. Regarding the liability for environmental pollution, there are a number of international agreements on this, and they are of the utmost importance for the question of liability. This study discusses first and foremost the 1969 International Convention on Civil Liability for Oil Pollution Damage (CLC), which has been a model and an inspiration for all other civil liability conventions for environmental pollution. Interestingly, it has also been used as a model for the international agreements on civil liability for passenger claims, despite the fact that the situations of coastal communities aggrieved by environmental pollution and of passengers injured in a maritime accident are very different at first glimpse. Still, the civil liability systems share many characteristics, which is also revealing of their place in the international civil liability framework for maritime incidents.

An analysis of the civil liability regimes would hardly be comprehensive without a thorough analysis of their scope of application and a discussion of the limits that there are. The accidents and cases in which the constraints of the relevant international agreements have been put to the test are often more revealing of the role of civil liability regimes than the cases in which the conventions' provisions have simply been applied. Of course, outside the framework of international harmonisation, claims are governed by the domestic law of interested States. Certain problems of jurisdiction and private international law must therefore already be discussed preliminarily in the chapters on civil responsibility regimes, whereas a more comprehensive and structured view of these matters is given in the third part of the study (Chaps. 6 and 7).

From a wider perspective, beyond a mere discussion of the provisions and principles governing civil liability for accidents at sea, this book also addresses the interplay between public international law and the law of the sea, the international agreements that essentially set out private law provisions, and the principles of domestic law that must be used in the absence of international law. The quest for the rules applicable to relationships that are not governed by a previous agreement between the parties—and in a geographical realm where States' territorial claims are reduced or inexistent—is meant to highlight one aspect of the wider question of global governance.

Part I
Maritime Accidents

Chapter 2
Maritime Accidents from a Legal Perspective

The purpose of this study is to provide an overview of the legal problems arising from maritime accidents in international law, from a civil law perspective but with the imperatives of public international law in mind. Shipping is an ancient business, and the hazards of maritime traffic have been recognised and dealt with for millennia now,[1] the dangers at that time having been considerably more important than today. The first rules on losses at sea and responsibility for them that have been recorded date back to antiquity. However, and notwithstanding the important developments in technology, maritime disasters continue to be a threat and a concern to modern societies and will also be inevitable in the future.[2] There has always been maritime trade, and there have always been unforeseen events.

This chapter of the study analyses the extent to which accidents involving ships have been subject to legislative regulation and the different approaches that have been adopted in this context. As undesirable and unforeseeable events, accidents are of course normally not directly subject to regulation—norms obliging maritime actors not to have accidents are hardly conceivable. It is in the context of avoiding accidents and of dealing with their consequences that the regulation takes place. Thus, a definition of what a maritime accident is must necessarily have regard to the purpose for which these rules are developed.

In this context, it may be hard to determine whether an occurrence should be regarded as an accident or not. It is easily conceivable that the grounding or sinking of a passenger ship or an oil tanker, resulting in loss of life or personal injury and considerable damage to the environment, with the ship being destroyed beyond repair, is an accident. The decision is harder to take with regard to a single container lost at sea—or deliberately dumped in order to ease the ship in bad weather—considering that shipping is a large scale business in which partial loss of cargo is a

[1] See J. A. Weiss, 'Maritime Disasters Through the Ages' (2001) 32, Journal of Maritime Law and Commerce, 215–40.

[2] See C. Schofield, 'Iceberg(s) Dead Ahead?' (2012) 8, Internationale Zeitschrift, 8–15.

S.F. Gahlen, *Civil Liability for Accidents at Sea*, Hamburg Studies on Maritime Affairs 32, DOI 10.1007/978-3-662-45555-5_2

common, yet undesirable, occurrence.[3] The role of intent, for instance in sacrificing a part of the cargo to save the ship, or even in sacrificing the vessel as such to bring the crew to safety, also needs to be taken into account.

This study adopts a rather broad approach as to the meaning of "accident", also including occurrences that are not in themselves disastrous. It will, however, be seen that regulation is of course effectuated to a certain end, and that accidents that do not cause the problem that is addressed by the regulation are left out of the legislative picture.

A. Maritime Accidents and Marine Pollution

Today, one of the major concerns related to maritime accidents is certainly the risk of environmental damage on a large scale, caused by the spill of bunker oil, cargo oil or other toxic substances. Still, regulation of maritime pollution first took place with regard to other practices of the shipping industry. It is only comparatively late that maritime accidents were recognised as a major source of contamination.

Even if environmental protection as such is a comparatively recent concern of the industrial world, the impact of shipping on the marine environment was recognised as early as 1954, when the government of the United Kingdom convened a conference meant to introduce measures against the deliberate discharging of oil and oily residues into the territorial sea of States. At that time, washing a ship's tanks with water and pumping the mixture of oil and water into the sea was a common shipboard operation,[4] and "waste disposal" of this kind was, outside the territorial sea of coastal States, actually considered as being covered by the freedom of the high seas.[5] Considering the growing size of tankers since the end of World War II, and the rise in oil transports due to the increasing number of motor cars worldwide, this practice started to raise concerns in some States,[6] namely the United Kingdom.[7] Although most of the negotiating States only reluctantly

[3] The available information as to the number of containers lost at sea greatly differs, with estimations sometimes rising up to 10,000 per year. The World Shipping Council, after surveying its members, claims that on average, only about 675 containers are lost at sea every year, including catastrophic events with losses of 50 containers or more; http://www.worldshipping.org/industry-issues/safety/Containers_Overboard__Final.pdf (last visited 6 March 2014).

[4] Z. O. Özçayir, *Liability for oil pollution and collisions* (London: LLP, 1998), p. 172; R. R. Churchill and A. V. Lowe, *The Law of the Sea*, 3rd edn. (Yonkers (NY), Manchester: Manchester University Press, 1999), p. 329.

[5] W. K. Bissell, 'Intervention on the High Seas: an American Approach Employing Community Standards' (1976) 7, Journal of Maritime Law and Commerce, 718–35 at 724.

[6] T. A. Mensah, 'Prevention of Marine Pollution: The Contribution of IMO', in J. Basedow and U. Magnus (eds.), *Pollution of the Sea: Prevention and compensation* (Berlin, New York: Springer, 2007), p. 43.

[7] Özçayir, *1998 "Liability for oil pollution"*, p. 172.

endorsed this initiative, the 1954 International Convention for the Prevention of Pollution of the Sea by Oil (OILPOL)[8] entered into force in 1958. It prohibited the discharging of oil or oily mixtures within the "prohibited zones" extending 50 miles from the mainland and compelled contracting States to enforce these rules. Furthermore, it encouraged coastal States to provide reception facilities in their ports. It was the first international treaty having environmental protection as its sole purpose.[9]

1958 also marked the first United Nations Conference on the Law of the Seas,[10] held in Geneva, which resulted in the adoption of the four Geneva Conventions: the Convention on the Territorial Sea and the Contiguous Zone, the Convention on the High Seas, the Convention on the Continental Shelf, and the Convention on Fishing and Conservation of the Living Resources of the High Seas. The first three conventions mirrored to a wide extent the customary international law as it had been established at that time and were ratified by an important number of States.[11] The Geneva Conference and these four conventions thus present a first, but important step in the codification and progressive development of the law of the seas, and they continue to be applicable to the States parties that did not ratify the 1982 Law of the Sea Convention (UNCLOS), such as the United States.

The 1958 Geneva Convention on the High Seas[12] did have provisions on marine pollution, but, here again, regard was only given to deliberate pollution. Art. 25 of the Convention called upon contracting States to take measures against the dumping of radioactive waste, and art. 24 dealt with the prevention of marine oil contamination by "discharge" from ships or pipelines. Even though art. 24 also refers to "pollution resulting from the exploration and exploration of the seabed and its subsoil", given that this wording does not refer as clearly to deliberate pollution as the first part of the sentence, it is fairly clear that maritime accidents were not yet recognised as a major source of marine pollution in the beginning of the codification process of the law of the seas.

It is remarkable that one single accident was enough to change this perception. In March 1967, the Liberian registered oil tanker *Torrey Canyon*, laden with 117,000 tonnes of crude oil, grounded on rocks off the Cornwall coast, just outside the British territorial sea which at that time only stretched 3 miles from the baselines. Following a puncture of the hull, the ship almost immediately released some 30,000 tonnes of its cargo into the sea. Unsuccessful attempts to rescue the ship resulted in a further loss of 30,000 tonnes of oil, the pollutant drifting towards both the British and French coasts. Eventually the British authorities considered that further salvage attempts would be useless and ordered the Naval and Air Forces to bomb the vessel

---

[8] 327 UNTS 3; BGBl. 1956 II, 379.

[9] Özçayir, *1998 "Liability for oil pollution"*, p. 172.

[10] See Churchill and Lowe, *1999 "The Law of the Sea"*, p. 15 for a historic outline.

[11] Ibid., p. 15.

[12] 450 UNTS 11; BGBl. 1972 II, 1089.

in order to set its remaining cargo on fire and avoid further damage.[13] The accident, tragic in its environmental[14] and economic[15] consequences, drew the attention of the international community to a number of open questions relating to maritime incidents of this kind. The two most important issues,[16] both of a legal character, were expressly identified by the Council of IMCO in an extraordinary session in May 1967: first, the question of whether a coastal State was entitled to take measures against vessels posing pollution threats to its coastlines and to what extent; and second, the very crucial question of responsibility and compensation for marine pollution damage. Both issues were addressed in discussions and negotiations,[17] eventually resulting in the adoption of two international treaties, which thus owe their existence more or less directly to the *Torrey Canyon* accident.

I. The 1969 Intervention Convention

The first convention influenced by the *Torrey Canyon* disaster was the 1969 International Convention Relating to Intervention on the High Seas in cases of Oil Pollution Casualties[18] (1969 Intervention Convention). It stipulated in art. I that every contracting State was entitled to take measures on the high seas in order to prevent or eliminate danger from oil pollution to their coastline or related interests following a "maritime casualty" or acts related to it. A 1973 Protocol,[19] which entered into force in 1983, extends the field of application to pollution by substances other than oil.

The term "maritime casualty" is defined in art. II, para. 1 as:

> a collision of ships, stranding or other incident of navigation, or other occurrence on board a ship or external to it resulting in material damage or imminent threat of material damage to a ship or cargo.

---

[13] K. Hakapää, *Marine pollution in international law: Material obligations and jurisdiction* (Helsinki: Suomalainen Tiedeakatemia, 1981), p. 264; A. T. Falkanger, *Maritime Casualties and Intervention - coastal state measures when casualties pose the threat of pollution* (Bergen: Fagbokforlaget, 2011), p. 20.

[14] V. P. Nanda, 'The "Torrey Canyon" Disaster: Some Legal Aspects' (1967) 44, Denver Law Journal, 400–25 at 400; A. Hovanesian, 'Post "Torrey Canyon": Toward a New Solution to the Problem of Traumatic Oil Spillage' (1970) 2, Connecticut Law Review, 632–47 at 633.

[15] Ibid., p. 633; Falkanger, *2011 "Maritime Casualties"*, p. 21.

[16] Hakapää, *1981 "Marine pollution"*, p. 265; Mensah, *2007 "Prevention"*, p. 403.

[17] See E. D. Brown, 'The Lessons of the "Torrey Canyon"' (1968) 21, Current Legal Problems, 113–36 at 114 et seq.

[18] 970 UNTS 211; BGBl. 1975 II, 137.

[19] 1313 UNTS 3; BGBl. 1985 II, 593.

1. The Requirement of Material Damage or Threat Thereto

The definition given in art. II, para. 1 of the Intervention Convention appears to be of a very wide character at first glance. A "collision of ships, stranding or other incident of navigation, or other occurrence on board a ship or external to it" seems to cover every possible event in relation to a ship outside the normal onboard routine. Still, the wording is open to interpretation considering the requirement of material damage or threat thereto in the first three situations "collision of ships, stranding or other incident of navigation".[20] The last part of the sentence "resulting in material damage" can be read as either referring to all the four alternative situations "collision, stranding, other incident of navigation or other occurrence" or only referring to the last one. This latter interpretation would mean that situations of collision, stranding or other incidents of navigation would be covered under the definition even if they did not result in actual or threatening damage.

It must be kept in mind that the Intervention Convention can only be invoked by States where there is a "grave and imminent danger to their coastline or related interests from pollution or threat of pollution of the sea by oil" (art. I para. 1) and that therefore, the number of cases where there is no perceptible damage to the ship and yet a danger to the coastline are limited, but not inconceivable.

It has been maintained that collisions and stranding will naturally entail damage to the ship and that the question can therefore only come into play for the "other incidents of navigation",[21] but this point of view depends heavily on the definition of "collision" or "stranding"—there can be collisions in which there is no actual contact between the vessels in question,[22] and depending on the circumstances of stranding, there may be exceptional situations in which the contact with the seabed does not result in substantial damage to the ship's bottom.

An example of an "incident of navigation" other than a collision or a stranding and not causing damage to the ship could consist in the "escape of oil by negligence",[23] meaning oily ballast, not cargo oil, or a mere alteration of a ship's course,[24] provided that this would really put the coastal State's interests at risk

---

[20] Hakapää, *1981 "Marine pollution"*, p. 266; Falkanger, *2011 "Maritime Casualties"*, p. 133.

[21] Hakapää, *1981 "Marine pollution"*, p. 266; Falkanger, *2011 "Maritime Casualties"*, p. 133, the latter on the one hand admitting that there can be collisions without physical contact between vessels, "but then an oil spill will rarely occur".

[22] Under the English law definition, it will be enough that one vessel forces the other to alter her course, English law putting the stress on the actionable fault (of navigation, or other) of one party, but requiring a damage as a cause of action in negligence, see R. G. Marsden, S. Gault, S. J. Hazelwood and A. M. Tettenborn, *Marsden on Collisions at Sea*, 13th edn. (London: Sweet & Maxwell, 2003), paras 1-20 and 4-04.

[23] Hakapää, *1981 "Marine pollution"*, p. 266, quoting *Scheffer*, the Head of the Netherlands Delegation to the Brussels Conference negotiating the Intervention Convention.

[24] Statement by the Irish Delegate *Quigley*, *Official records of the international Legal Conference on Marine Pollution Damage, 1969* (London, 1973), p. 318, giving an example of a situation that should not be covered by the definition in his opinion.

(art. I para. 1). It has been submitted, following a careful analysis of the provision, the wording of corresponding domestic laws and the coming into being of the provision in the Convention's *travaux préparatoires*, that this latter interpretation would render the Convention most effective and should thus be given preference.[25] This opinion is in line with the Canadian delegation's point of view maintained in the negotiations, that the Convention should be concerned with the danger to the coastline and the related interests only, regardless of the fate of the ship and the cargo.[26] On the other hand, the weaknesses of this interpretation from a rather pragmatic point of view can be put forward, pointing to the fact that the other conditions for intervention laid down in the convention "tend to exclude" measures against a vessel not experiencing "physical danger" itself.[27] It is also true that several authors seem to take for granted that the incident must result in a damage or a threat thereto[28] before the coastal State can intervene. The Soviet Delegation in the negotiations of art. 221 of the United Nations Law of the Sea Convention (UNCLOS)[29] was similarly undifferentiating when they defined the right granted in the 1969 Intervention Convention as restricted to vessels that had suffered damage.[30]

Considering the few and unconvincing examples of incidents of navigation threatening the coastline without putting the ship in actual danger, this attitude of not according too much thought to the question does not seem too surprising. Still, if it is considered carefully, the criterion of actual or threatening damage to the vessel or cargo is unsuitable for another reason—there may be damages on board a ship that do not directly endanger the ship's seaworthiness or its structural integrity or ability to perform[31] and that the coastal State would not need to be concerned with. It seems that actual or threatening damage to the ship in an "incident of navigation" is a rather void criterion and entails questions of degree that are hard to assess—the Canadian Delegate to the 1969 Conference was certainly right when he pointed out that there might be a problem with the burden of proof if damage to the ship was to be a decisive criterion to establish an "incident of navigation" as a "maritime casualty".[32]

---

[25] Falkanger, *2011 "Maritime Casualties"*, p. 139.

[26] *1973 "Official Records (CLC)"*, p. 314, also referring to the burden of proof for the damage to the ship or cargo that should not be borne by the coastal State.

[27] Hakapää, *1981 "Marine pollution"*, p. 266.

[28] Bissell, '1976 "Intervention"', 729, 730; Churchill and Lowe, *1999 "The Law of the Sea"*, p. 354; C. M. de la Rue and C. B. Anderson, *Shipping and the environment: Law and practice*, 2nd edn. (London: Informa Law, 2009), p. 900.

[29] 1833 UNTS 397; BGBl. 1994 II, 1798.

[30] See R. Platzöder, *Third United Nations Conference on the Law of the Sea: Documents, Volume XI* (Dobbs Ferry (NY): Oceana Publications, 1987), p. 358.

[31] Falkanger, *2011 "Maritime Casualties"*, p. 149.

[32] *1973 "Official Records (CLC)"*, p. 314.

The underlying discussion and the examples given of a mere alteration of a ship's course or the negligent discharge of a small quantity of waste oil reveal the core of the problem. A coastal State could feel threatened by the discharge or by a tanker altering its course and probably coming closer to the coast or passing through shallower waters than usual or than done by other ships,[33] and a State will certainly feel threatened by a collision or a stranding or other unforeseen events involving a tanker off its coast, even if the vessel in question did not—yet—suffer actual damage. Especially in chains of events, it will be very hard to assess whether the casualty has already taken place or whether there is only a significant likelihood of its coming about.[34] In any unforeseen event in shipping, there is naturally a certain probability of some kind of damage to the ship or cargo. Setting aside the requirement of actual or threatening damage in the first three situations, which already feature the outward appearances of a maritime accident,[35] only means giving a bigger discretion to the coastal State in judging the "grave and imminent" danger to its coastline—the latter being an indispensable condition for an intervention. The intervening State would thus only need to justify its assumption of danger to its coastline following the "incident of navigation" and would not need to be concerned with an assessment of the risks on board the ship. Especially in the cases of alteration of course, bringing the ship in shallower or more dangerous waters, it seems appropriate to grant the coastal State the right to intervention without having first to inquire into the reasons for it. As the case of the *Costa Concordia* has tragically shown, the cruise ship having run onto a rock and foundered with more than 4,200 people and about 2,400 tonnes of heavy fuel oil on board, a decision to deviate from a safe and established course can be taken deliberately and arbitrarily, outside of any emergency situation.

The same right to intervene without considering the events on board should be granted to the coastal State in the case of the negligent escape of oil. If the discharge is of consequence, to the degree that the coastal State can invoke the "grave and imminent danger to its coastline" (art. I para. 1 Intervention Convention), then it seems against the purposes of the Convention to require the State to establish a damage to the ship beforehand. With regard to the indispensable criterion of the danger to the coastline, set out by art. I para. 1 of the Convention, it seems impossible that an intervening State could invoke the "incident of navigation", even if damage does not need to be established, without good reason and outside of situations that the Convention is meant to cover. The setting aside of the requirement of damage thus does not widen the field of application of the Convention in a way that would be irreconcilable with the balance between the freedom of

[33] Hakapää, *1981 "Marine pollution"*, p. 267.

[34] Bissell, '1976 "Intervention"', 729; Falkanger, *2011 "Maritime Casualties"*, p. 131, giving a very descriptive example.

[35] Ibid., p. 132.

navigation and the coastal State's interests. Therefore, there should be no criterion of actual or threatening damage in the "incident of navigation".

For the sake of completeness, it should be added that actual or threatening damage to the ship does not need to be established either in collisions or strandings, already featuring the outward appearances of a maritime accident. The criterion of "material damage or imminent threat of material damage to a ship or cargo" only refers to "other occurrences on board a ship or external to it", the latter part of the sentence thus describing circumstances that do not appear like a maritime accident at first sight but threaten or disturb the ship's operational safety.

Thus, the Intervention Convention features a rather clear and unambiguous definition of the term "maritime casualty", even in exceptional cases. Still, it is a very wide definition, covering almost every anomalous event in relation to a ship. Considering that the field of application of the Intervention Convention is only opened if the coastal State can plead a "grave and imminent danger" of oil pollution to its coastline, the definition is workable in this context, because the respective events thus need to have consequences adverse enough to raise a coastal State's concern.

Still, it is not certain if art. II para. 1 of the Intervention Convention gives a definition of the term "maritime casualty" that could also be applied in other fields of maritime law. Following the wording of the provision ("for the purposes of the present Convention"), the understanding of the notion is closely linked to the purposes of the text and cannot be conferred to other rules of international law.

For the sake of clarity, it should be briefly recalled that—both among the contemporary[36] and modern[37] authors—the Intervention Convention is mostly, but of course not uncontestedly, seen not as creating new rights to intervention for the contracting States, but rather as codifying existing public international law in a spirit of clarification and certainty. It is maintained that beyond the field of application of the Intervention Convention, States can invoke customary law principles such as necessity[38] or, even if is more doubtful, self-defence[39] to intervene in situations threatening their vital interests, including cases in which an actual maritime casualty cannot or not yet be identified.[40]

---

[36] Brown, '1968 "The Lessons"', 124; D. M. O'Connell, 'Reflections on Brussels: IMCO and the 1969 Pollution Conventions' (1970) 3, Cornell International Law Journal, 161–88 at 177; Bissell, '1976 "Intervention"', 729.

[37] de la Rue and Anderson, *2009 "Shipping and the environment*", p. 899.

[38] Considered by Brown, '1968 "The Lessons"', 127 et seq.

[39] See Ibid., p. 126, denying that the principle of self-defence could have applied in the *Torrey Canyon* incident, self-defence generally requiring an unlawful act of another subject of international law.

[40] Falkanger, *2011 "Maritime Casualties*", p. 131.

2. The Exclusion of Intent

There is a far-reaching consensus among authors analysing the Intervention Convention that there can only be a "maritime casualty" where the reasons for the incident were unforeseeable and that deliberate pollution is not covered by the term.[41]

This may partly be due to the fact that deliberate pollution had been subject to international regulation before, namely, by the 1954 OILPOL Convention mentioned above. Shortly after the final drafting of the Intervention Convention, illegal intentional dumping was addressed by the 1972 Convention on Prevention of Marine Pollution by Dumping of Wastes and Other Matter (London Convention),[42] thus clearly continuing to separate the issue of deliberate and conscious pollution[43] from the accidental contamination covered by the Intervention Convention.

It is interesting to note in this context that under the 1954 OILPOL Convention, the prohibitions against discharging oil at sea was applicable neither to ships in distress which only did so in order to secure the safety of the vessel, crew or cargo (art. IV para. 1(a)), nor to accidental losses following damages to the ship or leakages (art. IV para. 1 (b)). Similar provisions can be found in the 1972 London Convention, excluding disposal of matter for other reasons than the actual disposal from the field of its application (art. III para. 1(b) ii)) and permitting in its art. VIII the dumping in emergency situations. Thus, the accidental loss or even the deliberate disposal of cargo in order to ease the ship or otherwise ensure its safety would not be covered.

a) The Differentiation According to MARPOL

In 1973, IMO took a new initiative for a convention concerning operational pollution, desiring to ban deliberate and operational contamination by pollutants other than oil. On 2 November, IMO adopted the International Convention on the Prevention of Pollution from Ships (MARPOL).[44] The Convention deals in its Annex I with oil pollution, incorporating most of the 1954 OILPOL rules. The other annexes concern regulations for the control of pollution by noxious liquid substances in bulk (annex II), the prevention of pollution by harmful substances carried in packaged form (annex III), by sewage (annex IV), by garbage from ships (annex V) and prevention of air pollution by ships (annex VI). MARPOL 1973 was

---

[41] Özçayir, *1998 "Liability for oil pollution"*, p. 176; P. W. Birnie, A. E. Boyle and C. Redgwell, *International law and the environment*, 3rd edn. (Oxford, New York: Oxford University Press, 2009), p. 426; Falkanger, *2011 "Maritime Casualties"*, p. 131.

[42] 1046 UNTS 120; BGBl. 1977 II 165.

[43] See Özçayir, *1998 "Liability for oil pollution"*, pp. 177 et seqq. for an overview of the London Convention.

[44] 1340 UNTS 184; BGBl. 1982 II 2.

originally meant to deal with deliberate pollution, regardless of the pollutant, other than dumping.[45] In 1978 and following a series of major tanker accidents, Annex I was altered by a protocol, requiring *inter alia* new oil tankers to have double hulls. Thus, provisions meant to reduce the risk of accidental discharges were introduced in the MARPOL legal framework. As MARPOL 1973 had not yet entered into force before the adoption of the protocol, the latter absorbed the parent convention and is now referred to as MARPOL 1973/78. The Convention therefore deals with both operational and accidental discharges of harmful substances[46] and is thus, at first glance, the first IMO instrument not to distinguish openly between deliberate and accidental pollution.

MARPOL's main goal is on the one hand to set out technical standards vessels have to comply with so that the risk of accidental discharges is minimised, and on the other hand, to set out requirements for the conduct of surveys and certification to ensure that operational discharges are only effectuated within certain limits.[47] MARPOL thus presents a framework of technical standards. It therefore features a slightly different approach, compared to the preceding instruments on marine pollution, being concerned with the ship's equipment and management in the first place, whereas 1954 OILPOL and the 1972 London Convention tried to intervene in the standard shipboard procedures at that time and thus tried to regulate human behaviour rather than technical questions. It is apparently MARPOL's goal not only to prohibit certain procedures such as discharges, but also to prescribe the technical facilities that render them unnecessary—such as oily water separators etc.—and also to facilitate monitoring of potentially harmful circumstances by certification and recording duties.

Considering the rules on accidents, the annex I exempts discharges of oil from its field of application when they are necessary for the safety of the ship or saving life at sea (Reg. 4.1), in almost the same way the 1954 OILPOL Convention did. The same applies to emergency discharges of noxious liquid substances in Annex II (Reg. 3). Thus, MARPOL still does not give up the separation between deliberate discharge for the sake of getting rid of the pollutants and the emergency dumping or discharge, the latter not being covered by the Convention.

Even if MARPOL's provision are significantly more far-reaching and technical than the OILPOL and the London Convention rules, and even if MARPOL obviously recognises accidents as a source of marine pollution, the Convention does not directly regulate casualties and exempts typical consequences of maritime accidents from its field of application. It thus seems that MARPOL continues to pursue the clear distinction between deliberate pollution—clearly covered by it—and

[45] Ibid., p. 181.

[46] Ibid., p. 182; Faure M. and Ying S. (eds.), *China and International Environmental Liability: Legal Remedies for Transboundary Pollution* (Cheltenham: Edward Elgar, 2008), p. 71; also see the IMO portrayal under www.imo.org/about/conventions/listofconventions/pages/international-convention-for-the-prevention-of-pollution-from-ships-(marpol).aspx (last visited 6 March 2014).

[47] See de la Rue and Anderson, *2009 "Shipping and the environment"*, p. 821 et seqq. for a comprehensive overview of the convention's terms.

accidental pollution, exempted from the field of application once it has occurred, and probably leaving it to other legal instruments such as the Intervention Convention.

This view is also reflected in Art. 12 MARPOL on investigation into casualties, one of the rare provisions in which the term "casualty" is actually used. The provision reads:

> Art. 12
> Casualties to ships
> (1) Each Administration undertakes to conduct an investigation of any casualty occurring to any of its ships subject to the provisions of the Regulations if such casualty has produced a major deleterious effect upon the marine environment.
> (2) Each Party undertakes to supply the Organization with information concerning the findings of such investigation, when it judges that such information may assist in determining what changes in the present Convention might be desirable.

It will be seen at a later point in this study that the duty to investigate casualties in order to ameliorate the existing legal framework is not genuine to MARPOL and was first codified with a view to improve the safety of life at sea. The current IMO guidelines on how an investigation under MARPOL is to be carried out actually also refer to the guidelines[48] adopted for this duty under the International Convention for the Safety of Life at Sea 1974 (SOLAS).[49] It will be seen *infra* under B. that SOLAS features a slightly different approach to the notion of maritime casualties.

Still, despite the fact that the MARPOL provision closely mirrors the wording of the corresponding SOLAS provision, MARPOL does seem to maintain its distinction between accidental and deliberate pollution in that it only orders States to investigate into "casualties" and not into "incidents". If "casualties" are only undesired events, whereas "incidents" also comprise intentional acts and are thus the broader notion (see art. II para. 6 MARPOL), it seems logical only to inquire into unintended pollution events and not into deliberate acts. As the underlying purpose of art. 12 MARPOL is the amelioration of the technical framework and the implementation of lessons learned from accidents, there is no interest in such an investigation where the source of pollution can easily be found in the violation of MARPOL provisions. Within the MARPOL framework, the distinction between deliberate pollution and accidents seems consistent. Still, the picture may be different if the other conventions are taken into account, even in the field of marine pollution only.

---

[48] IMO MEPC/Circ. 318 of 26 July 1996, available at http://www.imo.org/blast/blastDataHelper.asp?data_id=9576&filename=318.pdf referring to the Code for the Investigation of Marine Casualties and Incidents (Resolution A.849(20)), (last visited 6 March 2014).

[49] 1184 UNTS 2; BGBl. 1979 II, 141.

b) Difficulties and Borderline Cases

Art. 2 para. 1 of the Intervention Convention, discussed *supra* under A. I., does not explicitly require that there must be an outside reason for the "casualty" and that deliberate pollution is not covered. An "occurrence [. . .] on board a ship resulting in material damage" might also be a conscious and illicit decision taken by a member of the crew. Whereas it is admitted that the common understanding of "casualty" implies an element of chance or mishap, it seems surprising that almost all authors considering the scope of the Intervention Convention maintain that "casualties" do not cover intentional pollution,[50] without giving a more thorough reasoning or considering the consequences.

Furthermore, authors dealing with this matter surprisingly tend to draw a clear line between the accident as such and its consequences. All sources considering this problem maintain that a deliberate disposal of oil into the sea will be covered by the 1969 Intervention Convention if the discharge is done by a ship in distress, already experiencing a "maritime casualty", in order to ease the vessel's dangerous situation.[51]Wide-spread agreement as to this point of view was already obtained during the negotiations of the Convention.[52]

It is true that in most cases, the "maritime casualty" will already be established and damage to the ship will have taken place or will be imminent when the crew takes recourse to the rather extreme measure to dispose of the cargo that has been trusted to them. On the other hand, a grave misjudgement or clearly overhasty decisions by the master or the crew are not inconceivable. It is very uncertain whether a situation of this kind—where there is damage done to the cargo but consciously inflicted by the crew without actually being "intentional"—could be considered to fall under the notion of "maritime casualty". With regard to the purposes of the Convention and the interests of coastal States, which are exactly the same in this situation and in a "real" casualty, it seems that the criterion of "unintentional" distress is still fulfilled if the crew felt to some point compelled to act in this way. It is submitted that in some situations, especially in chains of events, it might be hard to assess for the coastal State whether the discharge was done on purpose or with regard to the ship's safety and that the criterion of a maritime casualty being an unforeseen and undesired event might thus prove to be problematic.

---

[50] Özçayir, *1998 "Liability for oil pollution"*, p. 176; Birnie, Boyle and Redgwell, *2009 "International law and the environment"*, p. 426; Falkanger, *2011 "Maritime Casualties"*, p. 131.

[51] Hakapää, *1981 "Marine pollution"*, p. 267; Falkanger, *2011 "Maritime Casualties"*, p. 131.

[52] *1973 "Official Records (CLC)"*, pp. 313 et seq.

II. The 1969 International Convention on Civil Liability for Oil Pollution Damage

The second convention directly owing its existence to the *Torrey Canyon* incident in 1967 is the 1969 International Convention on Civil Liability for Oil Pollution Damage (CLC).[53] It addresses the problem of responsibility for the damages resulting from oil pollution from tankers in the territorial waters of a contracting State, identifies the shipowner as the liable person ("channelling") and contains provisions on strict liability and mandatory insurance. In return, it confers the benefit of limited liability to the shipowner. A thorough analysis of the Convention's provisions will be given *infra* in Chap. 3, but in order to define the understanding of a maritime accident in public international law, this aspect of the CLC will be discussed here.

1. Events Covered by the Convention

Unlike the Intervention Convention, the CLC is only concerned with the consequences of a maritime incident resulting in oil pollution, not with the reasons that led to this event. Art. III para. 1 of the CLC simply states:

> 1. Except as provided in paragraphs 2 and 3 of this Article, the owner of a ship at the time of an incident, or where the incident consists of a series of occurrences at the time of first such occurrence, shall be liable for any pollution damage caused by oil which has escaped or been discharged from the ship as a result of the incident.

The field of application of the Convention is thus restricted to oil that escaped or was discharged from the vessel, which excludes pollution damage caused by fire or explosions.[54]

After the entering into force of the Protocol of 1992 to amend the CLC (1992 Protocol) the phrasing of art. III para. 1 was altered, but still refers to "incidents" and "occurrences":

> 1. Except as provided in paragraphs 2 and 3 of this Article, the owner of a ship at the time of an incident, or, where the incident consists of a series of occurrences, at the time of the first such occurrence, shall be liable for any pollution damage caused by the ship as a result of the incident.

The term "incident" is defined in art. I para. 8 CLC, as amended by the 1992 Protocol as:

---

[53] 973 UNTS 3; 1975 BGBl. 1975 II, 305.

[54] L. Lucchini, 'La pollution des mers par les hydrocarbures: Les Conventions de Bruxelles de novembre 1969 ou les fissures du droit international classique' (1970) 97, Journal du Droit International Privé, 795–827 at 815; C. Wu, *Pollution from the carriage of oil by sea*: *Liability and compensation* (London, Boston: Kluwer Law International, 1996), p. 47.

> any occurrence, or series of occurrences having the same origin, which causes pollution damage or creates a grave and imminent threat of causing such damage.

"Incident" is thus generally defined as the reason for pollution—additional requirements are not set out. It is, with regard to the purpose of the Convention to provide a simple and comprehensive liability regime to the victims of tanker oil pollution, a rather wide definition.[55]

The notion of "pollution damage" is explained, according to art. I para. 6 as amended by the 1992 Protocol, as "the contamination resulting from the escape or discharge of oil from the ship, wherever such escape or discharge may occur", and is defined as also including the costs of preventive measures, meaning "any reasonable measures taken by any person after an incident has occurred to prevent or minimize pollution damage" (art. I para. 7 CLC).

It is thus clarified that damages are only recoverable under the CLC once the "incident" has occurred, and the "incident" is clearly defined by its consequences: if there is pollution damage caused by a ship or a grave and imminent threat thereto, there has been an incident.

2. Deliberate Pollution Under the CLC

Considering that the CLC was negotiated and drafted together with the Intervention Convention, it is interesting to briefly consider how it addresses deliberate pollution.

First of all, the rather neutral term "incident" does not imply an element of chance or mishap in the way the word "casualty" does. Still, there has been discussion, during the negotiations and in the immediate aftermath, about the question of intentional pollution. In the draft articles submitted in September 1969, there was still a version of art. II explicitly stating that the shipowner should in any case be liable for deliberate pollution, except in cases of emergency discharge to save life at sea.[56] As the alternative wording of art. II was adopted for the final text of the Convention, this provision disappeared from the draft. Still, it does not seem that the mentioning of deliberate pollution played a role in the choice of the alternatives, as the main difference between them was a regime of presumption of fault in the first version as opposed to a regime of strict liability in the second,[57] and it seems a mere coincidence that the first version contained an explicit rule on deliberate pollution whereas the second was mute on this point. The negotiating parties' comments to the alternatives in the draft were not concerned with this question.[58] The only exception consisted in the United States' comment

---

[55] R. Altfuldisch, *Haftung und Entschädigung nach Tankerunfällen auf See* (Berlin, New York: Springer, 2007), p. 9.

[56] *1973 "Official Records (CLC)"*, p. 459.

[57] Lucchini, '1970 "La pollution des mers"', 816.

[58] *1973 "Official Records (CLC)"*, pp. 458 et seqq.

that the shipowner should be liable for deliberate discharge even in emergency situations if negligence had contributed to the emergency in the first place,[59] thus stressing the general principle of liability for intentional pollution. It does not seem that the question was considered beyond this.

After the adoption of the final text, it was argued that the word "pollution" implied a connotation of accidental discharge or release, as operational discharge could not be considered as "pollution".[60] A similar approach was revealed by the German Delegation in the negotiations to the 1971 International Convention on the Establishment of an International Fund for Compensation for Oil Pollution Damage 1971 (FUND),[61] the delegate *Herber* arguing that within the 1969 CLC, "incident" did not include operational discharge.[62] Still, the wording of the CLC is certainly ambiguous in this respect.[63]

It must first be kept in mind that at the time of the negotiations, operational discharge of oil and oily residues was still admissible if done within the limits of the 1954 OILPOL Convention, and that the regime on deliberate pollution only became considerably stricter with the adoption of MARPOL 1973/78.

Many authors, especially in later publications, take it for granted that the CLC applies both to accidental and intentional discharges, given the open wording that does not require a fortuitous element in the cause of the pollution,[64] and from the victims' point of view, it should certainly not make a difference whether the pollution incident was caused by mishap or intentionally.

Art. V para. 2 of the CLC provides for the loss of the shipowner's limitation of liability "if it is proved that the pollution damage resulted from his personal act or omission, committed with the intent to cause such damage, or recklessly and with knowledge that such damage would probably result". In the case of intentional pollution caused by the shipowner, he will be liable without limitation. The existence of this rule rather strongly argues for the application of the CLC to intentional pollution. Even if the provision refers to the behaviour of the shipowner, not of the crew, it can certainly be argued that where the shipowner is liable for intentional pollution, and even without limitation, the CLC should also apply where

---

59 Ibid., p. 465.

60 Lucchini, '1970 "La pollution des mers"', 815.

61 1953 UNTS 330; BGBl. 1975 II 301.

62 *Official records of the Conference on the Establishment of an International Compensation Fund for Oil Pollution Damage, 1971* (London: IMCO, 1978), p. 389.

63 Wu, *1996 "Pollution"*, p. 48; Altfuldisch, *2007 "Haftung und Entschädigung"*, p. 15.

64 A. L. Doud, 'Compensation for Oil Pollution Damage: Further Comment on the Civil liability and Compensation Fund Conventions' (1973) 4, Journal of Maritime Law and Commerce, 525–42 at 527; D. W. Abecassis and R. L. Jarashow, *Oil Pollution from Ships: International, United Kingdom and United States law and practice*, 2nd edn. (London: Stevens, 1985), para. 10–14; J. Rodríguez Docampo, 'Alcance de la canalización de la responsabilidad y de las daños ambientales en el convenio internacional sobre responsabilidad civil por contaminación de hidrocarburos: CLC/92', in J. L. García-Pita y Lastres, R. Quintáns Eiras and A. de La Díaz Rosa (eds.), *Estudios de Derecho Marítimo* (Cizur Menor: Thomson Aranzadi, 2012), p. 914.

the intentional act was caused by the crew. A detailed discussion of how art. V para. 2 CLC is to be understood, especially considering the consequences for the shipowner, can be found *infra* in Chap. 3, D.

3. Deliberate Pollution as a Reason of Exoneration

The CLC also contains provisions on deliberate pollution as a reason of exoneration for the shipowner. Art. III para. 2 (b) excludes the shipowner's liability if he can prove that the damage results from an act or omission of a third party with the intent to cause damage. From its wording, this provision only applies to situations in which not only the discharge or the pollution was caused consciously or intentionally, but where the actual damage was done intentionally. This would mean that art. III para. 2 (b) is only applicable where the people in question not only caused the discharge of oil on purpose, but consciously expected and wanted the ensuing damage. It is submitted that these cases will be very rare—almost inconceivable except with regard to insurance fraud. Most cases of deliberate pollution, where the wrongdoer consciously causes the discharge of oil in order to dispose of it, but rather hopes that there will be no conceivable damage, would thus still be covered by the CLC liability regime. It is furthermore very questionable whether crewmembers on board a vessel, hired and paid by the shipowner can be considered to be "third parties" in the relationship between the shipowner and a victim of pollution.[65]

Moreover, art. III para. 3 CLC offers a possibility of exoneration if the shipowner can prove that the victim caused the pollution damage intentionally or negligently. Both art. III para. 2 b) and art. III para. 3 remained unaffected by the 1992 Protocol, but it introduced an important change with regard to the channelling provisions of art. III para. 4 CLC. Whereas the original provision only protected "servants and agents of the owner" from claims against them, art. III para. 4 CLC now provides a list of persons benefitting from the channelling of liability unless they caused the damage intentionally or "recklessly and with knowledge that such damage would probably result". The scope of art. III para. 4 will be discussed *infra* in Chap. 3, D., also addressing the question of how these claims for intentional and reckless pollution can be brought.

From the existence of the provisions cited it can be inferred that the CLC is also concerned with deliberate pollution, by providing special rules where the exceptional character of intentional contamination requires a special consideration of the channelling and strict liability provisions. In view of that, it seems doubtful to maintain that the purpose of the CLC, to provide for a compensation mechanism for

[65] See M. N. Tsimplis, 'The Bunker Pollution Convention 2001: completing and harmonizing the liability regime for oil pollution from ships?' (2005), Lloyd's Maritime and Commercial Law Quarterly, 83–100 at 89, footnote 57, discussing the identical provision in the 2001 Bunker Oil Pollution Convention.

disastrous accidents, implicitly excludes intentional pollution from its scope of application.

It can furthermore be argued that it would constitute an inappropriate privilege to wrongdoers[66] if the CLC would not apply to deliberate pollution. It is certainly true that the damage caused by deliberate operational discharges of oil will in most cases be minor[67] and thus outside the typical field of application of the CLC, and that therefore there may be no victims interested in the protection offered by the regime. In any case, the inclusion of deliberate pollution would avoid the sometimes difficult differentiation between negligence, gross negligence and intent which would need to be contemplated in both the triggering event and the pollution as such.

III. The 1982 Law of the Sea Convention

The 1982 Law of the Sea Convention (UNCLOS), negotiated for nine years and finally adopted in 1982, is the most comprehensive multilateral agreement in existence after the UN Charter[68] and is also one of the most successful treaties in terms of global acceptance, currently having 166 States parties.[69]

1. Art. 221 UNCLOS

Environmental protection was a concern of the negotiating parties from the start.[70] The development of rules for preservation of the marine environment was trusted, among other tasks, to the Third Committee. The underlying goal in the negotiations to the provisions on marine environmental protection was to promote the preservation of marine resources without hampering legitimate maritime activities. Following the *Torrey Canyon* accident, the consequences of which were still perceivable in the mid-1970s[71] and considering the efforts made in the 1969 Conventions, maritime accidents were clearly recognised as a major source of marine pollution. The negotiating States were thus eager also to include provisions

[66] Altfuldisch, *2007 "Haftung und Entschädigung"*, p. 15. Also see Lucchini, '1970 "La pollution des mers"', 827, admitting this notwithstanding his opinion that the CLC does not apply to intentional pollution.

[67] Ibid., p. 827; Wu, *1996 "Pollution"*, p. 49.

[68] J. Kraska, *Maritime power and the law of the sea*: *Expeditionary operations in world politics* (Oxford, New York: Oxford University Press, 2011), p. 100.

[69] See the Status of the Convention as published by the Division for Ocean Affairs and the Law of the Sea (as at 10 January 2014) http://www.un.org/Depts/los/reference_files/status2010.pdf (last visited 6 March 2014).

[70] Abecassis and Jarashow, *1985 "Oil Pollution"*, para. 5–49.

[71] Hovanesian, '1970 "Post "Torrey Canyon""', 633.

on coastal States' intervention rights against ships posing a threat of pollution to their territorial waters and coastline. First proposals to address this issue were already made in the 1973 session to Sub-Committee III of the Seabed Committee, but it was only in 1976 that a first draft was included, as the then art. 31, in the negotiating text. It read:

1. Nothing in this Chapter shall affect the right of States to take measures, in accordance with international law, beyond the limits of the territorial sea for the protection of coastlines or related interests, including fishing, from grave and imminent danger from pollution or threat of pollution following upon a maritime casualty or acts related to such a casualty.
2. Measures taken in accordance with this Article shall be proportionate to the actual or threatened damage.[72]

Following the 1973 Protocol to the Intervention Convention 1969, the article was not only concerned with oil pollution, but provided a rule for all kinds of pollution. Furthermore, the wording clearly recognised existing international law on coastal States' intervention rights,[73] presenting a 'no prejudice'-clause from the start. Still, it must be noted that the extent of the reference is by no means clear—even during negotiations, delegates were at odds about what the customary international law on intervention should be.[74] It was even maintained by the Soviet Delegation that there was no customary right of intervention other than the one laid down in the Intervention Convention and that the only function of what was to become art. 221 UNCLOS was to give this right to the States that were not party to the Intervention Convention.[75] In their point of view, the coastal State was not given the right to intervene before a "maritime casualty" had been established or to intervene in cases where there was no damage to the vessel.

Considering that this interpretation of the text is at odds with the wording of the provision, this point of view can hardly be endorsed. The provision in question is explicitly referring to "customary international law" as opposed to treaty law. This is in accordance with a general understanding expressed in UNCLOS that there is general international law on the use of the sea beyond the treaty.[76]

A comprehensive study of the provenance and extent of a customary right to intervention following maritime casualties and the meaning to be given to the term

[72] A/CONF.62/WP.8/Rev.1/Part III; Official Documents of the UN Third Conference on the Law of the Sea, vol. 5, 179.

[73] R. Lagoni, 'Das Wrackbeseitigungs-Übereinkommen und das Völkerrecht' (2008), Transportrecht, 352–7 at 354.

[74] See M. H. Nordquist, S. N. Nandan and S. Rosenne, *United Nations Convention on the Law of the Sea, 1982. A Commentary*, Volume III (Dordrecht, Boston: Martinus Nijhoff; Distributors for the United States and Canada, Kluwer Academic Publishers, 2002), para. 221.4 et seqq. for a comprehensive overview on the debate.

[75] Platzöder, *1987 "UNCLOS Documents Vol. XI"*, p. 358.

[76] D. French, 'Treaty Interpretation and the Incorporation of Extraneous Legal Rules' (2006) 55, International and Comparative Law Quarterly, 281–314 at 294, footnote 49.

"casualties" in this context is beyond the scope of this work. The answer to the question will be found along the lines of the customary concept of necessity and will thus probably only give regard to the consequential pollution, not to the situation it arises from. Originally, necessity in international law could only be invoked where crucial interests of the State are at stake.[77] It can thus be argued that a customary right to intervention can only be invoked where there is a certain danger to the ship or the cargo, an interference with a ship carrying out operational discharges being rather of a punitive than of a protective nature.[78] When it is submitted that the State may invoke necessity in cases where the "casualty" cannot be established yet[79] it seems that at least the damage expected must be very important. Here again, the question of whether an actual or threatening "damage" to the ship or cargo is required would arise. Still, with regard to the uncertainties in the customary right to intervention, it will be almost impossible to define the notion of "casualty" in this context.

Art. 221 para. 2 UNCLOS as it now stands contains a definition of the term "maritime casualty". This appeared in the negotiation texts for the first time in 1978 in a French proposal.[80] The *Amoco Cadiz* accident in the same year, in which a tanker flying the Liberian flag and laden with 220,000 tonnes of light crude oil ran aground off the coast of Brittany, France, and released its whole cargo into the sea[81] (also see *infra*, Chap. 3, A. I. 1. a)), had prompted the debate on intervention during the UNCLOS negotiations. The French delegation, reviewing the then art. 222 of the draft, proposed to delete the referral to "grave and imminent" danger, reducing the threshold for intervention so that cases of "mere" danger would suffice.[82] The French proposal was rejected as too far-reaching[83] and only later gave rise to a slight alteration terming the conditions for intervention as

> pollution or threat of pollution following upon a maritime casualty or acts relating to such a casualty, which may reasonably be expected to result in major harmful consequences.

The adding of the definition was not retained in the first place, but reappeared in a suggestion of the Chairman of the Informal Meetings in May 1978.[84] It was included into the text from the seventh session on and later adopted as art. 221 UNCLOS.

---

[77] Brown, '1968 "The Lessons"', 128 et seq.

[78] Hakapää, *1981 "Marine pollution"*, p. 273.

[79] Falkanger, *2011 "Maritime Casualties"*, p. 131.

[80] Reproduced in R. Platzöder, *Third United Nations Conference on the Law of the Sea: Documents, Volume X* (Dobbs Ferry (NY): Oceana Publications, 1986), p. 218.

[81] E. Fontaine, 'The French Experience: "Tanio" and "Amoco Cadiz" incidents compared', in C. M. de la Rue (ed.), *Liability for damage to the marine environment* (London, New York: Lloyd's of London Press, 1993), p. 102.

[82] Hakapää, *1981 "Marine pollution"*, p. 271.

[83] Ibid., p. 272; Falkanger, *2011 "Maritime Casualties"*, p. 212.

[84] Reproduced in Platzöder, *1986 "UNCLOS Documents Vol. X"*, p. 236.

The provision, as such, mirrors closely the terms of the 1969 Intervention Convention.[85] Art. 221 para. 2 UNCLOS repeats the definition in art. II para. 1 of the Intervention Convention to the letter, apart from using the term "vessel" rather than "ship", which is in all likelihood only an editorial change,[86] having been made at the same time as parallel changes in other provisions of Part XII of the Convention.[87]

It can thus be seen that the provision in the UNCLOS draft text was meant to use the established vocabulary of the Intervention Convention and also copied the meaning of the term "casualty". On the other hand, it does not create new rights but merely refers to existing international law, the existence and scope of which is highly contested. Still, with regard to the term "casualty", it does not seem that UNCLOS features a substantially different understanding of the notion. The customary rights it refers to would therefore exist in situations similar to those that trigger intervention rights according to the 1969 Intervention Convention.[88]

This would mean that, here again, there must be an event outside the shipboard routine, either already featuring the outward appearances of a navigational incident such as a collision or a stranding, or another occurrence threatening the ship or cargo with imminent or actual damage, whereas intentional acts are ruled out. Considering the close relationship between the 1969 Intervention Convention and art. 221 UNCLOS, also pursuing the same aims in terms of protection of the marine environment, this interpretation is preferable to any other point of view. For the purposes of coastal States' intervention rights, there is therefore only one definition of the term "casualty".

2. Art. 211 para. 7 UNCLOS

Art. 221 UNCLOS is not the only provision of the convention in which the term "maritime casualties" is used. Art. 211, dealing with avoidance of pollution from vessels, compels the contracting States to adopt effective legislation to this end and provides in its para. 7 that States shall also have regard to rules of prompt information and communication to coastal States

> whose coastline or related interests may be affected by incidents, including maritime casualties, which involve discharges or probability of discharges.

The question of what a casualty would be in this provision has not been the subject of much debate. It has been argued that the definition in art. 221 para. 2 UNCLOS, based on the understanding as created by the 1969 Intervention Convention, should also be retained to interpret art. 211 para. 7: "Given the

---

[85] Hakapää, *1981 "Marine pollution"*, p. 271.

[86] Falkanger, *2011 "Maritime Casualties"*, p. 212.

[87] Nordquist, Nandan and Rosenne, *2002 "UNCLOS Commentary Vol. III"*, para. 221.7.

[88] Ibid., para. 221.9(f).

provenance and the generality of the definition, no clear reason exists for limiting the meaning of the expression to the purposes of art. 221."[89]

With a view to the wording of art. 211 para. 7 UNCLOS and the purpose of the provision, this appears right. The term "maritime casualties" is given as a sub-category of "incidents". "Incidents", following the definition in art. II para. 6 MARPOL, are events involving an actual or probable discharge of harmful substances into the sea, and maritime casualties can thus be qualified as the incidents in which the discharge is due to an unforeseen and undesirable event outside the shipboard routine, as opposed to deliberate pollution. Here again, the criterion of adverse consequences of the event is referred to in the provision, considering that the coastal States' interests need to be affected.

Furthermore, the term "marine casualties"—slightly different from the "maritime casualty" referred to in art. 211 para. 7 and 221 UNCLOS, but it is claimed, without special reason[90]—also appears in art. 94 para. 7 UNCLOS. The provision concerns the contracting States' duty to hold an inquiry "into every marine casualty or incident of navigation on the high seas involving a ship flying its flag and causing loss of life or serious injury to nationals of another State or serious damage to ships or installations of another State or to the marine environment."

The provision thus does not only concern marine pollution, but also the other categories of adverse consequences of a marine accident: loss of life or serious injury to humans or serious damage to material property. Following its history, it becomes obvious that the provision, based on rules in the 1958 Geneva Convention on the High Seas, first and foremost aims at the protection of life at sea[91] and that the reference to marine pollution was only made for the sake of completeness. The provision will thus be discussed in the second chapter of this part of the study.

IV. Conclusion

It seems ironic that in the field of marine pollution, regulation started out with the prohibition of deliberate pollution and that maritime casualties have only been recognised as a source of contamination at a comparatively late point in time. Still, most of the regulation only deals with the adverse consequences to the marine environment and approaches the topic from a pragmatic point of view, applying the respective regimes if there is damage to the environment threatening or caused, regardless of the reasons for it. This certainly holds true for the 1969 CLC, one of the most-applied conventions in the field of maritime accidents.

Other texts draw a sharp line between deliberate dumping or discharge of toxic substances and accidental losses. The distinction seems first and foremost to stem

[89] Ibid., para. 221.9(f).

[90] Ibid., para. 221.9(f).

[91] See ibid., para. 94.4 et seqq. and *infra* under B.

from the fact that deliberate contamination was recognised as a problem long before accidents such as the *Torrey Canyon* incident in 1967 raised concerns about the environmental impact of maritime accidents. Furthermore, it is clear that the deliberate pollution in standard shipboard operations and its adverse consequences was rather easy to identify and prohibit, even if effective control and enforcement has of course always been an issue.[92]

In contrast to that, effective regulation against unforeseen and undesired accidents must play on minimising the underlying risks, as it is undertaken by the 1973/78 MARPOL Convention. Still, MARPOL is not concerned by the consequences of possible accidents. It is a preventive instrument, but on a technical scale, meant to take effect long before a casualty can occur.

There is only one area in which it is necessary to define what a "maritime casualty" actually consists in: the regulation of preventive measures, against the consequences of a casualty that has already occurred, i.e. the coastal State's intervention. Not surprisingly, it is in this area that there are texts defining the term "maritime casualty"—art. II para. 1 of the Intervention Convention and art. 221 UNCLOS. Art. 211 para. 7 UNCLOS can also be included in this group since it deals with an accident's consequences by compelling States to have regard to the information rights of other, affected coastal States. Considering that in order to trigger these provisions, the accident as such must already have occurred, but not yet necessarily have entailed dramatic consequences—intervention is concerned with preventing the worst, after all—it seems logical first to require a situation that either has already entailed damage to the ship or cargo or at least features the outward appearances of a maritime accident—a stranding, collision or other incident of navigation, meaning a disturbance in the ship's function as a means of transport. If the interference with the proper functioning of the ship does not concern navigation—if, i.e. the ship is undermanned, or if it is threatened by external dangers such as underwater mines[93]—then it is necessary that these circumstances entail at least an imminent threat of danger to the ship or the cargo.

Deliberate acts by crewmembers leading to illicit marine pollution will be covered by the punitive legal framework of the 1972 London Convention and MARPOL, prohibiting these acts from the start, offering mechanisms of control to port States, coastal States and flag States and compelling them to sanction wrongdoers. Generally, coastal States therefore do not need to take recourse to the instrument of intervention to impede intentional pollution of this kind.

On the other hand, the 1969 Intervention Convention is the only written instrument conferring States the right of intervention against foreign-flagged ships on the

---

[92] G. Peet, 'The MARPOL Convention: Implementation and Effectiveness' (1992) 7, International Journal of Estuarine and Coastal Law, 277–95; de la Rue and Anderson, *2009 "Shipping and the environment"*, p. 825.

[93] Example given by Falkanger, *2011 "Maritime Casualties"*, p. 149.

high seas.[94] Neither MARPOL nor the 1972 London Convention would enable a State to undertake measures in this situation, let alone to intervene before damage has actually occurred.

It seems that thus, by restricting the notion of "maritime casualties" to unforeseen and undesired events, the contracting States of the 1969 Intervention Convention abstain from using an instrument that could easily be employed to fight deliberate pollution occurring on the high seas and, as it is mostly held, in their exclusive economic zone (EEZ),[95] but threatening their coastal waters.

It however seems that an extension of the right in this way would neither be in direct violation of the wording of art. II para. 1 of the Intervention Convention, nor pose an inappropriate threat to the freedom of navigation. An "occurrence on board a ship [. . .] resulting in material damage" could also be a deliberate decision of a member of the crew, and it does not seem that the traditional element of chance or mishap in the notion of "casualty" is too strong to be overridden. Especially in cases where the degree of intent is hard to assess and where the pollution is clearly illicit, priority should be given to an effective protection of coastal States' interests. Considering the balance between these interests and the principle of the freedom of navigation, it must be kept in mind that the Intervention Convention provides in its art. VI for a reasonable compensation if a State unduly invokes the intervention right. Secondly, the threshold of a "grave and imminent" danger to a States' coastline is a very high one that will, arguably, not be crossed in the typical cases of clandestine illicit pollution. Thirdly, States have so far only very reluctantly made use of the intervention right at all.[96]

It is admitted that the distinction between deliberate pollution—for the sake of disposal—and contamination for other reasons has a long tradition in the international regulation of maritime pollution and that the word "casualty" does bear a connotation of chance or mishap. For the purposes of this study, the term "casualty" will, as a rule, be considered in the way it is generally understood and thus, as not covering intentional action. Still, where there is an interest in scrutinising the impact that the exclusion of intent may have, the question will be addressed.

---

[94] P. Wendel, *State responsibility for interferences with the freedom of navigation in public international law* (Berlin, New York: Springer, 2007), pp. 48 et seq.; Falkanger, *2011 "Maritime Casualties"*, p. 105.

[95] The 1969 Intervention Convention, dating from a time before the concept of the EEZ was firmly established, in its wording only applies to measures "on the high seas" (art. I para. 1). Still, it is today widely accepted that the regime also applies to contracting States' EEZ, "on the high seas" meaning "beyond territorial waters" for the matters of the Convention, see Churchill and Lowe, *1999 "The Law of the Sea"*, p. 354; de la Rue and Anderson, *2009 "Shipping and the environment"*, p. 901; Falkanger, *2011 "Maritime Casualties"*, p. 125.

[96] Wendel, *2007 "State responsibility"*, p. 49, claims that there was, up to 2007, no incident in which a coastal State has relied on art. I para. 1 of the Intervention Convention.

B. Maritime Safety and the Protection of Life and Property at Sea

Considering that the roots of merchant shipping go back to antiquity and that maritime trade used to be even more of a bold venture in former times, the safety of ships has always been a concern to seafarers and shipowners. Unlike the idea of environmental protection that only arose during the last century when it became obvious that our planet's resources are limited, the necessity to safeguard life and property at sea has accompanied the development of merchant shipping from its very beginning. This was not only for humanitarian reasons, but also for the simple fact that severe accidents cause prejudice to the reliability and the cost of the worldwide trade.[97] It is in this area that the avoidance and the consequences of maritime casualties were considered at a much earlier point in time, but from a different point of view. In contrast to environmental law, which is concerned with the protection of a common, global good, the value of which is sometimes not easy to assess, the safety of ships and their crews have direct implications on the cost of maritime trade. On the one hand, safety measures almost always entail additional costs for shipowners; on the other hand, the consequences of accidents will in most cases be more costly than any preventive measure taken beforehand. Even if risks are insured, insurance rates may become prohibitively high if there are too many accidents, and vessel safety is thus also in the interest of insurers, who will demand certain standards to be met in order to properly assess the risk and keep it insurable.[98]

Again, regulation on the protection of life and property at sea centres on one hand on technical provisions, meant to minimise risks that may result in danger to the ship, its crew and its cargo. On the other hand, there are rules on how to mitigate the consequences once an accident has happened or is about to happen, i.e. the duty to render assistance to ships in distress and people in danger of being lost at sea.

I. Safety of Ships

Merchant shipping has always been a business heavily driven by profit and return considerations. As has been mentioned above, maritime safety has a direct impact on the cost of shipping, and this, both ways: enhanced safety measures tend to be costly, but can easily be offset by the cost of accidents, both to shipowners and insurers. In the seventeenth and eighteenth century, when marine insurance gradually became institutional, reliable information about ships' condition and nautical qualities were crucial to insurers in order to correctly assess the risk of a given

---

[97] Churchill and Lowe, *1999 "The Law of the Sea"*, p. 264.

[98] W. K. Talley, 'Vessel Safety and Accident Analysis', in C. T. Grammenos (ed.), *The Handbook of Maritime Economics and Business*, 2nd edn. (London: Lloyd's List, 2010), p. 521.

voyage. It was in this time that the first classification society, Lloyd's, emerged in London, first as a club collecting information and publishing it on a small scale, later in a more official context, and eventually in the framework of a shipping register providing neutral and faithful information,[99] coming close to the way classification societies work today. Still, there were no uniform standards to which a vessel's performance and seaworthiness were checked, criteria were often arbitrary and policy-driven[100] and it was left to the insurers and cargo-owners to interpret the information obtained. Despite the system's shortcomings, it created a viable market, the information available influenced the cost of insurance, and the vessel's seaworthiness or general condition thus became a factor in the shipowner's calculation. An original legal framework emerged around the maritime trade, in which liabilities were contractually distributed and in which each player was only concerned with the risks he was bearing[101] and tried to offset or minimise these rather than being concerned with the security of the undertaking as a whole. Furthermore, the reasons why ships were lost at sea were hardly ever known, so that there was no real means of enhancing safety by taking technical measures before the ships set sail.

In the nineteenth century, the development of the steam engine replaced the sailing ships, and vessels became more independent from adverse weather conditions. It seems that their general condition and performance was then even more closely scrutinised, considering that ships were expected to use their own power to master storms and keep away from rocks and other vessels, and that the length and the cost of sea voyages became more calculable. Risk assessment and safety measures still took place only between private players, as administrative regulation by States was considered to be misplaced in an industry governed by international competition.

Regulations concerning ship safety only very slowly emerged in the nineteenth century, and were very different from country to country.[102] Even in the beginning of the maritime passenger liner trade, suddenly exposing large numbers of people unacquainted with the maritime hazards to considerable risks, many lives were lost before the first steps towards an international framework of safety rules were made, such as the introduction of an International Code of Signals and the Rule of the Road at Sea as an international anti-collision agreement. It was only late in the nineteenth century that maritime casualties were finally perceived as avoidable

---

[99] See P. Boisson, 'Classification Societies and Safety at Sea' (1994) 18, Marine Policy, 363–77 at 363 et seqq. on the history of Lloyd's and other classification societies.

[100] Ibid., p. 365, referring to Lloyd's practice in 1797 to discriminate against ships not built on the river Thames.

[101] A. M. Antapassis, 'Liability of Classification Societies', in K. Boele-Woelki and S. van Erp (eds.), *General reports of the XVIIth congress of the International Academy of Comparative Law* (Bruxelles, Utrecht: Bruylant Eleven, 2007), p. 634.

[102] P. Boisson, 'Du "Titanic" au "Concordia": 100 ans de droit de la sécurité des navires de croisière' (2012) 64, Droit Maritime Français, 328–37 at 328.

events that could rather easily be prevented by technical precautions, and not as mere fate only to be dealt with in the aftermath and on a commercial basis.

Still, only after the devastating *Titanic* accident in 1912, also heavily influencing public opinion, maritime nations gathered under the guidance of the United Kingdom to develop a comprehensive and binding international agreement on the safety of life at sea.[103]

1. Maritime Casualties in the Framework of SOLAS

The first version of the International Convention of the Safety of Life at Sea (SOLAS)[104] was adopted in 1914 and constituted a compromise that did not win wide acclaim in the beginning, only obtaining five signatures. Moreover, World War I prevented the Convention from coming into effect in 1915 as it had originally been envisaged. Still, the new legal framework influenced domestic legislation into adopting more uniform standards, until another conference in 1929 led to the adoption of a new SOLAS Convention, already more successful in terms of ratification and which entered into force in 1933. A third version was adopted in 1948, in order to keep up with new technical developments, and a fourth version in 1960.

In 1974, another IMO conference revised SOLAS with the aim of consolidating it, including all the amendments made beforehand, and introducing a tacit amendment procedure which assumes that an amendment is favoured by the contracting States and thus adopted unless they explicitly object to it. The tacit amendment procedure has enabled the IMO to quickly adapt the Convention to technical developments. The last amendments in the current SOLAS version date of May 2011 and have entered into force on 1 January 2013.[105]

SOLAS does not mention the term "maritime casualties" often. The notion is only used in the context of the investigation into maritime casualties as provided for in the annex, chapter I, Part C, Regulation 21(a).

In addition to that, SOLAS only sporadically refers to "emergencies" or "distress", such as in the headline of art. V ("carriage of persons in emergencies") on the inapplicability of the Convention to ships evacuating persons in peril. In the Annex, Chapter II-1, Regulation 3 gives a definition of "emergency condition", and Chapter V, Regulation 31 and 33 regulate on "danger" and "distress" messages. This is certainly due to the fact that SOLAS is first and foremost concerned with technical provisions meant to minimise the likelihood of casualties or mitigate the consequences of an accident once it has occurred and therefore intervenes at an earlier point in time, not regulating on the accident as such. Even the provisions on

[103] See Boisson, '2012 "Du "Titanic" au "Concordia""'.

[104] 1184 UNTS 278; BGBl. 1979 II, 141.

[105] The list of amendments is available in the IMO's Knowledge Centre, http://www.imo.org/KnowledgeCentre (last visited 6 March 2014).

"emergency" or "distress" situations or conditions are still of a technical character and therefore feature a technical definition. According to the Annex, Chapter II-1, Regulation 3 (6), an "emergency" condition for instance is a condition in which the vessel's systems are not working properly due to the failure of the main source of electrical power. The definition is restricted to the description of a technical dysfunction and not concerned with the functioning of the ship as a whole or the outward appearance of what happened. This understanding of SOLAS is also mirrored in the notion of "distress". In the Annex, Chapter V Regulation 33 provides for the obligations to fulfil after receiving a distress message and simply starts from the point that there is such a message—so the senders themselves consider that they are in distress—or that the Master interprets it as a distress message. The provision does not feature a neutral definition of when people should be deemed to be "in distress" or what "distress" as opposed to a "casualty" is.

In contrast to that, in the SOLAS provision directly dealing with "maritime casualties", there is an element of hindsight. The Annex provision, Chapter I, Part C, Regulation 21 (a) reads:

> (a) Each Administration undertakes to conduct an investigation to any casualty occurring to any of its ships subject to the provisions of the present Convention when it judges that such an investigation may assist in determining what changes in the present regulations may be desirable.

The provision was already included in the 1948 version of SOLAS,[106] then as Regulation 20. The wording was almost identical, except for the fact that the 1948 version of the regulation compelled the administrations to inquire into any "major marine casualty" rather than merely "any casualty". It is interesting that therefore, the SOLAS convention features a notion of "casualties" that is effectively about 20 years older than the notion in the 1969 Intervention Convention, but the scope of which has never been analysed in order to overcome the uncertainties in the 1969 definition and the same term in the 1982 UNCLOS.

a) The Requirement of Adverse Consequences

It seems that this is partly due to the special character of SOLAS as a technical framework. The Annex provision of Chapter I, Part C, Regulation 21 (a) explicitly refers to the interest an investigation of casualties may have for the technical provisions of SOLAS. The provision is not really concerned with the casualty as such, but with the amelioration of the underlying technical framework.[107] It may

---

106 Text available under http://www.imo.org/KnowledgeCentre/ReferencesAndArchives/HistoryofSOLAS/Documents/SOLAS%201948%20UK%20Treaty%20Series.pdf (last visited 6 March 2014).

107 See J. Lang, 'The Investigation of Marine Accidents: A U. K. Accident Investigator's View' (2001) 75, Tulane Law Review, 1673–85 at 1676 on the U. K.'s Marine Accident Investigation Board criteria to choose the accidents that will be investigated.

therefore be possible that there is, even in a severe casualty, no real interest in an investigation, if for instance the reasons for it are outside the technical provisions of SOLAS or if the casualty is due to a violation of these rules, so that there is no lesson to be learned from the event that could improve the regulations. On the other hand, an incident may reveal that there are substantial shortcomings in the SOLAS regulations, even if there is no damage to the ship or if the incident does not even feature the outward appearance of a maritime accident.

SOLAS therefore cannot and does not need to provide an unambiguous definition of the term "maritime casualties"—it does not regulate on maritime casualties as such, but only on certain technical standards it considers to be at the source of casualties. A "casualty", for that matter, could be every event having undesirable consequences. Neither is personal injury to crewmembers or passengers nor damage to the ship or cargo a necessary condition to establish a "casualty", even if these adverse consequences will be at the root of the interest in the investigation in most cases.

Whereas the SOLAS Convention in the Annex provision of Chapter I, Part C, Regulation 21 (a) does not give a definition of "maritime casualty", a written definition can be found in the IMO's "Code for the investigation of marine casualties and incidents", adopted on 27 November 1997 by Resolution A.849 (20)[108] (1997 Code). The Resolution supersedes and revokes the earlier Resolutions A.637(16),[109] A.440(XI)[110] and A.173(ES.IV)[111] (the last dating from 28 November 1968) in which there was no definition of what a "casualty" is but which recall the importance of casualty investigation for the amelioration of the existing legal framework and compel States to cooperate in this task. As has been pointed out *supra* under A. I. 2. a), the provisions of the code also apply to investigations carried out under MARPOL, but it is obvious that the 1997 Code is rooted in the early provisions of SOLAS, and the reference to MARPOL was most likely only added for the sake of convenience.

According to the 1997 Code, art. 4.1,

> *marine casualty* means an event that has resulted in any of the following:
>
> .1 the death of, or serious injury to, a person that is caused by, or in connection with, the operations of a ship; or
> .2 the loss of a person from a ship that is caused by, or in connection with, the operations of a ship; or
> .3 the loss, presumed loss or abandonment of a ship; or
> .4 material damage to a ship; or

[108] Available under http://www.imo.org/blast/blastDataHelper.asp?data_id=22633&filename=A849(20).pdf (last visited 6 March 2014).

[109] Available under http://www.imo.org/blast/blastDataHelper.asp?data_id=26668&filename=A440(XI).pdf (last visited 6 March 2014).

[110] Available under http://www.imo.org/blast/blastDataHelper.asp?data_id=26668&filename=A440(XI).pdf (last visited 6 March 2014).

[111] Available under http://www.imo.org/blast/blastDataHelper.asp?data_id=28026&filename=A173(VI).pdf (last visited 6 March 2014).

.5 the stranding or disabling of a ship, or the involvement of a ship in a collision; or
.6 material damage being caused by, or in connection with, the operation of a ship; or
.7 damage to the environment brought about by the damage of a ship or ships being caused by, or in connection with, the operations of a ship or ships.

As opposed to "marine casualty", a "marine incident" is defined, in art. 4.4 of the Code for the Investigation of Marine Casualties and Incidents, as

> an occurrence or event being caused by, or in connection with, the operations of a ship by which the ship or any person is imperilled, or as a result of which serious damage to the ship or structure or the environment might be caused.

The definition of "marine casualty" clearly sets out from the results of the accident, i.e. the adverse consequences brought about by it. This also includes damage to cargo ("material damage") in point .6, if it is caused by the operation of the ship. In a broad interpretation of the wording, which is clearly desirable with regard to the purpose of the provision, "material damage" would also include loss of cargo.

Mirroring the definition in the Intervention Convention 1969 as advocated *supra* under A. I., the marine casualty is established if there is the outward appearance of a maritime accident, such as a stranding or collision, regardless of whether there is actual damage to the ship (point .5). The wording of this paragraph with its various alternatives makes it quite clear that there does not need to be damage to the ship in point .5—otherwise it would be redundant as included by point .4.

The definition in art. 4 of the Code for the Investigation of Marine Casualties and Incidents only concerns adverse consequences that are brought about by operational activities. This of course due to the fact that the SOLAS framework is meant to regulate on the operational performance of the ship and is therefore only concerned with casualties caused by operational dysfunctions that can be avoided in the future. Still, it seems that the term "operational" in the points .1, .2, .6 and .7 is quite broad, considering that it covers not only damage "caused by the operation" but also damage "caused in connection with the operation" of a ship. Bearing in mind the purpose of the Code, only occurrences completely unconnected with the function of the ship as a means of transport are ruled out, such as a death or injury due to a conflict within the crew. Concerning point .7, it actually seems difficult to imagine a situation in which a damage to the environment is brought about by a ship, but unconnected from its operational function.

According to the wording of art. 4.4, the difference between "casualty" and "incident" lies in the fact that a casualty has already effectuated graspable adverse consequences whereas in an "incident", the ship or any person, structure or the environment are only endangered. The notion of "incident" is therefore different from the one used in MARPOL and art. 211 para. 7 UNCLOS (see *supra*, A. III. 2.). Concerning the term "casualty", the 1997 Code for the Investigation of Marine Casualties and Incidents clearly defines maritime casualties by their adverse consequences, actually not requiring any other condition. Similar to the definition in the Intervention Convention 1969, the broad notion will be narrowed down by the purpose of the regulation: the rule does not apply to all maritime casualties but only to those where there is an interest in investigation, just like the intervention allowed by the 1969 Convention which will only be effectuated if a State can feel threatened to a reasonable degree.

Considering that it is the purpose of the investigation duty to establish the reasons for which the accident came about, it would of course not be helpful to hold up too many conditions before an event would fall into the category of "casualty". The presence of adverse consequences or danger to the ship or persons are therefore the only suitable conditions to trigger the investigation duty.

b) Intent

The definition in art. 4 of the Code for the Investigation of Marine Casualties and Incidents does not consider the question whether the event or its adverse consequences must have occurred accidentally. Only considering the wording of the 1997 Code, it does seem that the term "marine casualties" also covers cases in which the triggering event or the actual adverse consequences have been caused intentionally. Of course, it may only be found out in the course of the investigation that there was an element of intent in the development of the accident, but still, the provision does not even rule out the most obvious intentional acts. Within the framework of SOLAS, it is not necessary to draw a line between intentional doing and undesired accidents, which makes it considerably easier to define and detect a "marine casualty".

It therefore seems that at least within SOLAS, a "marine casualty" is only defined by its adverse consequences and that there are no additional conditions to restrict the term. This interpretation has the benefit of being all-embracing and offering the widest possible frame for an amelioration of the existing SOLAS framework. It further does away with the difficult distinction between intentional doing and actual accidents.

As has been mentioned (*supra* under A. I. 2. a)), the duty to investigate into casualties at sea by no means only exists in SOLAS. A corresponding duty exists under art. 12 MARPOL for casualties which have had "a major deleterious effect upon the maritime environment".

Considering that the notion of "casualties" under MARPOL does not include intentional doing, a deliberate act entailing oil pollution would not be a casualty and therefore not trigger the flag State's duty to investigate. On the other hand, under SOLAS and the 1997 Code, any damage to the environment caused by the operation, or in connection with the operation of a ship is considered a "marine casualty" and entails the theoretical obligation to investigate the case. SOLAS therefore seems to act as a "catch-all" and also covers the cases in which art. 12 MARPOL is not applicable, either because there was no environmental damage or because the harm was caused intentionally, always under the condition that there is an interest in the investigation. Considering that the reporting and investigation duties are similar under both conventions and that the IMO mandatory reporting system on

MARPOL investigations[112] even refers to the guidelines originally adopted for the SOLAS, and bearing in mind the impressive number of contracting States to both conventions,[113] it seems that the investigation obligations for States are comprehensive anyway, regardless of the restriction under MARPOL.

The investigation provisions seem to be another example that the exclusion of intentional doing out of the notion of "casualties" at sea does not help the purposes of regulation and rather seems to be a relic of a time when casualties and intentional pollution were treated very differently, either due to the lack of appreciation for a sound maritime environment or due to a failure in recognising that it was possible to avoid casualties by implementing technical administrative standards.

Of course, it must again be stressed that the annex provision of Part C, Chapter I, Regulation 21 is concerned with the investigation into maritime casualties and that therefore, in most of the cases, the sources of the accident will not yet have been established. Thus, it may be not possible to establish the reasons for the accident from the start, including the answer to the question whether the casualty was caused by mishap or on purpose. Still, it seems surprising that the definition in the 1997 Code does not rule out obviously intentional acts, but defines the term "marine casualty" from the adverse consequences only. An explanation to this may lie in the purpose of the Annex, Part C, Chapter I, Regulation 21 which ensures that States will only conduct investigations if they are of interest for the improvement of the legal framework—which will most likely not be the case if it is clear from the start that the damage was caused by intentional doing. Whereas it seems obvious that the investigation into a purely deliberate pollution incident would not be helpful to this end, there may be a considerable interest in inquiring into casualties that have been caused by gross negligence or even by simple ignorance of members of the crew—again, the line between carelessness and intent will be hard to draw. It therefore seems correct to follow the wording of the 1997 Code definition and to define a "marine casualty" by its adverse consequences only, not having regard to whether there may be an element of intent involved.

2. Art. 94 para. 7 UNCLOS

It has been shown that the flag State's duty to carry out investigations on maritime casualties, with regard to the goals and the legal framework of the existing legal texts is a comparatively old idea within the rather recent regulation of maritime safety, dating back to as early as 1948. It is therefore not surprising that it is

[112] IMO MEPC/Circ. 318, available at http://www.imo.org/blast/blastDataHelper.asp?data_id=9576&filename=318.pdf referring to the Code for the Investigation of Marine Casualties and Incidents (Resolution A.849(20)), (last visited 6 March 2014).

[113] As of 28 February 2014, 162 States are parties to SOLAS 1974 and 152 States are parties to the original MARPOL Convention 73/78, representing 98, 77 % and 99, 20 % of world maritime tonnage respectively, see http://www.imo.org/About/Conventions/StatusOfConventions/Documents/Summary%20of%20Status%20of%20Conventions.xls (last visited 6 March 2014).

introduced into many conventions in the field of maritime law. Apart from SOLAS and MARPOL, it is also provided for in art. 23 of the 1966 International Convention on Load Lines.[114] Still, there was no corresponding duty in the 1958 Geneva Convention on the High Seas, and the provision was only introduced into UNCLOS in the drafting process in the early 1970s.

A first attempt to introduce States' investigation duties into the UNCLOS draft was made in the second session of the Conference in 1974, when nine Western European Countries submitted a working paper on States' duties on the high seas.[115] It included a duty to cooperate in the inquiries of other States into maritime casualties if they had caused loss of life or serious injuries to nationals or damage to ships or other installations of that other State.[116]

This suggestion was developed into a consolidated text in the third session in 1975, in which the duty to cooperate was supplemented with an actual duty to States to investigate into casualties involving ships flying their flag, if these had caused loss of life or serious injury to nationals of other States "or serious damage to shipping or installations of another State or to the marine environment".[117]

The provision remained almost unchanged except for minor alterations in wording, replacing "shipping" by "ships" and was finally adopted in the final version:

> 7. Each State shall cause an inquiry to be held by or before a suitably qualified person or persons into every marine casualty or incident of navigation on the high seas involving a ship flying its flag and causing loss of life or serious injury to nationals of another State or serious damage to ships or installations of another State or to the marine environment. The flag State and the other State shall cooperate in the conduct of any inquiry held by that other State into any such marine casualty or incident of navigation.

It is questionable whether the "incident of navigation" in this provision should also be defined according to the definition in the 1997 Code as an event in which persons or the ship are imperilled. The term is neither mentioned in the 1948 version of SOLAS, nor in the 1974 version, there is no hint to it in the IMO Resolutions that preceded the Resolution A.849(20) and the 1997 Code, and it therefore seems that the UNCLOS term in Art. 94 para. 7 is prior to the 1997 definition. One could therefore be tempted to look at the 1969 Intervention Convention and the corresponding art. 221 UNCLOS, in which the term is employed as a subcategory of "maritime casualty",[118] namely, an event other than a collision or a stranding, outside the shipboard routine and affecting the ship in its function as a means of transport (see *supra*, at A. I. 1.). This definition is not too far from the

---

[114] 640 UNTS 63; BGBl. 1969 II, 249.

[115] Nordquist, Nandan and Rosenne, *2002 "UNCLOS Commentary Vol. III"*, para. 94.4.

[116] A/CONF.62/C.2/L.54, Official Documents of the UN Third Conference on the law of the sea, vol. 3, 229 et seq.

[117] C.2 Blue Paper No. 9, reproduced in R. Platzöder, *Third United Nations Conference on the Law of the Sea: Documents, Volume IV* (Dobbs Ferry (NY): Oceana Publications, 1983), p. 130.

[118] Nordquist, Nandan and Rosenne, *2002 "UNCLOS Commentary Vol. III"*, para. 94.8(k).

definition in point 4.4. of the 1997 Code, centring on the threat to the ship from, or in connection with, an operational cause.

Having considered the roots of the duty to investigation in the SOLAS Convention, it is fairly obvious that the provision's wording is close to the wording of the SOLAS annex, Part C, Chapter I, Regulation 21 formulation. The purpose of art. 94 para. 7 UNCLOS is formulated in a broader way, not being limited to the amelioration of an existing legal framework such as SOLAS or MARPOL, but attaching value to the investigation as such, if the casualty has had severe consequences. Still, in the context of art. 94 para. 1 and 3 UNCLOS, calling for safety at sea,[119] the provision aims for the same goals as the original SOLAS provision. It is therefore submitted that the notion of "marine casualties" as used in art. 94 para. 7 UNCLOS should be the same as in the early SOLAS provision. Indeed, the use of the term "marine casualties" rather than "maritime casualties" as in art. 221[120] most likely hints to the SOLAS roots of the provision, the term in the 1948 SOLAS version being "major marine casualty" before the allusion to "marine" was omitted in the 1974 convention, which only reads "casualty". It should therefore be accepted that the origin of art. 94 para. 7 is very different from the history of art. 221 UNCLOS and its definition, stemming from the field of marine pollution and intervention. UNCLOS is, after all, only an "umbrella convention" that sets out valid principles for the law of the sea, but that was implemented on top of already existing legal frameworks both in the area of marine environmental protection and maritime safety and is, in its vocabulary and its policies, inspired by the preceding conventions.[121]

It does therefore not seem completely right when it is submitted that "given the generality of that definition, there is no clear reason for limiting the meaning of the expression to the purposes of article 221 [. . .], and "marine casualty" and "maritime casualty" are taken to have the same meaning".[122] It should be kept in mind that for the art. 221 UNCLOS meaning, mirroring the definition in art. 2 para. 1 of the 1969 Intervention Convention, it is most often argued that intentional acts are not covered, seemingly as a result of the traditional separation between accidental and deliberate pollution. This is very questionable, as has been shown above, but still seems to be the understanding given to the term in most of the cases.

In contrast to this, the SOLAS provision is not concerned with the question of intent at all, as the causes of the accident and the events that allowed it to happen are still to be inquired into. The only conditions that render an incident a "casualty" under SOLAS are the adverse consequences referred to in the 1997 Code, featuring

[119] Ibid., para. 94.1.

[120] Ibid., para. 94.8(k), mentioning the difference and maintaining "that it is not clear why the English text maintained this minor linguistic discrepancy".

[121] See S. V. Scott, 'The LOS Convention as a Constitutional Regime for the Oceans', in A. G. Oude Elferink (ed.), *Stability and change in the law of the sea: The role of the LOS convention* (Leiden, Boston: Martinus Nijhoff, 2005), pp. 14 et seq.

[122] Nordquist, Nandan and Rosenne, *2002 "UNCLOS Commentary Vol. III"*, para. 94.8(k).

either material damage to the ship or crew or the outward appearances of a maritime accident. It therefore does not seem right to promote a common understanding for the whole of UNCLOS. If considered in detail, the understandings are different.

This does not mean that a common understanding of the term "maritime casualty" throughout UNCLOS and the other Conventions would not be desirable. Still, it seems more helpful to give up the outdated distinction between deliberate and accidental acts than to impose a modern understanding stemming from the area of marine pollution on a term that was already interpreted in a certain way in 1948.

For the purposes of this study, a "maritime casualty" will, in the following, rather be defined by the adverse consequences such as loss of life or injury to persons, damage to the ship, cargo or other property or the marine environment, without ruling out intentional acts from the start.

II. The Duty to Assist People in Distress at Sea

Keeping in mind that in the matter of the protection of life and property at sea a "casualty" only seems to be defined by the threat to the lives or the well-being of the people involved, the ship itself and the cargo, it seems worthwhile to have a look at regulation attempts concerning the general duty to assist people in distress at sea.

This problem is particularly interesting because the general obligation to offer help and rescue to persons in danger of being lost at sea has been recognised since time immemorial,[123] as a pragmatic solution to the extreme dangers of the maritime trade in its earliest beginnings.[124] Seafarers, who were, after all, engaged in the

[123] B. H. Oxman, 'Human Rights and the United Nations Convention on the Law of the Sea' (1998) 36, Columbia Journal of Transnational Law, 399–429 at 415; Nordquist, Nandan and Rosenne, *2002 "UNCLOS Commentary Vol. III"*, para. 98.1; J. E. Tauman, 'Rescued at sea, but nowhere to go: The cloudy legal waters of the "Tampa" crisis' (2002) 11, Pacific Rim Law & Policy Journal, 461–96 at 473; F. J. Kenney and V. Tasikas, 'The "Tampa" incident: IMO perspectives and responses on the treatment of persons rescued at sea' (2003) 12, Pacific Rim Law & Policy Journal, 143–77 at 148; R. Barnes, 'Refugee Law at Sea' (2004) 53, International and Comparative Law Quarterly, 47–77 at 49.

[124] Oxman, '1998 "Human Rights"', 414; Tauman, '2002 "Rescued at sea"', 473; K. X. Li and J. M. Ng, 'International Maritime Conventions: Seafarers' Safety and Human Rights' (2002) 33, Journal of Maritime Law and Commerce, 381–404 at 386; A. A. Severance, 'The Duty to Render Assistance in the Satellite Age' (2006) 36, California Western International Law Journal, 377–400 at 379; also see the early recognition of a humanitarian motive in "*Scaramanga v. Stamp*", 5 Law Reports, Common Pleas Division 295 (High Court (England), Common Pleas Division, 20 April 1880), p. 304: "The impulsive desire to save human life when in danger is one of the most beneficial instincts of humanity, and is nowhere more salutary in its results than in bringing help to those who, exposed to destruction from the fury of winds and waves, would perish if left without assistance. To all who have to trust themselves to the sea, it is of the utmost importance that the promptings of humanity in this respect should not be checked or interfered with by prudential considerations as to injurious consequences which may result to a ship or cargo from the rendering of the needed aid."

same enterprise, provided assistance to each other, regardless of the nationality of the distressed, the cost of rescue to the salvor and the juridical status of the sea area in which the distressed persons were found. It seems that the duty was widely recognised and apparently, as far as this can be assessed,[125] steadily fulfilled within the merchant marine, even though it was for a long time only considered a moral rather than a legal duty,[126] and means of enforcement were unclear in scope or inexistent.[127] Last but not least, its fulfilment generally did not result in a valid and quantifiable award for salvage,[128] but this of course also means that seafarers were not under an obligation to pay for their rescue.[129]

On an international level, this duty was first introduced into a written text in the 1910 International Convention for the Unification of Certain Rules Relating to Assistance and Salvage at Sea (1910 Salvage Convention).[130] Its Art. 11 reads:

> Every master is bound, so far as he can do so without serious danger to his vessel, her crew or passengers, to render assistance to everybody, even though an enemy, found at sea in danger of being lost.

Considering the political tensions between the major maritime nations at that time, the codification of this provision, reflecting the customary law in place,[131] can be seen as an important achievement. The Convention had a considerable success, especially in the first period after its adoption, obtaining more than 50 ratifications in the years before 1914,[132] among which figured most of the important maritime nations. The Convention however only placed an obligation on States and did not

---

[125] J. Z. Pugash, 'The Dilemma of the Sea Refugee: Rescue Without Refuge' (1977) 18, Harvard International Law Journal, 577–604 at 580, referring to the difficulties in controlling the fulfilment of this duty and in holding violators liable; M. Davies, 'Obligations and implications for ships encountering persons in need of assistance at sea' (2003) 12, Pacific Rim Law & Policy Journal, 109–41 at 1115 et seqq.; Severance, '2006 "The Duty to Render Assistance"', 397 et seq. concerning the ineffective enforcement mechanisms in more recent times.

[126] W. H. Smith, 'The duty to render assistance at sea: Is it effective or adrift?' (1971) 2, California Western International Law Journal, 146–63 at 146; Severance, '2006 "The Duty to Render Assistance"', 379.

[127] Smith, '1971 "The duty to render assistance"', 162, also referring to art. 11 of the 1910 Salvage Convention and art. 12 para. 1 (a) of the 1958 Geneva Convention on the High Seas.

[128] Ibid., p. 147; Severance, '2006 "The Duty to Render Assistance"', 380.

[129] Li and Ng, '2002 "International Maritime Conventions"', 387.

[130] RGBl. 1913, 66; the text of the convention (in English) is available at http://www.unidroit.info/program.cfm?menu=convention&file=convention&lang=en&msid=2 (last visited 7 March 2014).

[131] R. E. Fife, 'The Duty to Render Assistance at Sea: Some Reflections after "Tampa"', in J. Petman and J. Klabbers (eds.), *Nordic Cosmopolitanism: Essays in international law for Martti Koskenniemi* (Leiden, Boston: Martinus Nijhoff Publishers, 2003), p. 475.

[132] The status of the Convention, now widely absorbed by the IMO 1989 Convention on Salvage, can be found with the Belgian authorities, since Belgium is the depositary, http://diplomatie.belgium.be/fr/binaries/i2a_tcm313-79745.pdf (last visited 7 March 2014).

bind private actors,[133] and it did not address the problem that salvage remuneration was generally not awarded for saving life at sea, as opposed to the salvage of property.[134]

Since 1910, almost every international convention dealing with the matter of safety at sea comprises provisions that describe the duty to rescue life in distress at sea[135] or that mean to ensure the best possible preconditions to guarantee assistance at sea. Most of them are based on the wording of their predecessors.[136] Examples of this include the original 1914 SOLAS Convention,[137] containing in art. 12 the prohibition against the misuse of distress signals or the use of signals that could be mistaken for distress signals, and the explicit codification of the duty in art. 37. The provision became more refined in the 1948 SOLAS version, setting out in Chapter V, Regulation 10 a procedure to follow after the reception of a distress message. The duty was furthermore laid down in art. 12 (1) of the 1958 Geneva Convention on the High Seas and eventually in art. 98 UNCLOS, all of them conferring the obligation to enforce this duty on the flag State.[138] The 1979 International Convention on Maritime Search and Rescue (SAR)[139] obliges States to provide public search and rescue services[140] in the geographic areas they are responsible for.[141]

The 1910 Salvage Convention was replaced in 1989 by an IMO Convention, the 1989 Salvage Convention, which also refers to this duty but which is, after all, only applicable once judicial or arbitral proceedings relating to salvage matters are brought in a State Party (art. 2).[142]

All of these provisions do not refer to "accidents" or "casualties", but to "distress" or the "danger of being lost at sea". With regard to the humanitarian purpose behind this very ancient concept, it is clear that the notion does not centre on the reasons of why people find themselves in a perilous situation at sea, but on the situation of danger as such. The question of what "distress" or "danger of being

[133] M. Pallis, 'Obligations of States towards Asylum Seekers at Sea: Interactions and Conflicts Between Legal Regimes' (2002) 14, International Journal of Refugee Law, 329–64 at 332, footnote 13.

[134] Severance, '2006 "The Duty to Render Assistance"', 380 et seq.

[135] Pugash, '1977 "The Dilemma"', 579.

[136] Pallis, '2002 "Obligations of States"', 331.

[137] The text can be found at http://www.imo.org/KnowledgeCentre/ReferencesAndArchives/HistoryofSOLAS/Documents/SOLAS%201914.pdf (last visited 7 March 2014).

[138] Severance, '2006 "The Duty to Render Assistance"', 384.

[139] 1405 UNTS 97; BGBl 1982 II, 485.

[140] See Severance, '2006 "The Duty to Render Assistance"', 390 et seq. for more details.

[141] See Pallis, '2002 "Obligations of States"', 335 et seq.

[142] Davies, '2003 "Obligations and implications"', 114, also see F. Odier, 'L' assistance maritime - une institution en quête de reconnaissance', in *Droit de la mer: Études dédiées au doyen Claude-Albert Colliard* (Paris: Pedone, 1992), pp. 67 et seqq. for an overview on the differences between the 1910 and the 1989 Salvage Conventions.

lost at sea" means is not defined in international law,[143] at least not in this context. As opposed to that, definitions of "distress" have been elaborated in the similar situation of ships in distress seeking refuge in ports,[144] an occurrence in which the vessel's right to shelter is most often contested or denied because coastal States wish to protect their coastline from the floating hazards. In most of these modern cases, it is not the crew's lives that are at stake, but environmental or economic interests. Still, it is generally argued[145] that notwithstanding the complex problems arising in this situation, in any case, there is "distress" where peoples' lives are endangered.[146] In this context, the duty to rescue persons in distress can be considered to correspond to their right to life.[147]

In the matter of assistance to people in distress at sea, this question has been a subject of debate with regard to illegal immigration. Merchant ships' crews might be tempted to turn a blind eye on unseaworthy, stricken boats full of desperate migrants that will obviously not be welcome in the next or any other port of call,[148] and authorities of some coastal States have opposed disembarkation of these castaways in their ports,[149] in some extreme cases hinting at that the immigrants' dangerous situation at sea was to some degree self-inflicted.[150]

A full discussion of this problem is outside the scope of this study, but it is interesting that the discussion of when a ship or its crew or passengers are deemed to be in "distress" only seems to emerge in this context where, for policy reasons, States are reluctant to render the necessary assistance to castaways. It seems that outside this controversial debate, the duty to assist people in distress at sea is well-established and recognised and that notwithstanding the lack of a clear definition, there does not seem to be debate as to when people are considered to be "in distress".

This will most likely be due to the fact that customary law and the conventions pointed out above confer a wide discretion to the master and the crew of the vessel

---

[143] M. Pugh, 'Drowning not Waving: Boat People and Humanitarianism at Sea' (2004) 17, Journal of Refugee Studies, 50–69 at 58; S. Rah, *Asylsuchende und Migranten auf See* (Berlin: Springer, 2009), p. 102.

[144] Barnes, '2004 "Refugee Law at Sea"', 60; Rah, *2009 "Asylsuchende und Migranten"*, p. 102, claiming that the criteria are basically identical.

[145] Barnes, '2004 "Refugee Law at Sea"', 60, also see "*Phelps, James & Co. v. Hill*", 1 Law Reports, Queen's Bench Division 605 (High Court (England), Queen's Bench Division, 28 February 1891), p. 613: "So strict is the rule with regard to deviation, that, while the master may deviate to save life, he may not deviate to save property."

[146] Also see the detailed analysis of the problem in I. von Gadow-Stephani,*Der Zugang zu Nothäfen und sonstigen Notliegeplätzen für Schiffe in Seenot* (Berlin, New York: Springer, 2006), pp. 213 et seqq.

[147] Pallis, '2002 "Obligations of States"', 335.

[148] Davies, '2003 "Obligations and implications"', 141; Fife, *2003 "The Duty to Render Assistance"*, pp. 476 et seq.

[149] Tauman, '2002 "Rescued at sea"', 491 et seqq., quoting several examples.

[150] Pugh, '2004 "Drowning not Waving"', 56.

coming to the rescue.[151] This is obviously the only practicable solution, considering that both situations of danger and the appropriate means of assistance[152] may be manifold and that the rescuing ship's command is the only instance present that could judge the situation at close range.[153] In summary, it seems that there is "distress" where there are reasonable grounds to believe that without help the stricken vessel and its crew or passengers will not reach safety and be lost at sea,[154] which clearly points to an imminent threat of death,[155] thus creating a comparatively high threshold at first sight. If considered more carefully, it becomes obvious that in basically any situation in which a vessel, its crew or passengers meet with unfavourable conditions beyond a certain degree,[156] for either the lack of provisions, fuel, for technical or medical problems, there is a manifest risk of death considering that the oceans are an environment hostile to human life after all.

There is a strong opinion holding that the duty does not apply without restrictions where the situation of distress has been brought about intentionally by the stricken vessel's crew, with a view to obtaining access to the coastal States' institution and possibly circumventing immigration or trade laws[157] or even to lure ships closer to subsequently attack them.[158] Still, even in these cases of abuse of the right to assistance, the appropriate answer cannot be the point-blank denial of rescue if people are objectively in danger,[159] but must be found in the aftermath of the rescue, in the further proceedings of their cases, and thus, in either criminal law,[160] immigration law or in holding them liable for the rescue costs under civil law. It is maintained, and obviously rightly so, that it should not be the task of the salvors to assess the reasons for the perilous situation and act accordingly, but that they should unconditionally come to the rescue of those that they deem in distress.[161]

---

[151] Kenney and Tasikas, '2003 "The "Tampa" incident"', 151; Fife, *2003 "The Duty to Render Assistance*", p. 480; Pugh, '2004 "Drowning not Waving"', 59; Rah, *2009 "Asylsuchende und Migranten*", p. 102.

[152] Kenney and Tasikas, '2003 "The "Tampa" incident"', 151 et seq.

[153] Pugh, '2004 "Drowning not Waving"', 59, 64.

[154] D. R. Rothwell, 'The Law of the Sea and the MV Tampa Incident: Reconciling Maritime Principles with Coastal State Sovereignty' (2002) 13, Public Law Review, 118–27 at 123; Pugh, '2004 "Drowning not Waving"', 59.

[155] Churchill and Lowe, *1999 "The Law of the Sea*", p. 63.

[156] Also see "*Kate A. Hoff v. United Mexican States*", 4 UN Reports of International Arbitral Awards 444 (International Arbitral Tribunal, 02 April 1929), p. 447, holding that a ship can be considered to be in distress even when it can still manoeuvre or reach the port under its own power.

[157] A. Heiser, 'Border Protection; UNCLOS and the M. V. Tampa Incident 2001' (2002) 16, Australian and New Zealand Maritime Law Journal, 84–121 at 100; Fife, *2003 "The Duty to Render Assistance*", p. 481.

[158] Pugh, '2004 "Drowning not Waving"', 59.

[159] Fife, *2003 "The Duty to Render Assistance*", p. 481; Rah, *2009 "Asylsuchende und Migranten*", p. 103.

[160] Pallis, '2002 "Obligations of States"', 340.

[161] Fife, *2003 "The Duty to Render Assistance*", p. 482; Pugh, '2004 "Drowning not Waving"', 59.

There again, in this centuries-old duty to assist people in distress at sea, considerations about how the dangerous situation first came about are left aside, and intentional acts may lead to "distress" or "danger of being lost at sea" in the same way accidents can. Again, the situation that triggers the duty to act is simply defined by the adverse consequences on people's lives and well-being. In this area, one still mostly driven by humanitarian considerations, the reasons for defining the actionable event simply by the damage, danger or threat of damage become even more obvious.

Part II
Civil Liability Regimes

Chapter 3
Civil Responsibility Regimes and the Liability for Cargo Oil Pollution

The fact that the consequences of maritime accidents, especially the environmental damage, would not be confined to only one country, was recognised at an early point in time. In the *Torrey Canyon* incident, the coastlines of both England and France were affected, and it became obvious that an international regime to cope with pollution damage, bundling victims' claims and submitting them to a uniform regime, would be beneficial to both the victims and the shipping industry. Victims would be compensated faster, and according to a uniform regime, regardless of differences in domestic law, and the risk of environmental damage would become more calculable to shipowners and their insurers, also conferring on them the benefit of limited liability. The ensuing initiative eventually resulted in the adoption of the CLC, but the funds available from shipowners and their insurers soon proved to be insufficient. Not long after the adoption of the CLC, a second tier of compensation was established by the 1971 FUND Convention, which was paid for by the oil industry.

Other forms of environmental pollution have also been submitted to international civil responsibility regimes, some of them equally successful, such as the 2001 Bunker Oil Pollution Convention (BOPC),[1] some of them rather stuck in the process of coming into force, such as the 1996 International Convention on Liability and Compensation for Damage in Connection with the Carriage of Hazardous and Noxious Substances by Sea (HNS).[2]

In all of these later initiatives, the CLC, as the first and as a rather successful convention, has served as a model. In drafting the later regimes, the negotiating parties have shown a certain tendency to start from the already existing provisions of the CLC and to take them over or change them according to the necessities of the

[1] 40 ILM 1493; BGBl. II 2006, S. 578.

[2] 35 ILM 1415.

S.F. Gahlen, *Civil Liability for Accidents at Sea*, Hamburg Studies on Maritime Affairs 32, DOI 10.1007/978-3-662-45555-5_3

new regime in question.[3] As the CLC itself has been subject to quite important amendments, there is thus a certain development in which, it seems, the regimes influence each other. Still, the constant use of the already established provisions sometimes dating from as early as the 1960s has sometimes also prevented the adoption of new provisions that could arguably reflect the risks and challenges of shipping in the present days better.[4]

The early decision to create a responsibility regime for one pollutant, oil carried as cargo, has been at the root of a heavy fragmentation of the international law on liability for marine pollution.[5] Still, the ideas and concepts of the CLC have had a huge impact, and the civil liability regime for passenger claims that is discussed in Chap. 5 is perhaps the best example for the fact that civil liability for accidents at sea gives rise to very particular problems, but which tend to be the same for all the heads of damage.

Notwithstanding the fragmentation of the system, the different liability regimes thus influence each other. As will be seen, the interpretation of provisions in a given civil liability regime tends to be governed by the results that have been found under the corresponding provisions of other conventions, and despite the abundance of maritime civil liability conventions, there is a certain, and arguably very valuable, harmonisation between them.

However, the fragmentation of the regime following different pollutants in the conventions for the protection of the environment often creates overlaps,[6] and sometimes lacunae, where a claim or a case is left uncovered by all the international regimes. These latter claims will, in the last instance, need to be covered by domestic law, the difficulties of which will be analysed in the third part of this study (Chaps. 6 and 7). In the present and the two following chapters, the gaps as such will be highlighted, in order to consider whether there are ways to broaden the scope of application of the international regimes or whether there could be, in theory, a more universal regime.

With regard to the problem of uniformity, the conventions that will be analysed in this part of the study are generally applied by national courts, and according to domestic procedural rules. Depending on the respective mechanism to implement international conventions into national law, the provisions may either be applied directly, thus making them a part of the domestic legal system, or they may be transposed into national law by a separate act of law-making, rephrasing the convention's provisions into domestic law. If a convention is implemented in this latter way, there is of course a risk that the provisions of the national law will differ

---

[3] N. A. Martínez Gutiérrez, *Limitation of Liability in International Maritime Conventions* (London, New York: Routledge, 2011), p. 152.

[4] N. Gaskell, 'Compensation for Offshore Pollution: Ships and Platforms', in M. Clarke (ed.), *Maritime Law Evolving: Thirty Years at Southampton* (Oxford, Portland: Hart Publishing, 2013), p. 64.

[5] Lucchini, '1970 "La pollution des mers"', 827; Gaskell, *2013 "Compensation"*, p. 64.

[6] Ibid., p. 64.

from the original international convention, or will be more open to different understandings. Interpretations of provisions by national courts, even if the rules are directly applicable, always present the danger that the results of comparable sets of facts may differ a lot between countries. In the analysis of civil liability regimes in the following three chapters, some of the discrepancies will be presented.

International treaties are the result of a compromise between the negotiating parties. Especially with regard to the conventions regarding the liability for environmental pollution, the drafting of the texts was usually carried out to address a certain issue, and in the way the *Torrey Canyon* accident triggered the negotiation of the CLC, other accidents influenced the amendment of the regime or initiated the introduction of new regimes for other types of accidents. Considering that negotiation and drafting were often inspired by a certain disaster, the regimes as such tend to be rather incomplete in themselves.

With regard to the possible gaps in conventions, there may be "systemic" shortcomings on the one hand. The negotiating parties to a convention may have consciously chosen to exclude certain sets of facts from the scope of applicability, for a lack of unanimity in this question, or, not least, for policy reasons. On the other hand, there may be gaps in the regulatory framework that have simply been overlooked during the negotiations.

The CLC is a very successful convention, having been ratified by 132 States and covering 96, 86 % of the world tonnage in its amended form of 1992.[7] Together with its complementary conventions, the 1971 FUND Convention and the Supplementary Fund that was established by a Protocol in 2003, it creates a comprehensive regime, ensuring a comparatively quick compensation to victims and setting out uniform standards as to the liable parties and the amount of compensation.

The 1992 Protocol to the CLC amended the original version of 1969. Apart from factual amendments, it also contains a "denunciation" clause, by virtue of which States parties to the 1992 Protocol cease to be Parties to the 1969 CLC. The 1992 rules therewith replace the 1969 version, but the 1969 CLC remains applicable to States parties who did not sign the 1992 Protocol.

Considering the extent of damage that can be caused by modern tankers, the civil responsibility regime for cargo oil pollution would not be complete without the FUND Convention, originally adopted in 1971. The Convention creates a second tier of compensation, beyond the funds available under the CLC. It is exclusively funded by the oil industry, as contributions are collected from any person in a contracting State who has, within a year, received more than 150,000 tons of oil (art. X FUND). The monies are collected on a past-event basis, once the necessity to step in arises for the FUND Convention, and are then distributed to victims on the basis of claims made by them. The system is essentially meant to be triggered if the damages of a given oil spill exceed the amounts that are available under the CLC, or

[7] See the summary of the Status of IMO Conventions, http://www.imo.org/About/Conventions/StatusOfConventions/Documents/Summary%20of%20Status%20of%20Conventions.xls (last visited 7 March 2014).

in the case of the shipowner's or his insurer's insolvency. Furthermore, the FUND will also relieve the shipowner for the expenses he incurred voluntarily to prevent or minimize pollution damage (art. IV para. 1 FUND), even if the damage was avoided altogether,[8] thus encouraging shipowners to take all possible measures of this kind.[9] The Fund is a separate legal entity, independent from its contributors and legally represented by the Director (art. II para. 2 FUND). Liability under the FUND Convention is limited, too, but to a considerably higher amount than under the CLC. The limits under the 1971 original FUND Convention have been increased under a 1992 Protocol, actually establishing a new Fund that will only apply to incidents after this date.

The regime for cargo oil pollution furthermore features a third leg of compensation. Following a number of very disastrous oil spills by tankers in European waters, the European Commission came up with an initiative to create another Fund, that would provide for compensation beyond the means available under the FUND Convention, but which would only apply to incidents having taken place in European waters. It became clear that some contracting States of the CLC and FUND regime, and not only in Europe, considered the existing liability limits as inadequate for major oil spills, whereas other countries did not feel the need to increase the funds available and were afraid of the ensuing costs.[10] The regionalisation advocated by the European initiative was perceived as a major threat to the whole of the system,[11] and the only way out of this deadlock was the establishment of a voluntary third leg of compensation under a supplementary protocol to the FUND Convention. States that ratify it benefit from the higher liability limits and the additional financial means available under this Supplementary Fund, but for all other states the existing regime under the CLC and the FUND will remain in place. The 2003 Protocol establishing the Supplementary Fund has currently 30 contracting States, representing close to 19 % of world maritime tonnage.[12] The Supplementary Fund will distribute payments beyond the liability limits of the Fund and up to a higher limit wherever there is a risk that the damage may exceed the compensation available under the FUND Convention, thus being a

---

[8] R. H. Ganten, 'Developments in Oil Pollution Liability' (2008) 38, Environmental Policy and Law, 312–5 at 312.

[9] G. M. Gauci, 'Protection of the Marine Environment through the International Ship-Source Oil Pollution Regimes' (1999) 8, Review of European Community and International Environmental Law, 29–36 at 33.

[10] See M. Jacobsson, 'L'expérience française du FIPOL' (2007) 59, Droit Maritime Français, 968–84 at 972 et seq.

[11] See R. P. Balkin, 'Some Future Developments in Liability and Compensation for Environmental Damage at Sea', in M. H. Nordquist, J. N. Moore and S. Mahmoudi (eds.), *The Stockholm Declaration and Law of the Marine Environment* (The Hague, London: Martinus Nijhoff Publishers, 2003), pp. 440 et seqq. on the developments that led to the adoption of the Supplementary Fund.

[12] See http://www.imo.org/About/Conventions/StatusOfConventions/Documents/Summary%20of%20Status%20of%20Conventions.xls (last visited 7 March 2014).

potential benefit to claimants even before it can be stated that the means of the Fund will be exhausted by an incident. Whereas the Supplementary Fund is a separate legal entity from the Fund, featuring its own administrative organs, it operates in close cooperation with the Fund, only taking into account claims that have already been recognised by the Fund and relying on its claims procedure.

The CLC, the FUND and the 2003 Protocol establishing the Supplementary Fund thus create a comprehensive regime under which most claims for cargo oil pollution can be settled quickly and in an internationally uniform way.

A. Scope of Application as an International Regime

It goes without saying that an international convention can only bind the States parties to it. With regard to the difficulties to establish territorial claims at sea, a civil liability regime for marine environmental pollution must take into account the realm that its contracting States can submit to their jurisdiction. At the same time, pollution at sea does not respect borders, even less than pollution on land. With regard to the fact that shipping is an international business, and that there can be, apart from the affected States, quite a number of jurisdictions involved, it is obviously possible that claimants establish a judicial venue in a jurisdiction which is not a party to the CLC. The interplay of the Convention with the domestic choice of law rules of the forum can then lead to a number of difficulties.

Furthermore, the geographical scope of application causes difficulties in the borderline cases of pollution on the high seas, and, perhaps surprisingly, in cases where there is only one jurisdiction involved.

I. Geographical Applicability

According to art. II of the CLC 1969, the Convention applies exclusively to pollution damage caused on the territory including the territorial sea of a contracting State. This caused uncertainty as to the applicability to the Convention to States' EEZ when this concept was established by UNCLOS 1982. The 1992 Protocol clarified this point, altering art. II so as to explicitly include contracting States' EEZ into the field of application.

The CLC is therefore inapplicable to pollution damage caused in the jurisdiction of a non-contracting State. The United States, through whose waters pass a significant number of worldwide oil transports every year, both on the way from and to their ports[13] and in passage through their waters, are the most prominent example.

[13] H. Wang, *Civil liability for marine oil pollution damage: A comparative and economic study of the international, US and the Chinese compensation regime* (Alphen aan den Rijn: Kluwer Law International, 2011), p. 190.

Even apart from the U.S. and notwithstanding the impressive quota of ratification in terms of world tonnage, there is still an important absolute number of States that signed neither the parent Convention nor the 1992 Protocol,[14] and not all of them are land-locked. Especially among some of the Pacific Islands nations (Micronesia, Nauru) and some Caribbean States (Haiti, Cuba), there are gaps on the map of implementation of the CLC.

1. Applicability of the CLC in Non-contracting States

Apart from the fact that the CLC will not apply to damage caused on the territory of non-contracting States, not even when proceedings are started in a contracting State (art. II CLC), the CLC's provisions are not part of the domestic law of non-contracting States either and will possibly not be recognised in their courts. Thus, even if the oil spill took place in the waters of a contracting State, it is not evident whether a non-contracting State's court will apply the CLC.

An answer to this problem will necessarily entail an analysis of private international law. A non-contracting State's courts will treat an oil spill in the waters of a foreign jurisdiction first as a set of facts with a foreign element, which will lead to the application of conflict-of-law rules that differ from country to country. An outline of the conflict-of-law rules of every non-contracting State is clearly out of the scope of this study, but with regard to the importance of the United States as a market of transport of oil by sea and with regard to the *Amoco Cadiz* case, an analysis will be given of this problem as treated by the US courts.

a) The *Amoco Cadiz* Case

The *Amoco Cadiz* was a Liberian-flagged tanker, owned by the Liberian subsidiary of an US American company. The shipowner was, in effect to a wide extent, controlled by its parent. In March 1978, following a steering gear failure on her way from the Persian gulf to Rotterdam, the *Amoco Cadiz* ran aground off the coast of Brittany, while being fully laden with 220,000 tons of light crude oil. Neither the crew in its attempts to cope with the situation nor the ensuing salvage operations were successful and the ship released its entire cargo into the sea, resulting in a pollution of about 300 km of coastline and an estimated damage of about 800 million French Francs.

France, at that time, was a party to the CLC 1969, but not to the 1971 FUND Convention. Under the liability limits of the CLC 1969, the victims could have

[14] http://www.imo.org/About/Conventions/StatusOfConventions/Documents/status-x.xls (last visited 7 March 2014).

recovered about a tenth (77 million French Francs) of the actual damage[15] from the shipowner without having to prove his wrongful behaviour (art. III para. 1 CLC).

In effect, the shipowner immediately established a liability fund to this amount with the courts in Brest, Brittany. The aggrieved parties in France, among which were the French Republic in the first place, the *départements* of Finistère and Côtes-du-Nord and a number of municipalities in these regions, and a group of private interests consisting of private companies and associations, were of course interested in recovering their entire loss. In order to overcome the limitation of liability under the CLC and obtain compensation for the actual damage, it would have been necessary to prove the owner's actual fault or privity in the ship's stranding. This was considered as rather difficult,[16] not least because the managing owners of the ship were companies based in the United States which created difficulties with regard to obtaining evidence for proceedings in France.[17] Furthermore, it was considered easier to plead and prove the negligent conduct of an entire company group under US procedural law than under the corresponding French provisions.[18]

The number of different claims and parties that were involved in proceedings in the US American courts render the case rather complicated. First of all, it is important to mention that the French State—in contrast to the other claimants—did not sue the shipowner,[19] Amoco Transport Company (referred to as "Transport" in the judgments), because it considered that this claim should be governed by the CLC and could not be pursued in the American courts.[20] The defendants mainly targeted by the claimants were Transport's parent company Standard Oil Company Indiana ("Standard") who effectively operated the vessel, and Standard's parent company Amoco International Oil Company ("AIOC"). Transport was actually only sued by the *département* of Côtes-du-Nord. Still, this claim against Transport would have been the only one to which the CLC would have been applicable, even in the French courts, as the CLC then only provided for the strict liability of the shipowner and was not concerned with claims against other parties, as long as they could not be termed "servants or agents of the owner" (art. III para. 4 CLC 1969).[21] The legal situation has changed under the 1992 Protocol which includes "charterer, manager or operator" into the circle of persons against whom the action is

[15] Fontaine, *1993 "The French Experience"*, pp. 102 et seq.

[16] E. du Pontavice, 'Affaire "Droit de l'environnement versus Droit maritime" ou la décision rendue le 18 avril 1984 concernant "l'Amoco Cadiz"' (1985), Annuaire de droit maritime et aérien, 9–60 at 16 et seq.

[17] Ibid., p. 18.; Fontaine, *1993 "The French Experience"*, p. 103, A. de Raulin, 'L'épopée judiciaire de l'Amoco Cadiz' (1993), Journal du Droit International Privé, 63–96 at 88.

[18] L. Lucchini, 'Le procès de l'"Amoco Cadiz": Présent et voies du futur' (1985) 31, Annuaire français de droit international, 762–82 at 768; E. M. Rodríguez Gayan, 'Claves de derecho privado en el asunto "Prestige"' (2003), Revista Española de Derecho Internacional, 117–47 at 141.

[19] Emphasised by du Pontavice, '1985 "Affaire "Droit de l'environnement""', 11 et seq.

[20] Lucchini, '1985 "Le procès de l'"Amoco Cadiz""', 768; du Pontavice, '1985 "Affaire "Droit de l'environnement""', 11.

[21] See Lucchini, '1985 "Le procès de l'"Amoco Cadiz""', 772 for more details.

prohibited, and this problem will be considered at a later point in this study (see *infra* under E.).

Before considering the legal implications of the proceedings in the United States, it must therefore be kept in mind that after all, the only claim that can be addressed here as a CLC claim in proceedings in a non-contracting State is the claim against Transport, but it was bundled in the lengthy and complicated proceedings and was obviously not in the centre of attention. Most proceedings took place in the Northern District court of Illinois, which was designated as the competent court for all claims in connection with the accident by the Judicial Panel on Multidistrict Litigation in June 1979.[22]

An opinion on the grounds of the claims against the shipowner's parent company was issued on 18 April 1984.[23] In this opinion, the court clearly stipulated that since the damage had been sustained in French territorial waters or on the French coast, "the substantive law applicable to such claims would therefore have been French law".[24] The court, in designating the substantive law called to govern the case, followed the rule of *lex loci delicti*,[25] applying the law of the place where the damage was sustained. Whereas the localisation of the damage can be extremely problematic in other cases of maritime torts,[26] the case in question was only concerned with the pollution damage to the French coast, so that the place of the damage was rather obvious.[27] Still, the court did not apply French law, but maintained that "it was not proven different"[28] from that of the United States and that "Claimants Côtes du Nord, in open court, stipulated that United States law applied".[29]

The claimants thus somehow agreed to submit their case to United States law,[30] which overruled the conflict-of-law rule of *lex loci delicti*. This is possible under

[22] *"In Re Oil Spill by the Amoco Cadiz"* 1979 American Maritime Cases 1811 (Judicial Panel on Multidistrict Litigation, 04 June 1979).

[23] US Northern District of Illinois Court (Eastern Division), *"In Re Oil Spill by the Amoco Cadiz"* 1984 American Maritime Cases 2123 (US District Court, Northern District of Illinois (Eastern Division), 18 April 1984).

[24] Ibid., p. 2189.

[25] C. Kloss, 'Der Untergang der "Amoco Cadiz" vor amerikanischen Gerichten' (1989), Praxis des Internationalen Privat- und Verfahrensrechts, 184–8 at 186; with doubts, but, considering with regard to the decision's wording that the court applied the *lex loci delicti* rule rather than the rule of the most significant relationship; V. Tomljenović, 'Maritime Torts' (1999), Yearbook of Private International Law, 249–97 at 289.

[26] See, by way of example M. P. George, 'Choice of Law in Maritime Torts' (2007) 3, Journal of Private International Law, 137–71 and Part III of this study.

[27] N. J. Eskenazi, 'Forum Non Conveniens and Choice of Law In Re: The "Amoco Cadiz" Oil Spill' (1993) 24, Journal of Maritime Law and Commerce, 371–98 at 392.

[28] *US District Court, Northern District of Illinois (Eastern Division), 18.04.1984, "In Re Oil Spill by the Amoco Cadiz"*, p. 2189.

[29] Ibid., p. 2189.

[30] Tomljenović, '1999 "Maritime Torts"', 289; Rodríguez Gayan, '2003 "Claves de derecho privado"', 141.

United States procedural law,[31] which will apply the *lex fori* if the applicability of foreign law is not pleaded and proven, and which often fictitiously assumes that the foreign law is identical to the domestic rules. Considering the claim against Transport, the shipowning company, this assumption was slightly artificial, as the court subsequently referred very explicitly to the applicability of the CLC in French Law[32] to this claim.

Even if the court recognised that United States law was not called to govern the proceedings in the first place, and only applied it upon request of the parties, it did decide the case, to some extent, according to the domestic law of a non-contracting State. It has been rightly pointed out, and it is certainly interesting that in its reasoning as to the applicability of the CLC, the court only explicitly mentions the shipowner's mother companies and then declares that the CLC in any case does not apply to them, as they were not the shipowner's "servants or agents".[33] The question of whether the CLC applied to Transport, the shipowner, is not explicitly answered in the first instance judgment.[34] In its considerations as to the liability of Transport, the court came to the conclusion that in any case, Transport could not invoke the limitation of liability since the accident had been caused by its fault or privity.[35] The court however entirely omitted to declare whether it based its refusal to grant the limitation of liability on art. V para. 2 CLC 1969 or on the US general limitation statute for maritime claims. It however repeatedly considers the role of the CLC as a part of French law,[36] especially with regard to its non-applicability to defendants other than the shipowner and his servants or agents. It has consequentially been argued that the judgment implies that, as against Transport, the shipowner, the court implied the applicability of the CLC and actually decided this claim according to the Convention's provisions.[37] As refined as this observation is, it is at odds with the explications of the court in the second instance. The US Court of Appeal, affirming the District Court's judgment concerning jurisdiction and

[31] S. Symeonides, *American Private International Law* (Alphen aan den Rijn: Kluwer Law International, 2008), para. 161 et seqq.

[32] A. Kiss, 'L'affaire de l'"Amoco Cadiz": responsabilité pour une catastrophe écologique' (1985), Journal du Droit International (Clunet), 575–89 at 587; also see Eskenazi, '1993 "Forum non conveniens"', 397, but omitting the difference between the claim against Transport and the claims against the parent companies.

[33] *US District Court, Northern District of Illinois (Eastern Division), 18.04.1984, "In Re Oil Spill by the Amoco Cadiz"*, p. 2190.

[34] du Pontavice, '1985 "Affaire "Droit de l'environnement""', 25.

[35] *US District Court, Northern District of Illinois (Eastern Division), 18.04.1984, "In Re Oil Spill by the Amoco Cadiz"*, pp. 2193 et seq.

[36] P. Bonassies, 'La décision "Amoco Cadiz"' (1984) 36, Droit Maritime Français, 688–98 at 691; R. Kbaier and V. Sebek, 'New trends in compensation for oil pollution damage' (1985), Marine Policy, 269–79 at 272.

[37] D. A. Jacobsen and J. D. Yellen, 'Oil Pollution: The 1984 London Protocols and the "Amoco Cadiz"' (1984) 15, Journal of Maritime Law and Commerce, 467–88 at 485 et seq.; Lucchini, '1985 "Le procès de l'"Amoco Cadiz""', 771; du Pontavice, '1985 "Affaire "Droit de l'environnement""', 26.

liability in its entirety, clearly referred to the US general limitation statute when confirming that Transport could not benefit from a limitation of liability.[38]

It must however be mentioned that the proceedings concerning the compensation amounts in the *Amoco Cadiz* case, as opposed to the merits of the case, were decided under French Law.[39] Hence, in contrast to the common perception, it seems that it was the US procedural law[40] and the US courts' generosity in holding mother companies liable for the actions of their subsidiaries[41] that caused the claimants to be successful in their lawsuit on the merits of the claim, rather than the non-application of the CLC. Considering that it was French law that was applied to the question of which claims were compensable and to what amount, it does not seem that the foreign venue significantly contributed to higher awards, always assuming that Transport would have been found to have lost the limitation of liability under the CLC, too.

Still, it should not be forgotten that only a number of exceptional circumstances made this claim possible, namely the willingness of the French claimants to submit the proceedings to a different judicial system and the fact that the French State financially backed these proceedings, which would otherwise have been prohibitively expensive to private entities.[42] It must by all means be kept in mind that even in France, claims against the shipowner's parent companies Standard and AIOC would not have been governed by the CLC[43] and that a foreign oil company may, not least for policy reasons, also have incurred important liabilities in a tort claim in France following a disastrous pollution of the French coastline.[44]

Furthermore, in the United States proceedings, the defendants apparently could have defended themselves against the Illinois forum and the application of United States law by pleading *forum non conveniens*, considering that the proceedings concerned a casualty that occurred in France and mostly concerned French victims and in which pieces of evidence and witnesses were first and foremost found in France and in Italy where the vessel's crew came from.[45] On the other hand, all documents concerning the Amoco Parties were in the United States, and it has also been argued that the court may have considered the United States was the right

---

[38] *"In Re Oil Spill Amoco Cadiz"* 1992 American Maritime Cases 913 (US Court of Appeals for the Seventh Circuit, 24 January 1992), pp. 937 et seq.

[39] *"In Re Oil Spill by the Amoco Cadiz"* 1989 Il Diritto Marittimo 876 (US District Court, Northern District of Illinois (Eastern Division), 11 January 1988).

[40] du Pontavice, '1985 "Affaire "Droit de l'environnement"'", 37 et seq.

[41] de Raulin, '1993 "L'épopée judiciaire"', 87 et seq.; Kloss, '1989 "Der Untergang der "Amoco Cadiz"'", 185; T. Ballarino, 'Questions de droit international privé et dommages catastrophiques' (1990) 220, Recueil des Cours, 299–387 at 374.

[42] Fontaine, *1993 "The French Experience"*, p. 105.

[43] J. A. Zimmermann, 'Inadequacies of the Oil Pollution Act of 1990: Why the United States should adopt the Convention on Civil Liability' (2000) 23, Fordham International Law Journal, 1499–539 at 1527.

[44] But see de Raulin, '1993 "L'épopée judiciaire"', 88, expressing a more pessimist view.

[45] Eskenazi, '1993 "Forum non conveniens", 381 et seqq.

forum to judge an alleged tortious behaviour by American companies.[46] It can be presumed that the shipowner's mother companies preferred to defend themselves in proceedings in their own country, as they might not have been protected by the limits of liability under the CLC in France anyway.

b) The *Prestige* Case

The problem of the application of the CLC in a non-contracting State arose again, albeit in a slightly different context, in the US proceedings following the *Prestige* accident. The catastrophic sinking of this oil tanker took place off the coast of Galicia, Spain. The *Prestige* flew the Bahamian flag at the time, and she was owned by a Liberian shell company.[47]

On a voyage from Latvia to Singapore in November 2003 and laden with heavy fuel oil, the *Prestige* encountered heavy weather when passing the Spanish coast. Due to structural weaknesses in the 26-year-old tanker's hull, the *Prestige* began to list, drifted within about five miles of the Spanish coast and started to leak her cargo into the sea. Salvors tried over several days and apparently in several countries[48] to find a safe place of refuge for the ship, but without success. The *Prestige* was finally towed out to sea and sank in a distance of about 280 km from the Spanish coast and in considerable depth, the wreck still leaking oil. In total, 64,000 tons of oil were lost, contaminating 3,000 km of European coastline. Clean-up costs incurred by the Spanish government alone amounted to approximately 3 billion Euro, and greatly exceeded the amounts available under the CLC and the FUND Conventions.

Numerous proceedings were started in the Spanish courts,[49] but the Kingdom of Spain also decided to take the case to US courts, by suing the American Bureau of Shipping (ABS), the *Prestige's* classification society, in their home jurisdiction.[50] Spain pleaded, *inter alia*, that ABS had failed in its task to detect the deplorable condition of the vessel and was thus responsible for the ensuing accident and the oil spill. ABS invoked the provisions of the CLC 1992, which channelled the liability for oil spills to the shipowner and argued that the Convention barred claims against other persons. Still, the proceedings took place in the United States, and it was unclear whether the judge would admit ABS's defence, based on the provisions of a convention to which the US is not a party.

---

[46] Kloss, '1989 "Der Untergang der "Amoco Cadiz""', 185.

[47] See R. F. Hansen, 'Multinational Enterprise Pursuit of Minimized Liability: Law, International Business Theory and the "Prestige" Oil Spill' (2008) 26, Berkeley Journal of International Law, 410–51 at 419 et seqq. on the role of the corporations involved in the *Prestige* oil spill.

[48] Rodríguez Gayan, '2003 "Claves de derecho privado"', 119.

[49] See M. Michinel, 'The "Prestige" in the courts' (2007) 21, Australian and New Zealand Maritime Law Journal, 171–7 at 172 for more details.

[50] See M. Davies, 'Classification society liability in the United States', in D. R. Thomas (ed.), *Liability regimes in contemporary maritime law* (London: Informa, 2007), pp. 143 et seqq. on the rather complicated procedural background.

The case is highly interesting in more than one respect, especially with regard to the scope of the channelling provisions and the liability of classification societies as such (see *infra* under E. I. 4.), but the judges also pronounced themselves on the application of the CLC in non-contracting States.

The district court judge, in a decision of 2 January 2008, first considered ABS's defence that, according to art. IX (1) CLC, claims for compensation under the CLC can only be brought in contracting States.[51] The court held that if the CLC is applicable to the claim for damages brought against ABS, it deprives the US court of its jurisdiction so that the action must be dismissed. After having considered ABS's role with regard to the channelling provisions (see *infra* under E. I. 4.), the judge in the US District Court came to the conclusion that the classification society was protected by the CLC's provisions and that hence, the court lacked jurisdiction to decide the case.[52]

Spain, as the claimant in the case, had effectually sued ABS under American general maritime and Spanish law. It had, against ABS's defence, raised the argument that the CLC was not a part of the US law applicable in the case, the US not being a contracting State and that therefore, the court should not apply the Convention's provisions at all. The court, in a summary judgment relying solely on reasons of law, rejected this reasoning, basically arguing that Spain as a contracting State should be bound by the CLC's provisions in a US court, at least with regard to the rules on jurisdiction.[53] The court accordingly dismissed the Kingdom of Spain's claim.

The judgement has triggered considerable criticism in international doctrine, but mostly with regard to the decision to extend the channelling provisions to classification societies. It is surprising that the court's findings concerning the application of the CLC in the US as a non-contracting State have not been more thoroughly scrutinised, especially as they are hard to reconcile with the US courts' findings in the *Amoco Cadiz* case.

It should be emphasised that the court in its decision of 2 January 2008 does not consider whether the case has a foreign element triggering conflict-of-law rules,[54] and what should be the law applicable to the case. The judge simply started from ABS's defences and the law they were based on and only considered whether this was admissible.[55] Given the fact that the oil spill took place in Spanish waters, it is possible that the judge simply assumed the applicability of Spanish law and then came to the ensuing conclusions, considering the CLC to be a part of Spanish law. Still, with regard to the wording of the US District court's decision, referring to the

[51] *"Reino de España v. American Bureau of Shipping (The Prestige)"* 2008 American Maritime Cases 83 (US District Court, Southern District of New York, 02 January 2008), p. 87.

[52] Ibid., p. 91.

[53] Ibid., p. 91.

[54] I. Arroyo Martínez, 'Para esta viaje no se necesitan alforjas: El caso "Prestige"' (2009) 26, Anuario de Derecho Maritimo, 81–100 at 96 et seq.

[55] Ibid., p. 97.

"lack of jurisdiction", it does not seem that this was the court's inherent reasoning. The court clearly declared the action as inadmissible in the courts of a non-contracting State and therewith used a convention to which the US is not a party in order to rule out the competence of a US court.[56]

This reasoning was declared erroneous on appeal, the US Court of Appeals simply vacating the decision on the ground that the CLC should not have an impact on the procedural rules of a non-contracting State.[57] The court explained that this did not mean that the District Court was competent to hear the case and hinted to the fact that a dismissal for *forum non conveniens* might be possible, if there was a more appropriate forum at hand.[58] The Court of Appeals then explained that if a dismissal for *forum non conveniens* was not pleaded or granted, the case should be submitted to a conflict-of-law analysis in order to determine the applicable law,[59] but did not proceed itself to this analysis and remanded the case.

In a new decision of 3 August 2010,[60] the US District court for the Southern District of New York finally applied conflict-of-law rules. ABS, the defendant had pleaded that either the law of the Bahamas, the vessel's flag State, or the law of either the United Arab Emirates or China should apply, as the law of the places where the *Prestige's* last two surveys had been concluded.[61] Spain, on the other side, invoked the applicability of either the law of the United States as the place of ABS's headquarters or of Spanish law, but apparently Spanish domestic law, setting aside the CLC. The US court did not automatically apply the law of the place where the tort took place, also omitting to strike a difference between the place outside the Spanish territorial waters where the accident took place and the damage done to the Spanish waters and coastline.[62] Instead, it looked for the "proper law of the tort"[63] by considering a number of possible connecting factors following the precedents in federal admiralty jurisdiction of *Lauritzen v. Larsen*[64] and *Hellenic Lines Ltd.*

---

[56] J. Juste-Ruíz, 'Compensation for pollution damage caused by oil tanker accidents: from "Erika" to "Prestige"' (2010) 1, Aegean Review of the law of the sea and maritime law, 37–60 at 57.

[57] *"Reino de España v. ABS (The Prestige)"* 2009 American Maritime Cases 2402 (US Court of Appeals for the Second Circuit, 12 June 2009), pp. 2404 et seq.

[58] Ibid., p. 2404.

[59] Ibid., p. 2405.

[60] *"Reino de España v. American Bureau of Shipping (The Prestige)" (after remand)* 2010 American Maritime Cases 1877 (US District Court, Southern District of New York, 03 August 2010).

[61] Ibid., p. 1886.

[62] J. L. Goñi, 'L'arrêt de la Cour d'appel des États-Unis du 29 août 2012 dans l'affaire du "Prestige" et la responsabilité de la société de classification' (2012) 64, Droit Maritime Français, 1023–6 at 1025, footnote 6.

[63] See the discussion of the conflict-of-law rules with regard to torts under US State and Federal Law in Symeonides, *2008 "American Private International Law"*, pp. 107 et seqq., and Davies, *2007 "Classification society liability"*, p. 133 on the choice of law in maritime cases and Part III of this study.

[64] *"Lauritzen v. Larsen"*, 345 United States Reports 571 (US Court of Appeals for the Second Circuit, 25 May 1953).

v. Rhoditis.[65] The judge, after a long reasoning, finally came to the conclusion that United States law applied, mainly as the place of ABS's headquarters where the allegedly wrong class documents were issued, was Houston, Texas.[66] The court then decided that under the applicable US American law, ABS as the *Prestige's* classification society did not owe victims of oil pollution a duty to carefully perform its classification contract with the shipowner. The decision was affirmed on appeal,[67] the Court of Appeals explicitly declaring that the choice of US American law as argued by Spain, being the more generous legal system, was admissible.[68]

The facts of the case with regard to the choice of the law applicable to the tort were here certainly special, not least because of the policy considerations that a case brought by an European State against an US company necessarily entails.[69] Conflict-of-law problems with regard to tortious liability of classification societies are extremely hard to solve,[70] and a more thorough analysis of this will be given *infra*, in Chap. 7, D. of this study. Still, in this matter, the court did not apply the CLC to an oil spill that effectively and undeniably took place in the waters of a contracting State.

Here again, the question of applicability in a non-contracting State's court is closely linked to the question of whether the defendant was really faced with the application of the CLC to his case, even if the litigation had taken place in the courts of a contracting State. With regard to classification societies, this question is, to a certain degree, still open (see *infra* under E. I. 4.). However, by bringing the claim against ABS in the US forum and pleading for the application of US law, the Kingdom of Spain achieved its end to definitely opt out of the CLC provisions and the limitation of liability, even if the case failed for other reasons. By suing in a non-contracting State, the defendant at least triggers the application of conflict-of-law rules, which can point to the law of a non-contracting State, whereas in contracting States, the CLC is uniformly applied, without considerations of conflict-of-law rules.

---

[65] *"Hellenic Lines v. Rhoditis"*, 398 United States Reports 306 (US Court of Appeals for the Fifth Circuit, 08 June 1970).

[66] *US District Court, Southern District of New York, 03.08.2010, "Reino de España v. American Bureau of Shipping (The Prestige)" (after remand)*, pp. 1886 et seq.

[67] *"Reino de España v. ABS (The Prestige)"* 2012 American Maritime Cases 2113 (US Court of Appeals for the Second Circuit, 29 August 2012).

[68] Ibid., p. 2123.

[69] I. Naeemullah, 'A Decade Later, $1 Billion Saved: The Second Circuit Relieves a Maritime Classification Society of Unprecedented Liability for Environmental and Economic Damages in "Reino de España v. American Bureau of Shipping, Inc."' (2013) 37, Tulane Maritime Law Journal, 639–54 at 649.

[70] Antapassis, *2007 "Liability of Classification Societies"*, pp. 678 et seqq.

c) Conclusion

As both the *Amoco Cadiz* and the *Prestige* cases show, bringing proceedings in a non-contracting State can be a practicable way to opt out of the CLC regime, depending on the conflict-of-law rule in the chosen jurisdiction.

The problem of judicial standing outside the CLC system has not been too thoroughly addressed in this chapter, still, given the degree of globalisation in the oil business and the complicated constructions generally set up by players on the market in order to minimize liability and split liability risks,[71] it does not seem impossible to find a suitable target in a non-contracting jurisdiction. Considering the importance of the US as a place of incorporation on the oil market, it was certainly not difficult for the claimants in the *Amoco Cadiz* and *Prestige* cases to target US companies allegedly bearing some responsibility for the accidents.

In both cases, the court did not follow the *lex loci delicti* rule in determining the applicable law. In the *Amoco Cadiz* case, the parties to a certain degree agreed to apply the law of the court, by not pleading the applicability of any other law, and in the *Prestige* case, the court was faced with an impressive number of possibly applicable laws[72] and came to the conclusion that the *lex fori* was best placed to answer the case. It seems that the *Lauritzen/Rhoditis* approach in the US can lead to very uncertain results for maritime torts,[73] given that several connecting factors are taken into account and their importance judged on a case-by-case basis. It is thus perhaps not too bold to submit that it is not unlikely that a court in the US will, after a conflict-of-law consideration, come to the result that the law of the forum is applicable.

This possibility however causes the other, and seemingly often overlooked, problem of double proceedings. The phenomenon was highlighted in the *Amoco Cadiz* proceedings, in which the shipowner had, immediately after the accident, established a CLC limitation fund in the court of Brest, France. Whereas the French parties never claimed against the fund and pursued their proceedings in the United States, the UK government filed claims for the damages that the oil spill had caused to the English coast and to the Channel Islands. Considering the fact that the French parties had brought suit in the United States, invoking a fault of maintenance in the ship, the court refused to grant payment of the claims,[74] because the limitation would not subsist in the case that the shipowner would be found guilty of reckless conduct in the proceedings in the United States. On appeal, the *Cour d'Appel* of

[71] Hansen, '2008 "Multinational Enterprise Pursuit"', 418 et seqq.

[72] Davies, *2007 "Classification society liability"*, p. 146.

[73] See the case discussions in ibid., pp. 134 et seqq.

[74] Tribunal de Commerce de Brest, 8 June 1984, unreported. The judgment is referred to by du Pontavice, '1985 "Affaire "Droit de l'environnement""', 15; and Lucchini, '1985 "Le procès de l'"Amoco Cadiz""', 776.

Rennes[75] however declared that the American proceedings were carried out under a different law and on a different legal basis and could not have an impact on the proceedings in France. This reasoning however overlooks the problem that under these circumstances, a shipowner could be held liable twice, for the same incident, by claimants in two different countries. This may not have caused a problem in the present case as the Amoco Parties could not invoke their limitation of liability in the US anyway, and the limitation fund in the court of Brest could easily satisfy the UK government's claim. It is however not inconceivable that the courts may come to different results considering a conduct that would deprive the shipowner of the limitation of liability; and in a case such as the *Amoco Cadiz* accident, it seems arbitrary of the French court not to take into account the Illinois court's meticulous analysis of the defendants' conduct, coming to the result that they had acted with fault or privity.[76] For this reason, the view of the court in the first instance, that the proceedings should be stayed until there was a decision in the American courts, seems considerably more far-sighted on the international plan, even if a stay of proceedings in favour of a case in a foreign jurisdiction, to which there are even different parties, may be hard to justify under procedural law.

The possibility of double proceedings for the same accident should however also be considered by the shipowner, when deciding whether he will submit to the jurisdiction of a court in a non-contracting State or rather fight for proceedings in a jurisdiction where the CLC will be applied.

2. Is the CLC Part of the Domestic Law of Contracting States?

Even if the conflict-of-law rules of the non-contracting State point to the law of a contracting State, this does not mean that the CLC will be automatically applied.

The CLC is an international treaty containing uniform rules of law; its provisions do not need to be implemented into domestic law, but can apply directly. Still, implementation of international treaties is put into effect differently in different States, and jurisdictions following a dualistic system will not apply the provisions directly, but will enact national legislation reflecting the provisions of the convention, which can be a source for a different understanding of its contents.[77] Still, and either way, the provisions will generally be considered as national law in contracting States. It is however questionable whether a non-contracting State will be willing to apply a treaty to which it is not a party, even if its own conflict-of-law rules point to the law of a contracting State.

[75] *"Géniteau et Gouvernement Britannique v. Amoco"*, 38 Droit Maritime Français 99 (Cour d'Appel de Rennes, 03 October 1985).

[76] Lucchini, '1985 "Le procès de l'"Amoco Cadiz""', 777; M. Rémond-Gouilloud, 'Les surprises du "fonds"' (1986) 38, Droit Maritime Français, 89–98 at 95.

[77] M. Jacobsson, 'Uniform Application of the International Regime on Liability and Compensation for Oil Pollution Damage', in T. M. Ndiaye and R. Wolfrum (eds.), *Law of the sea, environmental law and settlement of disputes* (Leiden, Boston: M. Nijhoff, 2007), pp. 423, 426.

The court, in the *Amoco Cadiz* case, did consider this question, but addressed the matter rather implicitly. The defendants, the shipowner's parent companies, invoked the CLC, and namely the limitation of liability, since both France and the *Amoco Cadiz's* flag State Liberia were parties to it and as it was the only available international legal instrument in the field of oil pollution damage.[78] They implicitly argued that the applicability of the CLC should result in an absolute channelling of the liability to the shipowner, so that neither of Transport's parent companies would be liable.[79] The court did not follow the defendants' reasoning, but decided that with regard to the parent companies, they could not benefit from the channelling under the CLC as they were neither "agents" nor "servants" of the shipowner, so that claims could be directed against them outside the CLC, and the court also determined the scope of the CLC under French law:

> Even if French law, including the CLC, were applicable to this issue, none of the provisions of the CLC would bar suits against Standard or AIOC.[80]

The wording is very clear, supporting the view that the Illinois court considered the CLC to be a part of French law, to be applied by the United States if the case was governed by French law.

In the same court's judgment of 11 January 1988, on the compensation amounts payable, the court however maintained that it applied French law,[81] but completely disregarded the limitation of liability that would have applied under the CLC.[82] This is in line with the argument of the French claimants, who had, throughout the proceedings, invoked a provision of the French Civil Code as a starting point to calculate damages,[83] thus arguing domestic law excluding the CLC. The choice of French domestic law in the judgment on the compensation payable, presumably as the personal law of the victims[84] and the law of the jurisdiction they suffered their damages in, is however completely inconsistent with the court's earlier findings in the judgment on the merits of the case.

The problem of whether the CLC is part of the contracting States' domestic law was also addressed in the *Prestige* proceedings in which the Kingdom of Spain sued the *Prestige's* classification society. Spain, as the claimant, had openly pleaded the applicability of either US law or Spanish domestic law to the exclusion of the

---

[78] *US District Court, Northern District of Illinois (Eastern Division), 18.04.1984, "In Re Oil Spill by the Amoco Cadiz"*, p. 2190.

[79] Kbaier and Sebek, '1985 "New trends"', 273.

[80] *US District Court, Northern District of Illinois (Eastern Division), 18.04.1984, "In Re Oil Spill by the Amoco Cadiz"*, p. 2190.

[81] *US District Court, Northern District of Illinois (Eastern Division), 11.01.1988, "In Re Oil Spill by the Amoco Cadiz"*, p. 879 et seq.

[82] Ballarino, '1990 "Questions de droit international privé"', 381.

[83] *US Court of Appeals for the Seventh Circuit, 24.01.1992, "In Re Oil Spill Amoco Cadiz"*, p. 947.

[84] Ballarino, '1990 "Questions de droit international privé"', 381.

CLC,[85] obviously maintaining that even if the US conflict-of-law rule pointed to Spanish law, this did not include the CLC, and that the CLC, to which the US are not a party, remained inapplicable in the US courts.

The first decision of the US Court of the Southern District of New York seems to reject this reasoning by ruling that Spain was a party to the CLC and must submit its claim to the conventional rules including the provisions on jurisdiction,[86] possibly maintaining that there was no application of Spanish law outside the CLC. However, the court's complete reasoning has been annulled on appeal, and in the final judgment after appeal and the remand to the first instance and the second appeal, the court did not take this reasoning into consideration, but decided the applicable law to be the US maritime law, so that the question did not need to be considered.

It is therefore difficult to decide how the courts of a non-contracting State would deal with the argument that the CLC was not part of the domestic law of a contracting State, but meant for application between contracting States only. This would probably also include an analysis of whether the CLC has been implemented into the contracting State's national law by domestic statute or whether it is simply applicable as an international treaty.[87] There may be some authority in the *obiter dicta* in the judgment on the merits of the *Amoco Cadiz* case, that the CLC would be applicable as a part of the domestic law of contracting States. However, in the judgment on the compensation amount, French domestic law was applied without having regard to the CLC.

II. The High Seas

The CLC, according to its art. II as amended by the 1992 Protocol, only applies to pollution damage caused in the territory, including the territorial sea, of a contracting State, in the EEZ of a contracting State, or to measures meant to prevent or minimise such damage. There is therefore, even with regard to preventive measures taken on the high seas, a clear link to the impact on a State's coastline. The place where the incident occurred or where the oil was released into the environment is irrelevant, so that the CLC is inapplicable to an accident on the high seas if the pollution is not perceived on land, but will become applicable as soon as the oil drifts into the waters or the EEZ of a contracting State.

During the negotiations of the Convention, it was discussed whether the regime should be applicable to the high seas as well.[88] The proposal was not adopted,

[85] *US District Court, Southern District of New York, 03.08.2010, "Reino de España v. American Bureau of Shipping (The Prestige)" (after remand)*, p. 1882.

[86] *US District Court, Southern District of New York, 02.01.2008, "Reino de España v. American Bureau of Shipping (The Prestige)"*, p. 276.

[87] Jacobsson, *2007 "Uniform Application"*, p. 423.

[88] LEG/CONF/4/Add.2; LEG/CONF/C.2/WP.3, *1973 "Official Records (CLC)"*, pp. 660 et seqq.

mostly, it seems, for the reason that the drafting committees needed the localisation of the damage as an unambiguous criterion to trigger the application of the CLC.

Considering that the catastrophic consequences of an oil spill will in the first place be felt in coastal waters where they may have an impact on fishing and tourism and where victims could easily be identified, damages on the high seas were seemingly considered as negligible. It is perhaps not too bold to submit that the delegates were interested in coming to a quick understanding and reluctant to discuss borderline cases which would only lengthen the negotiations. This restriction to apply the CLC to the immediate victims of oil pollution only and not to take into account the long-term consequences of pollution on the high seas has been criticised,[89] and with good cause. It seems unlikely that the impact of oil pollution sustained on the high seas which is only perceived at a much later point in time by coastal States, could still be attributed to a given oil spill and compensated under the CLC,[90] even if the prescription provisions are left out of the picture.

In a pragmatic spirit, the importance of the cases in which the damage is solely sustained on the high seas is considered limited. Theoretically, oil pollution on the high seas could be detrimental to fishing in that zone,[91] it could damage private property brought to the high seas such as fishing gear, ships and their equipment and offshore installations.[92] However, the immediate damage resulting may in most cases not be substantive enough to justify the application of the CLC,[93] and this view was already endorsed during the negotiations of the Convention.[94] Many authors have maintained that such a claim could be settled under "domestic law"[95] or, as the delegates to the negotiation of the CLC put it, "under the existing rules".[96]

This study will, in its third part, give an overview on how and where a claim for damages on the high seas could be brought (see *infra*, Chap. 7, A.). The CLC's conventional regime does not offer solutions for this problem.

---

[89] Lucchini, '1970 "La pollution des mers"', 827.

[90] J. H. Bates and C. Benson, *Marine environment law* (London: Lloyd's of London Press, 1993), para. 4.19 maintain that oil emerging out of a wreck in later years "may" still be considered as a result of the original incident, but assume that both the original incident and causation could be established, and also refer to prescription.

[91] This argument was put forward by the Indian delegation during the negotiations, LEG/CONF/4/Add.2; LEG/CONF/C.2/WP.3, *1973 "Official Records (CLC)"*, pp. 660 et seqq.

[92] de la Rue and Anderson, *2009 "Shipping and the environment"*, p. 5.

[93] Ibid., p. 5.

[94] LEG/CONF/C.2/WP.3, *1973 "Official Records (CLC)"*, p. 563.

[95] de la Rue and Anderson, *2009 "Shipping and the environment"*, p. 5.

[96] LEG/CONF/C.2/WP.3, *1973 "Official Records (CLC)"*, p. 563.

III. A "National Exception"?

The CLC, according to its art. II as amended, explicitly applies, *inter alia*, "to pollution damage caused on the territory including the territorial sea of a Contracting State". The wording does not differentiate between cases having a national implication only and others in which there is a foreign element. The problem of the applicability of an international convention to a purely domestic case occurs of course with regard to almost all international conventions, and its solution seems to depend on the provisions, the characteristics and the purpose of the convention in question.[97]

From a theoretical point of view, some arguments can be drawn from the fact that the CLC's wording refers to pollution caused by "international trade", a term that is generally understood to mean trade between two different countries, or trade between two ports in the same country, but with a port of call in another country during the voyage.[98] Still, an accident during an oil transport between two ports of the same country can still cause pollution to another country, and if the victims there could not invoke the CLC, this would undermine the purpose for which the Convention was adopted. Bearing in mind that the purpose of the CLC is to ensure that victims can obtain damages without turning to a foreign jurisdiction, it might be argued that the Convention should not apply if the case has a national dimension only,[99] but this would rather be the case if the pollution was caused to one country only, irrespective of whether it was caused by a ship in international or in national trade.

The problem seems to be a very important one in China, where the relationship between an international convention and conflicting provisions in domestic law does not seem entirely clear. Whereas there is a rule setting out the general supremacy of the international agreement, some scholars argue that the international convention can only come into play where the case presents a foreign element that justifies application of these international rules.[100] Other authors however plead that conventions should also be applied to purely domestic cases.[101] Chinese

[97] See E. du Pontavice, 'Conventions Internationales et Droit Interne' (1980) 5, Annuaire de droit maritime et aérien, 17–81 and I. Arroyo Martínez, 'International Conventions and Domestic Law' (1986) 2, Yearbook of Maritime Law, 5–50 for a detailed discussion.

[98] See the extensive discussion in du Pontavice, '1980 "Conventions Internationales"', 30 et seq.

[99] Arroyo Martínez, '1986 "International Conventions"', 32 et seq.

[100] A thorough analysis of this problem of Chinese law is outside the scope of this work. See Wang, *2011 "Civil liability"*, pp. 327 et seqq.; Q. Chen, 'On the Application of CLC 92 in China', in M. Faure, L. Han and H. Shan (eds.), *Maritime Pollution Liability and Policy: China, Europe and the US* (Alphen aan den Rijn: Kluwer Law International, 2010), p. 347 et seqq.; L. Zhang, 'Compensation for the Domestic Oil Pollution in China's Coast: Which Law Shall Apply?', in M. Faure, L. Han and H. Shan (eds.), *Maritime Pollution Liability and Policy: China, Europe and the US* (Alphen aan den Rijn: Kluwer Law International, 2010), p. 359 et seqq. for a comprehensive outline of the difficulties arising out of the implementation of the CLC.

[101] Chen, *2010 "On the Application of CLC 92"*, p. 350.

courts have, in the past, mostly decided not to apply the CLC to an oil spill case having a national dimension only.[102]

It is clear that a State such as China with an impressive territory and long coastlines will be more likely to suffer an oil spill having a domestic dimension only than a smaller State. Within Europe, it seems highly unlikely that a major marine oil spill would affect the coastline of one country only. Not surprisingly, some European States have explicitly enacted domestic legislation by which the CLC provisions are implemented into national law and their applicability to domestic cases is made clear. Examples of this are the German *Vertragsgesetz*,[103] even if repealed and replaced by a more general reference to the CLC text,[104] and the UK Merchant Shipping Act 1995 which paraphrases in its Chapter III the provisions of the CLC.

Still, in France, even though the State has enacted legislation to implement the Convention's main provisions into its internal legal order,[105] it is sometimes still taken for granted that French domestic law, i.e. the respective provisions of the *Code Civil* will be applied to damages caused by a French vessel in French waters.[106] There are similar interpretations of the Convention in Spain, in which the applicability of the domestic law to cases without foreign elements is seen as a matter of course.[107] It is in this context submitted that the CLC, following its own text, is concerned with oil pollution on an international level and that the exclusion of purely domestic cases is to be concluded from this purpose.[108] The question was heavily debated in the case of the *Urquiola*, a tanker flying the Spanish flag, belonging to a Spanish owner, carrying cargo for a Spanish company and being insured with a Spanish insurer, which struck an underwater rock in Spanish territorial waters off La Coruña.[109] This problem in the case was however later superimposed by the question whether the Spanish state was liable for omitting to show the obstacle in official charts.[110]

---

[102] Wang, *2011 "Civil liability"*, p. 240; Zhang, *2010 "Compensation"*, p. 363.

[103] BGBl. 1975 II, 301.

[104] Altfuldisch, *2007 "Haftung und Entschädigung"*, p. 12, also referring to an "extension" of the Convention to domestic cases, even if apparently pleading that the CLC applies out of its own account.

[105] Loi du 26 mai 1977.

[106] P. Delebecque, 'Responsabilité et indemnisation des dommages dus à la pollution par des hydrocarbures' (2000), La Semaine Juridique Édition Générale, 125–7 at 125.

[107] J. L. Meilán Gil, *Problemas Jurídico-Administrativos Planteados por el "Prestige"* (Cizur Menor: Thomson Aranzadi, 2005), p. 169.

[108] Arroyo Martínez, '1986 "International Conventions"', 32 et seq.

[109] See I. Arroyo, 'The Application of the International Convention on Civil Liability for Oil Pollution Damage to the "Urquiola" Case' (1977), Lloyd's Maritime and Commercial Law Quarterly, 337–43.

[110] See J. L. Pulido, 'Compensation by Coastal States - The "Prestige" Disaster', in J. Basedow and U. Magnus (eds.), *Pollution of the Sea: Prevention and compensation* (Berlin, New York: Springer, 2007), p. 165.

It goes without saying that the successful functioning of a convention depends on its uniform implementation and application, and that a "domestic exception" will be detrimental to this aim[111] by withdrawing certain cases from the uniform regime. Even if the case in question may not have an international dimension in the first place, victims in another jurisdiction may only enter into the picture at a later point in time.[112] Not least, the wording of the provisions on the scope of application of the CLC runs counter to the "national exception", the applicability of the Convention being triggered if oil is spilled on the territory of a contracting State, without explicitly allowing for exceptions.[113] To the extent in which the FUND Convention is considered as a necessary and successful complement to the CLC, the "national exception" in the CLC risks to disturb the balance between the contributions of shipowners and the oil industry. This is due to the fact that the contributions to the Fund are levied on all oil transports, irrespective of whether it was a national or an international purchase.[114]

On the other hand, a State authority's wish to apply their own law to a case involving only domestic actors is to some extent understandable, especially if the compensation that can be obtained this way provides for higher liability limits or an easier procedure. The geographic position of a State or any other national particularity can in fact create specific circumstances that are perhaps best addressed by domestic law.[115] It also seems that the national exception is often promoted for policy reasons, in order to make sure that a disaster on a domestic scale is dealt with under the national rules, and possibly, in a more enclosed judicial environment and with more opportunities to take policy considerations into account than under the CLC. Even if, apart from China, the "national exception" does not seem to have much practical relevance yet, the doctrinal opinions taking the existence of such an exception for granted are revealing of a certain attitude towards the CLC—that it is useful as a lowest common denominator in cross-border cases, but that domestic legislation will be more favourable to domestic victims.

Considering the experiences that have so far been made with tort claims in domestic courts, it does not seem that this view is justified. Not least, it is submitted that the cases that truly only have a national dimension are very rare, as there may always be remote interests such as reinsurance in a third country. It does not seem

[111] Arroyo, '1977 "The Application"', 340.

[112] Ibid., p. 342, with regard to the direct action against insurers and possible claims against reinsurers, the Spanish insurer in the *Urquiola* case being reinsured by an English P & I Club.

[113] Ibid., pp. 339 et seq.; Rodríguez Docampo, *2012 "Alcance de la canalización"*, p. 914.

[114] M. Morin, 'Les rapports entre droit international public et droits internes: l'exemple du FIPOL' (1997) 49, Droit Maritime Français, 325–35 at 329.

[115] See M. Jacobsson, 'An ideal international scheme for compensation for marine pollution damage' (2011) 17, Journal of International Maritime Law, 263–73 at 266, discussing the possibilities of a system in which there would be a domestic first tier to the liability regime and an international second tier for damages that exceed the funds available under the national regime.

that it is very clear where the line should be drawn. The CLC already contains many systemic gaps and shortcomings, it should be avoided to add to their number by creating exceptions that are not even supported by the wording of the Convention's provisions.

B. Jurisdiction

As seen from the considerations under A., the quest for the exact outline of the CLC's scope of application heavily depends on considerations of international jurisdiction, triggering a court's decision to apply or not to apply the regime as a whole. However, the search for the competent jurisdiction within the CLC system, i.e., within the group of contracting States, will be just as important.

This is, first and foremost, due to the problem of uniform interpretation already mentioned in the introduction to this chapter. As undesirable as conflicting interpretations by courts from different contracting States may be, the occurrence of discrepancies is reality and may give rise to the phenomenon of forum shopping, claimants trying to obtain a judicial venue in a court that they perceive as favourable to their claim. It is therefore interesting to consider whether there can be a choice of this kind for the claimants in the CLC system or whether the localisation of a claim is clear-cut enough to render forum shopping unlikely.

Since this matter is to some extent linked to the question of the establishment of jurisdiction, it will also examined in this part if courts are, in their application of the CLC and the FUND, bound by precedents or authoritative interpretations of the Conventions, namely by the Fund's Claims Manual and the guidelines that will govern the Fund's decisions on whether to compensate a given claim or not.

In a second step, the problem of recognition and enforcement of judgments from one contracting State in other contracting States will be addressed. Even if there is not yet proof of a major difficulty in the application of the respective provisions in the CLC and the FUND, under certain circumstances, the system of uncomplicated recognition and enforcement of judgments may be heavily disturbed. The interplay with EU law in the member States of the European Union, that has become a problem for the more recent maritime civil liability conventions (see *infra* in Chaps. 4 and 5) is however less of a problem under the CLC.

As between member States of the European Union, questions of jurisdiction, recognition and enforcement of judgments in civil and commercial matters have been governed by uniform law since 1968, first by the Brussels Convention and, since 2001, by Regulation 44/2001 (Brussels Regulation). Possible conflicts between the general rules of these instruments and the rules on recognition and enforcement in international conventions thus needed to be considered from an early point in time and onwards. With regard to newer international instruments, the question of the interplay between European and international law can cause considerable difficulties, but the Brussels Regulation, in its art. 71, states that it does not affect any conventions to which the member States are parties. The CLC being the earlier instrument, its provisions can in any case not be affected by the Regulation, even if it applies as between EU member States to the extent in which its rules do not contradict arts. IX and X CLC. Problems could however still arise for member States ratifying the CLC now or in the future, after the coming into force of the Brussels Regulation.[116]

I. Judicial Venue

According to art. IX para. 1 CLC, "where an incident has caused pollution damage in the territory including the territorial sea of one or more Contracting States [...] actions for compensation may only be brought in the Courts of any such Contracting State or States." In the 1992 Protocols, the reference to the territory of a State was broadened, in order to also include the exclusive economic zone of contracting States or the equivalent area submitted to their jurisdiction.

1. Competent Court

It is therefore clear that generally, jurisdiction lies with the courts of the contracting State where the damage occurs, obviously with the aim to provide victims with a jurisdictional venue in the place where they have been aggrieved.[117] Following art. IX para. 1 CLC, a claim under the CLC can therefore not be brought in a State in which there has not been pollution damage. In the CLC system, the place of the damage is the decisive criterion, and other factors according to which jurisdiction may be attributed, such as the domicile of the defendant or the place where a vessel was arrested, have been ruled out at an early point in time.[118] However, in all cases where the damage affects more than one State, it can be discussed whether claimants can choose to bring their claim in any of these jurisdictions, or whether there can only be proceedings in each State for the damage sustained in that State.

The wording of art. IX para. 1 CLC ("the courts of any such Contracting State") clearly points to the first interpretation.[119] This means that if there is a pollution damage in several countries, claimants can address the courts of all of these countries, irrespective of their nationality or the exact localisation of the damage they bring their claim for.

---

[116] M. Jacobsson, 'Perspective of the Global Compensation Regimes: The Relationship between EU Legislation and Maritime Liability Conventions' (2012), European Journal of Commercial Contract Law, 63–76 at 71.

[117] Altfuldisch, *2007 "Haftung und Entschädigung"*, p. 45; de la Rue and Anderson, *2009 "Shipping and the environment"*, p. 128.

[118] Wu, *1996 "Pollution"*, p. 73.

[119] Altfuldisch, *2007 "Haftung und Entschädigung"*, p. 44; de la Rue and Anderson, *2009 "Shipping and the environment"*, p. 130.

There is no *lis pendens* rule in the CLC, according to which the proceedings in one contracting State could be stayed in favour of earlier proceedings in another contracting State.[120] Considering that art. 71 of the Brussels Regulation only rules out the application of its own rules where they are in conflict with the provisions of the special convention, it is therefore possible that the Regulation's *lis pendens* rules are still applied by member States' courts,[121] if the cases can be considered to concern the same subject matter. In any case, there will however be a final bundling of claims when it comes to the distribution of the limitation fund established by the shipowner. According to art. V para. III CLC, the shipowner has the right to set up the fund in any of the contracting States in which an action is brought, or, if no claim is brought, in any of the contracting States in which a claim could be brought.

Following art. IX para. 3 CLC, only the court of the place where the fund has been constituted is competent to decide on the apportionment and the distribution of the fund, which means that all claims for payments must in the end be addressed to this court. It is thus of course crucial that the court will easily recognise all judgments from other contracting States in order to grant all claimants a fair share in the distribution process.[122] Questions of recognition and enforcement will be dealt with *infra* under II. It should however already be mentioned that, since the shipowner can also exercise a certain choice where to establish his fund if the oil pollution was caused to more than one State, he may also exercise this choice with regard to his own interests.[123] To the extent that the court distributing the fund can be found, in deciding on the apportionment and the distribution of the fund, to have the authority to influence the extent of the shipowner's liability, it is actually possible that it is the shipowner who has the more important possibility of choice of forum.

Where there are damages in more than one State, claimants, but also the shipowner as a defendant thus have a choice between the courts of all the affected States. To the extent that there are differing interpretations of the Convention's provisions in neighbouring countries, claimants may consciously decide to bring their claim in the jurisdiction that will be more favourable. There is thus a certain danger of forum shopping[124] even within the CLC system. It has been submitted that the likelihood of a major disturbance of the system is however small, considering that many oil spills in the past have only affected one country, and that the

[120] R. Renger, 'Recht und Praxis der Haftung und Entschädigung für Ölverschmutzungsschäden auf See', in H. Koch and A. Willingmann (eds.), *Großschäden - Complex Damages*, (Baden-Baden: Nomos, 1998), p. 156; Altfuldisch, *2007 "Haftung und Entschädigung"*, p. 44.

[121] See J. Maseda Rodríguez, 'Responsabilidad civil por vertido de hidrocarburos en el mar: competencia judicial internacional y actividad procesal del naviero y víctimas', in J. M. Sobrino Heredia (ed.), *Mares y Océanos en un mundo de cambio: Tendencias jurídicas, actores y factores* (Valencia: Tirant lo Blanch, 2007), pp. 555 et seq.

[122] Wu, *1996 "Pollution"*, p. 74.

[123] Maseda Rodríguez, *2007 "Responsabilidad civil"*, pp. 550 et seqq.

[124] Altfuldisch, *2007 "Haftung und Entschädigung"*, p. 45.

discrepancies in the interpretation of the Convention are too difficult to foresee.[125] Indeed, there have so far not been many cases in which claimants consciously brought a case in another contracting State, most oil pollution victims clearly preferring their home jurisdictions. Still, the number of precedents of the application of the CLC has grown over time, and there are certainly tendencies of different interpretation in different countries, also supported by cases that have been decided under provisions from other conventions that however exactly mirror the wording of the CLC's rules. The possibility of forum shopping among contracting States renders the differences in the interpretation of the Convention's provisions a rather important problem, namely where they occur in countries that could very easily be affected by the same oil spill, such as England and France.

It is important to note that art. IX CLC understands the courts of the place or the places of the damage to be exclusively competent. A court in any other State party must declare itself incompetent. This can however cause distortions in cases where it is not entirely clear whether the claim that is brought is effectively a claim under the CLC. Shortly after the accident of the tanker *Erika* off the coast of Brittany in 1999, causing important oil spill damages to the French coastline, victims started to consider a possible liability of the Italian classification society RINA, which had certified the vessel to be in a sufficient state of repair for its last voyage. RINA therefore brought a negative declaratory action in the Italian courts, asking for a formal declaration that it could not be held responsible for the accident. The *Corte di Cassazione*, in the last instance, held that the Italian courts were not competent to decide on that matter, Italy being a member State to the CLC and art. IX CLC assigning exclusive jurisdiction to the courts of the State where the damage had occurred.[126] The *Corte di Cassazione* thus made sure that all proceedings would be held in France, avoiding the possibility of contradicting judgments from different States. The decision can however be criticised with a view to the fact that a claim against a classification society, not being one of the parties explicitly mentioned as beneficiaries of the channelling provisions (see *infra* under E. I. 4.) could also be considered as a claim outside the CLC, governed by domestic rules of jurisdiction and domestic tort law. The CLC does not offer a basis for a claim against a classification society, the convention may only be considered to apply to the case indirectly, by barring a claim under domestic law against a beneficiary of the channelling provisions. In this very case, it seems right to refer the case to the French courts, considering that there had already been some investigations against RINA in France and that the French courts were also the typically competent forum for a domestic tort claim for damages suffered in France, even outside the scope of the CLC. Still, to the extent that the applicable law of Italy and the EU allowed RINA to start proceedings in Italy, possibly creating a *lis pendens* in favour of the

---

[125] Ibid., p. 45.

[126] *"IOPC v. RINA"* 2003 Rivista di diritto internazionale 238 (Corte di Cassazione (Italy), 17 October 2002).

Italian courts, RINA was denied a trial in a jurisdiction that would have been competent for a claim outside the CLC.

Other than for claims against the shipowner under the CLC, the courts of the State or States in which the oil pollution damage occurred are also important for another reason. Claims against the Fund, in addition to or instead of a claim against the shipowner, are, according to art. VII para. 1 FUND, to be brought in the court that would be competent for a claim against the shipowner under art. IX CLC. As numbers from oil spills in the past have shown, the vast majority of claims addressed to the Fund are settled amicably, without any intervention of contracting States' courts.[127] The 1992 Protocols, effectively establishing a new Fund that would not be concerned with incidents having occurred before 1992, provided for an Assembly, a supreme organ, electing an Executive Committee that would, first and foremost control and approve the Fund's settlement claims. The 1992 Fund can also suggest a binding arbitration in order to come to a settlement with a given claimant.[128] In practice, the venue of art. VII para. 1 FUND is therefore more of a venue for judicial review of the initial decisions taken by the Fund, and the courts competent for the claims against the shipowner are also competent to judge on the final liability of the Fund.[129]

2. Authoritative Interpretation

The application of the international convention by national courts, according to their domestic procedural rules and within the framework of their legal system, can of course cause some discrepancies in the treatment of claims under the system. It goes without saying that international conventions can only fully serve their purpose of international harmonisation if they are applied in the same way in all the contracting States. Even beyond this, it seems that there is a more compelling reason for the uniform application of the provisions of the CLC and the FUND, considering that it is a system to which different players from different countries contribute. Not least, with the contributions of oil imports levied for the Fund, the oil industry in all contracting States compensates the aggrieved parties in one or at most a very few contracting States.[130] Disparities in the distribution of the funds, namely if they are rooted in decisions in these victim States could quickly lead to a lessened acceptance of the system.

In view of that, the Fund already called in 1993 upon the courts of the contracting States of the FUND to take the decisions of the organs of the FUND

[127] Jacobsson, '2007 "L'expérience française"', 969 et seq.

[128] L. C. Kappet, *Tankerunfälle und der Ersatz ökologischer Schäden* (Hamburg, Münster: Lit, 2006), p. 64.

[129] Renger, *1998 "Recht und Praxis"*, p. 159.

[130] Jacobsson, '2007 "L'expérience française"', 979.

on the interpretation and the application of the Convention into account.[131] Notwithstanding the emphasis on the importance of a uniform interpretation, it is obvious that the decisions of the Fund are not binding, considering that art. VII para. 1 FUND allocates judicial jurisdiction to the contracting States' courts, conferring the decision in the last resort on them.[132] None of the guidelines or decisions by which the decisions of the Fund's organs are governed can be considered to be authoritative for the contracting States' courts.

This circumstance has certainly been stressed in the decisions of national courts. Even if the number of domestic precedents on the liability of the Fund is modest,[133] namely in comparison with the number of cases that have been settled by the Fund's representing organs, there are still cases in a small number of jurisdictions in which the Fund's decision not to compensate a claim has been challenged by the victims. Most of these cases concern the question of what constitutes a "damage" that can be compensated, claimants having addressed the Fund for compensation of damage that was only a remote consequence of the actual pollution.

The *Braer* grounding off the Shetland Islands in 1993 caused one of the first claims where claimants from a region affected by an oil spill saw their claims denied by the Fund and subsequently challenged that decision in the domestic courts. The case of *Landcatch v. IOPC*[134] was brought by a local business that reared salmon from eggs in protected waters and then sold them for further ongrowing in sea water. The claimant's production was undisturbed by the oil spill, but due to a fishing ban and a prohibition to supply fish from the polluted waters, its customers were affected and did not take delivery of the smolt produced by the claimant. The Fund, when the claimant presented its claim, did not consider it compensable, due to the fact that the claimant was only affected by way of the impact on its customers, not in its own business. In the Scottish Court of Session, L. J. Clerk and McCluskey clearly stated that they did not consider themselves bound by the decisions of the Fund[135] and interpreted the notion of "damage" under domestic law. As a result, they however agreed with the Fund's argumentation that claimant's loss, resulting from the fact that it supplied other professionals whose business was in turn affected by the oil spill, was too remote to be compensated.

The English courts considered a very similar issue following the shipwreck of the tanker *Sea Empress* off Milford Haven in 1996. Following a fishing ban in the area after the oil spill, a seafood processing factory suffered losses, considering that it could not take delivery of the catches of local fishermen any longer. The Fund

[131] IOPC Fund, Resolution n° 8, http://www.iopcfunds.org/uploads/tx_iopcrulesandregs/Resolutions_92_e.pdf (last visited 7 March 2013).

[132] Morin, '1997 "Les rapports"', 333.

[133] M. Jacobsson, 'Dispute resolution in oil pollution cases: Are there alternatives to litigation in national courts?', in N. A. Martínez Gutiérrez (ed.), *Serving the rule of International Maritime Law: Essays in Honour of Professor David Joseph Attard* (Abingdon: Routledge, 2010), p. 332.

[134] *"Landcatch v. IOPC"* 1999 Scots Law Times 1208 (Court of Session (Scotland), 19 May 1999).

[135] Ibid., pp. 1217, 1221.

considered that this claim was too remote, as the factory was only a secondary business, providing services for the fishermen, who in turn directly depend on the affected maritime resource of seafood. When the refusal of the Fund was adjudicated in the English Court of Appeal,[136] L. J. Mance and Chadwick referred to the fact that, given the limitation of the Fund's liability, it could hardly cover all possible claims arising out of an oil spill.[137] Taking into account the Scottish *Landcatch* decision, they explicitly pointed out that the Fund's practice could not be taken as an unchallengeable interpretation of the act incorporating the FUND into English law,[138] and very carefully considered the case before coming to the conclusion that the damage suffered by the seafood factory was too remote, thus effectively upholding the Fund's decision.

In the courts of Galicia, Spain, a very similar statement was made concerning the case of the *Aegean Sea*, a tanker that grounded in the river of Ares, just outside the port of La Coruña. The proceedings were carried out in the criminal courts, following a criminal charge of the vessel's master and the representative of the port authority, but the question of the damages that were recoverable under the FUND Convention was also considered in a decision of the *Audencia Provincial de La Coruña*.[139] The court of La Coruña emphasised its independence from the restrictive criteria of the Fund, but at the same time set out criteria of its own, according to which damage should be economically calculable, the admission of unquantified ecologic damage having been one of the results of the decision in the previous instance.[140] Still, the *Audencia Provincial* overruled the Fund with regard to the cost of certain preventive measures that the Fund did not want to cover for lack of "reasonableness".

French courts were faced with the problem of the definition of "damage" following the accident of the tanker *Erika* off the coast of Brittany in 1999. The sinking of the tanker caused an important oil spill to a region that was known as one of France's holiday destinations, and many claims were brought by parties that had been indirectly affected by the consequences of the oil spill. The Fund, relying on its practice, rejected a number of demands for compensation in which the claimants were only affected "in the second degree", meaning that they did not carry out their business in the tourism industry directly, but provided goods and services to other professionals, who in turn offered these to holiday guests. Whereas these providers "in the second degree" were without doubt affected by the decline in tourist

[136] *"Algrete Shipping v. IOPC"* 2003 1 Lloyd's Law Reports 327 (Court of Appeal (England), 07 February 2003).

[137] Ibid., pp. 330, 338.

[138] Ibid., pp. 336, 338.

[139] 18 June 1997, only reported in Galician language. The judgment is presented in detail in M. Castillo Daudí, 'El siniestro del buque "Aegean Sea" y la indemnización de los daños' (2001) 18, Anuario de Derecho Maritimo, 261–301 at 281 et seq.

[140] M. Castillo Daudí, 'La responsabilidad civil por daños de contaminación por hidrocarburos y el siniestro del buque "Aegean Sea"' (1998) 50, Revista Española de Derecho Internacional, 69–92 at 86.

numbers that was felt by their business partners, the Fund did not consider these losses to be compensable. When claimants from Brittany seized the French courts in accordance with art. VII para. 1 FUND to obtain a binding decision on their claims against the Fund, the cases hinged on the question of whether the Fund's decisions and guidelines were binding on the jurisdictions of contracting States. In quite a number of decisions, the *Cour d'Appel* of Rennes decided that it was not bound by the Fund's criteria of compensation,[141] hinting at the fact that the question of remoteness of damage is not defined by the FUND Convention,[142] and that the Fund's Manual itself declares that it is not an authoritative interpretation.[143] However, the court rejected the claims of the aggrieved parties in all cases,[144] generally requesting a direct link of causality between the causes of the accident that was not established by the claimants.[145]

There is one decision of the Greek Supreme Court in which the decision of the Fund not to cover a claim was overruled (see *infra* under C. II.) but which was based on doubtful legal reasoning. In summary, the level of recognition of the Fund's decisions, leading to a rather uniform application of the FUND is thus rather high,[146] even if national courts are eager to stress that they still have their full discretion. The experience of the Fund Assembly with an equitable treatment of claims, guided by an impressive database of decisions, can hardly be overestimated, but it should be kept in mind that the text of the FUND confers jurisdiction to the courts of national States, competent under the CLC, very likely in the hope of achieving an alignment with the claims against the shipowner.

It is clear that with every domestic case in which the courts of the contracting States do not support the previous decision of the Fund, the incentive of claimants to seize national jurisdictions increases.[147] In this regard, it seems interesting that the courts of the Republic of Korea decided in a 2001 decision that the

---

[141] *"Lebaupain v. FIPOL"*, 57 Droit Maritime Français 529 (Cour d'Appel de Rennes, 25 May 2004); *"FIPOL v. Gouzer"*, 58 Droit Maritime Français 1014 (Cour d'Appel de Rennes, 24 May 2005); *"FIPOL v. Hôtel Celtique"*, 58 Droit Maritime Français 1019 (Cour d'Appel de Rennes, 24 May 2005).

[142] *Cour d'Appel de Rennes, 25.05.2004, "Lebaupain v. FIPOL"*, p. 532; *Cour d'Appel de Rennes, 24.05.2005, "FIPOL v. Gouzer"*, p. 1016.

[143] *Cour d'Appel de Rennes, 25.05.2004, "Lebaupain v. FIPOL"*, p. 532.

[144] *Cour d'Appel de Rennes, 24.05.2005, "FIPOL v. Gouzer"*, p. 1018, remanding the case to the former instance to inquire whether the claimants' loss of profit in his oyster culture was directly due to the consumers' disinclination to buy seafood after the oil spill; *Cour d'Appel de Rennes, 24.05.2005, "FIPOL v. Hôtel Celtique"*, p. 1021, remanding the case to the former instance to inquire whether the claimants' hotel's loss of profit was due to a decline in guest numbers after the oil spill, *"FIPOL v. Simenoff"*, 58 Droit Maritime Français 1021 (Cour d'Appel de Rennes, 24 May 2005), p. 1023, rejecting the claim of a supplier of machinery for cafés and restaurants.

[145] *"Boissons Morbihan v. FIPOL"*, 58 Droit Maritime Français 1030 (Cour d'Appel de Rennes, 27 June 2006), pp. 1030 et seq.; *Cour d'Appel de Rennes, 24.05.2005, "FIPOL v. Simenoff"*, p. 1023.

[146] Jacobsson, *2010 "Dispute resolution"*, p. 332.

[147] Morin, '1997 "Les rapports"', 334.

understanding of "damages" under the CLC and the FUND should be, in the interest of international uniformity, the same, so that under Korean law, damages other than those covered by the international conventions were not compensable.[148] In order to completely overcome the problem of differing interpretation and application of the conventions, it would be necessary to have an authoritative institution or tribunal, delivering a binding decision on any interpretative issue.[149] Out of the number of existing bodies of international jurisdiction, it would of course be the International Tribunal for the Law of the Sea (ITLOS) that could be considered,[150] bearing in mind that the international civil liability regimes have a legal basis in the law of the sea after all. It is however clear that major amendments to the civil liability conventions and the unconditional acceptance of all contracting States would be needed to this end, not to mention the change in the role of the ITLOS that this would cause.

II. Recognition and Enforcement of Judgments

An international civil liability regime would hardly be able to take any international effect if judgments from one contracting State would not be easily recognised in the other contracting States, in order to make them enforceable there.

In consequence, art. X CLC and the respective provisions in the 1971 FUND Convention are meant to create a regime in which all judgments by a competent court rendered in one jurisdiction can be recognised and enforced in all other contracting States.[151] The provision reads:

Art. X.

1. Any judgment given by a Court with jurisdiction in accordance with Article IX which is enforceable in the State of origin where it is no longer subject to ordinary forms of review, shall be recognized in any Contracting State, except:

(a) where the judgment was obtained by fraud; or

---

[148] Judgment in the case of the *Keumdong N° 5*, as presented by M. Jacobsson, 'L'indemnisation des dommages résultant des atteintes à l'environnement dans le cadre du régime international CLC/FIPOL' (2010) 62, Droit Maritime Français, 463–80 at 474.

[149] See C. Legros, 'L'intégration des conventions internationales dans le droit dérivé de l'Union Européenne: L'exemple du droit des transports', in J. Foyer, J. Meunier, F. Jault-Seseke, J. S. Bergé and I. Dauriac (eds.), *Le droit entre tradition et modernité: Mélanges à la mémoire de Patrick Courbe* (Paris: Dalloz-Sirey, 2012), pp. 381 et seq., also considering the role of the ECJ for matters where there is European regulation.

[150] N. Gaskell, 'Pollution, Limitation and Carriage in "The Aegean Sea"', in F. D. Rose (ed.), *Lex Mercatoria: Essays on International Commercial Law in Honour of Francis Reynolds* (London: LLP, 2000), p. 73; M. Requejo Isidro, 'Contaminación por hidrocarburos y responsabilidad civil: quién, cuánto, por qué', in J. M. Sobrino Heredia (ed.), *Mares y Océanos en un mundo de cambio: Tendencias jurídicas, actores y factores* (Valencia: Tirant lo Blanch, 2007), p. 493; Jacobsson, *2010 "Dispute resolution"*, pp. 338 et seq.

[151] Jacobsson, '2007 "L'expérience française"', 969.

(b) where the defendant was not given reasonable notice and a fair opportunity to present his case.

2. A judgment recognized under paragraph 1 of this Article shall be enforceable in each Contracting State as soon as the formalities required in the State have been complied with. The formalities shall not permit the merits of the case to be re-opened.

As a rule, however, judgments for claims under the CLC are not likely to require lengthy proceedings to be enforced. First of all, in most cases there will be a limitation fund that should recognise the claims that are filed against it. Secondly, the liability as against the shipowner will, as a rule, be covered by the shipowner's compulsory insurance, which can be directly invoked against the insurer.[152]

Here again, the rules of the EU Brussels Regulation, namely the procedural ones according to its art. 71 (2) (b), can apply to the extent in which they do not contradict the CLC as the earlier international convention.

1. Scope of the Recognition and Enforcement

As important as a general recognition of judgments without further inquiry in the recognising State is, art. X CLC can effectively create problems, namely with regard to the fact that liability is limited and that generally, the amounts available under the CLC are to be distributed between all the claimants. In the event that the total amount of claims exceeds the liability limits, all claims will be reduced in proportion. This means that every claim will have an impact on the interests of other claimants, possibly reducing the amounts that they can expect to receive under the Convention.

There is thus a very general difficulty in reconciling claims from different jurisdictions in the distribution of the fund without inquiring into their merits. Art. IX para. 3 CLC confers jurisdiction as to "all matters relating to the apportionment and distribution of the fund" on the courts of the State where the fund has been established, without however defining these matters more clearly. It has rightly been pointed out that this provision cannot only concern the procedural questions of distribution. General matters, such as the question of whether the shipowner can benefit from the limitation of liability at all or whether his conduct bars limitation, must be decided by one court authoritatively, if the goal of harmonisation of victims' rights is to be achieved, and there is some logic in considering that it should be the court at the place where the fund is established that decides on this question.[153] This would however mean that the shipowner, when having a choice as to the country in which he can establish the fund (see *supra* under I. 1.), can theoretically choose a jurisdiction in which it is very likely that the

[152] de la Rue and Anderson, *2009 "Shipping and the environment"*, p. 131.

[153] Advocated by S. Robert, *L'"Erika" : Responsabilités pour un désastre écologique*, Perspectives internationales (Paris: Pedone, 2003), p. 47; de la Rue and Anderson, *2009 "Shipping and the environment"*, p. 130.

limitation of liability will be granted, consciously avoiding jurisdictions in which a limitation can be easily denied.

The alternative solution would consist in a general denial of the limitation of liability, as soon as one decision from a court in a State party finds the shipowner liable without limitation. In this situation, the claimants of these proceedings would start to search for alternative assets of the shipowner, possibly anyway in other jurisdictions. It is submitted that with regard to the wording of art. X CLC, the question of whether a shipowner can limit his liability or not is quite substantial, and that it would form part of the "merits" of the case rather than being a question of the compensation amounts referred to in art. X CLC. It must however be kept in mind that it depends on the procedural law of the legal system in question whether this alternative solution, in which the unlimited liability would need to be recognised by the court of the limitation fund, is more beneficial to claimants or not. A shipowner having established a limitation fund will generally be able to refer claimants to this fund if they try to seize other assets, art. VI CLC, if he is entitled to limit his liability. Until it is found that he does not have the right to limitation, he may therefore secure his assets and "organise his insolvency". If, on the other hand, the right to establish the fund only arises when the right to limitation is established by judgment, a shipowner will find himself in a very insecure position until then. The CLC does not provide guidance on the procedure to follow, so that the decision must be dealt with by the courts under national law. Under the 1976 Convention on the Limitation of Liability for Maritime Claims (LLMC),[154] where there is the same problem, very different solutions can be found within one country already,[155] France being one of the very few jurisdictions in which there are precedents of a denial of the limitation under the LLMC (also see *infra* under D.).

With regard to the scope of art. X CLC, it must be pointed out that a recognition of judgments without inquiry into the merits of the case can cause a problem where there has been a judgment for an inflated claim in one contracting State. In general, with a view to the CLC's aim of unproblematic enforcement, other claimants will not be able to challenge the recognition of this claim in the courts of the place where the limitation fund is established. Therefore, there is a risk for them to see their own claims reduced by virtue of the recognition of another claim that they could not challenge, neither in the State of origin of the judgment, nor later on in the courts of the recognising State. It goes without saying that such a reduction of a parties' claim by virtue of another claim without a valid defence against this impairment of property rights may cause considerable difficulties with regard to human rights. The problem may be aggravated if the judgment to be recognised is at odds with the practice of the Fund. Recognition and enforcement of judgments against the Fund are governed by the same provisions as recognition and enforcement under the

[154] 16 ILM 606, BGBl. 1986 II, 786.

[155] See A. Vialard, '"L'apparence" de faute inexcusable comme cause de déchéance "provisoire" du droit à la limitation de responsabilité' (2000) 52, Droit Maritime Français, 787–93.

CLC,[156] so that the courts competent for the inflated claim against the shipowner will also be competent for the inflated claim against the Fund and will overrule the Fund if it refuses to satisfy the claim in the first place.[157]

One possible mechanism to mitigate this problem is of course the recognition of the Fund's practice as authoritative, or at least as a strong persuasive precedent for the courts of the contracting States. The more unanimous contracting States' courts are with regard the claims that are admissible, the smaller this risk. It has however been shown *supra* under I. 2. that there is no effectual way to force courts to recognise the Fund's practice and guidelines as authoritative.

It is to be hoped that it is not very likely that a domestic court will grant a claim that is so inflated that the property rights of other claimants in the distribution of the shipowner's limitation fund or even in the Fund's limitation are curtailed. Cases in which very large sums have been granted to claimants have, so far, been decided as cases of unlimited liability. Matters in which domestic courts decided against the Fund however concerned rather small claims. Still, to the extent where there is a risk that the conduct of one court may overturn the equity of the system, the ways in which this danger could be dealt with should be considered.

2. Exceptions to Recognition and Enforcement

It therefore seems that the exceptions to recognition and enforcement available under art. X para. 1 a) and b) CLC should be given a closer look, as a careful interpretation may be another way to mitigate this danger of claimants' property rights being curtailed by inflated claims.

At first glimpse, the exceptions of fraudulent claims and the defendant not having obtained the opportunity to present his case seem to be examples of major shortcomings in the proceedings,[158] and thus last resort exceptions that should in any case be interpreted restrictively. It does not seem that the exceptions have been used very often to rule out a claim. One of the only situations in which the exception of art. X para. 1 b) CLC has been used was the case of the *Amoco Cadiz* limitation fund in the French courts (see *supra* under I. 1. c)), where the court in the first instance ruled out the UK government's claim against the limitation fund. It held that the shipowner had not been properly given notice of the claim,[159] given that the main proceedings between the parties on the shipowner's side and the pollution victims from France were pursued in the US American courts.

[156] B. Vanheule, 'Oil Pollution: The International Liability and Compensation Regime' (2003) 38, European Transport Law, 547–76 at 562.

[157] See Gaskell, *2000 "Pollution, Limitation and Carriage"*, considering the problem more generally with a view to cases in which claims have been granted contrary to the provisions of the CLC and the FUND.

[158] Renger, *1998 "Recht und Praxis"*, p. 156.

[159] Unreported. The facts are mentioned by Rémond-Gouilloud, '1986 "Les surprises"', 92, footnote 14.

The notion of "fraud" seems to imply that the claimant needs to have some premeditation as to obtaining a judgment that he is not entitled to, and that the court errs as to the facts of the case rather than applying the Convention too generously while being fully informed about the facts. However, with a view to that art. X para. 1 a) and b) CLC are the only provisions by which a court could stop the enforcement of an inflated judgment harming other claimants, it is perhaps worth considering to interpret the provision in a broader sense, also covering judgments that have been obtained in direct violation of the principles applied in the cases of other claimants.

C. Scope of Application

The CLC's material scope of application was defined with regard to the oil tanker accidents that had triggered its negotiation. As previously mentioned, the negotiating parties focused on oil as a pollutant, due to the immediately visible damage that it causes to the marine environment. The extension of the Convention to other pollutants was ruled out at a very early point in the negotiations. Furthermore, fuel and other oil used in the operation of the ship was excluded from the coverage of the Convention, making it effectively only applicable to tankers and thus, to a smaller fraction of the world fleet.

I. Persistent Oil Carried as Cargo

The original version of the 1969 CLC, art. I para. 1, applied to "oil", meaning "any persistent oil such as crude oil, fuel oil, heavy diesel oil, lubricating oil and whale oil, whether carried on board a ship as cargo or in the bunkers of such a ship" (art. I para. 5). In the 1992 Protocol, the reference to whale oil, the only non-hydrocarbon oil which was initially included because of its viscosity and persistence similar to heavy oil,[160] was removed.

The restriction to "persistent oils" is due to the fact that all oils lighter than heavy diesel oil, such as light diesel oil, benzine, kerosene etc. will volatilise when released into the environment and are thus considered to be less harmful.[161] Considering the fact that crude oil accounts, by far, for most of the oil shipments in total,[162] the importance of other oil products is perhaps limited enough to justify the restriction.

[160] Abecassis and Jarashow, *1985 "Oil Pollution"*, para. 10-13, 10-120.

[161] However, the long-term effects of pollution by non-persistent oil have been discussed for a long time, see Jacobsen and Yellen, '1984 "Oil Pollution"', 479.

[162] UNCTAD, *Review of Maritime Transport 2011*. http://www.unctad.org/en/docs/rmt2011ch1_en.pdf (last visited 8 March 2014), pp. 13 et seqq.

The definition of what constitutes "persistent oils" is not to be found in the CLC 1969 and its Protocols, nor in the FUND Convention.[163] The Fund has however developed guidelines on this,[164] essentially describing experimental methods to determine whether a certain type of oil can be considered as "persistent".

As long as the ship in question was a tanker actually carrying oil, the original, 1969 CLC applied to the bunker oil as well, for the pragmatic reason that in the case of a major oil spill, it would be impossible to detect whether the oil released was cargo or fuel oil.[165]

In the 1992 Protocol to the CLC, bringing about some important amendments to the CLC regime, the definition of "ship" was extended to mean

> any sea-going vessel and seaborne craft of any type whatsoever constructed or adapted for the carriage of bulk as cargo, provided that a ship capable of carrying oil and other cargoes shall be regarded as a ship only when it is actually carrying oil in bulk as cargo and during any voyage following such carriage unless it is proved that it has no residues of such carriage of oil in bulk aboard.

This new definition, not too easily understandable in its wording, extends the applicability of the CLC Convention on the one hand to oily residues in the hull of oil tankers on voyages in ballast,[166] and on the other hand to the fuel oil of unladen tankers,[167] unless they have either never carried oil or it is proven that there are no oily residues on board.

The condition that the oil needs to be "carried" to trigger the application of the CLC presupposes that the oil is on board the ship in order to be transported. Some difficulties with this criterion have arisen in cases where the oil was spilt during the transfer onto or from a ship.[168] In general, the Fund seems to take the view that as long as the oil was first loaded onto the ship and then subsequently spilled, it has been "carried", even if the escape occurred before the vessel even moved. On the other hand, once the cargo is unloaded from the ship and then lost while being transferred to a terminal facility, the oil would no longer be considered as being "carried". The decisive factor to make the CLC apply seems to be whether the pipes

---

163 Özçayir, *1998 "Liability for oil pollution"*, p. 213.

164 See the "Non-technical Guide to the Nature and Definition of Persistent Oil", attached to the IOPC document FUND/A.4/11, available from the IOPC Document Service, http://documentservices.iopcfunds.org/decisions-database/standard-view/?category=4&subject=53&incident_type=all&parent=category (last visited 8 March 2014); also see Y. Baatz, *Maritime Law*, 2nd edn. (London: Sweet & Maxwell, 2011), pp. 430 et seq.

165 LEG/CONF/C.2/WP.7, *1973 "Official Records (CLC)"*, p. 565; Wu, *1996 "Pollution"*, p. 41.

166 M. Rémond-Gouilloud, 'The Future of the Compensation System as established by International Convention', in C. M. de la Rue (ed.), *Liability for damage to the marine environment* (London, New York: Lloyd' s of London Press, 1993), p. 91, attributing this initiative to the *Olympic Bravery* accident; Altfuldisch, *2007 "Haftung und Entschädigung"*, p. 15.

167 M. Jacobsson, 'Ship source pollution - Oil and hazardous and noxious substances', in D. R. Thomas (ed.), *Liability regimes in contemporary maritime law* (London: Informa, 2007), para. 1.9.

168 de la Rue and Anderson, *2009 "Shipping and the environment"*, pp. 96 et seq.

or hoses from which the oil is lost can be considered to belong to the ship.[169] This decision can however be just as difficult to take. Borderline cases in which oil is lost from a hose during pumping operations between a ship and a port facility may thus be very hard to solve. In any case, it is obvious that there may be cases not covered by the CLC.

II. Ships

According to its art. I para. 1, the CLC as amended by the 1992 Protocol applies to "any sea-going vessel and any sea-borne craft of any type whatsoever constructed or adapted for the carriage of oil in bulk as cargo". The definitions of "cargo oil" and "ship" are therefore to some extent inter-dependent, as the Convention is meant to apply to ships that can carry the pollutant targeted.

The Convention's definition of "ship" is thus rather wide and does not specify characteristics such as the outward appearance or the means of propulsion, only centring on whether the ship is capable of the carriage of oil. Problems of interpretation however arose with this very criterion. Whereas it is reasonably clear that the CLC and the FUND do not apply to offshore installations for the exploration and exploitation of the marine subsoil,[170] the line can be more difficult to draw between a ship and a storage facility, namely if the latter is theoretically capable to be used for transport purposes.

In 2006, the Assembly of the Fund decided that vessels used as storage facilities or storage facilities capable of transporting oil should only be covered by the CLC and FUND regime if they carried oil to or from a port or terminal outside the location where they normally operate, not if they are only used as the "long arm" of a terminal or other port facility.[171]

This view was seriously challenged by the Greek courts in the *Slops* case, concerning an accident that took place in the harbour of Piraeus in 2000. The *Slops* was a former tanker that was still registered in the Greek Shipping register. It had however undergone major conversion during which its means of self-propulsion had been removed, its propeller had been taken away and the engine had been deactivated. Even though the *Slops* was still registered as an oil tanker, it did not carry the mandatory insurance that it would have needed under the CLC,

---

[169] R. B. K. Kiran, 'Liability and Compensation for Oil Pollution Damage: An Examination of IMO Conventions' (2010) 3, National University of Juridical Sciences Law Journal, 399–421 at 402.

[170] J. Harrison, 'Conflicting Interpretations - The "Slops" Incident and the Application of the International Oil Pollution Liability and Compensation Regime to Offshore Storage and Transfer Operations' (2008) 20, Journal of Environmental Law, 455–64 at 457.

[171] IOPC Document 92FUND/A.11/35, para. 32.8., available from the IOPC's Document Service, http://documentservices.iopcfunds.org/decisions-database/standard-view/?category=5&subject=289&incident_type=all&parent=category (last visited 8 March 2014).

and the insurance requirement was never enforced by the Greek authorities. Waste oil from ships arriving in the harbour of Piraeus was collected on board. At the time of the incident, the *Slops* had been used as an oily water reception facility, permanently at anchor, for more than five years. On 15 June 2000, the *Slops* caught fire and exploded, subsequently spilling a considerable amount of oily water, which damaged port facilities and the environment outside the harbour. With a view to the shipowner's insolvency, the aggrieved parties addressed the Fund, claiming that the damage caused by the accident was recoverable under the FUND Convention. The Fund, hinting at its guidelines and the fact that the *Slops* had effectively not moved since 1995 and was only used as a stationary reception facility, refused to compensate the claimants. In accordance with art. VII FUND, the claimants filed thus suit against the Fund before the Greek courts. The claim was successful in the first instance, but the decision was reversed by the Court of Appeal.[172] In the final proceedings in the Greek *Areios Pagos* however, it was held that the *Slops* was a ship for the purposes of the CLC, considering that it had been a tanker and was still generally in a condition to carry oil as cargo.[173]

The decision has been criticised as policy-driven,[174] and it is certainly true that the reasoning of the court is at odds with the decision of the Greek authorities not to enforce the requirement of mandatory insurance under the CLC against the *Slops*, if the vessel was to be considered a "ship" under the Convention.[175] Looking at the wording of art. I para. 1 CLC, it has rightly been pointed out that the term of "carriage of oil" presupposes that the oil is moved with the watercraft it is loaded on, and that the prospective move at another point in time and with another means of transportation can hardly be enough to consider that the oil is "carried" by a reception facility such as the *Slops*.[176]

The case is certainly worth discussion for the problem of authoritative interpretation of the Convention, more than for the interpretation of the term of "ship" as

---

[172] "*Marine Environmental Services MC and Environmental Protection Technical SA v. International Oil Pollution Compensation Fund 1992*", (2004) 26 Peiraki Nomologia 188 (Piraeus Court of Appeal, 16 February 2004). An English summary and translation in part can be found in 2004 International Law in Domestic Courts 855 (GR 2004).

[173] "*Marine Environmental Services MC and Environmental Protection Technical SA v. International Oil Pollution Compensation Fund 1992*" (final appeal judgment), (2006) 54 Epitheorisi Emporikou Dikaiou 1797 (*Areios Pagos*, 6 July 2006). An English translation can be found in 2006 International Law in Domestic Courts 856 (GR 2006). Also see Z. Peplowska, 'What is a ship? The Policy of the International fund for Compensation for Oil Pollution Damage: the effect of the Greek Supreme Court judgment in the "Slops" case' (2010) 1, Aegean Review of the law of the sea and maritime law, 157–64 at 160.

[174] Harrison, '2008 "Conflicting Interpretations"', 462 et seq. referring to the interest in compensation for environmental damage; Peplowska, '2010 "What is a ship?"', 162, considering the interests of otherwise uncompensated victims.

[175] T. A. Mensah, 'Can the "Slops" be considered as a ship for the purposes of the 1992 Civil Liability Convention and the 1992 Fund Convention?' (2010) 1, Aegean Review of the law of the sea and maritime law, 145–55 at 151; Peplowska, '2010 "What is a ship?"', 163.

[176] Mensah, '2010 "Can the "Slops" be considered as a ship?"', 148.

such. After the final judgment by the Greek Supreme Court, the Fund did not have any further means of judicial review and effectively paid the sum claimed by the aggrieved parties. It has however insisted on its view and implicitly stated that it would treat future cases according to the guidelines, regardless of the Greek Supreme Court's decision.[177]

It goes without saying that the directly conflicting interpretations by the Fund and the Greek courts can in the future be a source of legal uncertainty[178] and possibly inequality between claimants from different countries.[179] In the very case, the absence of insurance for the *Slops* caused the Fund to compensate an accident that would otherwise have been covered—entirely or at least to the biggest extent—by the shipowner's insurer. Under these circumstances, the decision of course harms the balance between contributions from the shipping interests and the oil industry that has been established by the CLC and FUND system.[180]

With regard to the risk that an uninsured shipowner may be insolvent, so that the whole of the damage needs to be compensated by the FUND, it should however not be forgotten that the mandatory insurance applies only to ships carrying more than 2,000 tonnes of oil in bulk as cargo (art. VII para. 1 CLC). Thus, even considering the modern trend of ships increasing in size and tonnage, a significant number of oil tankers escapes the insurance obligation. According to statistics quoted by the European Maritime Safety Agency (EMSA),[181] oil and chemical tankers carrying less than 500 tonnes represented more than 7 % of the world merchant fleet in 2012.

D. The Limitation of Liability and the Loss of the Right to Limitation

Limitation of liability is rather the rule than the exception in maritime law, and the conventions on civil liability for environmental pollution do not form an exception to this. During the negotiation of the CLC, the limitation of liability that would allow shipowners to calculate their maximum financial risk was perceived as the counterweight to the strict liability, by virtue of which shipowners would be liable for oil pollution damage irrespective of their actual fault. Considering the FUND, it must be borne in mind that the oil industry financing it contributes to the compensation for oil pollution claims in other countries, in which they may only have a marginal interest. It is rather unrealistic to expect that States will require their oil industry to contribute to the compensation of these events to an unlimited

[177] IOPC Document 92FUND/A.11/35, para. 32.12.

[178] Harrison, '2008 "Conflicting Interpretations"', 463.

[179] Peplowska, '2010 "What is a ship?"', 162.

[180] Ibid., pp. 162 et seq.

[181] http://emsa.europa.eu/implementation-tasks/equasis-a-statistics/items/id/472.html?cid=95 (last visited 7 March 2014).

amount,[182] and the same can of course be argued with regard to the marine insurance industry and the shipowners bearing the burden under the CLC.

The liability limits in the CLC and the FUND are calculated in accordance with the ship's tonnage (art. V para. 1 CLC, art. IV para. 4 FUND), so that the limits are higher for bigger ships that are likely to carry more important cargoes. The initial limits in the 1969 CLC and the 1971 FUND have been significantly increased by the 1992 Protocols.

The shipowner's limitation of liability may be far-reaching, but it is by no means absolute. In the original version of the CLC, according to art. V para. 2, shipowners lost their right to limit their liability if the pollution damage was found to be the result of their "actual fault or privity". In the 1992 Protocol, the wording was changed to rule out the limitation of liability only where the pollution damage was proved to result from the owner's "personal act or omission committed with the intent to cause such damage, or recklessly and with knowledge that such damage would probably result".

The original wording of art. V para. 2 CLC, referring to the shipowner's "actual fault or privity" was directly taken from the 1957 Convention relating to the Limitation of the Liability of Owners of Sea-Going Vessels.[183] This latter Convention was replaced by the 1976 Convention on the Limitation of Liability for Maritime Claims (LLMC), which, in addition to a substantial increase in the liability limits, changed the condition for the loss of the limitation privilege to the wording later employed by the CLC 1992. The Convention is said to give the owner a "virtually unbreakable"[184] right to limitation of general maritime claims.

By virtue of the 1992 Protocol, art. V para. 2 CLC was changed to mirror the new art. IV LLMC—both Conventions now refer to the "personal act or omission committed with the intent to cause such damage, or recklessly and with knowledge that such damage would probably result". Given the fact that the provision has been drafted in accordance with the corresponding article of the LLMC, it is often claimed that the shipowner's right to limit his liability for oil pollution damage under the CLC is equally difficult to overcome and "practically unlimited".[185] Obviously, one of the negotiating parties' strongest motives for changing the provision were the problems recognised in the old version—the owner's "actual fault or privity" was often invoked and had first to be considered and to be decided on, which lengthened the compensation process and created uncertainties for the insurance interests.[186]

---

[182] Jacobsson, '2011 "An Ideal International Scheme"', 268.

[183] 1986 UKTS 52; BGBl. 1972 II, 672.

[184] See, not least, the explanation on the IMO homepage, http://www.imo.org/About/Conventions/ListOfConventions/Pages/Convention-on-Limitation-of-Liability-for-Maritime-Claims-(LLMC).aspx (last visited 7 March 2014).

[185] Vanheule, '2003 "Oil Pollution"', 552.

[186] Wu, *1996 "Pollution"*, p. 174.

Considering that both Conventions make use of the same wording and that the CLC provision was drafted after the model of the LLMC provision, it suggests itself that the notion of "recklessly and with knowledge that such damage would probably result" should be understood in the same way.[187] The expression has, effectively, also been used in other maritime—and other transport—conventions,[188] before and after the LLMC and the CLC. Furthermore, similar concepts of a reproachable behaviour that is either driven by intent or almost equals intentional wrongdoing can also be found in national legislations.[189]

The term is, effectively, very open to interpretation and its wording gives rise to difficulties in more than one way. Domestic jurisdictions sometimes rely rather heavily on their national understanding of the norm. Hence, it seems that art. V para. 2 CLC is one of the provisions of the CLC most influenced by the appreciation of the national courts, with very different outcomes. It has already been mentioned that the provision was, during the negotiations of the 1984 and 1992 protocols, expected to render the shipowner's right to limit his liability almost absolute, but this is definitely not always the case in certain jurisdictions.

I. The Shipowner's "Personal" Act or Omission

When considering the wording of art. V para. 2 CLC as amended, it first refers to the shipowner's "personal act or omission", which clearly points to the fact that the deed must have been committed by the shipowner himself and that there should be limits to attribute another person's behaviour to him. Therefore, wrongful behaviour by the a vessel's crew or captain cannot, generally, trigger the exception of art. V para. 2 CLC against the shipowner, unless the latter can be blamed with a behaviour that fostered or made possible the wrongful behaviour.

The notion of the shipowner's "personal" act or omission of course causes some problems with regard to shipowning companies which only act through their directors and employees. It is very questionable which person needs to have privity or knowledge of the act or omission for it to be attributable to the company as a

---

[187] Altfuldisch, *2007 "Haftung und Entschädigung"*, p. 18; but the uniform interpretation of the provision is questioned by A. Vialard, 'L'évolution de la notion de faute inexcusable et la limitation' (2002) 54, Droit Maritime Français, 579–85 at 580, advocating that there should at least be different understandings with regard to the differences between material and corporeal damage.

[188] Art. XXV of the 1929 Warsaw Convention for the Unification of Certain Rules Relating to International Carriage by Air, as amended by the 1955 Protocol; Art. IV para. 5 of the Convention for the Unification of Certain Rules concerning Bills of Lading, as amended by the 1968 and 1979 Protocols (Hague Visby Rules); arguably Art. 6 of the BOPC, by reference to the provisions of the LLMC, and Art. 9 (2) of the HNS Convention.

[189] See J. A. Schaar, 'The Shipowners' Limitation of Liability Act: Still Afloat or Sinking Fast?' (2000) 24, Tulane Maritime Law Journal, 659–704 on the US Shipowners' Limitation of Liability Act.

whole. The attribution of representatives' fault to the company is a common problem of company law in every legal system, and here again, the CLC does not provide guidance, so that the decision will effectively be left to the national courts. These will consider the question almost certainly with the domestic company law provisions in mind.

Very generally, the person whose knowledge is considered will need to be a member of the company's administrative staff. Nautical personnel such as vessels' masters or engineers is generally not considered to be high enough in the company hierarchy to render the corporation aware of the wrong.[190] Arguably, art. V para. 2 CLC, by differentiating between the shipowner's and the nautical staff's acts and omissions is also drawing a distinction between the conduct of staff that is actually employed in the nautical operation, and thus perhaps more prone to spontaneous or rash decisions in a given situation,[191] and administrative staff which is expected to take decisions in a planned and objective way. Since the implementation of the International Safety Management Code (ISM Code) into the SOLAS Convention (Chapter IX) in 1994, it may have become easier to establish that the management of a shipowning company had knowledge of shortcomings in safety matters,[192] the ISM Code requiring the company to designate a person having access to the top management as a person responsible for safety matters. By virtue of the Code, there is thus the theoretical possibility to trace non-compliance with safety measures from the ship to a responsible person within the shipowning company.[193] Since the coming into force of the ISM Code, shipowning companies will at least no longer be able to claim that they did not sufficiently assess the implementation of safety practices on board, as this could in itself be an omission of their obligations under the ISM Code. It will however still be a matter of a case-to-case consideration whether a non-compliance with safety practices in a given situation can really be attributed to the management.[194]

Existing precedents and doctrine in domestic law most often deal with the problem under the LLMC, but the results are generally considered to be the same

---

[190] Altfuldisch, *2007 "Haftung und Entschädigung"*, p. 20, on English and German law; D. Veaux and P. Veaux-Fournerie, 'La pénétration en droit français de la thérorie de la faute inexcusable en matière aérienne et maritime sous l'influence des conventions internationales', in J. Béguin and P. Bourel (eds.), *L'internationalisation du droit: Mélanges en l'honneur de Yvon Loussouarn* (Paris: Dalloz, 1994), p. 398, on French Law; Schaar, '2000 "The Shipowners' Limitation of Liability Act"', 696, considering the problem under the US Shipowners' Limitation of Liability Act and referring to respective case law.

[191] Y. Tassel, 'Le droit maritime - un anachronisme?' (1997), Annuaire de droit maritime et océanique, 143–57 at 151.

[192] Gauci, '1999 "Protection of the Marine Environment"', 29; Altfuldisch, *2007 "Haftung und Entschädigung"*; H. Honka, 'Modern Comprehensive Approach to Safety at Sea', in U. Karpen, D. Lukic and S. Heinze (eds.), *Maritime Security - Current Problems in the Baltic Sea*, 1st edn. (Baden-Baden: Nomos, 2008), p. 110.

[193] See C. H. Allen, 'The ISM Code and Shipowner Records: Shared Safety Goals vs. Industry's Privacy Needs' (1998) 11, University of San Francisco Maritime Law Journal, 1–46.

[194] See Martínez Gutiérrez, *2011 "Limitation of Liability"*, pp. 70 et seq.

under the CLC.[195] Giving a broad summary of the general tendencies, it seems that the act or omission must be found in a person that can be considered to be the *alter ego* of the company, having an overview of the management of the ship and the authority to take decisions on behalf of this ship—seemingly regardless of their actual function within the company.

With regard to the normal tasks of a shipowner and the practical possibilities of controlling and monitoring a ship at sea, there is a certain tendency to consider that the shipowner has a general obligation to ensure the vessel's seaworthiness and equipment, and that shortcomings in the technical equipment or omissions in monitoring the normal operational procedures on board can constitute a "personal act or omission" of the shipowner. Mere nautical faults by the ship's crew or sudden technical problems should, in accordance with the wording, be excluded.[196] A nautical mistake can become a shipowner's personal act or omission where the technical problems were due to organisational mistakes or insufficient manning for which the shipowner can be held responsible.[197]

II. The Degree of Negligence and the Foreseeability of the Damage

Apart from determining the person whose knowledge or privity triggers the exception, it is questionable which actions can actually be considered to be committed "with the intent to cause such damage or recklessly and with knowledge that such damage would probably result" in the sense of art. V para. 2 CLC.

With regard to the term "intent", it should be noted that it is actually not only the act, but also the damage that must be caused intentionally—which is hard to imagine in oil pollution cases except perhaps for matters of insurance fraud. The alternative "recklessly and with knowledge that such damage would probably result" obviously describes a state of mind in which the circumstances by which the damage is caused are clearly known to the wrongdoer and the ensuing damage could be foreseen not only as possible, but rather as probable. From a general point of view and bearing in mind the intention of the negotiating parties when art. V para. 2 CLC was changed to its current wording, it is fairly clear that the exception triggering the shipowner's unlimited liability should be reserved for faults that are

[195] See, for example de la Rue and Anderson, *2009 "Shipping and the environment"*, p. 791 as to the approach under English law, but mostly with regard to the 1957 convention, Altfuldisch, *2007 "Haftung und Entschädigung"*, p. 19, as to German law and I. Corbier, 'La faute inexcusable de l'armateur ou du droit de l'armateur à limiter sa responsabilité' (2002) 54, Droit Maritime Français, 403–17 at 411 for an overview on French maritime law.

[196] Ibid., p. 408; Altfuldisch, *2007 "Haftung und Entschädigung"*, p. 20.

[197] See Corbier, '2002 "La faute inexcusable"', 411 et seq. giving examples under French law.

committed in a state of mind very close to the intentional doing, which is to be distinguished from mere negligence.

The unanimity about how to interpret "recklessly and with knowledge that such damage would probably occur" however stops there. Given that the expression is used in many maritime and other transport conventions, and to a certain extent in a similar way in domestic legal systems, it is almost impossible to give a comprehensive overview of the possible understandings. Only an outline of the prevailing interpretations of art. V para. 2 CLC can therefore be given in this place, most of them derived from the understanding of the term in the LLMC. The comparison with the interpretation of this term in transport conventions, namely the Hague and Hague Visby rules will not be considered here. Even if the legal terms are very similar or the same, transport conventions only concern the relationship between carrier and shipper, whereas general limitation frequently occurs between the shipowner and third parties.[198] This is also the situation with regard to the CLC, so that there is another similarity between the two legislative frameworks, apart from the legislative history of art. V para. 2 CLC as amended by the 1992 Protocols, aligning the provision with the LLMC.

The English courts have, to this day, shown themselves extremely reluctant to break the shipowner's limitation of liability under the LLMC. It has, in a case concerning very careless behaviour, explicitly stressed[199] that the breaking of the limitation is an absolute exception, to be reserved for very extraordinary cases, without however being entirely inconceivable.[200] The case of *Margolle v. Delta Maritime* concerned a collision between a fishing vessel and a freighter. The fishing vessel had navigated against the Traffic Separation Scheme in the Dover Strait at night with only one deckhand on watch, contravening rule 10 of the Convention on the International Regulations for Preventing Collisions at Sea 1972 (COLREG). It was under the command of the shipowner who had consciously directed this course for the only motive to overtake rival trawlers on the way to the fishing grounds. The English court still seems to have regarded this as a borderline case, denying the claim for limitation of liability without a hearing and allowing the case to go on trial. The case was however settled before a court could decide on the possibility to limit the claim.[201]

There is a lot of abstract discussion about the degrees of negligence and foreseeability of the damage under English law, and in one decision,[202] the Court

---

[198] A. Vialard, 'La limitation de responsabilité, clé de doute pour le droit maritime du 21e siècle' (2009) 61, Droit Maritime Français, 21–8 at 24, footnote 4; S. Miribel, 'L'affaire "Rosa Delmas": Limitation et faute inexcusable, une nouvelle approche?' (2011) 63, Droit Maritime Français, 863–9 at 869.

[199] *"Margolle v. Delta Maritime Co. Ltd."* 2003 1 Lloyd's Law Reports 203 (High Court (England), Queen's Bench Division, 11 November 2002), p. 209.

[200] Martínez Gutiérrez, *2011 "Limitation of Liability"*, p. 75.

[201] M. N. Tsimplis, 'Liability in respect of passenger claims and its limitation' (2009) 15, Journal of International Maritime Law, 125–50 at 143.

[202] *"The Leerort"* 2001 2 Lloyd's Law Reports 291 (Court of Appeal (England), 27 June 2001).

of Appeal stressed that the defendant must have foreseen "the very loss that actually occurred, not merely the type of loss that occurred".[203] The findings were partly based on the interpretation of the term "such loss". The case concerned a collision between two ships in Colombo harbour, the *Leerort*, which was lying berthed, and the *Zim Piraeus*. The crew of the latter craft entered the harbour at excessive speed, then failed to observe the fact that at this speed it would not be possible to carry out an astern movement in automatic control mode, and failed to put the engine immediately full astern when they had finally engaged the manual control mode. Even though the parties responsible for the *Zim Piraeus* admitted this conduct, the court did not consider that it defeated their right to limitation. Not least, the court stressed that it is hardly conceivable that the crew acted recklessly while having the knowledge that this manoeuvre was likely to involve the still new-built *Zim Piraeus* in a collision. This is however to some extent at odds with the findings of the court in the *Margolle v. Delta Maritime* case, where it was held that the likelihood of the fishing vessel being involved in a collision resulted from the presence of other vessels in the same area,[204] irrespective of the fact that the shipowner had repeatedly infringed the COLREG in that area without any accident beforehand, and doubtlessly expected that everything would go well again this time.

Under English law, the shipowner's right to limit liability is virtually never broken in practice.[205] There is however one rather recent Canadian case in which limitation of liability was denied to a shipowner. The facts of it are rather peculiar and do not contribute significantly to a better understanding of how the provision is to be understood. In the case of *Société Telus v. Peramoco Inc.*,[206] a telephone cable lying at the bottom of the St. Laurent River was cut by the defendant, a self-employed fisherman, when his gear became entangled in the cable. The defendant believed the cable to be abandoned and unused, even though the cable's position had been duly published and the defendant ought to have known of it. When the claimants challenged the defendant's right to limit his liability, the court considered the scope of art. IV LLMC rather thoroughly. Harrington J. held that the defendant acted intentionally when he cut the cable, and that the mere fact that he believed the damage to be negligible as he deemed the cable to be abandoned did not change his premeditation.[207] It was only as an additional argument that the judge considered whether the defendant acted recklessly, so the findings as to this are mere *obiter dicta*. He stressed the fact that the defendant failed to inform himself about the existence of the cable, even though the knowledge was readily available, which

---

[203] Ibid., p. 295.

[204] Martínez Gutiérrez, *2011 "Limitation of Liability"*, p. 75.

[205] As asserted by Steel J, *"MSC Mediterranean Shipping v. Delumar BVB et al. (The MSC Rosa M)"* 2000 2 Lloyd's Law Reports 399 (High Court (England), Queen's Bench Division, 14 July 2000), p. 400: "there are no examples in English law of the defence being successfully run in the maritime context."

[206] *Société Telus v. Peracomo Inc.* 2011 Federal Court (CanLII) 494 (Federal Court (Canada), 06 May 2011).

[207] Ibid., para. 74 et seqq.

rendered the conduct reckless. As to the probability of the loss, the judge deemed it as certain, considering that the defendant wanted to cut the cable. The decision was affirmed on appeal.[208] With a view to the fact that the case was actually decided as a case of intent barring the right to limit liability, the findings as to recklessness are only of limited help. The scarcity of precedents in English and Canadian[209] law also makes it rather difficult to establish the degree to which the shipowner must foresee the damage and the degree to which the damage foreseen must correspond to the damage that actually occurs. It is however certain that the breaking of the shipowner's liability under English law will be reserved for exceptional cases.

This view is in considerable contrast with the interpretation of art. IV LLMC that has been upheld by the French courts for a considerable time, even if the general position seems to be in the process of change. Still, there are precedents in which the limitation of liability was denied to the shipowner, not only concerning defaults in the maintenance or technical equipment of a vessel or the choice of a master known for his lack of responsibility,[210] but also shortcomings in the operation of the ship[211] that rather resemble nautical mistakes. In the case of the *Johanna Hendrika*, the accident in question can actually be considered to have been caused by the negligence of the persons on board. The *Johanna Hendrika* was a dredger, employed in digging operations in port. The vessel operated on ground that eventually fell dry. As the dredger was not properly anchored, it slid towards the quayside and damaged two vessels that were moored there. It was obvious that the master of the *Johanna Hendrika* had committed a mistake, but the courts also detected reckless conduct of the shipowner without any precise statement as to the actual act or omission they blamed him for. This was why the decision was finally quashed by the *Cour de Cassation*.[212] After remand, the *Cour d'Appel* of Caen stated that the *Johanna Hendrika's* anchorage system had been insufficient for the task and that there were only two persons on the dredger at the time of the accident, whereas it should have been manned by at least eight crewmembers.[213] The decision can certainly be criticised for determining the matter on rather vague criteria,[214] omitting to establish whether the anchorage system could have

[208] *Peracomo Inc. v. Société Telus (Appeal)* 2012 Federal Court of Appeal (CanLII) 199 (Federal Court of Appeal (Canada), 29 June 2012).

[209] See *Federal Court (Canada), 06.05.2011, Société Telus v. Peracomo Inc.*, para. 62, emphasising that this is the first recognised case which is decided under art. IV LLMC.

[210] *"Brescou"*, 57 Droit Maritime Français 713 (Cour d'Appel de Montpellier, 04 November 2004).

[211] A. Vialard, 'Observations CA Bordeaux 31.05.2005' (2005) 57, Droit Maritime Français, 846–50 at 859.

[212] *"Johanna Hendrika"*, 49 Droit Maritime Français 976 (Cour de Cassation (France), 20 May 1997).

[213] *"Johanna Hendrika"*, 53 Droit Maritime Français 981 (Cour d'Appel de Caen, 02 October 2001), p. 987.

[214] O. Cachard, 'Observations CA Bordeaux 14.01.2013' (2013) 65, Droit Maritime Français, 212–20 at 218.

been sufficient if the dredger had anchored in another spot, whether the accident was really caused by the insufficient crew or whether the two moored vessels should rather have been moved out of the *Johanna Hendrika's* reach to avoid the accident.

In the case of the *Stellaprima*,[215] the *Cour d'Appel* of Montpellier argued that the omission to check project cargo before lifting it could constitute reckless behaviour of the shipowner, apparently attributing the mistake of the persons involved in the loading process to him. The *Cour de Cassation* merely stated that the shipowner failed to establish a control procedure that the crew would have had to follow prior to loading.[216] In the same line of argumentation, the shipowner was denied the limitation of liability in the case of the *Multitank Arcadia*,[217] for not having given stricter instructions and carried out stricter controls concerning the maintenance procedures on board. It must however be pointed out that in other cases where the accident was caused by rather unexpected defects in otherwise well-maintained machinery,[218] limitation of liability was granted.

Thus, certain authors, relying on respective case law, have identified a general *obligation de resultat*,[219] that is, an absolute duty to provide a seaworthy vessel with an adequate crew and the suitable equipment for the very task the vessel is employed in. Any problem in the employment of the ship can of course be possibly traced back to an omission of the shipowner, who, depending on the circumstances, could also have foreseen the probability of the ensuing damage. In view of the precedents cited, French jurisprudence can be considered to circumvent the limitation provisions to a wide extent,[220] increasing the risk of shipowners to face unlimited liability to a considerable degree. In more recent jurisdiction, there is however a tendency to interpret art. IV LLMC in a more restricted way.

It is rather ironical that one of the most famous cases in this matter that was first considered as revolutionary in its denial of the limitation,[221] turned out to be one of the most important precedents for the opposite position after more than 20 years of proceedings. In the case of the *Heidberg*, a vessel collided with a landing stage when leaving port, an accident which was caused by a six minutes' absence of the

---

[215] *"Stellaprima"*, 52 Droit Maritime Français 813 (Cour d'Appel de Montpellier, 07 December 1999); *"Stellaprima"*, 54 Droit Maritime Français 460 (Cour de Cassation (France), 03 April 2002).

[216] Ibid., p. 462.

[217] *"Multitank Arcadia"*, 54 Droit Maritime Français 150 (Cour d'Appel d'Aix-en-Provence, 10 October 2001).

[218] *"Zulu Sea"*, 54 Droit Maritime Français 125 (Cour d'Appel d'Aix-en-Provence, 05 November 1998); *"Moldavia"*, 54 Droit Maritime Français 132 (Cour d'Appel d'Aix-en-Provence, 08 June 2000).

[219] Corbier, '2002 "La faute inexcusable"', 413; K. Le Couviour, 'Erika: l'arrêt salvateur de la Cour de Cassation' (2012), La Semaine Juridique Édition Générale, 2095–9 at 2098.

[220] Corbier, '2002 "La faute inexcusable"', 416.

[221] A. Vialard, 'L'affaire "Heidberg": gros temps sur la Convention de Londres 1976 sur la limitation de responsabilité en matière de créances maritimes' (1993) 45, Droit Maritime Français, 706–11 at 707.

master from the bridge. In the first instance,[222] the court held that the master should not have left the bridge to the pilot and the helmsman to attend other business in the engine room, and the fact that he did so was revealing of an undermanning of the vessel, for which the shipowner was to blame. The *Heidberg*, flying the German flag, was effectively only manned by two officers, in accordance with the German regulation then in force. Whereas one can certainly acclaim this French initiative to discourage shipowners from operating their ships under the absolute minimum safety standards,[223] the judgment can however be considered to have some weaknesses especially in establishing the causality between the master's absence, his decision not to summon the second officer to the bridge and the circumstances of the accident.[224] Still, the decision was affirmed on appeal,[225] the *Cour d'Appel* of Bordeaux explicitly holding that the shipowner was liable for reckless behaviour because he knowingly accepted that the *Heidberg*'s master did not trust his sailors with the normal onboard tasks, even though they were formally competent.[226] In this way, the master's error of judgment in the minutes before the allision was traced back to the shipowner's decision as to the crewing of the vessel. This decision was however, more or less for procedural reasons, quashed in the next instance. After remand, the *Cour d'Appel* of Bordeaux finally re-established the shipowner's right to limit his liability.[227] The court, after meticulously considering the facts of the case held that the master's decision to leave the bridge during the critical moments of the *Heidberg* leaving port was certainly a nautical mistake, but not induced by a management or operational mistake of the shipowner. The *Heidberg's* manning complied with the regulation in place at the time of the accident, and even if the problems in the collaboration of the crew had been established, the shipowner could hardly foresee that the master would commit the mistake to leave the bridge in a critical moment without ensuring that the vessel was properly attended to.[228] The decision thus stresses that the degree of probability with which the damage would occur should be assessed objectively, to the standards of a reasonable shipowner.[229]

---

[222] *"Heidberg"*, 45 Droit Maritime Français 731 (Tribunal de Commerce de Bordeaux, 23 September 1993).

[223] Vialard, '1993 "L'affaire "Heidberg""', 709 et seqq.

[224] See T. Clemens-Jones, '"Heidberg": malfaiteur ou victime d'une injustice?' (1993) 45, Droit Maritime Français, 712–7 at 716 et seq.

[225] *"Heidberg"*, 57 Droit Maritime Français 839 (Cour d'Appel de Bordeaux, 31 May 2005).

[226] Ibid., pp. 843 et seq.

[227] *"Heidberg" (after remand)*, 65 Droit Maritime Français 201 (Cour d'Appel de Bordeaux, 14 January 2013).

[228] Ibid., p. 210.

[229] Cachard, '2013 "Observations CA Bordeaux 14.01.2013"', 220.

These findings, namely the objective appreciation of the foreseeability of the damage[230] are supported by another, earlier decision[231] concerning an allision between a public transport ferry and the facilities of a fish farm off Toulon. The court stated that the master of the ferry certainly committed a serious nautical mistake, but that in the absence of any wrongful conduct of the shipowning transport company, the limitation of liability would not be denied.

In their recent cases, the French courts also require a close link of causality between a shipowner's act or omission and the ensuing damage, even if the mistake was a serious one. In the collision case of the *Antares* and *Audrey Marie*[232] it was held that the shipowner and master's omission to train his crew to use a newly installed radar and his decision to leave the bridge for a few moments without setting the anti-collision alarm did not constitute a reckless behaviour, because the radar was only one of the means of monitoring the vessel's surroundings. Similarly, the fact that the crew had seriously exceeded their working hours could not constitute reckless behaviour as it could not be found that the collision was caused by fatigue.[233] In another case decided by the *Cour de Cassation*, it was held that the owner of a pleasure craft seriously exceeding the speed limit could still invoke the limitation of liability following an accident with loss of life, in the absence of other facts that would constitute recklessness.[234] The court pointed out that the victim of the accident, a diver, also failed to ensure his visibility in the water.[235] Both cases can also be understood as to require a higher degree of foreseeability in the interpretation of the term "with knowledge that such damage would probably occur", considering that it is the courts' opinion that the accidents could still have been avoided if other precautions had been taken. Notwithstanding the shipowners' obvious nautical mistakes, the courts thus seem to consider that the shipowners were somehow justified in their expectations that everything would go well.

Even if there is, thus, a recent development as to a stricter appreciation of the shipowner's conduct barring limitation, it is by no means sure that the same considerations will apply to accidents that are caused by defaults in ships' maintenance or security equipment.[236] In the *Erika* case, in which reckless conduct was heavily debated, the notion of "recklessly and with knowledge that such damage will probably result" also being the requirement for overcoming the channelling provisions (see *infra* under E. II.), charterers, managers and the classification

[230] P. Bonassies, 'Observations CA Aix-en-Provence 11.10.2012' (2013) 65, Droit Maritime Français, 224–7 at 226.

[231] *"Creux Saint Georges III"*, 65 Droit Maritime Français 221 (Cour d'Appel d'Aix-en-Provence, 11 October 2012).

[232] *"Antares" and "Audrey Marie"*, 65 Droit Maritime Français 875 (Cour d'Appel de Rouen, 14 February 2013).

[233] Ibid., p. 880.

[234] *"Dizzy"*, 65 Droit Maritime Français 142 (Cour de Cassation (France), 11 December 2012).

[235] Ibid., p. 157.

[236] Distinction pointed out by Cachard, '2013 "Observations CA Bordeaux 14.01.2013"', 218.

society of the vessel were held to have acted recklessly in operating and classifying a ship that was obviously in bad state.

There has always been considerable debate among French authors whether the foreseeability must be given as an objective criterion—so, if the standard is the one of a reasonable person—or if it is subjective and must have been envisaged by the actual shipowner in question. Most cases in which this problem was explicitly considered, including the earlier decisions, argue for an objective appreciation,[237] maintaining that the probability of the damage was foreseeable for a professional fisherman[238] or a professional of a society providing nautical services.[239]

In summary, French courts are apparently still inclined to find recklessness with knowledge that the damage would probably occur in a shipowner's behaviour. Even if it seems that jurisdiction at least in the courts of the second and third instance moves away from the extreme cases such as the *Johanna Hendrika*, it is still considerably more likely that the limitation of liability will be denied in the French courts than in the courts of another jurisdiction. It is submitted that the French courts' tendency to appreciate the foreseeability of the damage objectively is partly responsible for it. By requiring that the actual shipowner in question must have consciously foreseen the damage, "recklessness" actually becomes constructive intent, considering that even the most careless wrongdoer will, as a rule, still hope for everything to go well.[240] It is thus not surprising that the only case in which reckless behaviour was found even under the subjective appreciation, the Canadian case of *Société Telus v. Peramoco Inc.*, was actually a case of intent. It is true that at first glimpse the subjective approach is easier to reconcile with the wording of art. IV LLMC and art. V para. 2 CLC, rather hinting at the actual shipowner's knowledge. However, it is almost impossible to judge the actual thoughts of a person before the accident,[241] and denying the limitation of liability only to those shipowners who admit that they foresaw the damage would amount to punishing the sincere.

It must however be stressed that especially in the past, French courts have gone very far, and arguably too far, in denying the limitation of liability to careless shipowners. The exception of art. IV LLMC and art. V para. 2 CLC are still meant to be exceptions, not to be invoked for behaviour that only constitutes negligence. An interpretation of this kind certainly contradicts the negotiating parties' original

---

[237] Ibid., p. 226.

[238] *"L'Inglais" and "François Delphine"*, 57 Droit Maritime Français 708 (Cour d'Appel de Montpellier, 18 November 2003), pp. 710 et seq.

[239] *"Moheli"*, 56 Droit Maritime Français 727 (Cour d'Appel de Rouen, 09 November 2004), p. 732.

[240] See C. Haddon-Cave, 'Limitation against passenger claims: Medieval, unbreakable, unconscionable' (2000), CMI Yearbook, 234–43 at 237 et seq., in the context of passenger claims.

[241] Y. Tassel, 'Observations CA Rouen 09.11.2004' (2005) 57, Droit Maritime Français, 734–7 at 736.

intention to strengthen the shipowner's position with regard to the limitation by amending art. V CLC.[242]

III. The Burden of Proof

It is argued by a considerable number of authors that art. V para. 2 CLC as amended shifts the burden of proof for the wrongdoer's premeditation to the claimants, who, in order to overcome the limitation, have to establish that the shipowner acted with a higher degree of negligence ("the owner shall not [. . .] if it is proved").[243] It seems that in the view of the English courts, the difficulties to establish a "reckless" behaviour are partly due to this distribution of the burden of proof, as the claimants face the difficulty to prove both the defendants' reckless conduct and their state of mind, i.e. the knowledge that the relevant loss would probably result.[244]

Considering that the appreciation of the shipowner's behaviour is in the majority held to be done according to objective standards, this view is not shared by French authors. Many consider the limitation of liability as a privilege, only accorded to shipowners who can show that their behaviour was at least not reckless.[245] It seems that in most cases the courts proceeded to a certain assumption of the conscious mind of the defendants seeking to limit their liability, claiming that the probability of the damage was obvious. With the courts' recent tendency to be stricter in finding a shipowner's behaviour reckless, this result may be mitigated in practice. Examples of this can be found in the *Heidberg* case where the court in the last instance decided that according to the facts of the case it was highly unlikely that the shipowner would have foreseen the chain of events that led the master to quit the bridge and eventually caused the collision. In the cases in which the shipowner is on board, such as the judgments that concern fish trawlers under the command of their owners, an objectively dangerous course of action, in violation of established maritime law and practice, can very easily trigger the denial of limitation of liability, as a reasonable fisherman or shipowner ought to have foreseen the adverse consequences of a breach of the rules. In these cases, it would then rather fall to the shipowner to prove that he never foresaw the possibility of the accident.

Whereas it is true that the wording of art. V para. 2 CLC rather explicitly hints at the onus of proof being on the person who challenges the limitation of liability, the

---

[242] P. Bonassies, 'Observations CA Caen 02.10.2001' (2001) 53, Droit Maritime Français, 989–93 at 993.

[243] de la Rue and Anderson, *2009 "Shipping and the environment"*, p. 803 with regard to the LLMC; Wang, *2011 "Civil liability"*, p. 147; Baatz, *2011 "Maritime Law"*, p. 441, footnote 126.

[244] *High Court (England), Queen's Bench Division, 11.11.2002, "Margolle v. Delta Maritime Co. Ltd."*, p. 207; *High Court (England), Queen's Bench Division, 14.07.2000, "MSC Mediterranean Shipping v. Delumar BVB et al. (The MSC Rosa M)"*, p. 403.

[245] P. Bonassies, 'Problèmes et avenir de la limitation de responsabilité' (1993) 45, Droit Maritime Français, 95–107 at 106; Corbier, '2002 "La faute inexcusable"', 413.

other paragraphs of art. V CLC are not entirely clear in this respect. The limitation of liability is also alluded to as an "entitlement" (art. V para. 2 and 11 CLC) or a "benefit" (art. V para. 3 CLC).

Certainly, the proof of a defendant's guilty mind is always difficult to establish, and that in most cases the claimants and the courts will need to have recourse to secondary evidence that allows the drawing of conclusions as to the state of mind of the defendant. It does not seem inadmissible to consider an obviously extremely dangerous and careless behaviour as a hint to that the person in question consciously took the risk that the ensuing damage would arise. It does however seem that the French courts have gone too far in this in certain cases,[246] and arguably, rather often for policy reasons. This is clearly undesirable as it renders the provision legally uncertain and impairs the compensation system that is heavily based on the limitation of liability as a general principle.

IV. Insurance

It has been argued that the concept of the owner acting "recklessly and with knowledge that such damage would probably result" equals "wilful misconduct" which may be, under insurance legislation, a reason for the insurance to reject the claim.[247] The shipowner may thus lose his insurance cover, so it would not be in the interest of claimants to invoke that he acted recklessly. Art. VII para. 8 CLC, in the fourth sentence, in fact, allows the insurer to avail himself of the defence that the owner acted with wilful misconduct. Art. V para. 11 CLC, however, in stipulating that the insurance fund may be constituted to the liability limits even if the owner is not entitled to invoke the limitation, actually does not presuppose that insurance cover is necessarily totally lost in situations of reckless behaviour.

The answer to the problem will depend on the terms of the insurance contract, the law governing it, the definition of "wilful misconduct" under this law[248] and the degree of proof required to establish the wrongful behaviour.[249] A clear distinction

[246] Considered by A. Vialard, 'La faute de l'assuré dans le droit français des assurances maritimes' (2008), Annuaire de droit maritime et océanique, 605–19 at 614: "unmoderated".

[247] Wu, *1996 "Pollution"*, p. 181; P. Wetterstein, 'Shipowners' Environmental Impairment Liability - Developments and Challenges', in H. Tiberg, M. Clarke, A. Beskow, M. Carlsson, R.-M. Lundström, T. Utterström and S. Arvmyren (eds.), *Essays on Tort, Insurance Law and Society in Honour of Bill W. Dufwa: Volume II* (Stockholm: Jure, 2006), p. 1252.

[248] See E. Gold, *Gard Handbook on P & I Insurance*, 5th edn. (Norway: GARD, 2002), para. 7.2.3. on the question under Norwegian law; D. Damar, *Wilful misconduct in international transport law* (Berlin, Heidelberg: Springer-Verlag, 2011), pp. 35 et seqq. on English law and Vialard, '2008 "La faute de l'assuré"' on French law.

[249] See H. Tiberg, 'Fraud and Sinking Ships in Marine Insurance', in H. Tiberg, M. Clarke, A. Beskow, M. Carlsson, R.-M. Lundström, T. Utterström and S. Arvmyren (eds.), *Essays on Tort, Insurance Law and Society in Honour of Bill W. Dufwa: Volume II* (Stockholm: Jure, 2006), p. 1181 on the proof of wilful misconduct and gross negligence, in the context of the proof of actual intent.

must however been drawn between contracts for hull and cargo insurance, meant to indemnify the shipowner for damage done to his vessel and its cargo, and P & I insurance, meant to indemnify the shipowner for any liabilities he incurs as against third parties. Only the latter is contemplated here, as it is the insurance covering liability for pollution damages.[250] The question of wilful misconduct in P & I insurance contracts is very widely considered as a matter of English law, which governs most of these contracts.[251]

Considering the importance of the owner's fault breaking the limitation of liability in the French courts, it appears surprising that P & I insurances have seemingly abstained from pleading the *faute inexcusable* or, following the newer and more precise terminology,[252] *faute de témérité*, of their assured in most cases.[253] Considering that P & I insurance is generally organised as a mutual self-insurance between shipowners, in which all of them are insurers and beneficiaries at the same time, it may not be in their interest to deny payments on the ground of reckless conduct and thus to create precedents. Another possible explanation could be found in the fact that within the insurance contract relationship, the burden of proof for establishing the shipowner's fault is clearly on the insurer refusing to pay,[254] and it is perceivable that courts may require other standards of proof in this relationship than in the tort relationship between a shipowner and an oil pollution victim under the CLC, or even between a shipowner and a third party under the LLMC. Furthermore, under French domestic insurance law, it may not be easy for the insurer to make the shipowner answerable for the reckless conduct of his agents or employees.[255]

Still, if the owner's reckless conduct is invoked by victims to overcome the limitation of liability under the CLC, the possible impact on the insurance contract should however not be overlooked, as victims may face the possibility of the shipowner's losing the insurance cover. It is however submitted that in most cases in which victims aim for breaking the liability limits to recover unlimited compensation from the shipowner, they will expect him to have additional funds at his disposal in any way, so that they most probably do not need to rely on the limited insurance cover at all.

---

[250] See however T.-L. Wilhelmsen, 'Misconduct of the Assured and Identification' (2003), CMI Yearbook, 540–72 on the problem under hull and cargo insurance, giving a very comprehensive overview from a comparative perspective.

[251] Vialard, '2008 "La faute de l'assuré"', 610.

[252] Cachard, '2013 "Observations CA Bordeaux 14.01.2013"', 225.

[253] Vialard, '2008 "La faute de l'assuré"', 609.

[254] Ibid., p. 609.

[255] Ibid., pp. 615 et seq.

E. Overcoming the Channelling Provisions

The channelling of liability to the shipowner and the limitation of liability in the CLC is arguably not absolute. Art. III CLC, in its original and in its amended version, only enumerates certain persons to be protected from claims by third parties, and the situation is not too clear with regard to other actors that are in a rather remote relationship to the shipowner. Protection of these parties by the channelling provisions could lead to a total exemption from liability, and could thus result in barring an additional source of compensation for victims.[256]

Moreover, the channelling only works in relation to third parties. According to art. III para. 5 CLC, nothing prejudices the shipowner's right to recourse, so that the channelling definitively does not amount to an exemption of liability for the beneficiaries of the provision.

It should, thirdly, not be forgotten that parties engaged in the maritime business may have the right to limit their liability by virtue of other legislative instruments, such as the LLMC. The Convention applies to all "maritime claims", defining these as, generally, claims arising in direct connection with the operation of the ship (art. II LLMC). It can be invoked by a defined class of beneficiaries comprising shipowners—the term covering owners, charterers, managers and operators—and salvors (art. 1 LLMC), and also the persons "whose neglect, act or default" the shipowner or salvor is responsible for.[257]

The LLMC obtained a number of ratifications that equals 54, 55 % of the world maritime tonnage, and 45, 13 % with regard to the 1996 Protocol to the Convention[258] that, in the first place, was meant to increase the liability limits. The Convention generally applies to all claims for the limitation of liability brought before a State party's courts (art. XV LLMC), with the *caveat* that domestic legislation of a State party may deny the application to defendants who do not have their habitual residence or principal place of business in the State in question. This serves to restrict shipowner's forum shopping for jurisdictions that allow the limitation of liability.

Defendants who will not benefit from the CLC's channelling provisions may thus try to invoke the LLMC. The limits of liability under the LLMC are, however, different from the CLC limits, being calculated according to the tonnage of the ship and considerably higher for claims for loss of life or personal injury than for other claims. Art. II para. 1 LLMC sets out the limitation of liability for

---

[256] Juste-Ruíz, '2010 "Compensation for pollution damage"', 45.

[257] See P. Griggs, R. Williams and J. Farr, *Limitation of Liability for Maritime Claims*, 4th edn. (London: LLP, 2005), pp. 7 et seqq. on the scope of art. 1 LLMC.

[258] See http://www.imo.org/About/Conventions/StatusOfConventions/Documents/Summary%20of%20Status%20of%20Conventions.xls (last visited 7 March 2014).

(a) claims in respect of loss or life or injury and loss or damage to property (including damage to harbour works, basins and waterways and aids to navigation), occurring on board or in direct connection with the operation of the ship or with salvage operations, and consequential loss resulting therefrom;
(b) claims in respect of loss resulting from delay in the carriage by sea of cargo, passengers or their luggage;
(c) claims in respect of other loss resulting from infringement of rights other than contractual rights, occurring in direct connection with the operation of the ship or salvage operations;
(d) claims in respect of the raising, removal, destruction or the rendering harmless of a ship which is sunk, wrecked, stranded or abandoned, including anything that is or has been on board such ship;
(e) claims in respect of the removal, destruction or the rendering harmless of the cargo of the ship;
(f) ...

The provisions of the LLMC, the more general convention which is applicable to all kinds of "maritime claims", could therefore also be applicable to claims for oil pollution, arising clearly in direct connection with the operation of the ship.

However, according to art. III (b) LLMC, "claims for oil pollution damage within the meaning of the International Convention on Civil Liability for Oil Pollution Damage, dated 29 November 1969 or of any amendment or Protocol thereto which is in force" are excluded from the scope of applicability of the LLMC. Hence, claims for oil pollution will be governed by the more specialised provisions of the CLC rather than the general provisions of the LLMC and the two conventions do not interfere with each other. It is therefore questionable whether a claim for oil pollution damage directed against a person other than the shipowner and therefore not subject to the CLC channelling and limitation provisions is "a claim for oil pollution damage within the meaning" of the CLC.

During the negotiations of the 1976 LLMC, the problem of the interplay between the two conventions was debated in some length. The delegates were faced with the problem to word the LLMC's provision in a way that would be acceptable to all States, irrespective of whether they were parties to the CLC or not.[259] Considering that the liability limits under the LLMC would be lower than those under the CLC, it was feared that oil pollution claims under the LLMC could not be sufficiently satisfied. The delegates obviously preferred the solution to withdraw oil pollution claims from the LLMC in any case, irrespective of whether the CLC would apply to them in the actual case,[260] and thus effectively opting for unlimited liability where oil pollution claims would not be covered by the CLC.[261] The problem can be

[259] LEG/CONF5/C.1/W.P19 and LEG/CONF.5/C.1/WP.21, Berlingieri F. (ed.), *The Travaux Préparatoires of the LLMC Convention, 1976 and of the Protocol of 1996* (Antwerpen: CMI, 2000), pp. 96 et seq.

[260] Position of the Chairman of the Working Group on art. III (b) LLMC, ibid., p. 101.

[261] J. L. Gabaldón García and J. M. Ruiz Soroa, *Manual de Derecho de la Navegación Marítima*, 3rd edn. (Madrid, Barcelona: Marcial Pons, 2006), p. 831; Martínez Gutiérrez, *2011 "Limitation of Liability"*, p. 185.

mitigated by domestic legislation giving effect to the LLMC,[262] but from its wording, art. III (b) LLMC is generally understood to bar the application of the Convention to all oil pollution claims,[263] regardless of whether the CLC actually applies to them or not.

However, in the proceedings in the British courts following the loss of the oil tanker *Aegean Sea*, in which the owner brought a recourse claim for the oil pollution damage he had had to pay against the charterer, the court clearly stated that art. III (b) of the 1976 Convention only ruled out claims against the person seeking the right to limit that were actually made under the CLC; otherwise a party other than the shipowner would face unlimited liability for pollution claims and that could not have been intended.[264] It must, however, be kept in mind that the English law expressly addresses the problem in its national law and does away with the ensuing consequence of unlimited liability for persons other than the shipowner.

It does seem that in most cases in which the benefit of the channelling provisions of the CLC was denied to defendants, they were not successful in limiting their liability by virtue of other instruments. The LLMC can of course only apply if it is in force in the jurisdiction in question and if the defendant is an "owner" in the sense of art. II LLMC. However, the limitation of liability for maritime claims is a principle to be found in many pieces of domestic legislation, even if they are not a party to the LLMC. This holds true for the United States,[265] where, in the *Amoco Cadiz* proceedings, several defendants sought to limit their liability under the national law, without success in the case of the two main parties.[266]

Of course, however, under the LLMC, and possibly under the domestic legislation, too,[267] the right to limit liability is also subject to the beneficiary not having acted with intent or recklessly and with knowledge that the damage would result, so that this right is lost under the same circumstances under which it is lost by virtue of the CLC (art. III para. 4 CLC), too (see *infra* under II.)

---

[262] See V. Looks and D. Sinkus, 'Die Kosten für Ölbekämpfungsmaßnahmen und die Haftungsbeschränkung unter der Londoner Konvention von 1976' (1994), Transportrecht, 263–6 at 264 et seq. on § 486 III HGB in the then German law. The contents of the provision has been left unamended by the 2013 German maritime law reform but is now § 611 III HGB. Also see B. Soyer, '1996 Protocol to the 1976 Limitation Convention: A More Satisfactory Global Limitation Regime for the Next Millennium?' (2000), Journal of Business Law, 153–72 at 169; Griggs, Williams and Farr, *2005 "Limitation of Liability"*, p. 28; setting out the English domestic legislation that "solves" the problem by rephrasing art. III (b) under the respective national law; Martínez Gutiérrez, *2011 "Limitation of Liability"*, pp. 48, 185.

[263] Looks and Sinkus, '1994 "Die Kosten für Ölbekämpfungsmaßnahmen"', 264; Martínez Gutiérrez, *2011 "Limitation of Liability"*, p. 185.

[264] *"The Aegean Sea"* Lloyd's Law Reports 39 (High Court (England), Queen's Bench Division, 07 April 1998), p. 54.

[265] See Schaar, '2000 "The Shipowners' Limitation of Liability Act"', 665 et seqq. for more details.

[266] Ibid., pp. 676 et seq.

[267] See Ibid., p. 692 on the "owner's privity or knowledge" under the US Shipowners' Limitation of Liability Act.

I. Claims Against Persons Outside the CLC Regime

Even if the channelling of liability to the shipowner is an essential characteristic of the CLC regime, there is quite an important number of defendants who are arguably not protected by the channelling provisions and could therefore be sued outside the CLC regime.[268] This responsibility of third, non-protected parties has, especially by French authors, already been identified as a "fourth leg" of responsibility[269]—after the responsibility guaranteed by the shipowner's mandatory insurance under the CLC regime, the guarantee by the 1971 FUND Convention and the shipowner's unlimited liability if the claimants can establish intent or reckless behaviour (art. V para. 2 CLC). It must be stressed that by bringing a claim in domestic tort law against a person outside the CLC, the claimants may overcome both the channelling provisions and the limitation of liability: not only will there be another defendant, other than the shipowner, but this defendant may possibly also be liable to the full extent of the claim,[270] if he cannot invoke a limitation regime such as the LLMC. Putting the scope of the channelling provisions to the test has therefore been a strategy often followed by oil pollution victims faced with insufficient compensation under the CLC and FUND Conventions.

The channelling provision of art. III para. 4 CLC has undergone some change with the adoption of the 1992 Protocol. The changes were not only of a clarifying nature but altered the subject matter of this provision. It will therefore be interesting to analyse the scope of art. III para. 4 CLC and consider whether certain maritime operators, typically involved in maritime accidents, will benefit from the channelling provisions.

In its original 1969 version, art. III para. 4 CLC reads:

> 4. No claim for compensation for pollution damage shall be made against the owner otherwise than in accordance with this Convention. No claim for pollution damage under this Convention or otherwise may be made against the servants or agents of the owner.

Following the wording of the original provision, the channelling benefitted the owner's "servants or agents" only. As the Convention did not—and even in its amended form still does not—offer a definition to decide which persons should be regarded as agents or servants, the interpretation is left to domestic law as applied by the national courts.[271] Considering that the concepts of agency or service may be

---

[268] M. Rémond-Gouilloud, 'Leçons d'un naufrage' (1979), Recueil Dalloz Chronique, 133–8 at 135; P. Bonassies, 'Après l'Erika: Les quatre niveaux de réparation des dommages résultant d'une pollution maritime par hydrocarbures' (2000), Revue de Droit Commercial, Maritime, Aérien et des Transports, 140–5 at 140; Meilán Gil, *2005 "Problemas Jurídico-Administrativos"*, p. 169; Requejo Isidro, *2007 "Contaminación por hidrocarburos"*, p. 477; de la Rue and Anderson, *2009 "Shipping and the environment"*, p. 109.

[269] Bonassies, '2000 "Après l'Erika"', 140.; M. Ndendé, 'L'accident de l'Erika' (2007), Revue de Droit des Transports, 11–7 at 15.

[270] Rémond-Gouilloud, '1979 "Leçons d'un naufrage"', 135.

[271] Abecassis and Jarashow, *1985 "Oil Pollution"*, para. 10-25.

understood differently, the provision may be another starting point for different understandings and court decisions in contracting States. However, it was to some extent clear that the provision only benefitted persons who stood in a certain relationship to the shipowner, by either being employed by him and thus subject to his directions, or by a similar agreement.

The original purpose of the channelling provisions was the bundling of claims against the shipowner as the mandatorily insured party, in order to benefit oil pollution victims. The provision aims at channelling all claims to a single defendant, who is easily identified and has the limited liability insurance fund at his disposal, so that victims will not need to decide which party to sue and can obtain compensation quickly and easily. The main goal was to create a single and solvent defendant against whom victims' claims would be directed. It seems that the scope of the channelling provision and the question if and how other parties could be targeted by claimants under tort law or otherwise, was not too thoroughly envisaged during the drafting of the CLC.[272] However, it has been pointed out that the channelling supports the limitation of liability. If there is more than one defendant, and if defendants are inter-dependent as parts of a group of companies, it is possible that the mother company will need to pay twice for the same damages in the end,[273] in the case that the total of the victim's claims exceeds the limitation fund established in the first place.

In the immediate aftermath of the *Amoco Cadiz* decision, during the negotiations to the 1984 Protocol to the CLC, amendments to art. III CLC were already discussed. The 1984 Protocol was eventually superseded by the 1992 Protocol, and art. III para. 4 was changed to the following wording:

> 4. No claim for compensation for pollution damage may be made against the owner otherwise than in accordance with this Convention. Subject to paragraph 5 of this Article, no claim for compensation for pollution damage under this Convention or otherwise may be made against:
> (a) the servants or agents of the owner or the members of the crew;
> (b) the pilot or any other person who, without being a member of the crew, performs services for the ship;
> (c) any charterer (howsoever described, including a bareboat charterer), manager or operator of the ship;
> (d) any person performing salvage operations with the consent of the owner or on the instructions of a competent public authority;
> (e) any person taking preventive measures;
> (f) all servants or agents of persons mentioned in subparagraphs (c), (d) and (e);
> unless the damage resulted from their personal act or omission, committed with the intent to cause such damage, or recklessly and with knowledge that such damage would probably result.

The provision is now wider in scope and covers persons who could by no means have been described as the "agent or servant" of the shipowner. Still, the provision

[272] Wang, *2011 "Civil liability"*, pp. 83 and seqq.

[273] Robert, *2003 "L'"Erika""*, p. 37.

does leave room for interpretation,[274] especially with regard to subparagraph (b) and the question of the criterion of "performing services for the ship".

It is occasionally argued that the concept of channelling in art. III para. 4 CLC should be interpreted as broadly as possible, so that basically all parties that could have contributed to an oil spill should be allowed to refer to the shipowner's liability and be shielded from liability themselves. Admittedly, the possibility to sue outside the regime can damage the balance of rights and liabilities that has been found between the shipowner, the oil industry and the insurers, considering that the limitation of liability is considered as the trade-off for the shipowner's responsibility for oil spills irrespective of actual faults. This repartition of risks can be endangered where third parties are held civilly liable without limitation, but then possibly have a right of recourse against the shipowner.[275]

It must however be stressed that the wording and the structure of art. 4 para. III CLC by no means support the idea of an absolute channelling without any possibility to target parties outside the regime. There must be some persons who cannot invoke the channelling, otherwise there would not have been a need to meticulously set out specific parties in the list.[276]

1. Crewmembers, Servants and Agents of the Owner

"Servants and agents of the owner" were the only persons explicitly addressed in the original 1969 version of the CLC, and it goes without saying that the "servants or agents" of the owner that the negotiating parties originally had in mind were the polluting vessel's crew members, effectually taking part in the nautical operation of the ship, and the master, representing the owner. There is certainly in maritime law a privilege to crewmembers and sometimes even the ship's master, having to decide matters of navigation on the spot.[277] Whereas it has been pleaded that the provision served to protect these crew members, who could most likely not financially answer for an oil pollution claim anyway,[278] it must be kept in mind that art. III para. 4 CLC does only shield from third party claims, and that the recourse of the shipowner against his own employees remains perfectly possible.[279] Of course, in this relationship, the employees may invoke labour law provisions or other safeguards

[274] P. Delebecque, 'Aspects civils de l'arrêt Erika' (2010), Droit Maritime Français, 878–85 at 882.

[275] B. D. Daniel, 'Potential Liability of Marine Classification Societies to Non-Contracting Parties' (2006) 19, University of San Francisco Maritime Law Journal, 185–296 at 229.

[276] F. Berlingieri, 'Les sociétés de classification peuvent-elles bénéficier de la canalisation prévue à l'article III.2 de la CLC 1992?' (2012) 64, Droit Maritime Français, 1015–9 at 1019.

[277] Li and Ng, '2002 "International Maritime Conventions"', 390.

[278] Kiss, '1985 "L'affaire de l'"Amoco Cadiz""', 588.

[279] Kbaier and Sebek, '1985 "New trends"', 274; T. A. Mensah, 'International Regime on Liability and Compensation for Marine Environmental Damage from Ships: Genesis and Achievements', in R. Lagoni and M. Paschke (eds.), *Seehandelsrecht und Seerecht: Festschrift für Rolf Herber zum 70. Geburtstag* (Hamburg: Lit, 1999), p. 421; Vanheule, '2003 "Oil Pollution"', 553.

against the shipowner, and may thus find themselves in a more favourable position than against a third party directly affected by the oil pollution incident. It seems likely that the shipowner in his recourse action, if he undertakes it at all, will be considerably more scrupulous with regard to possible outcomes than a given third party[280] claiming under domestic tort law and aiming for an official condemnation of a responsible party.[281] Furthermore, the shipowner's recourse will generally be limited to the amount set out by the CLC, the loss that the owner actually suffered, whereas a third party tort claim could in theory be unlimited. To a certain extent, the channelling provision therefore does protect the persons benefitting from it, but the channelling does by no means amount to a complete exemption of responsibility for them.

In the amendments brought about by the 1992 Protocols, the "members of the crew" are now explicitly mentioned in art. III para. 4 (a) CLC. This begs the question for the understanding of "servants and agents", as opposed to crewmembers. In general, it seems that a "servant" is hired for the ongoing exercise of duties to the shipowner, such as an employee,[282] in contrast to an independent contractor who is paid for achieving a given result. An agent is generally a person acting on behalf and for the account of the principal, without having independent interests as to the matter.[283]

The difficulties to interpret these terms were highlighted in the *Amoco Cadiz* case, still concerning the 1969 original provision only protecting the "servants and agents of the owner". Whereas the US American courts in the matter did not apply the CLC, the applicability was considered and the judges came to the conclusion that the CLC would not bar the proceedings against the shipowner's mother companies, as they could not be termed "agents or servants" of the single-ship company actually owning the *Amoco Cadiz*.[284] The experiences from this case were largely responsible for the following changes to the CLC.[285]

Admittedly, at first sight it does not seem too unfair or socially undesirable that the solvent mother companies should step in for their single-ship subsidiary deprived of means apart from the insurance fund, especially if they can be reproached with a behaviour giving rise to a claim in tort.[286] On the other hand, and considering that means-deprived single-ship subsidiaries of prosperous mother

[280] P. Bonassies, 'Aspects nouveaux de la responsabilité du capitaine' (2002) 54, Droit Maritime Français, 3–13 at 3.

[281] de Raulin, '1993 "L'épopée judiciaire"', 65, claims that one of the reasons for the claimants in the *Amoco Cadiz* proceedings to sue in the US was to obtain a verdict against the Amoco parties in their home jurisdiction.

[282] Berlingieri, '2012 "Les sociétés de classification"', 1018.

[283] Ibid., p. 1018.

[284] *US District Court, Northern District of Illinois (Eastern Division), 18.04.1984, "In Re Oil Spill by the Amoco Cadiz"*, pp. 2190 et seq.

[285] de la Rue and Anderson, *2009 "Shipping and the environment"*, p. 108.

[286] See Kbaier and Sebek, '1985 "New trends"', 275, expressing sympathy for the findings of the court.

companies are a common occurrence today,[287] the Illinois court's interpretation of art. III para. 4 CLC, allowing the claim against the closely interested mother companies, did overturn to a certain extent the CLC regime. Unlimited liability tort claims against the deep-pocket controlling companies, and possibly charterers and cargo owners, would have become the rule rather than the exception, and the limitation of liability granted to the shipowner and his mandatory insurance would have become ineffectual.[288] It was considered that in most of these cases, the shipowner would probably have been legally required to reimburse the defendant mother company, charterer or cargo owner for the liability they incurred, and the strict liability CLC regime would have been replaced by an unlimited liability tort regime. Hence the interest in adapting the channelling provisions and widening its scope in the ensuing 1992 Protocol.[289]

It must be kept in mind that in the States that did not sign the 1992 Protocol, the original provision remains applicable and that therefore it might be easier for claimants to target solvent mother companies controlling the shipowner in these jurisdictions.

The problem re-emerged in the proceedings following the accident of the *Erika*, a 23-year-old oil tanker, which had frequently changed her name and her flag and had been managed by several management companies. On her last voyage, she was flying the Maltese flag and chartered by a company of the Total group, a major French oil company. On 12 December 1999, laden with 30,000 tonnes of heavy fuel oil, she encountered heavy weather in the Atlantic, off the coast of Brittany. Due to corrosion damages in the hull leading to structural weaknesses, she finally broke in two, spilling about 20,000 tonnes of her cargo, which subsequently polluted about 400 km of French coastline, and resulting in a damage equalling an amount of about 1 billion € in total.[290] The limited liability fund available under the CLC 1992 was entirely insufficient to cover this amount, and even the means available under FUND Convention as they were set at that time, seemed only partly adequate.[291] Furthermore, under the procedural obligations set out by the FUND Convention, claimants have to keep close documentation of the damages they incurred and present consistent accounts to prove their economic losses. This was in the instance not easy for very many small enterprises in fishing and tourism in the Breton

[287] K. Le Couviour, 'Responsabilités pour pollutions majeures résultant du transport maritime d'hydrocarbures' (2002), La Semaine Juridique Édition Générale, 2269–77 at 2271; see J.-P. Beurier, 'Le transport maritime, le droit et le désordre économique international', in Y. Daudet and P. M. Eisemann (eds.), *La mer et son droit* (Paris: A. Pedone, 2003), pp. 88 et seq. giving reasons for this development.

[288] de la Rue and Anderson, *2009 "Shipping and the environment"*, p. 107.

[289] C. B. Anderson and C. de la Rue, 'Liability of Charterers and Cargo Owners for Pollution from Ships' (2001) 26, Tulane Maritime Law Journal, 1–60 at 7.

[290] See the outline of the facts in Ndendé, '2007 "L'accident de l'Erika"', 11.

[291] Bonassies, '2000 "Après l'Erika"', 143; Anderson and de la Rue, '2001 "Liability of Charterers"', 4; Ndendé, '2007 "L'accident de l'Erika"', 17.

regions affected by the spill, and the municipalities faced similar problems when quantifying the damages to their reputation as agricultural and touristic regions.[292]

The French authorities started criminal proceedings against several persons suspected of having contributed to the catastrophe. French law allowing for criminal proceedings against legal persons, the *Erika's* owner, the ship management company, the chartering oil company and the Italian classification society were targeted on the one hand. On the other hand, proceedings were also started against the master and several employees of the shipping companies.

Whereas the case was therefore a criminal case in the first place, many victims declared themselves *parti civile* and triggered an adhesive procedure in the courts, in which the criminal law court decided on compensation to be paid to the victims. An analysis of how adhesive procedures of this kind can influence the existing liability regime will be given *infra* under H. I., but in this case, the *Tribunal Correctionnel* carefully took the CLC provisions into account when deciding on the merits of the civil law claim.

The court was thus called to decide on the scope of the channelling provisions in more than one respect, and namely with regard to Mr. Savarese, the individual who was the beneficial owner of the *Erika's* registered owner, Tevere Shipping. The *Tribunal Correctionnel* on the one hand repeatedly mentioned that he had exercised control over the ship in directing to restrict repair works on the vessel to an extent that rendered the ship unsafe for service,[293] and on the other hand held that he was "neither the owner, nor the agent or the servant of the owner".[294] These findings have been criticised as contradictory,[295] and in effect, the *Cour d'Appel* in the next instance overturned the decision by deciding that Mr. Savarese was a "servant" of the *Erika's* registered owner, by being the company's beneficiary[296] and the director of the company who officially owned the *Erika*. Even if he did not escape unlimited liability due to the fact that the court found his personal fault in directing the last repairs "inexcusable" in the sense of art. V para. 2 CLC as amended, the result found by the *Cour d'Appel* is to a certain extent paradoxical. Admittedly, as a director, Mr. Savarese was an employee of the registered shipowner, Tevere Shipping, but on the other hand, he exercised immediate control over the company and was found, by the *Cour d'Appel*, to have personally directed the extent of the *Erika's* last and insufficient repairs.

The solution is consistent in that Mr. Savarese has been found to be liable according to art. V para. 2 CLC, notwithstanding the channelling of liability that he could otherwise have invoked, and it is considered to be in line with the French

[292] Bonassies, '2000 "Après l'Erika"', 141; Ndendé, '2007 "L'accident de l'Erika"', 17.

[293] *"Erika"*, 26 Annuaire de droit maritime et océanique 701 (Tribunal correctionnel de Paris, 16 January 2008), pp. 708,709, 720.

[294] Ibid., p. 727.

[295] de la Rue and Anderson, *2009 "Shipping and the environment"*, pp. 109, 110.

[296] *"Erika"*, 62 Droit Maritime Français 857 (Cour d'Appel de Paris, 30 March 2010), p. 867.

domestic law on employee responsibility.[297] The decision has, in this respect, explicitly been upheld by the *Cour de Cassation* in the next instance.[298]

However, and considering the set of facts abstractly, it does seem strange that the director of a shipowning company which is probably only a shell company or a sham, and who is directly responsible for every decision taken on behalf of the company, should be able to invoke the protection of the channelling provisions and be shielded by the same company that he effectively directs. It does seem that the CLC's channelling provisions do not work too well in these employment situations, but of course, there will generally be little interest in invoking the responsibility of an individual. The *Cour d'Appel's* meticulous consideration of Mr. Savarese's role within the shipowning company, could be carried out due to the fact that the case was brought as a criminal case, where responsibilities were thoroughly inquired into, and the civil liability of all persons held responsible was scrutinised in a second step.

From a more general point of view, it is of course questionable what the applicable criteria are to establish an "agency" or "service relationship". It can of course be argued that an employment contract should be enough to prove agency, but the outcome is questionable if the "agent or servant" is, in the way Mr. Savarese was, the main decision-taker in the shipowning entity.

The answer to this question seems widely to depend on how the purposes of the channelling provisions are considered. If they are seen as provisions at least partly protecting weaker parties that will not be able to answer for the claim, they should only be applied to persons who effectively work on behalf of the company and follow directions from a higher level. The provisions would then not be applicable to members of the management, such as Mr. Savarese who effectively took all decisions concerning the repairs of the *Erika* himself. The US district court in the *Prestige* proceedings seems to have at least partly adopted this view, when they argued that the *Prestige's* classification society could not be considered as a servant of the owner—even though there was a service contract for the classification of the ship—because the company was independent and did not work "at the direction of or on behalf of the owners".[299]

As has been pointed out above, the US court's judgment was held to be erroneous for applying the CLC, but the reasoning with regard to the role of "servants" and "agents" seems to be a valid one. If "servants" and "agents" are found to be rather down the line of command, being subject to directions or at least powers of control from a higher level, this could mitigate some of the difficulties oil pollution victims may experience when they are faced with company frameworks in which decisions are generally not made by the entities who officially present

---

[297] Delebecque, '2010 "Aspects civils"', 883.

[298] *"Erika"*, 64 Droit Maritime Français 985 (Cour de Cassation (France), 25 September 2012), p. 992.

[299] *US District Court, Southern District of New York, 02.01.2008, "Reino de España v. American Bureau of Shipping (The Prestige)"*, p. 273.

themselves in their capacity as shipowners. The channelling provisions would then only benefit the executing organs of a company group, not the actual decision-makers who try to shield themselves from liability by creating subsidiaries.

If, on the other hand, the channelling provisions should simply be seen as a clear fingerpoint to the responsible person and if the internal relationship between the shipowner and his employees is not given regard to, the extension to all persons simply being bound in contract to the shipowner seems admissible.

Since the change of wording of art. III para. 4 CLC, the channelling provisions do not only apply to "servants and agents" of the owner, but to "servants and agents of persons mentioned in subparagraphs (c), (d) and (e)" (art. III para. 4 (f) CLC), i.e. to servants and agents of, firstly, charterers, owners and managers, secondly, salvors and thirdly, persons taking preventive measures. The subparagraph explicitly does not refer to art. III para. 4 (a) CLC. As the *Cour d'Appel* of Paris in its decision in the *Erika* case rightly observed, the referral is thus restricted in a way not to cover agents or servants of persons providing services for the ship,[300] and it is hardly conceivable that the omission of this group should not have been consciously made. The *Cour d'Appel* used this reasoning to hold Mr. Pollara, the director of the *Erika's* technical management company, civilly liable in tort, maintaining that the management company performed services for the ship and that Mr. Pollara was a mere agent of a service provider, hence, not protected by the CLC provisions.

The *Cour d'Appel*'s reasoning is refined from a legal point of view, yet possibly debatable with regard to the facts. Panship Management, the company of which Mr. Pollara was director, was in fact in charge of the nautical management of the *Erika*[301] and could thus have been considered to be a "manager" and covered by art. III para. 4 (c) CLC, and M. Pollara thus by art. III para. 4 (f) CLC. It will be seen *infra* that the scope of the term "charterers and managers" is not easy to determine.

However, the *Cour d'Appel's* observation regarding art. III para. 4 (f) CLC not referring to para. 4 (b) CLC is an accurate one and can possibly be used to determine the scope of art. III para. 4 (b) CLC in more detail (see *infra* under 4.).

2. Charterers and Managers

As previously mentioned, the question whether charterers could be held liable for oil pollution damage outside the CLC regime emerged for the first time in the *Amoco Cadiz* proceedings. Considering that in most cases, at least in the bareboat and time charter situations, the charterer will exercise more control over the ship than the registered owner, it does seem surprising that the question did not come up at an earlier point in time.

In the carriage of oil by sea, charterers are often prosperous oil majors or subsidiaries of oil companies, and therefore a good target for claims not subject

[300] *Cour d'Appel de Paris, 30.03.2010, "Erika"*, p. 867.

[301] Ibid., p. 867.

to limitation of liability, whereas the registered owners are often destitute single ship companies.[302] Under the wording of the CLC 1969, where the only parties benefitting from the channelling are the "agents or servants" of the owner, claims against the charterer could—and can—be easily brought, considering that there will be hardly any contractual relationship in which the charterer could be termed the "agent or owner" of the shipowner.[303]

In the *Amoco Cadiz* proceedings, the court's findings considering the responsibility of the shipowner's parent companies who effectively operated the vessel were considered revolutionary, a "departure from the classical rules of maritime law"[304] or "an important step forward".[305] However, the international community embraced this new idea of charterers' and managers' responsibility rather quickly. In the *Tanio* proceedings following the shipwreck of this Malagasy tanker off the coast of Brittany in 1980, damages were paid by the shipowner's insurance under the CLC and by the IOPC to an extent of 70 %, but the French State, together with the IOPC also started proceedings against other parties in tort.[306] Among these were the bare boat charterer, who had effectively managed the ship, both in his capacity as a charterer and through a management company, and the management company itself.[307] It is true that these proceedings were widely considered as recourse actions of the IOPC, aiming at recovering the amounts paid out to victims, but the French State effectively sued for a "completion of the compensation"[308] already obtained, and thus for funds beyond the limitation.

However, the possibility to sue charterers outside the CLC regime has certainly changed with the coming into force of the 1992 Protocols. Under subparagraph (c), "any charterer (howsoever described, including a bareboat charterer), manager or operator of the ship", and, under subparagraph (f), "any agent or servant" of these persons, benefit from the channelling provision and cannot be directly sued by oil pollution victims.

It is therefore clear that claims against the "nominal" charterer, whose name figures on the charterparty, are not admissible. Still, problems arise with regard to persons or entities who exercise a power of control over a chartered vessel without being officially given legal authority to do so. This was the situation in the *Erika* proceedings with regard to the charterer's mother company and general group mother, Total, which had directed and instructed the nominal charterer, a mere shell company.[309] The parent company having assessed the vessel's condition and

[302] Beurier, *2003 "Le transport maritime"*, pp. 90 et seq., attributing this development to, *inter alia*, the CLC's channelling provisions.

[303] Anderson and de la Rue, '2001 "Liability of Charterers"', 7.

[304] Kbaier and Sebek, '1985 "New trends"', 273.

[305] de Raulin, '1993 "L'épopée judiciaire"', 88.

[306] Zimmermann, '2000 "Inadequacies"', 1515.

[307] Fontaine, *1993 "The French Experience"*, p. 106.

[308] http://www.cedre.fr/fr/accident/tanio/tanio_indem.php (last visited 7 March 2014).

[309] *Tribunal correctionnel de Paris, 16.01.2008, "Erika"*, p. 721.

performance in a vetting procedure that was internal to the company and thus, basically voluntary, was held to have failed in discovering the vessel's actual state. The mother company was thus reproached with an omission without having a corresponding obligation to do so.[310] Whereas the *Tribunal Correctionnel* in the first instance, and possibly for policy reasons,[311] held Total liable as they were not the *Erika's* nominal charterers,[312] the *Cour d'Appel* in the next instance overturned this decision. The court first considered the wording of art. III para. 4 CLC and declared that it had a very wide meaning, hinting at the expression "charterer (howsoever described [...])", and should therefore also cover the "real" charterer ("*véritable affréteur*")[313] as opposed to the "nominal" charterer only figuring in the charterparty but actually being subject to the instructions of the parent company.

Whereas the *Cour de Cassation* upheld this qualification of Total as the "charterer", who would normally have been able to benefit from the channelling, it however pointed out that Total exercised considerable control over the choice and the operation of the ship by the nominal charterer, by virtue of the vetting procedure in which Total assessed the state of maintenance before allowing it to be chartered. The fact that Total ignored the bad state of repair of the vessel and the circumstance that the class documents would expire in two months and were not yet renewed, together with other particularities that pointed to the *Erika's* incapability to withstand bad weather,[314] caused the *Cour de Cassation* to hold that Total had acted "recklessly and with knowledge that such damage would probably occur", so that it could not invoke the benefit of the channelling provisions. The *Cour de Cassation* however did not overturn the *Cour d'Appel's* decision that Total was a "charterer", notwithstanding the fact that it did not figure in the *Erika's* charterparty.[315]

Whereas the *Cour d'Appel's* decision as to this starts from the wording of art. III para. 4 CLC and takes into consideration the reality of modern shipping, i.e. a business environment in which it is easy to create tailor-made subsidiaries who can act as nominal charterers and shipowners while they are under the control of the parent company,[316] the solution is just as debatable as the opposite result found by the *Tribunal Correctionnel* in the first instance. As has been pointed out above, the channelling mechanism must apply to more entities than just to the "nominal" shipowner in order to render the CLC regime workable, but on the other hand, the

---

[310] K. Le Couviour, 'Catastrophe de l'Erika: premiers commentaires d'un jugement déjà exemplaire' (2008), La Semaine Juridique Édition Générale, 3–5 at 4; B. Parance, 'Note TGI Paris, 16.01.2008' (2008), La Semaine Juridique Édition Générale, 36–9 at 38.

[311] de la Rue and Anderson, *2009 "Shipping and the environment"*, p. 111.

[312] *Tribunal correctionnel de Paris, 16.01.2008, "Erika"*, p. 721.

[313] *Cour d'Appel de Paris, 30.03.2010, "Erika"*, p. 867.

[314] *Cour de Cassation (France), 25.09.2012, "Erika"*, p. 994.

[315] P. Delebecque, 'Aspects civils de la décision "Erika"' (2012) 64, Droit Maritime Français, 1000–6 at 1003.

[316] Hansen, '2008 "Multinational Enterprise Pursuit"', 422 et seqq.; L. Neyret, 'L'affaire "Erika": moteur d'évolution des responsabilités civile et pénale' (2010), Recueil Dalloz, 2238–46 at 2240.

CLC never intended to create a system of absolute channelling, considering that the original version started out from the mere inclusion of "servants and agents".[317]

One problem that seems to be at the core of this debate is the fact that the extension of responsibility to controlling parent companies and their being held responsible for the acts of their subsidiaries amounts to a "piercing of the corporate veil" and an exception to the principle of legal personality of limited liability companies.[318] Questions of company law are outside the subject of this study. However, it must be mentioned that the channelling of liability to the shipowner in art. III para. 4 CLC leads to the rather paradoxical result that a shipowner's parent company or "real" charterer will benefit from the subsidiaries' limitation of liability if the corporate veil is pierced during the proceedings and the court establishes that the parent exercised a power of control or instruction over the "nominal" charterer or shipowner.[319] If, on the other hand, this link of control between parent company and subsidiary cannot be established, the parent company can hardly be held to have acted in the capacity of a "charterer" and could then not benefit from the channelling provisions. Of course, it may then be hard to establish the parent company's responsibility in tort without proving some exercise of power or control in the operation of the ship, but depending on the domestic tort law regime, it does not seem impossible. In the *Amoco Cadiz* proceedings, the US district court of Illinois held that Standard, the highest parent company of the Standard Oil combine, was liable for the accident not only under tort law for having acted negligently, but also as the entity controlling her subsidiaries, the charterers. It seems that this power of control was mostly detected with regard to the close cooperation and the common administration between the companies, whereas it does not seem that Standard actually took part in the nautical management of the vessel.[320] In its controlling capacity, Standard was held liable for the wrongful behaviour of the charterers and shipmanagers as if it had acted itself.[321] There have also been legislative initiatives of this kind in France, aiming at establishing the responsibility of parent companies for wrongful behaviour of their subsidiaries,[322] and this is, under certain

[317] P. Bonassies, 'Sociétés de classification et Convention de 1969/1992 sur la responsabilité pour les dommages dus à la pollution par les hydrocarbures' (2008) 60, Droit Maritime Français, 691–6 at 693 hints at that the original wording was adopted from the 1968 Protocol to the Hague Rules, extending the limitation of liability to the carrier's "servants or agents".

[318] M. Ndendé, 'Les enseignements du jugement pénal de Paris sur les responsabilités des acteurs mis en cause dans la catastrophe de l'"Erika"' (2008) 26, Annuaire de droit maritime et océanique, 253–72 at 265, even claiming that the standards for piercing the corporate veil are easier to establish under civil than under criminal law; Delebecque, '2012 "Aspects civils"', 1003.

[319] Delebecque, '2010 "Aspects civils"', 881; K. Le Couviour, 'Erika: décryptage d'un arrêt peu conventionnel' (2010), La Semaine Juridique Édition Générale, 804–7 at 806.

[320] See K. A. Hofstetter, *Sachgerechte Haftungsregeln für multinationale Konzerne* (Tübingen: Mohr Siebeck, 1995), pp. 33 et seq. on the criteria that the court considered to establish the link of control.

[321] *US District Court, Northern District of Illinois (Eastern Division), 18.04.1984, "In Re Oil Spill by the Amoco Cadiz"*, p. 2194.

[322] See Parance, '2008 "Note TGI Paris, 16.01.2008"', 38 for further details.

circumstances, also possible under Spanish law.[323] It is thus possible that under company group law, a parent company may be held liable for a pollution incident without actually having taken part in the operative business, but it is of course a matter of interpretation in each given case whether a national court will be ready to establish a sufficient degree of control. As the decision by the *Cour de Cassation* in the *Erika* case shows, the establishment of control can also be detrimental to the parent entity, if its conduct is found to have been reckless and with knowledge that the damage would probably result.

However, the wide understanding of the term "charterer" and the extension of the notion to controlling parent companies is consistent with the underlying purpose to effectively establish the limitation of liability on the shipowner's and ship management's side. On the other hand, the solution may not be too handy in that a court will need to scrutinise whether a defendant entity had power of control over the ship or was effectively running or managing it.[324] Considering that there is no comprehensive definition of when a company is considered to take part in the chartering of a vessel, courts may be tempted to take recourse to national company law to decide whether there was a link of control or effective exercise of power between the defendant and the "nominal" charterer. Especially in cases of "exercise of power" over the subsidiary without an actual legal right to justify it that the *Tribunal Correctionnel* detected in the *Erika* case,[325] the problem may be a very complicated one and either decision seem arbitrary in a given case.

Furthermore, it does not seem that this problem could be solved by a change of wording of art. III CLC or by a legal definition of what a "charterer" is. Shipping companies, especially in the oil business, have been very adroit in creating tailor-made subsidiaries for their purposes to repartition responsibility without giving up control,[326] and the fact that tort proceedings against a parent company under domestic law will generally be costly, lengthy and complicated is perhaps enough of a safeguard to guarantee the workability of the CLC regime. Many victims will be reluctant to embark on this undertaking, and content themselves with the financial means offered by the shipowner's insurance fund and possibly by the IOPC and the Supplementary Fund. It does not seem too damaging to the regime that victims who are not satisfied with it may still be tempted to invoke a charterer's parent company under domestic law, running however the risk that a sufficient degree of control over the nominal charterer may not be established.

[323] Meilán Gil, *2005 "Problemas Jurídico-Administrativos"*, pp. 183 et seqq.

[324] Delebecque, '2012 "Aspects civils"', 1004, pointing out that the corresponding findings in the *Erika* decisions were possible due to the inquiries in the criminal case.

[325] *Tribunal correctionnel de Paris, 16.01.2008, "Erika"*, p. 721.

[326] A. Vialard, 'Faut-il réformer le régime d'indemnisation des dommages de pollution par hydrocarbures?' (2003) 55, Droit Maritime Français, 435–49 at 438; Hansen, '2008 "Multinational Enterprise Pursuit"', 431; Le Couviour, '2008 "Catastrophe de l'Erika"', 4.

3. Cargo Owners

Cargo owners do appear in the framework of the civil liability conventions, as contributors to the fund established by the 1971 FUND Convention. Apart from that, the Conventions do not confer liability on them for any given accident.[327] Still, they may be responsible if the oil in their possession causes harm to the environment or to third parties. Considering that neither the CLC in its 1969 version nor the amendments brought about by the 1992 Protocol mention cargo owners as beneficiaries of channelling provisions, a claim against them remains possible in theory.[328]

However, a claim in tort under any given jurisdiction may be difficult to establish, considering that cargo owners have normally conferred the care for the cargo to the shipper and do not exercise sufficient control over the goods in shipment to be reproached with a wrongful behaviour.[329] If the cargo owners, in their capacity as charterers, have tortiously chosen an unseaworthy ship, it is in their capacity as charterers that they will be held liable, not as cargo owners.

Civil liability could still arise where the given tort law of a State provides for strict liability, regardless of fault.[330] Major oil companies seem to have circumvented liabilities of this kind by making sure that the title in the goods is not held by the parent company, but by a subsidiary or another controlled entity, which may again induce courts to set aside constructions of this kind and pierce the corporate veil. Considering on the other hand that the market in crude oil is mostly run by a limited number of big players who tend to control a cargo in more than one respect, i.e. not only as cargo owners, but possibly also as charterers or even shipowners, it does not seem that a claim against a cargo owner in this capacity only is likely. Cargo owners will most likely enter the proceedings at a later point in time, when shipowners and charterers try to distribute the cost incurred between the parties to the shipping contract.

However, a claim against cargo owners by third parties may still be a valid recourse if victims have a claim against cargo owners under a given jurisdiction and a given set of facts. Considering that cargo owners are not mentioned under art. III para. 4 CLC as amended by the 1992 Protocol, choosing the cargo owner as opposed to the charterer may even be a safer way to overcome the CLC's channelling provisions.

From a practical point of view victims trying to hold cargo owners liable may face the same difficulties that made the parties abstain from conferring liability on cargo interests during the negotiations to the CLC: ownership in oil cargoes is most

---

[327] Anderson and de la Rue, '2001 "Liability of Charterers"', 6.

[328] Arroyo Martínez, '2009 "Para esta viaje"', 8, however criticising the wording of the provision.

[329] Anderson and de la Rue, '2001 "Liability of Charterers"', 15, with reference to liability under the US 1990 Oil Pollution Act.

[330] Ibid., p. 21, giving examples of US States in which this is the case.

likely to change several times during a shipment,[331] and different parts of one and the same bulk cargo may belong to different owners.[332] The ownership in a given cargo may thus be hard to establish for a given moment, and the identity of cargo owner may not always be documented.

4. Classification Societies

One of the most-debated problems with regard to art. III para. 4 of the CLC is the question of whether classification societies can invoke the benefit of the channelling provisions, or, more accurately, whether they fall under the category of "the pilot or any other person, who, without being a member of the crew, performs services for the ship" in the sense of art. III para. 4 (b).

Classification societies, as the private entities controlling and monitoring ships in terms of safety and performance, exercise a very important task in the maritime business. The concept of classification societies certainly started out as a purely private enterprise, enabling the shipowner to insure his vessel at a reasonable cost and encouraging interested third parties to transact with the ship by giving a neutral account of its performance and safety (see *supra*, Chap. 2, B. I.). Still, the obligation owed by classification societies slowly became important as a public task, and classification societies today generally take over the flag State's control obligations to a very wide extent.[333] At the same time, they have a contractual relationship with the shipowner to carry out the necessary surveys and issue the class documents without which the ship cannot take part in international trade. It is thus very often considered that the classification societies' task can be split into a private and a public function, the first being performed for the shipowner, the second for the authorities of the flag State or even for the public benefit. It is however very different to determine the line between the two functions in a given country,[334] let alone on an international level.

Following a maritime accident that is wholly or partly caused by poor maintenance or technical or structural shortcomings in a given vessel, the classification society having issued the class certificates may be a rewarding target for a claim in contract or tort. Classification societies are generally sound and wealthy businesses,

---

[331] Lucchini, '1970 "La pollution des mers"', 821; S. M. Carbone and L. Di Schiano Pepe, 'Uniform Law and Conflicts in Private Enforcement of Environmental Law: The Maritime Sector and Beyond' (2009) 111, Il Diritto Marittimo, 50–80 at 52.

[332] Ibid., p. 52.

[333] Antapassis, *2007 "Liability of Classification Societies"*, pp. 635, 640 et seqq.

[334] P. F. Cane, 'The liability of classification societies' (1994), Lloyd's Maritime and Commercial Law Quarterly, 363–76 at 364, concerning England.

continuously generating profits.[335] Depending on the degree to which it can be established that the classification society omitted to detect the ship's real condition, it may even be easy to prove a tort, once the causes of the maritime accident have been established.

The tortious liability of classification societies towards aggrieved third parties such as oil pollution victims is problematic in more than one respect and cannot be displayed here.[336] Suffice it to summarise that classification societies can incur liability in tort under a number of domestic legislations,[337] and that there are cases in which they have been successfully sued by third parties.

The question whether classification societies are covered by art. III para. 4 (b) CLC and can be considered to "perform services for the ship", so that they are shielded from liability for oil pollution damage was a theoretical one for a considerable time, up to 2008 when two courts in the United States and in France decided the question almost simultaneously and with directly opposite results.

The Paris *Tribunal Correctionnel*, in the first instance decision of the *Erika* proceedings, held the Italian classification society RINA civilly responsible for the pollution damage. The court rejected RINA's argument that it had, by issuing the classification certificates for the *Erika*, performed services for the ship and should thus benefit from the channelling provisions. The *Tribunal Correctionnel* centred its argumentation on the fact that art. III para. 4 (b) CLC should benefit "persons, who, without being members of the crew, perform services for the ship by directly participating in the nautical operation of the ship",[338] which was not the fact with regard to RINA. This interpretation was confirmed by the *Cour d'Appel* of Paris in early 2010,[339] but quashed again by the *Cour de Cassation* in its final decision on the case.[340] The court however decided that RINA had acted recklessly and with knowledge of the likelihood of the damage, so that it could not invoke the channelling benefit that it would enjoy in its capacity as a person performing services for the ship. The outcome for RINA, that had been found civilly liable without limitation under domestic tort law by the *Cour d'Appel*, was therefore not

---

[335] F. Siccardi, 'Pollution Liability and Classification Societies: Is the System a Fair One?' (2005) 107, Il Diritto Marittimo, 691–710 at 697; W. Wurmnest, 'Third Party Liability of Classification Societies in the Context of Shipping Accidents', in M. García Pérez and F. J. Sanz Larruga (eds.), *Seguridad marítima y medio ambiente* (Oleiros: Instituto Universitario de Estudios Marítimos, 2006), p. 112.

[336] See H. Honka, 'The Classification System and its Problems with Special Reference to the Liability of Classification Societies' (1994) 19, Tulane Maritime Law Journal, 1–36; J. Basedow and W. Wurmnest, *Third-party liability of classification societies* (Berlin, New York: Springer, 2005); Wurmnest, *2006 "Third Party Liability"*; N. Lagoni, *The Liability of Classification Societies* (Berlin, New York: Springer, 2007) p. 59 et seqq.

[337] See the comparative analysis in Antapassis, *2007 "Liability of Classification Societies"*, pp. 657 et seqq.

[338] *Tribunal correctionnel de Paris, 16.01.2008, "Erika"*, p. 128.

[339] *Cour d'Appel de Paris, 30.03.2010, "Erika"*, p. 867.

[340] *Cour de Cassation (France), 25.09.2012, "Erika"*, p. 992.

altered by the decision in the last instance. Still, the *Cour de Cassation's* interpretation of art. III para. 2 CLC is remarkable, especially considering the fact that it may be putting an end to a question that has been decided very controversially in two catastrophic oil pollution cases in different jurisdictions.

Two weeks before the first decision in the *Erika* case by the *Tribunal Correctionnel* of Paris, the United States District Court of the Southern District of New York was called to decide on the civil liability of the US classification society ABS ("American Bureau of Shipping"), which had issued the classification documents for the oil tanker *Prestige*. ABS pleaded, *inter alia*, that the US courts were not competent to hear the case, and, in the subject matter, that it had performed services for the *Prestige* by surveying it and issuing the class certificates and that it should therefore benefit from the channelling provisions of the CLC. Notwithstanding the problem of the competence of the US court to hear the case and the application of the CLC in a State that is not a party to it (see *supra* A. I. 1. b)), the court considered that it had power to "examine Defendants' contention that Spain's claims can only be litigated in the courts of a state signatory to the CLC."

In its discussion of whether the CLC was applicable to the Kingdom of Spain's claims against ABS, the court followed ABS's reasoning, holding that it was a person within the meaning of art. III para. 4 (b) CLC, against whom the action outside the CLC was barred,[341] and that all actions under the CLC regime, according to art. IX para. 1 CLC, have to be brought in a contracting State.

It is true that the US court did not explain its decision thoroughly, the judge actually referring to that the "undisputed factual record, even when read in light most favorable to Spain, clearly"[342] indicated that ABS was a person within art. III para. 4 (b) CLC. Still, neither did the *Tribunal Correctionnel* of Paris explicitly state why it considered that persons would only fall within the meaning of art. III para. 4 (b) CLC when they participated directly in the nautical operation of the ship.

For the sake of completeness, it should be repeated that in the ensuing appeal proceedings in the US, the courts refrained from applying or interpreting the CLC to any further degree, finally considering that it was not a part of US law and as such could not deprive a domestic court of judicial competence. All following decisions, including the most recent decision of the US Court of Appeals were made under federal US American Law (see *supra* under A. I. 1. b)), the case then centring on the question whether and under which circumstances ABS could be liable towards the Kingdom of Spain as a third party to the classification contract.[343]

Still, even if the question was thus set aside in the *Prestige* proceedings in the US, the question of interpretation of art. III para. 4 (b) CLC remains an important one, as it may determine the role of classification societies in the future.[344]

---

[341] *US District Court, Southern District of New York, 02.01.2008, "Reino de España v. American Bureau of Shipping (The Prestige)"*, pp. 274 et seq.

[342] Ibid., p. 275.

[343] See Goñi, '2012 "L'arrêt"', 1024 et seq.

[344] Wurmnest, *2006 "Third Party Liability"*, p. 119.

Multiple attempts have been made to interpret art. III para. 4 (b) CLC with regard to the negotiations between contracting parties on the wording of the 1992 Protocol.[345] It must be retained that the negotiating parties were at odds about whether specific maritime actors should be listed in the provisions or whether there should be a "general" channelling clause, whether the finally adopted list should be understood broadly or narrowly. It is perhaps not too speculative to emphasise the fact that the wording of art. III para. 4 CLC contains clauses open to interpretation, such as the notion of "servants and agents" and the provision of letter (b), and that the negotiating parties trusted that these notions would be interpreted by national courts according to the meaning that they favoured. Neither the explicit inclusion nor the exclusion of classification societies from the list of art. III para. 4 CLC have been addressed during the negotiations,[346] so the *travaux préparatoires* to the 1992 Protocol cannot really give a hint as to whether negotiating parties wanted to see them included in art. III para. 4 (b) or not. The fact as such that the list sets out specific actors does not point to a wide or narrow understanding in this respect,[347] even if there is some value in the argument that art. III para. 4 CLC comprises a positive list of persons enjoying the benefit of channelling, so that in doubt, persons not mentioned on the list can be expected to be excluded from the concept.[348]

With regard to the wording of art. III para. 4 (b) CLC, it has been argued that the wording "the pilot or any other person, who, without being a member of the crew, performs services for the ship" hints to a narrow understanding, because the pilot is effectually present on the ship and takes an active part in the nautical operation,[349] two requirements that are typically only fulfilled by crewmembers. Considering that crewmembers are already covered by art. III para. 4 (a) CLC,[350] the reference to the crew in art. III para. 4 (b) must however have some meaning, and the argument that by this reference the application is restricted to persons effectively on board the vessel, such as the pilot and the crew, is a logical one. Following this interpretation, art. III para. 4 (b) CLC could theoretically cover other independent contractors directly assisting in the nautical operation of the ship,[351] and perhaps

[345] Bonassies, '2008 "Sociétés de classification"', 693; P. Boisson, 'La société de classification bénéficie-t-elle de l'exclusion prévue par l'article III (4) de la Convention CLC?' (2008) 60, Droit Maritime Français, 696–703 at 699 et seqq.

[346] Bonassies, '2008 "Sociétés de classification"', 693; Carbone and Di Schiano Pepe, '2009 "Uniform Law and Conflicts"', 62, however referring to "a number of elements that can be derived from the relevant *travaux préparatoires*".

[347] Arroyo Martínez, '2009 "Para esta viaje"', 89.

[348] Meilán Gil, *2005 "Problemas Jurídico-Administrativos"*, p. 171; Rodríguez Docampo, *2012 "Alcance de la canalización"*, p. 923.

[349] Lagoni, *2007 "The liability of classification societies"*, p. 290; Bonassies, '2008 "Sociétés de classification"', 694; Arroyo Martínez, '2009 "Para esta viaje"', 95; Delebecque, '2012 "Aspects civils"', 1002; Rodríguez Docampo, *2012 "Alcance de la canalización"*, p. 924.

[350] Berlingieri, '2012 "Les sociétés de classification"', 1019.

[351] Siccardi, '2005 "Pollution Liability"', 707; de la Rue and Anderson, *2009 "Shipping and the environment"*, p. 111; Rodríguez Docampo, *2012 "Alcance de la canalización"*, p. 924.

persons performing auxiliary services, such as staff on board the pilot cutter.[352] The scope of application of the provision would then be very narrow, but it must be kept in mind that art. III para. 4 (b) focuses on the pilot and only extends its applicability to persons that can at least be compared with him. The function of the pilot being a very exceptional one, it may merit its own provision under the letter (b).[353]

Another argument in favour of a narrow understanding of art. III para. 4 (b) CLC can actually be drawn from the French *Cour d'Appel's* exact observation that the channelling provisions do not benefit the servants and agents of service providers,[354] but the servants and agents of all other beneficiaries (see *supra* under 1.). If one considers that the omission of letter (b) from the groups addressed in letter (f) was consciously made, this also justifies a very narrow understanding of the term of "service providers" in letter (b), a group that was not meant to cover servants and agents, but only the persons directly acting on behalf of the ship—probably only the persons present on the pilot's boat, and very possibly not even legal persons. If the provision of art. III para. 4 (b) is thus given a closer look with having company law provisions in mind, it actually seems problematic that a classification society as such—possibly a very large business having several hundred employees—should benefit from the channelling provisions, but not its agents or servants. It can hardly be the purpose of art. III para. 4 (b) CLC to protect classification societies from third party claims, but explicitly omitting its employees from this protection.

The argumentation can furthermore also be based on the expression "performs services for the ship" in art. III para. 4 (b) CLC. It can be argued that a classification society does not perform services for the *ship*, but rather for the shipowner,[355] following the classification contract concluded between them. What is more, it can even be argued that the assessment of the vessel's compliance with the SOLAS and MARPOL provision is not even carried out for the shipowner, but rather on behalf of the flag State.[356] In contrast to the argumentation of the *Cour de Cassation*, that referred to the contract by virtue of which RINA had to issue classification documents,[357] it is rightly pointed out that RINA's responsibility was not invoked as a breach of the classification contract, which would basically only regard the shipowner, but as a shortcoming in its performance as a delegate of the flag State.[358] In this capacity it could have benefitted from State immunity, but there are several problems as to this assumption of a delegation of flag State's duties. Even if the classification society disposed of a discretion as to granting the classification on the

[352] P. Bonassies, 'Réflections d'un juriste français sur le jugement "Erika"' (2008) 110, Il Diritto Marittimo, 249–57 at 254 et seqq.

[353] Siccardi, '2005 "Pollution Liability"', 707.

[354] *Cour d'Appel de Paris, 30.03.2010, "Erika"*, p. 867.

[355] Bonassies, '2008 "Sociétés de classification"', 694.

[356] Berlingieri, '2012 "Les sociétés de classification"', 1015.

[357] *Cour de Cassation (France), 25.09.2012, "Erika"*, p. 992.

[358] Berlingieri, '2012 "Les sociétés de classification"', 1015 et seq.

one hand, it did not have the power to revoke a vessel's class on the other hand, as the Maltese authorities had reserved this power for themselves. In view of that, it can be considered that RINA was not given genuine State power, but only acted as an administrative manager for Malta, the State not having delegated the sovereign right entirely.[359] RINA's possibility to invoke State immunity was ruled out both by the *Cour d'Appel*[360] and the *Cour de Cassation*[361] for the rather procedural reason that the classification society was considered to have waived its immunity by participating in the proceedings without prior protest. The *Cour de Cassation's* argumentation is slightly inconsistent when it argues that RINA benefits from the channelling provisions due to its contract with the shipowner, but could otherwise have invoked State immunity for having acted on behalf of Malta, had it only protested in time. If this latter argument is accepted, it means that classification societies could only be held civilly liable in the vessels' flag States where they do not enjoy immunity.[362]

In any case, if it is considered, by pursuing the *Cour de Cassation's* argument further, that a person is covered by art. III para. 4 (b) for "performance of services for the ship" as soon as he is in a contractual relationship with the shipowner, the number of parties shielded by the channelling provisions would become overly large, also comprising parties such as shipyards, ship repair services and brokers.[363]

For the inclusion of classification societies in art. III para. 4 (b) CLC on the other hand, it can be argued the classification does have an impact on the nautical management as it would be impossible to operate it without valid class documents.[364] It can certainly be argued that in this respect, classification societies are comparable to pilots, who also fulfil the task of ensuring maritime safety and are also paid by the shipowner, and that therefore, a differentiation between pilots and classification societies cannot be justified.[365] Furthermore, it can be argued that the differentiation between persons taking part in the nautical operation of the ship and

[359] Neyret, '2010 "L'affaire "Erika""', 2239.

[360] *Cour d'Appel de Paris, 30.03.2010, "Erika"*, p. 867.

[361] *Cour de Cassation (France), 25.09.2012, "Erika"*, p. 988.

[362] Neyret, '2010 "L'affaire "Erika""', 2239; S. Kopela, 'Civil and Criminal Liability as Mechanisms for the Prevention of Oil Marine Pollution: The "Erika" Case' (2011) 20, Review of European Community and International Environmental Law, 313–24 at 319; S. G. Kaufmann, 'Reformbedarf auf internationaler Ebene für die rechtliche Aufarbeitung von Tankerkatastrophen: Das Erika-Urteil des französischen Kassationsgerichtes vom 25. September 2012' (2013), Zeitschrift für ausländisches öffentliches Recht und Völkerrecht, 509–41 at 523.

[363] Delebecque, '2012 "Aspects civils"', 1002; Rodríguez Docampo, *2012 "Alcance de la canalización"*, p. 925.

[364] Daniel, '2006 "Potential Liability"', 227; Boisson, '2008 "La société de classification"', 698; Carbone and Di Schiano Pepe, '2009 "Uniform Law and Conflicts"', 63; Arroyo Martínez, '2009 "Para esta viaje"', 95. The argument was also put forward in the *Prestige* decision, *US District Court, Southern District of New York, 02.01.2008, "Reino de España v. American Bureau of Shipping (The Prestige)"*, p. 274.

[365] P. Boisson, 'La responsabilité de la société de classification dans l'arrêt "Erika"' (2010) 62, Droit Maritime Français, 897–902 at 901.

other persons is discriminatory,[366] and that pilots may also provide their services by radio from a point other than the ship.[367]

Regard can again be given to the referral to persons performing services, but not being members of the crew who should benefit from the channelling of liability to the shipowner. In effect, especially with regard to the shipowner's point of view it seems fair to have this benefit only conferred to persons who are under the scrutiny of the shipowner, and possibly only to those performing services mostly on or around the ship[368] so that he can control their performance.[369] More independent contractors, who are not subject to the shipowner's orders, should perhaps not benefit from the channelling,[370] which amounts to an exemption of liability in all cases in which the shipowner does not have a right of recourse against them. It is true that this argument slightly mingles a semantic interpretation with policy considerations in wanting to uphold tortious liability against independent players in order to ensure their acting with due diligence. Even if this is a very valid and pragmatic thought, it does not seem that this policy reason is at the heart of art. III para. 4 CLC, because other persons benefitting from the channelling such as the charterers, managers or operators may likewise not be subject to the orders of the shipowner and still escape liability.[371] Even beyond that, and without scrutinising any further into the relationship between pilots and masters,[372] the pilot is certainly not subject to the shipowner's power of control, so that the benefit conferred to him would not be justified.

In conclusion, it does seem that considering the wording of art. III para. 4 (b) CLC, the comparison with pilots and crewmembers and the systematic structure of the whole paragraph, the best arguments are found to be against the inclusion of classification societies into the channelling provisions. A narrow understanding of art. III para. 4 (b) CLC that restricts this benefit to persons working on board or in the close vicinity of the ship and participate in its actual operation is in line with the underlying idea of the list that there must be parties who are not protected by the channelling provisions, and better corresponds to the wording of the provision.

Many further arguments for or against this are policy-driven, and there, the argumentation heavily depends on the way the role of classification societies is considered.

---

[366] Ibid., pp. 900 et seq.

[367] Carbone and Di Schiano Pepe, '2009 "Uniform Law and Conflicts"', 63.

[368] Ibid., p. 62.

[369] Vanheule, '2003 "Oil Pollution"', 553; Boisson, '2008 "La société de classification"', 698.

[370] In this sense, Baatz, *2011 "Maritime Law"*, p. 438: "...a purposive analysis of the provision...".

[371] Vanheule, '2003 "Oil Pollution"', 553.

[372] See Bonassies, '2008 "Sociétés de classification"', 694: "the pilot, who, even if he is, legally, only an advisor to the master", and, more generally, K. Woonings, "The Plot of the Pilot': Pilotage and Limitation of Liability in Maritime Law' (2010) 24, Australian and New Zealand Maritime Law Journal, 126–37.

It has rightly been pointed out that the concept of channelling in connection with the limitation of liability leads to the somewhat paradoxical result that victims of catastrophic oil spills, where damages exceed the compensation amounts available, will try to find solvent third parties who do not benefit from the channelling provisions nor from the limitation of liability and may thus be faced with unlimited, arguably uninsurable, claims.[373] This may be deplored, especially with a view to the fact that these persons may not be the mainly responsible parties whose behaviour is at the root of the accident.[374] It has rightly been observed that even in cases of severe omissions, the classification society will, as a rule, not be the only instance who should have known about a vessel's bad condition.[375]

However, monitoring a vessel's condition and performance, to the extent to which they undertook to do so,[376] is the core task of classification societies and one for which they are the best placed instance, disposing over the necessary expert knowledge[377] and acting within a framework of applicable standards. Parties on the shipowner's side, such as the charterer or the cargo owner generally do not have comparable means of controlling the vessel's condition.[378] The shipping industry relies to a considerable extent on the certificates issued, and entirely exempting classification societies from liability amounts to questioning the system of classification as a whole.[379] With regard to the general responsibility of classification societies, there should certainly be a system striking a balance between the need to hold them liable to a certain extent and the interest of keeping classification workable and the liability of classification societies insurable.[380]

In the absence of a valid regulation on the liability of classification societies, it seems more appropriate to hold them liable for oil pollution claims in tort outside the CLC regime. Correctly assessing the condition of a ship and the likelihood of a major maritime catastrophe is not more than what they actually contractually undertake,[381] and in the case of an inadequate or even blatantly wrong class

[373] Rémond-Gouilloud, *1993 "The Future of the Compensation System"*, p. 95; Siccardi, '2005 "Pollution Liability"', 697; Daniel, '2006 "Potential Liability"', 186.

[374] Robert, *2003 "L'"Erika""*, p. 123; Siccardi, '2005 "Pollution Liability"', 697, referring to them as "mere side players".

[375] Vialard, '2003 "Faut-il réformer le régime"', 446.

[376] See Daniel, '2006 "Potential Liability"', 195 et seqq., pointing out that a survey by a classification society is by no means a detailed inspection of the whole vessel.

[377] Antapassis, *2007 "Liability of Classification Societies"*, p. 682.

[378] Vialard, '2003 "Faut-il réformer le régime"', 447.

[379] See Honka, '1994 "The Classification System"', 33: "In order to keep classification societies from acting as mere formal administrators, some type of liability in damages must be allowed."

[380] In this sense, ibid., pp. 33 et seqq.; Wurmnest, *2006 "Third Party Liability"*, p. 118; L. Han and P. Yu, 'New developments regarding the liability of classification societies' (2006) 12, Journal of International Maritime Law, 243–50 at 249 and Lagoni, *2007 "The liability of classification societies"*, pp. 316 et seqq., respectively suggesting an international convention to this end.

[381] K. Krüger, 'Fault Liability for Classification Societies towards Third parties?', in L. Gorton, J. Ramberg and J. Sandström (eds.), *Festskrift till Kurt Grönfors* (Stockholm: Norstedts, 1991), p. 286.

examination oil pollution victims should have the possibility to hold them liable to the extent that the incident can be traced back to their negligence. It is true that this may lead to a certain shift in the liability system, entailing the need for classification societies to either take out comprehensive insurance or to proceed to more "defensive" classification and probably charge higher costs for their services.[382] However, it does not seem that an authority that is, strictly speaking, independent from the shipowner and whose only task is the inspection of the ship's condition, should be allowed to hide behind the shipowner's limitation of liability. Furthermore, the limitation of liability granted to persons involved in shipping is historically based on the incalculable risks of this business, which classification societies are not directly confronted with.[383] Unlike shipowners, they will not be liable for the failure of the maritime venture, but only for bad performance of the service contract for classification. It is, in this context, interesting to keep in mind that classification societies generally do not benefit from the limitation of liability under the LLMC either.[384]

Having regard to the overall cost of oil pollution incidents, not only to the environment and the victims on shore, but also to cargo owners, insurance companies and the maritime industry as a whole, it is submitted that with regard to the cases of the *Prestige* and the *Erika*, a stricter class examination would perhaps after all not be too disadvantageous for the maritime business.[385]

5. Shipyards

In their capacity as independent contractors that are contractually bound to deliver a safe and secure vessel, or to carry out repairs to this end, shipyards may have considerable responsibility in a shipwreck that causes oil pollution. In the *Amoco Cadiz* proceedings, the shipowner filed a recourse claim against the shipyard that had built the vessel, and the yard was found liable for the construction faults that had contributed to the accident.[386] Considering that this was a recourse claim, brought by the shipowner, the application of the CLC's channelling provisions was not argued.

It does not seem that there are many cases in which oil pollution victims would have tried to hold a shipyard liable for contributing to or causing an accident, so that

---

[382] Siccardi, '2005 "Pollution Liability"', 701, referring to the disproportion between the liability incurred and the general cost of classification and to US case law in this question.

[383] Vialard, '2009 "La limitation de responsabilité"', 23.

[384] Delebecque, '2012 "Aspects civils"', 1002; Rodríguez Docampo, *2012 "Alcance de la canalización"*, p. 923. See however Daniel, '2006 "Potential Liability"', 233, hinting at the shipowner's non-delegable duty to provide a seaworthy vessel that may enable classification societies to be covered as a "person for whose act he shipowner is responsible".

[385] Le Couviour, '2002 "Responsabilités pour pollutions"', 2275.

[386] *US District Court, Northern District of Illinois (Eastern Division), 18.04.1984, "In Re Oil Spill by the Amoco Cadiz"*, p. 2195.

the question of whether shipyards can invoke the channelling benefit of art. III para. 4 CLC would have been put to the test. Authors considering the problem from a theoretical point of view are however very unanimous in maintaining that shipyards having built or having performed repairs on a vessel are not protected by art. III para. 4 (b) CLC.[387]

It is even maintained that a protection of shipyards by the channelling provisions is virtually unthinkable.[388] In this context, a comparison with classification societies is often drawn, and most authors who oppose the protection of classification societies argue that shipyards are not, and cannot be protected either.[389]

A shipyard will, as a rule, of course be an independent contractor working for the shipowner, and not be involved with the operation of the ship as a means of transport. Considering that the shipyard is not, strictly speaking, on the shipowner's side, but carries out its task independently from the nautical operation, it does seem right that the shipyard should not be protected by the CLC channelling provisions, but should be held responsible for negligence to the extent that this negligence contributed to an oil pollution accident. Following the interpretation of art. III para. 4 (b) CLC that has been advocated *supra* under 4., it must be repeated that the meaning of "the pilot, or any other person who, without being a crewmember, performs services for the ship" must be a person that is in a situation comparable to that of the pilot or a crewmember, i.e. a person involved in the actual nautical operation of the ship and carrying out tasks on board or in the close vicinity of the ship. This means that shipyards can clearly not be covered by it. Even more so than classification societies, they are independent contractors carrying out their task before the ship undertakes its nautical operations.

6. Salvors

The reference to salvors benefitting from the channelling provisions was also added to art. III para. 4 CLC by the 1992 Protocols. The amendment was also due to the experiences of the *Amoco Cadiz* accident, considering that in the ensuing proceedings the salvors were sued for allegedly having contributed to the oil pollution by their unsuccessful salvage attempts. In the instance, the salvage company was

[387] H. Smets, 'Indemnisation des dommages exceptionnels à l'environnement causés par les activités industrielles', in R.-J. Dupuy (ed.), *L'avenir du droit international de l'environnement* (Dordrecht: Nijhoff, 1985), p. 317; P. Bonassies, 'Note Tribunal de Commerce de Saint-Nazaire, 16.12.2000' (2002) 54, Droit Maritime Français (Hors série n° 6), 43–5 at 43; Ndendé, '2007 "L'accident de l'Erika"', 16; Requejo Isidro, *2007 "Contaminación por hidrocarburos"*, p. 477; M. N. Tsimplis, 'Marine pollution for shipping activities' (2008) 14, Journal of International Maritime Law, 101–52 at 112; de la Rue and Anderson, *2009 "Shipping and the environment"*, p. 109; Delebecque, '2010 "Aspects civils"', 882; Baatz, *2011 "Maritime Law"*, p. 438.

[388] Bonassies, '2008 "Réflections d'un juriste français"', 255: "an excess".

[389] Ibid., p. 255.

not found liable because the *Amoco Cadiz's* dimensions exceeded the length and tonnage of any other ship that had been towed in bad weather up to this point in time,[390] and the salvage tug's crew was held not to have acted grossly negligent or wilfully in any case, so that under US law, they could not be held liable.[391] Still, under different circumstances and in a different jurisdiction, there could have been another outcome, and it was apparently felt that salvage companies should be protected from third party claims in order not to discourage salvors from intervening in favour of ships in distress.[392]

However, it is important that according to the wording of art. III para. 4 d) CLC, the channelling can only apply if the salvor either acted with the owner's consent or on the instructions of a competent public authority. Considering that shipowners or ship managers may be reluctant to order salvage services if they do not deem the situation hopeless, the reference to the competent public authority's orders is certainly advantageous to all parties possibly involved in the incident.[393]

Salvage operations are still widely performed on "no cure, no pay" contracts and that salvors, as a rule, therefore have a strong interest in bringing the salvage operation to a successful end.[394] With regard to salvage operations only benefitting the environment or other values not directly calculable in money, the 1989 Salvage Convention[395] provides for adequate remuneration. It is not strikingly successful in terms of ratification, but at least over 50 % of world tonnage is covered by its provisions.[396] Generally, and unlike classification societies working for a fixed fee, salvors have therefore enough incentive to act in the most effective way. Considering the risks involved in the salvage business, it is clearly appropriate to protect salvors from third party claims, even if they generally work entirely independently from the shipowner's orders, are not involved with the normal nautical operation of the ship and only intervene once the vessel is in distress.

Following the amendments in the 1992 Protocol, the CLC considers the impact of wrongful behaviour of the parties that are benefitting from the channelling provisions. In the original version of art. III para. 4 CLC, the channelling benefit

---

[390] Kiss, '1985 "L'affaire de l'"Amoco Cadiz""', 579.

[391] *US District Court, Northern District of Illinois (Eastern Division), 18.04.1984, "In Re Oil Spill by the Amoco Cadiz"*, p. 2189.

[392] du Pontavice, '1985 "Affaire "Droit de l'environnement""', 24; Kbaier and Sebek, '1985 "New trends"', 275; A. Bishop, 'The liability of salvors for pollution', in D. R. Thomas (ed.), *Liability regimes in contemporary maritime law* (London: Informa, 2007), p. 100.

[393] Kbaier and Sebek, '1985 "New trends"', 275.

[394] But see S. Girvin and T. Stephens, 'Liabilities of salvors', in D. R. Thomas (ed.), *Liability regimes in contemporary maritime law* (London: Informa, 2007); A. Bishop, 'Places of refuge - Environmental salvage', in N. A. Martínez Gutiérrez (ed.), *Serving the rule of International Maritime Law: Essays in Honour of Professor David Joseph Attard* (Abingdon: Routledge, 2010), p. 357.

[395] 1953 UNTS 193; BGBl. 2001 II, 510.

[396] See http://www.imo.org/About/Conventions/StatusOfConventions/Documents/Summary%20of%20Status%20of%20Conventions.xls (last visited 7 March 2014).

was granted to servants and agents of the owner unconditionally, so that they were shielded from third party claims even if the incident had occurred because of their fault.

II. Loss of the Channelling Benefit in Cases of Wrongful Behaviour

According to art. III para. 4 CLC as amended by the 1992 Protocols, the beneficiaries of the channelling provisions will only be protected from third party claims if the damage did not result from "their personal act or omission, committed with the intent to cause such damage, or recklessly and with knowledge that such damage would probably result."

As has already been pointed out, the wording mirrors the conditions under which, according to art. V para. 2 of the CLC, a shipowner will lose the right to limit his liability. The provision starts from the idea that the benefits of maritime law will be lost for reckless behaviour beyond a certain point, and it is generally considered that art. V para. 2 and art. III para. 4 CLC should be understood in the same way.

The admission of claims against persons that would otherwise be protected by the channelling provisions begs the question whether these claims can be considered as CLC claims or whether they are claims under domestic law. The answer to the question has some consequences on the claims against the parties mentioned in art. III para. 4 CLC. First and foremost, if these claims do not have a footing in the CLC, but are only allowed by it, it means that the claims in question must be entirely justified and proven under their basis in the national law. It is obvious that this will at first glimpse not be a difficulty in practice, as a conduct that constitutes reckless behaviour with knowledge that damage will probably occur will most likely give rise to a claim in almost all legal systems. It however means that the claim would also need to be admissible according to the other rules of the forum, on general jurisdiction,[397] the capacity to be made a defendant, prescription, and so on. If the claim is not governed by the rules of the CLC, domestic law may provide for a different forum or a different procedure. The exclusive jurisdiction clause of art. IX CLC could then actually not be used, and there could be proceedings in courts of States parties where there has not been pollution damage (see *supra* under B. I. 1.). This could in the end mean that the proceedings against the shipowner and the proceedings against the person not enjoying the channelling benefit due to his conduct would not take place in the same court, even if the shipowner was also guilty of reckless behaviour endangering his limitation of liability.

[397] Robert, *2003 "L'"Erika""*, p. 118.

Up to now, it has simply been assumed that a claim against persons that would be shielded by the channelling provisions but for their reckless behaviour were claims under domestic law,[398] and this was clearly the position of the French courts in the *Erika* proceedings.[399] The wide use that the *Cour d'Appel* made of the exception to the channelling provisions was even extended by the *Cour de Cassation*.

The *Cour d'Appel* held Mr. Savarese, the *Erika's* owner's director, personally liable, reasoning that he had acted recklessly when restricting the vessel's last repairs before the accident to a minimum, to the point that the vessel was no longer fit for service, and to then charter the vessel.[400] The court decided that as an employee of the shipowning company, Mr. Savarese would normally have been shielded by the channelling provision of art. III para. 4 (b) CLC, but due to his reckless behaviour, he could not invoke that benefit.[401]

Looking at the set of facts, it does seem that, Mr. Savarese's behaviour and the way he took decisions during the last repairs of the *Erika* did appear to be reckless, completely ignoring the consequences that this would have on the seaworthiness of the ship. The repairs carried out on the *Erika* were originally planned to be of a much wider extent, and Mr. Savarese explicitly decided that only a small part of the originally planned works should be effectuated, and this, for economic reasons only.[402] Furthermore, Mr. Savarese could be blamed with having obtained classification certificates of which he knew that they did not correspond to the actual condition of the ship.[403]

Interestingly, with regard to Total's, the *Erika's* actual charterer's, behaviour in the vetting process, the *Cour d'Appel* held that the company did not receive enough information on the vessel's state to render the chartering a "reckless" act.[404] The *Tribunal Correctionnel* in the first instance had held Total liable, considering that it could not benefit from the channelling provisions because it was not the *Erika's* nominal charterer,[405] the charterparty having been signed by a mere subsidiary of Total (see *supra* under I. 2.). The *Cour d'Appel* opposed this view and considered Total as a charterer, then holding that Total committed errors of appreciation in the vetting process and failed to detect the condition of the ship, but that it did not lose the channelling benefit. The *Cour d'Appel* did not consider that the charterer had the "knowledge" that pollution damage would result from the decision to charter the *Erika* despite her doubtful condition.

---

398 Ibid., p. 113; Delebecque, '2010 "Aspects civils"', 882.

399 Juste-Ruíz, '2010 "Compensation for pollution damage"', 47 et seq.

400 *Cour d'Appel de Paris, 30.03.2010, "Erika"*, pp. 867 et seq.

401 Ibid., p. 868.

402 Also see the findings of the court in the first instance, *Tribunal correctionnel de Paris, 16.01.2008, "Erika"*, p. 708.

403 Ibid., p. 709.

404 *Cour d'Appel de Paris, 30.03.2010, "Erika"*, p. 86.

405 *Tribunal correctionnel de Paris, 16.01.2008, "Erika"*, p. 721.

In the final judgment on the case, the *Cour de Cassation* overturned this view of the *Cour d'Appel*. According to the *Cour de Cassation*, Total caused the oil spill recklessly, given the fact the persons in charge were aware of certain peculiarities, such as the circumstance that even after recent repairs, the *Erika's* classification certificates were only issued for two months.[406] From the facts stated in the judgments in all instances, it is true that there were signs that should have rendered every responsible charterer suspicious as to the real condition of the ship, but still, the *Cour de Cassation* based its decision rather on the facts of which Total was aware, less on the consideration what they should have known if they had acted responsibly.[407] With a view to the debate whether the knowledge that the damage would probably result should be appreciated objectively or subjectively (see *supra* under D. II.), it can be argued that the *Cour de Cassation* here rather inclines to a subjective appreciation,[408] considering whether the possibility of an accident with an important oil spill was foreseeable to Total's representatives even given their limited interest in the vessel's state. Under an objective appreciation of the "knowledge that the damage would probably occur", it would have been enough to state that a reasonable charterer would have taken more interest in the ship's state of repair. Even if it theoretically retains a stricter degree of appreciation of the "knowledge", the judgment can certainly be criticised for in fact merely assuming that Total's representatives had the knowledge,[409] without establishing whether the possibility of an accident was actually on their minds, and without discussing whether this fault was "personal" to Total.[410] The decision in this respect rather runs counter the current state of jurisdiction in the French courts, that have become much more reluctant to find for a "personal act, committed recklessly and with knowledge that such damage would probably occur" under art. IV LLMC[411] (see *supra* under D. II.).

Within the *Erika* decision, the *Cour de Cassation* was however equally ready to hold RINA, the vessel's classification society liable for reckless behaviour. Whereas the *Cour d'Appel* had decided that RINA was, as a classification society, not protected by art. III para. 4 (b) CLC, the *Cour de Cassation* overturned that finding, but in return decided that even though RINA was benefitting from the channelling provisions, it had acted "recklessly and with knowledge" as to the consequences of the accident.[412] The classification society's wrongful conduct, following the criminal inquiry the court refers to, mainly consisted in the omission to check the thickness of the steel plates in the *Erika's* hull[413] and to issue

[406] *Cour de Cassation (France), 25.09.2012, "Erika"*, p. 991.

[407] Le Couviour, '2012 "Erika: l'arrêt salvateur"', 2098.

[408] Ibid., p. 2098.

[409] Delebecque, '2012 "Aspects civils"', 1004.

[410] Ibid., p. 1004.

[411] Cachard, '2013 "Observations CA Bordeaux 14.01.2013"', 215.

[412] *Cour de Cassation (France), 25.09.2012, "Erika"*, p. 992.

[413] *Tribunal correctionnel de Paris, 16.01.2008, "Erika"*, p. 711.

classification certificates notwithstanding the insufficient survey. The court does not clearly establish the facts from which it derives that RINA's representatives knew or ought to have known that a structural failure of the hull was probable.

The *Cour de Cassation* used the exception of art. III para. 4 CLC twice in the *Erika* proceedings to overcome a channelling benefit that it considered otherwise granted to the parties, however, it is unclear whether the findings can be of value as a precedent. The entire *Erika* case is heavily influenced by the fact that the proceedings started as a criminal trial, in which evidence against the accused was firmly established before the civil liability was decided on in a second step. It can even be argued that the *Cour de Cassation* felt compelled, by the convictions of the criminal trial,[414] to find similarly for the civil liability.[415]

It does not seem that the exception of reckless conduct in art. III para. 4 CLC had ever been considered in any other case. Especially with a view to the strict interpretation of the term "recklessly and with knowledge that such damage would probably occur" under art. IV LLMC and art. V para. 2 CLC in the English courts, it does not seem that the *Erika* case could be considered as a good precedent there.

Moreover, even with regard to the identical formulation in art. III para. 4 CLC, art. V para. 2 CLC and art. IV LLMC, it must be kept in mind that the loss of the channelling benefit and the loss of the right to limitation do not exactly have the same consequences, and that the persons whose conduct is considered are in rather different situations. The persons listed in art. III para. 4 CLC are almost all involved in the actual nautical operation and to some extent exercising control over the ship, therefore the likelihood of them causing a "personal act or omission" on the ship is bigger than with regard to the shipowner, and the conduct that causes the loss of the benefit may be of a completely different character.

Under the exception to the channelling provisions, virtually all persons involved in the accident could become possible defendants, if they can be blamed with reckless behaviour. In all cases of oil pollution from substandard ships—of whose deplorable condition at least the crew must be aware—and in all cases of classification certificates obtained without proper examination of the ship, of which shipowners and probably charterers must be aware, the exception to art. III para. 4 CLC could be a practicable way to invoke the responsibility of other, and possibly more solvent, parties on the shipowner's side.[416]

There is however a very simple, procedural reason for which claimants may be reluctant to invoke the exception. In order to hold any beneficiary of the channelling system liable for reckless behaviour, there must first be proceedings against them,

414 See S. Grunvald and C. Saas, 'Regards croisés sur les catastrophes pétrolières: De l'Amoco Cadiz à l'Erika, le regard du pénaliste' (2009), Annuaire de droit maritime et océanique, 379–96 at 388 et seq. on the question of the fault in the criminal proceedings.

415 Delebecque, '2012 "Aspects civils"', 1002; H. Jessen, 'Neues zur Dritthaftung privater Klassifikationsgesellschaften nach "Erika" und "Prestige"?' (2013), Recht der Transportwirtschaft, 387–93 at 393.

416 Bonassies, '2002 "Aspects nouveaux"', 12, with regard to masters.

otherwise the exception to the channelling cannot be pronounced by the court. As mentioned earlier, these proceedings, in that they are not CLC claims, may even need to be brought under different rules on jurisdiction and procedure and may yet be struck out very quickly if the court considers that the exception of art. III para. 4 CLC does not apply. It is one of the problems of this exception to the channelling—even more so than in the limitation of the shipowner's liability according to art. V para. 2, see *supra* under D.—that it can only be decided upon once a claim against the person in question has been brought, proceedings that are at first sight inadmissible.

III. Recourse

The CLC clearly states in its art. III para. 5 that "nothing [...] shall prejudice any right of recourse of the owner against third parties." It is not the Convention's aim to provide for a complete exoneration of the persons benefitting from the channelling provisions, but only to protect them from an unmanageable number of claims by different claimants, possibly in various foreign jurisdictions.[417] They are, on the other hand, not protected from recourse claims by the shipowner, so that action can be taken by the shipowner against, for instance, salvors, or even port authorities.[418] Moreover, recourse claims for damages paid can also be brought by the Fund, and there are effectively some examples of cases where the Fund recovered damages from parties who had contributed to an accident. One example is the Fund's recourse claim against the Milford Haven Port Authority for the *Sea Empress* accident, in which the port authority was blamed for wrongful conduct.[419] The case however highlighted the difficulties in establishing the responsibility of the port authority under domestic law and was finally settled in mediation. In other cases, the Fund refrained from bringing a recourse claim, supposedly for political reasons, even if the claim would have been more than justified. The most prominent example[420] is perhaps the case of the *Slops* (see *supra* under C. II.) in which the Greek authorities were clearly to blame for their omission to enforce the CLC's mandatory insurance provisions against the *Slops's* owners, considering that they considered the floating oil storage device as a "vessel".

The CLC does not offer provisions on the law or the procedure applicable to the recourse claim, which will therefore be governed by domestic law in a given jurisdiction. The claim may also be decided under alternative dispute resolutions,

[417] Bishop, *2007 "The liability of salvors"*, p. 100 with regard to salvors.

[418] A. Mandaraka-Sheppard, *Modern Maritime Law and Risk Management* (London: Informa, 2009), p. 964.

[419] Jacobsson, *2010 "Dispute resolution"*, p. 337.

[420] Peplowska, '2010 "What is a ship?"', 163.

such as arbitration or mediation.[421] The shipowner may have different causes of action, depending on the relationship he has with the defendant to the recourse claim. Especially in cases in which there is a contractual relationship, there may be other circumstances or legal provisions allowing the defendant to escape liability partly or entirely and thus making his liability calculable and possibly insurable. Examples for this may be labour law provisions governing seamen's pay contracts or contractual provisions in charterparties or management contracts. Between different actors that have been held liable, outside of the CLC or to without limitation, recourse may entail questions of contributory negligence.[422]

As to questions of limitation in recourse claims, it must again be stressed that the recourse claim is not a claim under the CLC and may be governed by a limitation regime of its own. The limitation amounts under the CLC can however be of importance to the extent that they present the damage that the shipowner suffered and for which he turns back to the defendant of the recourse claim. Whether the recourse claim will be subject to limitation again depends on the circumstances of that case.

This was highlighted in the *Aegean Sea* case[423] in the English courts. The *Aegean Sea* was a tanker carrying nearly 80,000 tonnes of crude oil, which grounded on rocks when approaching the port of La Coruña. The vessel's owners faced CLC claims for the clean-up, damage to property destroyed by the oil spill and loss of profit of fishermen and local businessmen. They turned back against the *Aegean Sea's* charterers and claimed, *inter alia*, indemnity for these claims, on the basis of a breach of the charterparty, the charterers having ordered the *Aegean Sea* to proceed to La Coruña. The charterers invoked the limitation of liability under the LLMC as against the owners, but the court refused to grant it, pointing to the fact that the limitation of liability benefitted charterers to the extent that they acted as owners, i.e. managing and operating the ship, but not in the internal relationship between charterers and owners where they were only charterers.[424] The court very adroitly considered the fact that under the applicable limitation regime, one fund would be established for the event, and if the owners had a claim against the charterers for which the charterers could limit, the owners would share the limitation fund that they had themselves established, also diminishing the amounts available for other claimants. Moreover, there was of course the problem that the limitation amounts under the CLC and the LLMC were manifestly different. Under the LLMC, the charterers, even if they were held liable for the full damage towards the owners, would get a considerable discount in comparison to the amount the owners had paid under the CLC.

The distinction that was drawn in the *Aegean Sea* between claims incurred by the charterers as owners and claims incurred as against the owners was later overruled

---

[421] Jacobsson, *2010 "Dispute resolution"*, p. 336.

[422] Delebecque, '2012 "Aspects civils"', 1003.

[423] *High Court (England), Queen's Bench Division, 07.04.1998, "The Aegean Sea"*.

[424] Ibid., pp. 49 et seq.

by the Court of Appeal in the case of the *CMA Djakarta*.[425] That case concerned a recourse claim by the owners against the charterers for a breach of the charterparty, and on appeal, the court granted the charterers the limitation of liability for all claims, except those for the damage done to the ship itself, which are not covered by the wording of art. II LLMC.[426] The Court of Appeal claimed that by this interpretation of the LLMC, the problem of the owners sharing their own limitation fund would be solved, as the owner's main claim would always be the one for the damage done to the ship.[427] This argumentation holds true for the *CMA Djakarta* case, in which both owners and charterers invoked the LLMC, but it does not present a satisfactory solution for the problem of the different limitation funds under the CLC and the LLMC. If, in the *Aegean Sea* case, the charterers had been granted the limitation under the LLMC, they would have needed to either establish a second limitation fund for the event, that would arguably only have been available to the owners, or the owners would have needed to integrate the funds provided by the charterers into their CLC fund. It therefore seems that the limitation issue in CLC recourse claims continues to cause difficulties, even if there is a general tendency, at least in the English courts, to consider recourse claims separately and assess the limitation of liability independently for every claim.

The court in the *Aegean Sea* commented on this matter in *obiter dicta*, when considering whether a claim for pollution damage such as the one put forward by the owners against the charterers could be subject to limitation at all. Thomas J. considered it was, theoretically, covered by art. II (1) (a) LLMC.[428] This statement was explicitly approved by Longmore L. J. in the *CMA Djakarta*.[429] In the same spirit, it was stressed in *The Breydon Merchant*[430] that cargo owners who had incurred costs during salvage services and filed a recourse claim against the shipowner for having caused the emergency, still claimed for breach of contract, not for salvage. Therefore, the shipowner as the recourse defendant could limit his liability, even if the salvage liability that the cargo-owners had incurred as against the salvors was not subject to limitation.

It is therefore possible that the defendant in the recourse claim, depending on whether he is a beneficiary of the LLMC provisions, may be allowed to limit his liability under the maritime law applicable or the LLMC,[431] and possibly to marginal amounts.[432]

[425] *"The CMA Djakarta"* 2004 1 Lloyd's Law Reports 460 (Court of Appeal (England), 12 February 2004).

[426] Ibid., p. 467.

[427] Ibid., pp. 463, 468.

[428] *High Court (England), Queen's Bench Division, 07.04.1998, "The Aegean Sea"*, p. 52.

[429] *Court of Appeal (England), 12.02.2004, "The CMA Djakarta"*, p. 468.

[430] *"The Breydon Merchant"* 1992 1 Lloyd's Law Reports 373 (High Court (England), Queen's Bench Division, 14 November 1991).

[431] Martínez Gutiérrez, *2011 "Limitation of Liability"*, p. 188.

[432] de la Rue and Anderson, *2009 "Shipping and the environment"*, p. 98, referring to a collision between a tanker and a fish trawler.

The shipowner's right to recourse to some extent impairs one of the purposes of the channelling system, the avoidance of double or multiple insurance.[433] One of the aims in establishing the channelling principle was to determine the shipowner as the one and only defendant, who would be the only person to have to take out insurance. If, under a possible right to recourse, the persons benefitting from the channelling provisions can be confronted with a claim by the shipowner, they still need to have adequate insurance to face this claim, the amount also depending on whether they can limit their liability themselves.[434]

Again, however, the possible right to recourse may encourage the beneficiaries of the channelling provisions to act carefully and thus presents an incentive to prevention.[435] It seems that it would hardly be beneficial to the system as a whole if all parties protected by the channelling provisions were entirely exonerated from liability, the shipowner being the only entity liable.[436]

IV. Conclusion

The channelling provisions in the CLC are a subject of ongoing debate, and most of the arguments in favour of a wider or narrower interpretation of the channelling provisions are made with a view to policy or economic considerations. The costs of a catastrophic oil spill, especially if they can be claimed outside the CLC and thus unlimited in amount, can be considerable and could in the long run put the responsible entities' existence at stake. There is hence a strong interest in allocating the damage between given players and imposing limits of liability.

The question of how the limits of liability are to be calculated, how they have been adapted over the years and how they can be further adapted under the existing regime has not been scrutinised into in this study and is a question of calculation and of economics rather than a question of law. It must of course be mentioned that the costs of many average and minor oil spills will probably not exceed the liability limits under the CLC and the amounts available under the FUND Convention. In these cases, there will be little incentive for victims to start lengthy and costly tort proceedings in domestic jurisdictions against persons not benefitting from the channelling provisions.

---

[433] Wang, *2011 "Civil liability"*, p. 87.

[434] Gaskell, *2000 "Pollution, Limitation and Carriage"*, pp. 84 et seq.

[435] An economic analysis can be found in Wang, *2011 "Civil liability"*, p. 323 et seq.

[436] Wu, *1996 "Pollution"*, p. 170, also referring to the discussion between the negotiating parties to the Convention.

However, as long as there are limits to liability in a given civil responsibility regime, there will be cases in which the damage exceeds the caps[437]—if all cases were covered by the available insurance funds in any event, there would be no need for the limitation of liability. In fact, even when the limitation amounts of the 1992 Protocols were agreed on, it was already obvious that they would not be high enough to cover the damages of an accident such as the *Amoco Cadiz* oil spill that had occurred four years earlier.[438] Thus, there will always be cases in which victims will look for alternative ways to recover a surplus. There may also be cases in which there are sufficient funds available, but where victims face difficulties in obtaining compensation for procedural or other reasons.

Where victims see the necessity to start proceedings outside the CLC regime, there will be a parallelism between the Convention and domestic tort law. It is submitted that this coexistence of the shipowner's strict liability under the CLC and the tort claims outside the Convention is inherent in the compensation system and could only be avoided if the channelling provisions were absolute. It is often argued, and certainly rightly, that the channelling of liability benefits victims and that the system should not be disturbed or impaired by allowing too many claims outside the system.

The system of channelling has, after all, also been supported by the insurance industry,[439] as avoiding double or multiple insurance for different actors in the same case is beneficial to the industry as a whole. Nevertheless, the possibility of the shipowner's recourse impairs the system to a certain degree—however, it seems that this risk has been managed well so far. Entities like shipyards that cannot claim to be protected by the CLC's channelling provisions, have apparently been able to cover their risks by obtaining insurance or other measures.

There may of course be cases in which the unlimited liability of a person outside the CLC regime could exceed the funds available under the CLC and even the FUND Convention, which are after all expressly established to compensate for oil pollution.[440] It is of course debatable whether it is "fair" to hold a negligent side player responsible to this amount. It is also admitted that national courts may be driven by policy considerations when declaring a defendant liable, especially if they feel the need to protect victims in their own jurisdiction. This legal uncertainty is the price that must be paid for not having a more comprehensive regime than the CLC covering more actors, or not having higher liability limits that will cover a greater number of catastrophic spills.

---

[437] Smets, *1985 "Indemnisation des dommages"*, p. 312; A. Vialard, 'Responsabilité limitée et indemnisation illimitée', in A. Knops (ed.), *Liber Amicorum Roger Roland* (Gent: Larcier, 2003), p. 574.

[438] R. Renger, 'Haftung und Entschädigung für Ölverschmutzungsschäden auf See' (1993), Transportrecht, 132–5 at 132.

[439] Wu, *1996 "Pollution"*, p. 170.

[440] Wurmnest, *2006 "Third Party Liability"*, p. 119, referring to the *Prestige* case in which the FUND initially only offered monies to compensate for 15 % of the damage and the vessel's classification society could have been held liable for the remaining 85 %.

However, major oil spills constitute high risk incidents that are difficult to insure for everyone. In the absence of a target entity that could be held liable to the entire amount, the damage will be borne by the States affected, and, of course, by the victims. Whether the maritime transport of oil is in the interest of the world population as a whole and whether there is a general interest to keep the transport and insurance cost low and rather accept consequential accidents from time to time is a policy question; in any case, it does not seem fair that the consequences should in the first place be borne by some coastal States and their littoral populations, who do not benefit more from maritime oil transport than the rest of the world. From this point of view, it is not explainable why the consequences of catastrophic spills should be borne to a wider extent by them than by anyone else, and it does make sense to keep legal ways open by which oil pollution victims can at least theoretically obtain comprehensive compensation.

It must be stressed that domestic tort claims against persons outside the CLC regime will in all likelihood be lengthy, costly and bring about significant uncertainty as to the outcome. Very generally, oil pollution victims, especially private entities and individuals, will consider very carefully whether they have chances to recover a surplus under domestic law. It seems that this should be enough of a determent to them not to bring abusive claims, and not to start proceedings in too many countries. Of course, the sheer existence of different legal systems in which different outcomes can be expected, adds to the unfairness of the regime, but this is a problem the CLC cannot do away with as long as it is not of universal applicability.

With regard to the persons not covered by the channelling provisions, such as parent companies not involved in the actual operation of the vessel, shipyards and, arguably, classification societies, the possibility of unlimited third party claims in tort could actually acts as an important incentive to exercise due care in the performance of their duties. Recognising the fact that these actors also need to assume responsibility for catastrophic oil spills—to the extent that they have been caused by a given behaviour or by their power of control over subsidiaries—may be one step to incite them to take preventive measures. The CLC system has been frequently criticised for encouraging decision-takers to save costs and rely on the limitation of liability rather than investing into risk minimisation,[441] but it does seem that there are considerable gaps by which the limitation of liability and the channelling can be circumvented.

If the existence of these systemic gaps—and possibly, their justification with regard to victim's interests—were recognised to a wider degree, this could help to assess the overall cost of prevention and risk minimisation and perhaps contribute to an increase in maritime safety.

[441] Le Couviour, '2002 "Responsabilités pour pollutions"', 2270; Vialard, '2003 "Faut-il réformer le régime"', 439.

F. The Shipowner's Possibilities of Exoneration

Art. III para. 2 and 3 of the CLC give the shipowner the possibility to escape the otherwise strict liability if he can prove that the damage was caused by an outside cause. The provisions read:

2. No liability for oil pollution damage shall attach to the owner if he proves that the damage:
(a) resulted from an act of war, hostilities, civil war, insurrection or a natural phenomenon of an exceptional, inevitable and irresistible character, or
(b) was wholly caused by an act or omission done with intent to cause damage by a third party, or
(c) was wholly caused by the negligence or other wrongful act of any Government or other authority responsible for the maintenance of lights or other navigational aids in the exercise of that function.
3. If the owner proves that the pollution damage resulted wholly or partially either from an act or omission done with the intent to cause damage by the person who suffered the damage or from the negligence of that person, the owner may be exonerated wholly or partially from his liability to such person.

The list in art. III para. 2 and 3 CLC was, it seems, mostly dictated by the insurance interests, as the enumerated risks are of a kind that is generally not borne by insurers.[442] The exoneration clauses have, in the past, only been invoked very rarely.[443] It should briefly be mentioned that in the cases of damages caused by acts of God, intentional acts of a third party and wrongful acts of public authorities, damages will be available under the 1971 FUND Convention. Art. IV para. 1 (a) FUND stipulates that the Fund will step in in cases in which no liability arises under the CLC, but sets out exonerations of the FUND for war risks and ships on government service (art. IV para. 2 (a) FUND) and the victim's contributory fault (art. IV para. 3 FUND).

I. Act of War or Hostilities

According to art. III para. 2 (b) CLC, the shipowner is exonerated if the damage "resulted from an act of war, hostilities, civil war, insurrection or a natural phenomenon of an exceptional, inevitable and irresistible character". The first alternative refers to typical war risks that are normally excluded from maritime insurance cover.

[442] Abecassis and Jarashow, *1985 "Oil Pollution"*, p. 204; de la Rue and Anderson, *2009 "Shipping and the environment"*, p. 99. D. J. Bederman, 'Dead in the Water: International Law, Diplomacy, and Compensation for Chemical Pollution at Sea' (1986) 26, Virginia Journal of International Law, 485–514 at 488 claims that due to these exceptions, the strict liability was insurable with the same limits that would have applied to a liability for fault with the burden of proof reversed.

[443] Jacobsson, '2011 "An Ideal International Scheme"', 267.

Considering the wording of art. III para. 2 (b) CLC, the shipowner is already exonerated if the damage "resulted" from act of war or hostilities, the link of causality can therefore be weaker than in the other cases of art. III para. 2 CLC where the damage needs to be "wholly caused" by the exonerating set of facts.[444] It therefore seems that both the shipowner and his insurance have good chances to escape liability as soon as there is some war risk at the root of the oil pollution damage, without necessarily being the only cause for it. Whereas this is certainly in the interest of shipowners and crews who find themselves in an extreme situation facing war risks or hostilities, it should not be forgotten that terrorism and violence at sea have become a rather frequent occurrence recently and that pollution damage resulting from a deliberate attack can be very considerable. With regard to piracy, it is generally considered that an "act of war" presupposes a political motivation, which pirates, acting for private ends, according to the definition in public international law, generally do not have. The line between a political and a private motivation can however be very difficult to draw in some cases,[445] and where ransoms are claimed to support a political cause, or where pirates differentiate between victims from different States, the attack could be considered as an act of war.

It does seem problematic that there should be no compensation on an industrial scale available, considering that the damage caused by act of war is not compensated by the FUND either, constituting the only valid exemption for the FUND. It is not clear at all who could possibly be liable for an incident of this kind—even if the persons answerable for the attack could be held responsible,[446] which is problematic under public international law and for very practical reasons[447]—they could hardly cover a claim of this kind. A shipowner's marine war risk insurance will generally cover the loss sustained by the shipowner—i.e. the loss of the vessel that may be caused by a government intervention against the pollution hazard[448]—but not the damage to the environment, given the fact that the shipowner is not answerable for that damage in the first place.

The fact that the shipowner is exonerated from liability leads to the result that pollution damage following a terrorist or piracy or armed robbery attack will not be paid for at all, and the victims would be left without compensation. This is, with regard to the fact that attacks of this kind normally take place in politically unstable regions,[449] a very unsatisfying result.[450] Whereas the targeting of oil tankers in past

---

[444] de la Rue and Anderson, *2009 "Shipping and the environment"*, p. 99.

[445] L. Zhu, A. K. Y. Ng and M. Zhang, 'Some thoughts on civil liability for maritime piracy-caused oil pollution' (2012) 18, Journal of International Maritime Law, 306–15 at 313.

[446] Possibly in this sense, Wu, *1996 "Pollution"*, p. 83 referring to the "risk resulting from human actions" and the corresponding discussion during the negotiations of the FUND Convention.

[447] Gauci, '1999 "Protection of the Marine Environment"', 32.

[448] See M. D. Miller, *Marine War Risks*, 3rd edn. (London: LLP, 2005), para. 32.46 et seqq.

[449] Zhu, Ng and Zhang, '2012 "Some thoughts"', 307 et seq.

[450] Gauci, '1999 "Protection of the Marine Environment"', 32.

conflicts, due to the mode of attack with radar-guided missiles, did not lead to major pollution incidents,[451] piracy in certain parts of the world certainly causes a major risk of pollution. Incidents like the hijacking of the *Sirius Star*, a supertanker carrying about two million barrels of crude oil, in late 2008, could easily result in a catastrophic pollution incident, both during the piracy attack and during possible rescue operations by State authorities.

II. Natural Phenomena

The shipowner will furthermore be exonerated from liability if the damage "resulted from . . . a natural phenomenon of an exceptional, inevitable and irresistible character". Here again, it is sufficient that the phenomenon is at the root of the event, it does not need to be "wholly caused" by it and the causal link between the event and the damage can thus be rather weak.

However, the threshold for the event to be qualified as "exceptional, inevitable and irresistible" is extremely high, probably even higher than the "perils of the sea" exception in contracts of carriage.[452] It must in any case be shown that the damage could not have been avoided under any circumstances.[453] However, the mere foreseeability of a natural phenomenon by weather forecasts and other warnings does not prejudice the shipowner's exoneration if the fatal consequences cannot be avoided, considering that the provision does not require the phenomenon to be "unforeseeable".[454] Even very extreme weather conditions, therefore, do not suffice to trigger the exoneration, because they can normally be provided for or avoided. It is therefore submitted by certain authors that tidal waves would be the only phenomenon covered.[455]

According to art. IV para. 1 (a) FUND, notwithstanding the shipowner's exoneration, compensation is available under the 1971 Convention.

III. Intentional Acts of Third Parties

The exception under art. III para. 2 (b) available to the shipowner for "acts or omissions done with intent to cause damage" can cover terrorist and other malicious attacks like sabotage, effectuated with the third party's intent to cause damage. The

[451] de la Rue and Anderson, *2009 "Shipping and the environment"*, p. 99.

[452] Ibid., p. 99.

[453] Abecassis and Jarashow, *1985 "Oil Pollution"*, p. 205.

[454] Tsimplis, '2005 "The Bunker Pollution Convention"', 88, footnote 56 with regard to the identical provision in the BOPC.

[455] Abecassis and Jarashow, *1985 "Oil Pollution"*, p. 205; de la Rue and Anderson, *2009 "Shipping and the environment"*, p. 100.

applicability of the provision to persons who are clearly outsiders to the operation of the ship is obvious, but problems arise with regard to malicious acts committed by crew members, passengers or independent contractors.[456] Answers to this problem can only be found in domestic legal systems,[457] and courts faced with the problem will proceed to a legal appreciation under the applicable law.

The causal link between the pollution damage and the act or omission of the third party needs to be immediate, the damage must be "wholly caused" by the action. This can lead to very problematic situations where the attack was at the root of the pollution incident but other factors contributed to it[458]—the shipowner's strict liability would here not be excluded. Before 2001, the risk would generally not have been covered by P & I cover, but the clubs decided to adapt their rules after the lacuna became visible following the terrorist attacks on the World Trade Center.[459]

With regard to the causal link that can be weaker under the art. III para. 2 (a) exception for act of war or hostilities,[460] it may be in a shipowner's interest rather to invoke the war exception than the "intentional acts of third parties" or terrorist exception under art. III para. 2 (b) CLC. Whereas very generally, a piracy attack will not be carried out with the intent to cause damage in the first place as pirates usually want to take possession of the ship or the cargo in an undamaged condition,[461] their intentions can however change during the attack. In the hijacking of the *M/V Pacific Express*, pirates reportedly set the ship ablaze and left the ship after unsuccessful attempts to take over control of the vessel.[462] In cases of this kind, there may be considerable uncertainty as to whether the shipowner's exoneration of pollution resulting from the damage done to the ship is covered by art. III para. 2 (a) or (b) CLC, and this may make a difference, having regard to the distinction in the degree of causation.[463] It is perhaps in view of this difficulty that the introduction of an express exoneration clause for terrorist acts has been discussed after the events of 2001.[464]

---

[456] Tsimplis, '2005 "The Bunker Pollution Convention"', 89, footnote 57, with regard to the identical provision in the BOPC.

[457] See D. M. Gunasekera, *Civil liability for bunker oil pollution damage* (Frankfurt am Main: Peter Lang, 2010), pp. 208 et seq., giving a comprehensive overview of the problem mainly with regard to English law.

[458] M. Jacobsson, 'Bunkers Convention in force' (2009) 15, Journal of International Maritime Law, 21–36 at 26.

[459] de la Rue and Anderson, *2009 "Shipping and the environment"*, pp. 100, 747 et seq.

[460] Gauci, '1999 "Protection of the Marine Environment"', 32.

[461] Zhu, Ng and Zhang, '2012 "Some thoughts"', 313.

[462] See the NATO-ACO report on the rescue operation, http://www.aco.nato.int/nato-warship-rescues-mv-pacific-express-crew.aspx (last visited 7 March 2014).

[463] Jacobsson, '2009 "Bunkers Convention in force"', 26, also giving examples.

[464] N. Gaskell and C. Forrest, 'Marine Pollution Damage in Australia: Implementing the Bunker Oil Convention 2001 and the Supplementary Fund Protocol 2003' (2008) 27, University of Queensland Law Journal, 103–65 at 121, footnote 120.

IV. Wrongful Acts of Public Authorities

The exceptions of art. III para. 2 (b) CLC, for wrongful acts of public authorities, and art. III para. 3, the victim's contributory fault, are closely linked to each other. Considering that it will generally be the coastal States bearing the damage of oil pollution on a large scale, but may on the other hand have contributed to the occurrence or the gravity of an accident, the shipowner can in some cases raise both exceptions against a claim under the CLC. The provision is thus meant to be an expression of a certain fairness,[465] since it avoids to render the shipowner answerable for a damage that is essentially caused by the coastal State's shortcomings in maintaining its navigable waters. The wording of the provision does not require any qualified degree of negligence or premeditation by the State or public authority, mere negligence is enough to trigger the exception.[466]

However, the damage needs to be "wholly caused" by the shortcomings of the coastal State. The typical cases in which a shipowner would want to blame coastal authorities for having contributed to a maritime accident, by not reacting to a distress situation in time or not providing a place of refuge for a stricken ship,[467] therefore do not give rise to the exemption. It could be debated whether a public authority's omission to safeguard or control maritime zones with an increased piracy risk are enough to exonerate the shipowner under art. III para. 2 (b),[468] but especially in this case, the incident can hardly be considered to have been caused by the shortcomings in control only. The scope of the provision is rather restricted after all.[469]

The exception of art. III para. 2 (b) CLC refers to the "maintenance of lights or other navigational aids" and thus clearly points to obligations of the coastal State. Considering the progress in navigation supported by satellite positioning and other technical equipment on board the ship, the importance of lights and other means of guidance from the shore is perhaps declining. The provision has been invoked by shipowners with regard to the updating of maritime charts in the *Tsesis* case[470] before the Swedish courts. It concerned a Soviet tanker that ran aground in Swedish waters while being under pilotage and in the leading sector of a lighthouse. The tanker in effect hit a rock that had been discovered by the Swedish Maritime Authority as an obstacle eight years earlier but had, due to the procedure for selecting and handling information,[471] never been entered into the official charts.

465 Abecassis and Jarashow, *1985 "Oil Pollution"*, p. 206.

466 Zhu, Ng and Zhang, '2012 "Some thoughts"', 313.

467 See Rodríguez Gayan, '2003 "Claves de derecho privado"', 141 et seq. on the *Prestige* incident.

468 Zhu, Ng and Zhang, '2012 "Some thoughts"', 313.

469 de la Rue and Anderson, *2009 "Shipping and the environment"*, p. 101.

470 *"Tsesis"* 1983 Nordiske domme i sjøfartsanliggender 1 (Svea Court of Appeal, 06 June 1981).

471 See H. Tiberg, 'Oil pollution of the sea and the Swedish "Tsesis" decision' (1984), Lloyd's Maritime and Commercial Law Quarterly, 218–26 at 218 et seq. for a more detailed description of the causes of the accident.

The Swedish authorities tried to hold the shipowner responsible for the oil pollution caused, and the latter raised the defence of art. III para. 2 (b) CLC. The Swedish court accepted the motion and effectively also sentenced the Swedish State to pay damages to the shipowner.

It is, of course, debatable what kind of "navigational aids" can be considered to be covered by the provision, and what the meaning of "maintenance" as opposed to new installations or improvement should be.[472] Problems especially arise with regard to maritime charts.[473] There may in effect be considerable difficulties in determining whether a chart is correct or not.[474] With regard to the upkeep of information, it will be impossible to determine whether a government knew or should have known about a changed fact that renders the chart imprecise or wrong[475]—at the end of the day, an authority not effectuating too much survey of their waterways may be in a more favourable situation than a more careful authority. Furthermore, it will most likely be determined by the coastal State's national laws whether the government had an obligation—violation of which may give a right to remedy or not—to the maintenance of the navigational aids, so that obligations may be very different in different States,[476] unless the observance of international standards is advocated.[477]

With regard to the understanding of "navigational aids", it has been held, also by the Swedish courts in the *José Marti* case,[478] that a pilot error does not entitle the shipowner to exoneration, considering that a pilot cannot be considered a "navigational aid". Even if this outcome may be harsh to the shipowner, it can be considered right, the relationship between the pilot and the master being of a special kind (also see *supra* under E. I. 4.). Furthermore, as a human being acting on the ship and deciding immediately and independently on his actions, the pilot is clearly prone to human mistakes, in a way that may not be comparable with the administrative processes in a government authority responsible for the maintenance of lights and navigational aids.

Another difficulty may arise with regard to the reciprocal claims between the shipowner and the government authority. As long as the shipowner only raises art. III para. 2 (c) CLC as a defence against the wrongdoing government only and does not claim damages himself, the provision does not cause problems, but there can be

---

472 See P. Wetterstein, 'Complete Freedom of the Seas or Polluter Pays for Everything?', in J. Schelin (ed.), *General Trends in Maritime and Transport Law 1929-2009* (2009), p. 116, footnote 266, hinting at this problem.

473 See the abundant discussion in Tiberg, '1984 "Oil pollution"', 220 et seq. on whether the term should cover maritime charts.

474 K. Pineus, 'L'échouement du "Tsesis"' (1983) 35, Droit Maritime Français, 248–9 at 249.

475 de la Rue and Anderson, *2009 "Shipping and the environment"*, p. 103.

476 Tiberg, '1984 "Oil pollution"', 224.

477 Suggested by de la Rue and Anderson, *2009 "Shipping and the environment"*, p. 103, but considered "impracticable" by Tiberg, '1984 "Oil pollution"', 224.

478 *"José Martí"* 1987 Nordiske domme i sjøfartsanliggender 64 (Svea Court of Appeal, 26 November 1986).

difficulties with regard to third party claims and the claims of the shipowner for the losses he incurred. Certainly with a view to not leaving third parties without compensation, it can been argued that the shipowner should only be exonerated if the coastal State's domestic law provided for a remedy by which compensation could be obtained from the wrongdoing State instead. However, the wording of the provision does not imply a condition of this kind. If the exoneration possibility is meant to be fair to shipowners, it must be effective in every set of circumstances, notwithstanding national law provisions. It must however be mentioned that since the coming into force of the FUND Convention, compensation for damages that have been caused by the wrongful act of a public authority is available to victims under the FUND.

The CLC does not set out an obligation for coastal States to compensate the damage caused to any other party. Considerations of this kind are left to domestic law. Whereas it seems possible for claimants, both shipowners and third parties, to obtain compensation from the public authorities in some jurisdictions, this may not be the case in others where public authorities or other State entities categorically refuse to grant compensation. The CLC is, however, an instrument that is only concerned with the strict liability regime imposed on the shipowner. It can hardly set out a default responsibility regime of States to overcome these injustices.

V. The Victim's Contributory Fault

As previously mentioned, the exception of art. III para. 3 CLC mostly serves shipowners to defend themselves against government claims for pollution damages that have partly been caused by negligence of public authorities. It is hardly conceivable that any other victim than a State entity, especially a private entity, could have contributed to the pollution damage and then suffered from it.

Art. III para. 3 CLC is more generous towards shipowners in that it allows exoneration for damages that have only partially been caused by the victim's contributory fault, and that the "contributory fault" can be of any kind. The provision may thus be often invoked by shipowners, whose vessels have been denied assistance or refuge in a port, but the result is uncertain—firstly, art. III para. 3 CLC does not order the exoneration, but only states that it "may" be envisaged,[479] and secondly, there may only be exoneration in part. Thirdly, there are some difficulties in establishing causality, depending on the link that is required by the jurisdiction in question, as it may be argued that the omission to provide assistance only occurred when the ship was already experiencing difficulties.[480] It does not seem that the CLC very strongly encourages victim States to assume

[479] Abecassis and Jarashow, *1985 "Oil Pollution"*, para. 10-41.

[480] S. Hetherington, '"Prestige" - Can the Law Assist?' (2003), CMI Yearbook, 361–74 at 368 et seq.

responsibility for their contributory fault, considering that the wording of the provision does not oppose a blank refusal of exoneration. Furthermore, the burden of proof for the contributory fault of the other party is still on the shipowner,[481] and the corresponding information may be hard to get from public authorities.

The provision is, like art. III para. 2 (c) CLC, only invoked as a defence. It can in this context be of use in situations in which art. III para. 2 (c) CLC does not apply, arguably because the State did not provide for "navigational aids" or markings of underwater obstacles at all, as opposed to omissions in maintenance,[482] for collisions with State-owned vessels[483] or for inappropriate action in fighting the pollution damage,[484] and possibly also for serious mistakes of pilots in mandatory pilotage.[485]

The provision obviously does not give rise to state responsibility for negligent behaviour of State representatives, but it may be a gateway to make domestic courts consider the behaviour of States before, during and after a maritime incident involving their coasts.

G. Outside the CLC Provisions

In its original version, the CLC defined "pollution damage", in its art. I para. 6 as "loss or damage caused outside the ship carrying oil by contamination resulting from the escape or discharge of oil from the ship, wherever such escape or discharge may occur, and includes the costs of preventive measures and further loss or damage caused by preventive measures".

On a first glance, the provision is almost all-encompassing. By referring to "loss" it also comprises subsequent losses such as economic losses by the fishing or tourism industry that are only caused by the damage already done to the environment, as opposed to the direct damage of clean-up costs and damaged property. During the negotiations to the CLC, the notion of "damage" was consciously worded in an open way, leaving questions of the admissibility of merely ecological damage aside, i.e. the mere fact that the environment is impaired without having regard to financial consequences of it.[486]

The provision in its original form however soon gave rise to discussions about whether purely economic damage should be covered, especially after attempts of certain States to put forward claims for the mere impairment of the environment in

[481] Robert, *2003 "L'"Erika""*, p. 34.

[482] Abecassis and Jarashow, *1985 "Oil Pollution"*, p. 206.

[483] de la Rue and Anderson, *2009 "Shipping and the environment"*, p. 105.

[484] Ibid., p. 105, referring to the *Patmos* incident in which the Messina Port Authority was held partly liable for having used dispersants in an improper way.

[485] Ibid., p. 104.

[486] Kappet, *2006 "Tankerunfälle"*, p. 69.

the following years.[487] The issue was clarified in the 1992 Protocol to the CLC, retaining the original definition of "pollution damage" only to some extent. The provision now reads:

> Pollution damage means
> (a) loss or damage caused outside the ship carrying oil by contamination resulting from the escape or discharge of oil from the ship, wherever such escape or discharge may occur, provided that compensation for impairment of the environment other than loss of profit from such impairment shall be limited to costs of reasonable measures of reinstatement actually undertaken or to be undertaken;
> (b) the costs of preventive measures and further loss or damage caused by preventive measures.

The paragraph thus now gives some guidance to the problem and allows compensation for the impairment of the environment, as long as efforts to restoration are actually carried out and can be considered as "reasonable".[488] Still, the provision is far from being unambiguous. Problems arise with regard to the question of what measures are "reasonable" within the meaning of art. I para. 6 CLC, especially with regard to efforts undertaken by governments.[489] Furthermore, numerous problems arise with the compensation of economic loss. It is clear that oil pollution damage in a given coastal region can have an impact on secondary, tertiary and even more remote business, and it is in effect very difficult to draw a line and decide which damage should still be covered. The problem of striking a balance between claims that are "oil pollution claims" and others that are considered too remote has already been mentioned *supra* under B. I. 2. when discussing whether the Fund's criteria for admitting a claim could be considered binding on the competent domestic court. Even if the domestic courts have generally accepted these criteria, without however declaring themselves bound by them, the question of admissibility of claims is still far from certain. The fact that claimants have, over the years, become more and more confident in presenting borderline claims[490] has

---

[487] See the outline in M. Mason, 'Civil liability for oil pollution damage: examining the evolving scope for environmental compensation in the international regime' (2003) 27, Marine Policy, 1–12 at 3 et seq.

[488] W. Oosterveen, 'Some recent developments regarding liability for damage resulting from oil pollution - from the perspective of an EU Member State' (2004) 6, Environmental Law Review, 223–39 at 231.

[489] See the very comprehensive outline of this problem in de la Rue and Anderson, *2009 "Shipping and the environment"*, pp. 375 et seqq.; also see Rodríguez Gayan, '2003 "Claves de derecho privado"', 135, footnote 75; M. Jacobsson, 'How clean is clean? The concept of "reasonableness" in the response to tanker oil spills', in G. Berlingieri, A. Boglione, S. M. Carbone and F. Siccardi (eds.), *Scritti in onore di Francesco Berlingieri. Volume I*, (Genova: Il Diritto Marittimo, 2010), pp. 570 et seqq.; and, from a French point of view, M. Ndendé, 'La politique d'indemnisation des victimes des marées noires devant le FIPOL' (2008), Revue de Droit des Transports, 27–35 at 28 et seq.

[490] Renger, *1998 "Recht und Praxis"*, p. 165; Gaskell and Forrest, '2008 "Marine Pollution Damage"', 116; also see the historic outline in de la Rue and Anderson, *2009 "Shipping and the environment"*, pp. 438 et seqq.

contributed to the problem. On the other hand, claimants, apart from materially establishing the link between the contamination and their loss of income, may face problems of proving the likelihood of the expected earnings, especially for very small businesses that operate on a seasonal basis.[491]

I. Compensation for Oil Pollution Claims Outside the CLC

Depending on the rules in the domestic legal system that will decide on the claim, plaintiffs may therefore be tempted to bring actions outside the CLC and FUND system, under domestic law, by submitting that they do not constitute claims for "pollution damage", but for damage of a different kind, that is only caused by the intermediate of the pollution as such. The question of whether such a claim for "pure economic loss" is admissible in the tort law of a given legal system and to which degree compensation can be obtained is one of the most complicated questions of this field of law. It is, even in civil law systems, mostly a matter of case-by-case considerations,[492] and a thorough comparative analysis is outside the scope of this analysis. This part of the study only considers whether such a domestic tort claim, assuming that it is possible, would be reconcilable with the provisions of the CLC.

The possibility of bringing claims for damages that have been remotely caused by an oil spill, but are not covered by the CLC, has been advocated in French doctrine very often.[493] This is probably due to the wording of the CLC in French, referring to its applicability to "*dommage par pollution*", which bears a connotation of that the damage has to be caused by the pollution, and not by more remote secondary effects of it. This is not reflected in the English term of "pollution damage". However, the existence of claims outside the CLC and FUND regime is also considered and asserted elsewhere.[494] It is thus often submitted that the damage done to the image or the trademark of a region as a holiday destination or a particular natural reserve could not be compensated under the FUND Convention.[495] In effect, the only claims that have so far been admitted under the Convention are the costs of particular campaigns to overcome the negative image caused by

---

491 Ndendé, '2007 "L'accident de l'Erika"', 17.

492 An outline on this problem under English and US law can be found in de la Rue and Anderson, *2009 "Shipping and the environment"*, pp. 417 et seqq., also drawing a comparison to civil law systems.

493 Bonassies, '2000 "Après l'Erika"', 145; Delebecque, '2010 "Aspects civils"', 881; P. Bonassies and C. Scapel, *Droit maritime*, 2nd edn. (Paris: Librairie Générale de Droit, 2010), p. 329.

494 Anderson and de la Rue, '2001 "Liability of Charterers"', 8; P. Griggs, *International Convention on Civil Liability for Bunker Oil Pollution Damage, 2001*, p. 3, available from http://www.bmla.org.uk/; Requejo Isidro, *2007 "Contaminación por hidrocarburos"*, p. 477; de la Rue and Anderson, *2009 "Shipping and the environment"*, p. 112.

495 Ndendé, '2007 "L'accident de l'Erika"', 17.

a given oil spill,[496] addressed to targeted markets,[497] that were started after and because of the incident. In contrast to that, the immaterial damage done to an already established trademark has not been recognised, and the economic value may be hard to quantify. Where a legal system allows immaterial claims of this kind, claimants may think it worth their while to bring actions. Depending on the entity that established or owns the trademark, the claim can benefit municipalities or other communities that could, unlike the hotel owners or fishers, not put forward a claim for economic losses,[498] because they are not carrying out an economic business themselves.

In the *Erika* proceedings, the *Tribunal Correctionnel* distinguished very clearly between immaterial and material damages.[499] Compensation for material damages such as the cost of the clean-up, losses of local businesses and the cost of restoration of the environment, such as the treatment of oiled birds was granted to an extent that appears to be in line with the policy of compensation by the Fund. With regard to the "immaterial damage" (*préjudice moral*), the affected municipalities effectively obtained between 100,000 and 300,000 €, some up to 1.5 million €[500] as a compensation for the damage done to their trademark and corporate identity as a holiday destination, natural and agricultural region. The amounts were granted on the basis of objective criteria, such as the extent to which the coastline was polluted, the character of the coast as a natural landscape as opposed to heavily industrialised zones, and the gravity of the pollution and its impact on the region.[501] In addition to the municipalities, certain associations for the protection of the environment were granted compensation for their immaterial damage, consisting of the damage caused to the values and resources they were meant to protect.[502]

However, the *Tribunal Correctionnel's* decision did not concern the registered owner, but the beneficial owner, the director of the nautical management company, the nominal charterer's mother company and the classification society, who did not benefit from the channelling provisions (see *supra* under E. I.) and therefore found themselves outside the CLC regime anyway. Strictly speaking, the *Tribunal Correctionnel's* decision was rendered under French domestic law exclusively, once it was decided that the defendants could not invoke the CLC.

On appeal, the *Cour d'Appel* overturned the decision with regard to Total, the charterer's mother company, holding that it was the "charterer" in the sense of the

---

[496] Renger, *1998 "Recht und Praxis"*, p. 167; de la Rue and Anderson, *2009 "Shipping and the environment"*, p. 467. Also see Jacobsson, *2010 "How clean is clean"*, p. 574 for examples.

[497] Ibid., p. 574.

[498] de Raulin, '1993 "L'épopée judiciaire"', 69; Grunvald and Saas, '2009 "Regards croisés"', 393.

[499] L. Neyret, 'Naufrage de l'"Erika": vers un droit commun de la réparation des atteintes à l'environnement' (2008), Recueil Dalloz Chronique, 2681–9 at 2683.

[500] *Tribunal correctionnel de Paris, 16.01.2008, "Erika"*, pp. 738 et seq., the amounts having been confirmed in the second instance, *Cour d'Appel de Paris, 30.03.2010, "Erika"*, pp. 875 et seq.

[501] Neyret, '2008 "Naufrage de l'"Erika""', 2683 et seq.; M. Rémond-Gouilloud, 'Sur le préjudice écologique' (2012) 64, Droit Maritime Français, 1020–2 at 1021.

[502] *Tribunal correctionnel de Paris, 16.01.2008, "Erika"*, pp. 731 et seq.

CLC and therefore protected by the channelling provisions, and consequentially declared that Total was not liable at all. With regard to Mr. Savarese, the *Cour d'Appel* however decided that he was the registered owner's employee and as such protected by art. III para. 4 (a) CLC, but could not invoke the benefit of the channelling provisions due to his reckless behaviour. The *Cour de Cassation* undertook the same reasoning regarding the charterer and the classification society, holding them liable as beneficiaries of the channelling system who lost the benefit due to their conduct.

It has been set out *supra* under E. I. that the French courts considered the claims in the *Erika* proceedings as claims under domestic law, the legal basis of which was found outside the CLC in French domestic law. The argumentation of the courts, namely the *Cour d'Appel*, to discharge Total from all claims against them, including the compensation of the *préjudice moral*, thus clearly means that the court decided the channelling of art. III para. 4 CLC would be absolute in covering all claims,[503] regardless of their basis or of whether they could have been compensated under the CLC. This interpretation is of course in line with the wording of art. III para. 4,[504] but it is however remarkable as the possibility to obtain compensation for immaterial damage to the environment is well established in French law, the *Erika* case being by no means the first or the only case to recognise that head of damage.[505]

The French courts in the *Erika* case thus recognised the possible parallel application of domestic law and the CLC, but at the same time stated that the CLC opposed that parallelism to the extent that its channelling provisions apply.

In the same context, the claimants in the *Landcatch* case in the Scottish Court of Session, bringing a claim against the Fund for compensation of a loss that was considered too remote, had argued that their claim for damages was not a claim under the legislation implementing the CLC, but was a different claim under common law that ought to be admissible. L. J. Clerk however rejected this argument, holding that the definition of "damage" under the CLC and the domestic law were the same.[506] The reasoning implies that the judge considers that the inadmissibility of the claim under common law is due to the rules of common law,[507] rather than due to an effect of the CLC legislation that would bar claims outside the CLC.

From what is set out in legal literature and in limited case law available, there is a certain tendency to recognise the possibility of claims outside the CLC system, as long as the proceedings do not interfere with the provisions of the Convention. Hence, as soon as there is a possibility to claim against a party not protected by the channelling provisions, this party not only faces a liability which is not limited by

[503] Kopela, '2011 "Civil and Criminal Liability"', 319.

[504] Meilán Gil, *2005 "Problemas Jurídico-Administrativos"*, p. 171.

[505] See Rémond-Gouilloud, '2012 "Sur le préjudice écologique"', 1020.

[506] *Court of Session (Scotland), 19.05.1999, "Landcatch v. IOPC"*, p. 1218.

[507] Gaskell, *2000 "Pollution, Limitation and Carriage"*, p. 78.

the CLC provisions, and which, additionally, could also be invoked for damages that are not compensated under the CLC.[508]

II. Limitation of Liability for Claims Outside the CLC

If it is considered that there can be claims for pollution damage outside the CLC, this begs, in a next step, the question whether these claims are "maritime claims" under the LLMC, to the extent that the defendant is accorded a limitation of liability under that convention. The question was not considered in the *Erika* proceedings, but with a view to the fact that the court there established that the defendants acted "recklessly and with knowledge that the damage would probably occur", limitation of liability under the LLMC would have been denied by virtue of art. IV LLMC in any case.

1. The Interplay Between CLC and LLMC

The admission of claims against the shipowner for damages other than direct pollution damage, but which actually based on an accident again raises the problem of coherence between the LLMC and the CLC. According to art. III (b) LLMC, "claims for oil pollution damage within the meaning of the International Convention on Civil Liability for Oil Pollution Damage, dated 29 November 1969 or of any amendment or Protocol thereto which is in force" are excluded from the scope of applicability of the LLMC. Hence, claims for oil pollution will be governed by the more specialised provisions of the CLC rather than the general provisions of the LLMC and the two conventions do not interfere with each other. However, any claim for oil pollution outside the scope of the CLC is not covered by art. III (b) LLMC and therefore again subject to the general rule of art. 2 para. 1 (c) LLMC, rendering it a claim subject to the general limitation of liability.

The problem of coherence between the LLMC and the CLC has already been debated with regard to the liability of other parties than shipowners (see *supra* under E. I.) who cannot limit their liability under the CLC and may therefore find themselves outside the CLC regime. It is generally understood that these claims, notwithstanding the inapplicability of the CLC limitation of liability regime, are still claims under the CLC and therefore not subject to the LLMC, unless there is domestic legislation that solves the problem of potentially unlimited liability for these claims.[509]

---

[508] Neyret, '2010 "L'affaire "Erika""", 2242.

[509] Griggs, Williams and Farr, *2005 "Limitation of Liability"*, pp. 27 et seq. describing this as the general understanding of the provision and referring to English domestic legislation solving the problem.

It is however debatable whether this argumentation also holds true for claims for damages other than oil pollution that are not covered by the CLC at all. It has been argued that these claims would not trigger the exclusion clause of art. III (b) LLMC,[510] presumably because they are simply not covered by the wording, not being "claims for oil pollution damage within the meaning" of the CLC.

Claimants could therefore, even if they obtain compensation for a claim outside the CLC regime, be faced with another limitation of liability regime if the defendant can invoke the LLMC, it being repeated that the LLMC limits are generally lower than the CLC limits. This fact certainly adds to the uncertainties of an action regarding damage other than oil pollution.

2. Oil Pollution Claims Under the LLMC

The LLMC is, as a general convention for the limitation of maritime claims, not concerned with the particularities of pollution claims. As will be seen *infra* in Chap. 4, A. IV. 1., the question whether bunker oil pollution claims would be covered by the provisions of the LLMC has given rise to considerable debate. It however seems that the problems encountered there do not apply to the oil pollution claims outside the CLC that are discussed here. The application of the LLMC to cargo oil pollution claims has been rather unproblematic in the few cases in which this has been envisaged.

Art. II para. 1 of the LLMC reads:

> Art. II
> Claims subject to limitation
> 1. Subject to Articles III and IV the following claims, whatever the basis of liability may be, shall be subject to the limitation of liability:
> (a) claims in respect of loss or life or injury and loss or damage to property (including damage to harbour works, basins and waterways and aids to navigation), occurring on board or in direct connection with the operation of the ship or with salvage operations, and consequential loss resulting therefrom;
> (b) claims in respect of loss resulting from delay in the carriage by sea of cargo, passengers or their luggage;
> (c) claims in respect of other loss resulting from infringement of rights other than contractual rights, occurring in direct connection with the operation of the ship or salvage operations;
> (d) claims in respect of the raising, removal, destruction or the rendering harmless of a ship which is sunk, wrecked, stranded or abandoned, including anything that is or has been on board such ship;
> (e) claims in respect of the removal, destruction or the rendering harmless of the cargo of the ship;
> (f) claims of a person other than the person liable in respect of measures taken in order to avert or minimize loss for which the person liable may limit his liability in accordance with this Convention, and further loss caused by such measures.

[510] Ibid., p. 28.

In summary, the LLMC applies to claims arising on board or out of the "operation" of the ship. The provisions that could possibly apply are para. 1 (a) concerning the damage to property, para. 1 (c) concerning infringements of other rights and para. 1 (f) concerning damage arising out of preventive measures. This means that claims for remote damages would be covered by para. 1 (a), the term of "consequential loss" being wide enough to cover losses in the secondary or tertiary sector. Considering the claims for damage done to the environment, such as the damage done to a region's trademark or reputation would, as infringements of immaterial rights, most likely be covered by para. 1 (c).

The typical damage of the cost of cleaning beaches and maritime zones near the coast will hardly ever occur as a claim outside the CLC provisions. Still, before the coming into force of the CLC, the *Rechtbank Rotterdam* decided in its decision on the *Torrey Canyon* accident that the shipowner could invoke the limitation of liability for UK's and France's claims for the cost of cleaning up the oil that was spilled and burned in the accident.[511] The judgment was rendered under the 1957 Limitation Convention, the predecessor of the LLMC, but very similar to the LLMC, since it provided for limitation of liability for "loss of, or damage to, any other property".

In the *Aegean Sea* proceedings in England, the Commercial Court considered very thoroughly whether a claim for oil pollution damage that was not submitted to the CLC could however be limited under the LLMC. The question arose with regard to the shipowner's recourse claim against the charterers, for, *inter alia*, the liabilities he incurred under the CLC. The court's findings were *obiter dicta* as it decided the LLMC could not apply in the relationship between the shipowner and the charterer at all (see *supra* under E. III.). However, Thomas J. clearly stated that the claims in respect of property polluted by oil would be covered by art. II para. 1 (a) LLMC, including the claims for the cleanup, which are, in his view, claims for consequential loss.[512] It does not seem that he drew a distinction between damage done to private property and the damage done to the coastline and natural resources.

In the cases discussed, the pollution claims for oil pollution outside the CLC were still considered as "maritime claims" under the LLMC. It therefore seems that there is a certain likelihood that claims for damages caused by oil pollution outside the CLC will however be limited by the lower limits of the LLMC, to the extent that the defendant can invoke the Convention's provisions to his benefit. A claim for oil pollution damage outside the CLC against parties typically protected by the LLMC, such as the shipowner or the charterer of the vessel would therefore only appear promising where there is also a chance of breaking the liability limits of the LLMC.

---

[511] *"Torrey Canyon"* 1969 Nederlandse Jurisprudentie 1333 (Rechtbank Rotterdam, 08 April 1969).

[512] *High Court (England), Queen's Bench Division, 07.04.1998, "The Aegean Sea"*, p. 52.

H. The Role of Criminal and Administrative Law

Whereas it is fairly clear that the channelling provision of art. III para. 4 CLC has a direct influence on civil cases by blocking claims outside the CLC regime, the impact it has on other claims is less obvious. Oil spills may, on the one hand, give rise to proceedings in criminal law against the responsible persons. Furthermore, and especially in the cases in which the entity incurring the damage is a municipality or some other representative of a State, the person responsible for the damage can be subject to obligations under administrative law in some States. The cases of the *Prestige* and the *Aegean Sea* highlighted the difficulties in the interplay of criminal law with the international civil liability regime, whereas the *Erika* oil spill can be quoted as an example for both the impact of criminal and administrative law claims.

I. Criminal Law

Damage done to the environment can constitute a crime in many legal systems in the world, and maritime accidents can thus often be subject to proceedings in both criminal and civil courts.[513] However, criminal law normally only gives rise to criminal sanctions against natural persons and the conviction to a fine is usually not understood as a means of compensation to victims and not paid to them, but to the State.[514] Unlike civil law, criminal procedure generally focuses on the accused's conduct rather than on the damage, declaring that an act or omission was violating the applicable standards of behaviour.[515] The important question of coastal and flag States' criminal jurisdiction for the safety at sea and the protection of the maritime environment,[516] especially with regard to the reduced degree of sovereignty in the EEZ and the lack of sovereignty on the high seas, is outside the scope of this study.

Still, certain legal systems feature mechanisms to incorporate civil proceedings into criminal proceedings, in a spirit of establishing criminal responsibility and civil liability at the same time. There are also States in which a person held criminally liable for an event will automatically incur civil responsibility for the consequences as well.[517] Victims can thus obtain compensation during or in the immediate

[513] Jacobsson, *2007 "Uniform Application"*, p. 428.

[514] See P. K. Mukherjee, 'The Penal Law of Ship-Source Marine Pollution: Selected Issues in Perspective', in T. M. Ndiaye and R. Wolfrum (eds.), *Law of the sea, environmental law and settlement of disputes* (Leiden, Boston: M. Nijhoff, 2007), pp. 480 et seqq. on the characteristics of criminal law in the field of marine pollution.

[515] Robert, *2003 "L'"Erika""*, p. 132.

[516] See Grunvald and Saas, '2009 "Regards croisés"', 381, treating the question from a French point of view.

[517] See Castillo Daudí, '2001 "El siniestro del buque "Aegean Sea""', 274 et seq. referring to the Spanish rules on criminal procedure; Pulido, *2007 "Compensation"*, p. 154; Jacobsson, *2007 "Uniform Application"*, p. 435.

aftermath of the criminal proceedings. It is therefore interesting to consider whether the channelling provisions, and, possibly, the limits of liability set out by the CLC, can be defeated or influenced in criminal proceedings of this kind.

The relationship between the civil liability regime and a domestic criminal law regime was certainly put to the test in the *Erika* proceedings. The limited liability fund available under the CLC 1992 was, as in many other cases, entirely insufficient to cover this amount, and even the means available under the FUND Convention as they were set at that time, seemed only partly adequate.[518] As has already been shown, the victims were particularly interested in the responsibility of certain actors who claimed to be protected by the CLC's channelling provisions. Hence the victims' interest to declare themselves *partie civile* in the ensuing criminal proceedings against—on the one hand—the *Erika's* owner, the ship management company, the chartering oil company, the Italian classification society and—on the other hand—the master and several employees of the shipping companies.

As French law allows both criminal proceedings against legal persons and the merger of civil proceedings on behalf of the victims with the corresponding criminal proceedings, the forum was in this respect advantageous to the French claimants,[519] even though a first decision was only rendered by the *Tribunal Correctionnel* of Paris in 2008, more than eight years after the incident. A thorough analysis of the legal bases in French criminal law[520] cannot be given in this study, but it must be stressed that the *Erika* decisions were first and foremost criminal cases, and that the courts in all instances carefully distinguished between the criminal proceedings and the question of civil responsibility that was decided in a second step. In the adhesive procedure however, the courts did not act differently from what the civil courts would have decided, considering the applicability of the CLC and deciding the claims against the defendants according to domestic law (see *supra* under G.).

Still, this may not be the case in other countries. There have been cases in which masters have been declared personally liable for oil pollution damages, after having been held criminally liable for the same event.[521]

Reference can be made here, *inter alia*, to Spain. There have been at least two cases in the Spanish courts in which civil liability claims were considered in criminal proceedings. The first is the case of the *Aegan Sea*, the Greek oil tanker that ran aground at night and in bad weather off La Coruña in North-West Spain. The fault for the accident was considered to be both with the master, who was held to have been careless in attempting the manoeuvre with regard to the weather

---

[518] Bonassies, '2000 "Après l'Erika"', 143; Anderson and de la Rue, '2001 "Liability of Charterers"', 4; Ndendé, '2007 "L'accident de l'Erika"', 17.

[519] Furthermore, French procedural law allows, in cases for environmental pollution, even claims for "indirect", "non-pecuniary loss", see D. Papadopoulou, 'The Role of French Environmental Associations in Civil Liability for Environmental Harm: Courtesy of "Erika"' (2009) 21, Journal of Environmental Law, 87–112 at 100 for more details.

[520] See Grunvald and Saas, '2009 "Regards croisés"', 382 et seq. for an overview.

[521] Jacobsson, *2007 "Uniform Application"*, p. 435.

conditions, and with the state-employed pilot.[522] Spain, at that time, was a member to both the 1969 CLC and the 1971 FUND Convention. Still, the criminal court of La Coruña, following the respective provisions under the Criminal Procedure Act, also decided on civil claims and held that civil liability should be apportioned between the master and the pilot,[523] apparently disregarding the position of the master as an employee, or agent, of the owner that would bar the claim for civil liability according to art. III para. 4 CLC. Furthermore, by stating that a domestic Spanish rule on the limitation of liability of professionals did not apply in this case, the court of La Coruña indirectly stated that the master and the pilot would be liable without limitation, whilst, somehow incoherently also pronouncing that the shipowner's insurer and the FUND were civilly liable according to the limits of the CLC.[524] The court, in accordance with Spanish domestic law, also declared the shipowner and the Spanish state subsidiarily liable, as being respectively answerable for the actions of the master and the pilot as their representatives. In Spanish law, civil liability for torts can be entirely independent from civil or commercial law, resulting in a non-application of the CLC.[525] On appeal, there were no substantial changes of the decision by the tribunal in the next instance, neither concerning the criminal judgment, nor the civil responsibility of the parties in question.[526] Even if the responsibility of the master and the pilot were in fact not invoked,[527] the decision is at odds with the principles of the CLC regime.[528] The mere existence of these judgments may be detrimental to the Convention's uniform application, there being a certain risk that it might be considered as a precedent in other cases.

A similar situation arose in the criminal proceedings which were caused by the loss of the *Prestige* some 130 miles off the coast of Galicia. The claims for the accident as calculated and requested by the Spanish government exceeded the mark of 1 billion US$.[529] Here again, victims of the oil pollution damage took the

---

[522] *Juzgado de lo Penal número 2 de La Coruña*, 30.04.1996, unreported. The decision is presented by A. Crespo Hernández, 'Nota a la Sentencia del Juzgado de lo Penal núm. 2 de La Coruña de 30 abril de 1996' (1996) 47, Revista Española de Derecho Internacional, 372–5; Castillo Daudí, '2001 "El siniestro del buque "Aegean Sea""'; Pulido, *2007 "Compensation"*, p. 154.

[523] de la Rue and Anderson, *2009 "Shipping and the environment"*, p. 105, see also IOPC document 71FUND/A.23/14/1, para. 8.3., available from the IOPC's Document Service, http://documentservices.iopcfunds.org/meeting-documents/search-results/titles-only/1/start-year/2000/end-year/2000/sort/newest/fund-session/session-9/agenda_item/3/ (last visited 8 March 2014).

[524] Castillo Daudí, '2001 "El siniestro del buque "Aegean Sea""', 275.

[525] Rodríguez Gayan, '2003 "Claves de derecho privado"', 130, footnote 54.

[526] Judgment of the *Audiencia Provincial de La Coruña* of 18th of June 1997; see Castillo Daudí, '2001 "El siniestro del buque "Aegean Sea""', 282 et seqq. for the details.

[527] See IOPC document 71FUND/A.23/14/1, para. 8.4 on the ensuing distribution of liabilities between the Spanish State, the IOPC and the shipowner's insurer.

[528] Crespo Hernández, '1996 "Nota Juzgado de lo Penal de La Coruña 30.04.1996"', 374 et seq.; Castillo Daudí, '1998 "La responsabilidad civil"', 82; Gaskell, *2000 "Pollution, Limitation and Carriage"*, p. 73; Meilán Gil, *2005 "Problemas Jurídico-Administrativos"*, p. 171; Requejo Isidro, *2007 "Contaminación por hidrocarburos"*, p. 493, footnote 31.

[529] Michinel, '2007 "The "Prestige""', 173.

opportunity of having their civil claims adjudicated in the penal proceedings, and the criminal court of Concurbión confirmed both the criminal and the civil responsibility of, *inter alia*, the *Prestige's* master, her first officer, her chief engineer and her owner.[530] The representatives of the Spanish authorities, who had taken the decision to refuse the vessel access to ports in Spain and ordered the *Prestige* out to sea were, on the other hand, acquitted. Even if the judgment is a decision in criminal law only, it was perceived that the decision could prepare a civil liability trial.[531] Spanish law actually allows to undertake civil action not only at the same time with the criminal trial, but also at a later point in time, once the decision in the criminal courts is made.[532] Not least, this decision and the ensuing criminal proceedings in Spain have been heavily criticised for being unfair towards the accused and using them as scapegoats for an accident that had multiple causes after all. So far, the criminal charges have mostly been abandoned, the criminal court holding in late 2013 that the only charge to be retained was the captain's failure to obey authorities, all other persons accused having been entirely acquitted. It must however be kept in mind that to the extent in which civil proceedings remain possible against persons found guilty under criminal law, the criminal proceedings already flawed by allegations of arbitrary action could have had even more severe consequences.

Apart from France and Spain, other civil law legal systems, such as Germany,[533] know the adhesive procedure, the concept of merging civil proceedings for compensation to the victims with the criminal proceedings for the damaging event. In many common law countries, including England, criminal courts can order restitution to the victims in the criminal proceedings, and the court-ordered restitution is then part of the sentence. It is perhaps worth considering that the risk of decisions that are severe on the defendants may be higher in criminal law courts, taking a different approach to questions of fault and responsibility and being naturally reluctant to pronounce an exemption or exoneration in civil law for an accused that has been found guilty in criminal law. In view of that, criminal law courts may not be the right forum for the application of the CLC's rules.[534]

This risk will be more real in jurisdictions in which legal persons can be held criminally liable, or where restitution obligations incurred in a business environment must be paid by the accused's employer.[535] Whereas the cases of masters found civilly liable seem alarming, it does however not seem that victims of oil

[530] The judgment is presented by I. Arroyo Martínez, 'El caso "Prestige". Consideraciones sobre el auto de 18 de Marzo de 2009 del Juzgado de Concurbión, A Coruña (España)' (2009) 26, Anuario de Derecho Maritimo, 205–16.

[531] Ibid., p. 213.

[532] Michinel, '2007 "The "Prestige""', 173.

[533] *Adhäsionsverfahren*, §§ 403ff StPO, though of limited importance in practice.

[534] Castillo Daudí, '1998 "La responsabilidad civil"', 88.

[535] de la Rue and Anderson, *2009 "Shipping and the environment"*, p. 1106, also referring to the risk that criminal liability may be found with a clear view of facilitating subsequent civil proceedings to victims.

pollution will, as a rule, try to sue natural persons if the liability limits set out by the CLC and the FUND Convention are inadequate to cover the damage.

Given the procedural differences between criminal and civil proceedings, the question whether the adhesive procedure is a benefit to victims depends very much on the set of facts in a given case, and, of course, the procedural rules of the forum.[536] Whereas it is, generally, the public prosecution that will investigate the case and probably save the this effort to the victims who can then easily present the facts of the case to support their claim,[537] the threshold to find the accused guilty may be more difficult to reach than the threshold to establish civil liability.[538] Still, the start of criminal proceedings and a possible adhesive procedure may in some cases and in some fora be a very practicable way to obtain compensation for pollution damage.

II. Administrative Law

Victims of oil pollution may take recourse to other mechanisms to obtain compensation for the loss they incurred. Concerning the oil spill caused by the *Erika*, the Breton municipality of Mesquer tried to be refunded for its efforts in cleaning up the spilled oil by invoking a French administrative law, the *loi du 15 juillet 1975*. Its provisions oblige producers of waste, if it is likely to cause harm to the environment, to dispose of it or to assure that it will be disposed of. The municipality argued that the fuel oil transported by the *Erika* amounted, at least after the spill that rendered it unfit for further use, to waste, and that therefore Total as the entity responsible for this "waste" was liable to pay for the disposal. The law in question is a so-called *loi de police*, establishing liability of the polluter as against the State and allowing a public authority to enforce this obligation or to undertake it itself and then charge the polluter. Furthermore, the municipality of Mesquer also invoked the respective French tort law by which persons are liable for damage caused by a thing under their control (art. 1384 *Code civil*).

The case was in the first instance decided by the *Tribunal de Commerce* of Saint-Nazaire in Brittany.[539] The decision largely dealt with the notion of "waste" in the sense of the law in question, and whether the fuel oil spilt by the *Erika* can be considered to be waste, both before the accident and after the dispersal in the water, and came to the final result that it was not.[540] Considering the domestic French law

---

[536] See Ndendé, '2007 "L'accident de l'Erika"', 14, 16 on the possible disadvantages of the adhesive procedure to victims under French procedural law.

[537] Ibid., p. 12.

[538] Delebecque, '2000 "Responsabilité et indemnisation"', 125; P. Bonassies, 'Note TGI Paris 16.01.2008' (2008) 60, Droit Maritime Français (Hors série n° 12), 19–29 at 24, referring to the benefit of doubt in criminal law.

[539] *"Erika"* (2001) 53 Droit Maritime Français 259 (Tribunal de Commerce de Saint-Nazaire, 16 December 2000).

[540] Ibid., pp. 269 et seqq.

invoked by the claimant municipality, the *Tribunal de Commerce* dismissed the claim, because Total was the cargo owner only and did not exercise control over the fuel oil. The court also referred to the CLC and to the fact that strict liability was incurred by the shipowner only.[541] The question of whether Total benefitted from the provision of art. III para. 4 CLC, either in its capacity as "charterer" in a wide sense or in any other capacity, was not directly addressed by the court. As has been pointed out,[542] the court probably omitted this discussion in order not to decide on the applicability of the 1992 Protocol, to which France was already a member in 1999, but Malta, the flag State, was not.

On appeal, the *Cour d'Appel* of Rennes confirmed the decision regarding the administrative law provisions, and the municipality of Mesquer took the case to the highest French court, the *Cour de Cassation*. Considering that the French administrative law in question was based on the European directive 75/442 as amended, the *Cour de Cassation* suspended the procedure and addressed the question of how the notion of "waste" in the directive should be understood to the European Court of Justice (ECJ).

It should be emphasised that the European directive, even if maritime pollution was not excluded from its field of application, had its roots in terrestrial pollution law and was heavily inspired by the polluter-pays-principle, which is difficult to reconcile with the strict liability of the shipowner under the CLC.[543]

In its decision,[544] the ECJ considered the relationship between the directive and the CLC, to which almost all EU member States are parties. However, the court's reasoning centred on the interpretation of the notions of "waste" and "discard" as employed in the EU directive and came to the conclusion that fuel oil accidentally spilled in a maritime accident can be seen as "waste" and that in circumstances such as the *Erika* accident, the seller of the oil or the charterer of the ship could be considered as "producer" or "holder" of the waste and thus liable for the full clean-up costs, provided that they had contributed to the risk if the accident taking place.[545] The ECJ emphasised that it was not bound by the provisions of the CLC, the EU itself not being a party to the Convention, and it considered that EU law obliged the member States to ensure that spilled oil, considered as "waste",

[541] Ibid., pp. 279 et seq.

[542] Bonassies, '2002 "Note Tribunal de Commerce de Saint-Nazaire, 16.12.2000"', 45.

[543] M. Le Bihan-Guénolé, 'Une cargaison d'hydrocarbure accidentellement déversée au cours de son transport maritime peut-elle être qualifiée de déchet?' (2007) 59, Droit Maritime Français, 740–9 at 749; O. Cachard, 'Observations CJCE 24.06.2008' (2008) 60, Droit Maritime Français, 712–9 at 714; R. Caddell, 'Expanding the Ambit of Liability for Oil Pollution Damage from Tankers: The Charterer's Position under EU Law' (2009) 15, Journal of International Maritime Law, 219–23 at 221; E. Somers and G. Gonsaeles, 'The Consequences of the Sinking of the M/S "Erika" in European Waters: Towards a Total Loss for International Shipping Law?' (2010) 41, Journal of Maritime Law and Commerce, 57–83 at 69.

[544] *Commune de Mesquer*, (2008) 60 Droit Maritime Français 704 (European Court of Justice, 24 June 2008).

[545] Ibid., p. 709.

would be cleaned up at the expense of the "producer". The court, with this decision, did not show much respect for the liability mechanism in the CLC, obviously very consciously, by stressing[546] that the EU was, unlike its member States, not a party to the CLC and FUND Conventions.[547] The reasoning has, by some authors, been criticised for indiscriminately extending terrestrial norms to the marine environment and impairing the functioning of the oil pollution regime.[548]

It is true that the ECJ's reasoning is more refined than one may, at first blush, perceive. It has rightly been pointed out[549] that the court in its argumentation first and foremost considers the cases in which the amounts available under the CLC or even the FUND Convention do not suffice to cover the clean-up costs. The court emphasises that EU law, of course, does not prohibit that these costs are generally borne by the shipowner or by the Fund,[550] that EU member States only need to ensure the responsibility of the waste producer to the extent in which the costs are not covered by the Conventions, and moreover, only under the conditions of the EU directive, i.e., if the producer contributed to the risk of the pollution taking place.[551]

However, in cases such as the *Erika* accident, in which the charterer is actually the cargo owner and has exercised a certain degree of control over the vessel, the ECJ's reasoning and the application of the EU directive will clearly prejudice the liability mechanism set out by the CLC. Both the limitation of liability and the channelling will be overridden if the charterer is, under the directive or the respective national law, held liable for the clean-up costs exceeding the liability limits. In effect, the subsidiary liability of the "producer" or the "holder" of the waste, can be seen as an additional tier to the CLC and FUND system.[552]

Following the ECJ's decision, the *Cour de Cassation* considered the interpretation given by the court and remanded the case to the *Cour d'Appel* of Bordeaux to inquire into the facts of the case in order to establish whether Total contributed to the risk of maritime pollution. If this fact is confirmed, in application of the EU directive, Total will effectively have to bear the clean-up costs, regardless of the CLC's channelling provisions and regardless of the limitations on liability set out

---

546 Ibid., p. 710.

547 A. Cudennec, 'La politique communautaire de préservation du milieu marin: Quelle place pour le droit international?', in R. Casado Raigón and G. Cataldi (eds.), *Mélanges de droit de la mer offerts à Daniel Vignes* (Brussels: Bruylant, 2009), p. 222.

548 Le Bihan-Guénolé, '2007 "Une cargaison d'hydrocarbure"', 747; Cachard, '2008 "Observations CJCE 24.06.2008"', 716 et seqq.; Somers and Gonsaeles, '2010 "The Consequences of the Sinking"', 76.

549 P. Bonassies, 'Observations CJE 24.08.2008' (2009) 61, Droit Maritime Français, 17–21 at 19; Cudennec, *2009 "La politique communautaire"*, p. 223; N. de Sadeleer, 'Liability for Oil Pollution Damage versus Liability for Waste Management: The Polluter Pays Principle at the Rescue of the Victims' (2009) 21, Journal of Environmental Law, 299–307 at 306; Wetterstein, *2009 "Complete Freedom"*, p. 128.

550 *European Court of Justice, 24.06.2008, Commune de Mesquer*, p. 709.

551 Baatz, *2011 "Maritime Law"*, p. 464.

552 de Sadeleer, '2009 "Liability for Oil Pollution Damage"', 307.

by the Convention. It must however be observed that according to the *Cour de Cassation's* final ruling in the matter, Total is, notwithstanding its position as a charterer, liable under art. V para. 2 CLC, having acted "recklessly and with knowledge that such damage will probably occur". Following this decision, the CLC does not oppose Total's unlimited liability for the clean-up costs.

If the loss of the channelling benefit is left aside, the decision to hold a "holder" or "producer" of spilled oil can generally however be in conflict with the liability for oil pollution as determined by the application of the CLC. In cases of oil pollution such as the *Erika* accident, "victims" will often be municipalities or other public entities, whose oil pollution damage coincides with the costs that the State has to bear for the cleaning of beaches and the marine environment. It is obvious that administrative law—especially when backed up by an EU provision considered to have wider authority than a merely domestic provision[553]—can here be used to override both the provisions on channelling and on the limitation of liability and obtain full compensation from a parent company or another solvent player on the cargo owner's side. Therefore, not only would the State not be a secondary creditor, as seems desirable according to some authors,[554] but it could actually obtain unlimited compensation outside the limitation fund. Depending on the administrative law of the country in question, the compensation could possibly even be requested by a public act, without need to take recourse to the courts in the first place.

However, the relationship between the CLC and the administrative law provisions can also be seen as a parallelism rather than a conflict. It has been argued, and not without cause, that the CLC only applies to "mechanisms of civil responsibility" and that administrative law, or *lois de police* such as the waste disposal directive are not concerned with responsibility at all, but rather with a public duty.[555] It can be argued that the administrative obligation is not a debt of compensation to victims in the first place and thus not a "civil responsibility", but meant to relieve the State for the efforts incurred[556] following an illicit act of a private entity.[557] To a certain extent, a State—or a public entity such as a municipality—has, out of its duty towards its citizens, the obligation to preserve a clean and viable environment, and must carry out clean-up operations and restore the environment

[553] See Cachard, '2008 "Observations CJCE 24.06.2008"', 716 et seqq. and Cudennec, *2009 "La politique communautaire"*, pp. 222 et seq. on the conflict between EU law and international conventions.

[554] Altfuldisch, *2007 "Haftung und Entschädigung"*, p. 172.

[555] Bonassies, '2002 "Note Tribunal de Commerce de Saint-Nazaire, 16.12.2000"', 43; Y. Tassel, 'Observations Tribunal de Commerce de Saint-Nazaire 16.12.2000' (2001) 53, Droit Maritime Français, 281–4 at 283; Wetterstein, *2009 "Complete Freedom"*, p. 131.

[556] Robert, *2003 "L'"Erika""*, p. 131, Baatz, *2011 "Maritime Law"*, p. 465.

[557] Also see F. Stroink, 'The relation between private law and administrative law in view of globalization', in M. Faure and A. van der Walt (eds.), *Globalization and private law: The way forward* (Cheltenham: Edward Elgar, 2010) on the characteristics of administrative law and its impact on civil law.

even if there is no full compensation for these efforts. In cases where these costs exceed the funds available under the Conventions, the difference will often be borne by public actors.[558]

From this point of view, it is perhaps convincing to grant these public actors a claim against the "polluters"—if their contribution to the actual pollution can be established[559]—without having regard to the distribution of risk, insurance and liability as set out by conventions as the CLC.[560] However, it goes without saying that claims of this kind will damage the balance of responsibilities and liabilities established by the CLC if they are put forward on a large scale. The claim of a single municipality, such as Mesquer in the *Erika* case, asking for about 50,000 €, will clearly not endanger the system as a whole, nor the economic survival of a company such as Total. However, the decision clearly threatens to create an erosion of the system, considering that the expenses for removing spilled oil and restoring the environment are typically a substantial, if not the most important part, of the overall cost of an oil tanker accident.[561] It is up to States and public entities to either refrain from using administrative law in a way that jeopardises a system initially created by their own accord[562] or adapt their administrative law accordingly, if this is possible within the framework of EU law.[563]

I. Questions of Limitation of Liability in Time

According to art. VIII CLC, the rights of compensation are subject to extinction if a claim is not brought within the time limits set out. The provision reads:

> Rights of compensation under this Convention shall be extinguished unless an action is brought thereunder within three years from the date when the damage occurred. However, in no case shall an action be brought after six years from the date of the incident which caused the damage. Where this incident consists of a series of occurrences, the six years' period shall run from the date of first such occurrence.

[558] Hansen, '2008 "Multinational Enterprise Pursuit"', 417, referring to the *Prestige* oil spill.

[559] See Caddell, '2009 "Expanding the Ambit"', 223, on the difficulties to establish this and the degree of proof required.

[560] In this sense, Vialard, '2003 "Faut-il réformer le régime"', 442, claims that the question of compensation to third parties should not be considered as a matter of maritime, but as a matter of environmental law.

[561] Le Bihan-Guénolé, '2007 "Une cargaison d'hydrocarbure"', 749; Caddell, '2009 "Expanding the Ambit"', 223; Somers and Gonsaeles, '2010 "The Consequences of the Sinking"', 70.

[562] Vialard, '2003 "Faut-il réformer le régime"', 436.

[563] See F. Smeele, 'International Civil Litigation and the Pollution of the Marine Environment', in J. Basedow, U. Magnus and R. Wolfrum (eds.), *The Hamburg Lectures on Maritime Affairs 2007 & 2008* (Berlin, Heidelberg: Springer, 2010), p. 86, maintaining that EU Member States could properly implement the Waste Directive by targeting the oil manufacturer.

As Art. VI para. 1 FUND mirrors the wording of art. VIII CLC, the time limits under the FUND Convention are the same. The CLC and FUND thus set out rather clear time limits to the actions that can be brought. Still, claimants need to be aware of the fact that the start of proceedings against the shipowner may not be enough to also keep the prescription period as against the Fund.[564]

Where claimants need more time to assess their damages, they must still bring their claims within the three years, only backed up by provisional data if necessary. Following the *Erika* accident, a great number of claimants brought claims towards the very end of the prescription period, and this even on the recommendation of the Fund and the *Erika's* P & I Club.[565] Unlike prescription periods in some domestic legal systems like Germany that can be either interrupted or re-started, the time limits in art. VIII CLC do not provide for periods of prolongation or interruption.[566] This issue arose in the *Aegean Sea* criminal law proceedings in the courts of La Coruña, Spain. A number of claimants who had joined the action in adhesive procedure explicitly reserved their civil law claims until the question of criminal responsibility would be decided, which was a valid cause for interruption of a civil law prescription period under Spanish law.[567] The Fund, however, considered that the time limits in art. VI para. 1 FUND are absolute and not subject to interruption of any kind, the provisions of the convention overruling domestic law in any case. The Fund's point of view is certainly in line with the wording of the Conventions, and is also beneficial for practical reasons, enabling the Fund to close affairs after a certain time. However, claimants may be disappointed in trusting provisions of domestic procedural law, which may lead to harsh results in individual cases.

The prescription provisions have also given rise to some case law considering the exact interpretation of the limitation in time. Following the shipwreck of the tanker *Braer* off the Shetland Islands on 5 January 1993 and the spill of its cargo into the sea, heavy winds carried oil spray across the islands, contaminating meadows, farmland and buildings. This caused, in a rather unusual way, considerable damage on land. Numerous claims were brought for the clean-up costs and economic loss of farmers and livestock owners, and in some cases property rights in the affected assets created problems. In the case of *Eunson v. The Braer Corporation*,[568] a timely claim had been brought for damage caused to buildings by one spouse. The property, however, was jointly ("pro indiviso") owned by the couple, and after the three years' time limit had lapsed, the claimant tried to add his wife as

---

564 Gaskell, *2000 "Pollution, Limitation and Carriage"*, p. 74.

565 See J.-S. Rohart, 'Les critères du FIPOL sur l'admissibilité des demandes d'indemnisation' (2006) 58, Droit Maritime Français, 970–6 at 973; Ndendé, '2008 "La politique d'indemnisation"', 32.

566 Renger, *1998 "Recht und Praxis"*, p. 156; Altfuldisch, *2007 "Haftung und Entschädigung"*, p. 44.

567 Gaskell, *2000 "Pollution, Limitation and Carriage"*, p. 75; Castillo Daudí, '2001 "El siniestro del buque "Aegean Sea""', 293.

568 *"Eunson v. The Braer Corporation"* 1999 Scots Law Times 1405 (Court of Session (Scotland), Outer House, 30 July 1998).

a second party to the action, contending that she pursued the same action as he did, so that the claim had actually already been brought and could not be considered as extinguished. The claimant argued that art. VIII CLC and the corresponding national law provision transposing it concerned extinction of the remedy only and not of the right. The Court of Session rejected this reasoning, holding that claims were individually brought by individual claimants and that neither could the husband be considered to have brought the claim on behalf of his wife, nor could the wife appropriate her husband's claim. Whereas the decision rather seems to be concerned with property interests than with the scope of art. VIII CLC at a first glimpse, the conclusion that can be drawn from it is rather important. The court's reasoning shows a tendency to consider individual person's rights, in a scope that is determined by the claim itself. Under the Scottish court's reasoning it will not be possible to put forward a wide claim for damage to property to which other persons with proprietary interests can then, after the lapse of the limitation period, adhere. The prescription is therefore reinforced.

Another decision from the same court and following the same accident concerns the relationship between the limitation periods of three and six years mentioned in art. VIII CLC. In *Gray v. The Braer Corporation*,[569] the partners of a fishing firm sought compensation for the damage done to their business by the *Braer* oil spill. They first did so as individuals within the time limits, and then, after a restructuring of their business as a partnership, in the name of the partnership, but only lodged the second claim after the three years' time limit. The claimants maintained, however, that as a fishing firm directly affected by the pollution and the interdiction by the authorities to fish in their usual waters, they incurred ongoing losses in the following years. They pleaded that therefore, the six year period should be applicable to their claims, as art. VIII CLC provided for two periods, one for immediate losses incurring only once of three years, and another for ongoing losses, of six years.

The Scottish court was very clear in declaring that the two time limits were in effect only just one period, and that they applied cumulatively, not alternatively. The three years' time limit applied generally to all claims, the period starting when the damage, i.e. the oil pollution, first occurred. However, six years after the incident that caused the damage, that is, normally, the shipwreck,[570] all claims are absolutely barred.

Whereas this strict interpretation is in line with the provision's wording, it is harsh on victims if one considers claims that only occur or are possibly discovered some years after the incident. The same Scottish court, following the *Braer* incident and its catastrophic consequences for the landscape of the Shetland Islands, decided that a psychological condition developed due to the impact of the pollution on the patient's surroundings and life, can be a damage requiring compensation under the

569 *"Gray v. The Braer Corporation"* 1999 Scots Law Times 1410 (Court of Session (Scotland), Outer House, 29 December 1998).

570 Ibid., p. 1413.

CLC and the FUND Convention.[571] It does, therefore, not seem inconceivable that a claimant only develops—or only discovers that he has developed—such a condition six years after the incident. The same applies to possible remote consequences for a claimant's property, too, such as a degradation of farmland or the impact on the condition of livestock that only appears after some years. The CLC and the FUND Conventions here favour an unambiguous administrative procedure, enabling the authorities to close a case after a given time, over the unconditional compensation for all claims. In comparison to time bars in domestic legal systems, especially for personal injury or damage to real estate, prescription periods under the CLC are thus rather short.[572] On the other hand, very long limitation periods would certainly lead to longer proceedings before the definite amounts to be paid can be determined, and claimants would find themselves in a very uncertain position until that point in time, which can effectively be criticised with regard to their right to property.[573] With regard to the fact that, according to art. III para. 4 CLC, claims for compensation for oil pollution against the owner or the persons benefitting from the channelling provisions can only be made under the Convention, it may be in the interest of claimants to maintain that such a late claim for remote damages is not a claim for "pollution damage" but to find a legal basis in domestic law (see *supra* under G.).

Another problem in this context has been pointed out by way of example by the Scottish judge[574]: The escape of oil from a shipwreck only some years after the sinking, for which no compensation would be available. The prescription periods can thus cause some problems in this case. Considering that the CLC applies to oil pollution by "ships", meaning sea-going craft (art. I para. 1, art. III para. 1 CLC), it can be maintained that oil pollution caused by wrecks that are damaged beyond repair[575] do not fall under the CLC,[576] but the differentiation would make tremendous problems in practice and has, so far, not been effectuated in the administration of the CLC/FUND regime.[577] In the *Prestige* accident, for instance, a considerable part of the ship's cargo sank with the vessel. Only parts of it leaked very slowly

---

[571] *"Black v. The Braer Corporation"* 1999 Scots Law Times 1401 (Court of Session (Scotland), Outer House, 30 July 1998).

[572] Gaskell, *2000 "Pollution, Limitation and Carriage"*, p. 74.

[573] See A. Gouritin, 'The International Regime for the Compensation of Oil-pollution Damage: A Good Candidate to Have a Human Rights Law Approach?' (2011) 20, Review of European Community and International Environmental Law, 194–207.

[574] *Court of Session (Scotland), Outer House, 29.12.1998, "Gray v. The Braer Corporation"*, p. 1413.

[575] See H. Rak, 'Liability for Pollution from Shipwrecks', in H. Rak and P. Wetterstein (eds.), *Shipwrecks in International and National Law - Focus on Wreck Removal and Pollution Prevention* (Turku, 2008), p. 53 on the difficulties to define the term "wreck".

[576] Ibid., p. 61.

[577] H. Ringbom, 'Wrecks in International Law', in H. Rak and P. Wetterstein (eds.), *Shipwrecks in International and National Law - Focus on Wreck Removal and Pollution Prevention* (Turku, 2008), p. 17; Jacobsson, '2010 "L'indemnisation"', 472 et seq.

from the sunken hull, until the Spanish authorities decided to remove the entire cargo from the wreck in a complicated and costly salvage operation. The Fund partly refused to indemnify them for it, holding that the complete removal was not a "reasonable" measure under the FUND Convention as the operation was disproportionately costly in relation to the harm that could be caused by the oil remaining in the wreck.[578] It is however impossible to predict the environmental impact of oil caught in shipwrecks on a long-term basis, and the prescription period as set out in the CLC may become a problem one day.

Setting out rules for the civil liability for the removal of wrecks is one of the objectives of the 2007 Nairobi International Convention on the Removal of Wrecks (WRC)[579]. The rules of the Convention generally confer liability for locating, marking and removing wrecks on the shipowner and require him to take out compulsory insurance to guarantee the fulfilment of these obligations. It must however be kept in mind that "locating, marking and removing" a wreck may not necessarily include compensation for removing oil from an area in some distance from the wreck, so the WRC and the CLC do not apply in exactly the same circumstances.[580] The WRC generally applies to all wrecks located in the EEZ of contracting States, so that coastal States can take the measures to facilitate the removal of wrecks (art. IX WRC), even if the wreckage was caused before the entry into force of the Convention. However, there is a practical problem as the rules on compulsory insurance will only be enforced once the Convention becomes binding on its contracting States, and victims of pollution by shipwrecks may thus be left without compensation. The shipowner will, additionally, not be liable to the extent to which he is already liable under the CLC (art. XI WRC).

Furthermore, the owners of older wrecks may not exist as legal entities any longer or they may even be unknown.[581] For these latter cases the Convention, in art. IX para. 7 WRC only provides that the State affected by the wreck may remove it, which does not change the state of affairs in most cases. Wrecks in a coastal State's EEZ that pose an environmental hazard can, under public international law, always be removed by the State in question, considering the coastal State's jurisdiction regarding environmental protection in its EEZ (art. 56 para. 1 (iii) UNCLOS). Art. IX para. 7 WRC may therefore, arguably, only make a difference where there is a wreck in a coastal State's EEZ that poses an environmental threat or a navigational danger to another State.

[578] See the outline of the problem in M. Jacobsson, 'The International Liability and Compensation Regime for Oil Pollution from Ships - International Solutions for a Global Problem' (2007) 32, Tulane Maritime Law Journal, 1–33 at 22, and Jacobsson, *2010 "How clean is clean"*, pp. 576 et seqq., the author having been the director of the Fund at the time the decision was taken.

[579] 46 ILM 694.

[580] Lagoni, '2008 "Das Wrackbeseitigungs-Übereinkommen"', 355, stressing that the hazard within the scope of application of the WRC must emerge from the actual wreck; Tsimplis, '2008 "Marine pollution"', 111, footnote 90; Baatz, *2011 "Maritime Law"*, p. 437, footnote 96.

[581] Rak, *2008 "Liability for Pollution"*, p. 60.

A more thorough analysis of the WRC is outside the scope of this study. It should however be noted that the WRC itself sets out a limitation of liability in time, of three years from the occurrence of the hazard by the shipwreck, barred by an absolute limit of six years after the casualty that caused the wreckage. Considering that these provisions mirror the prescription rules of the CLC, the WRC can thus hardly be used to obtain compensation beyond the time limits of the CLC.

It seems, therefore, that the WRC, after coming into force, will not present a full solution to the problem of long-term damage by oil pollution or other contamination from shipwrecks either. Rules on civil liability for long-term damage have seemingly consciously been left out of the compensation regimes, possibly with regard to the fact that it will be very hard to reconcile claims of this kind with the limitation of liability and the immediate coverage of all costs by an insurance fund that is meant to be exhausted and then wound up. Claimants desiring compensation for damages that are only perceived after the lapse of the prescription periods will thus need to find a footing in domestic law, and outside the scope of application of the CLC provisions in order to avoid the short limitation periods under the Convention.

J. Conclusion

The initiative resulting in the adoption of the CLC directly stemmed from the traumatic *Torrey Canyon* accident, kindling public pressure for the first time with regard to the severe consequences of oil pollution. This does not mean that policy makers and scientists were not yet aware of the dangers of other pollutants than oil at that time,[582] but during the negotiations to the CLC, the elaboration of a responsibility regime for oil pollution from tankers was thought to be most urgent. It was considered that additional negotiations for other pollutants would slow down the negotiation process or make it altogether impossible to come to terms.[583] Certainly, oil is a major pollutant, transported on a large scale, directly visible and perceivable, and following the former practice of operational discharge, its detrimental characteristics could already be detected on a minuscule scale.[584] There was never a serious initiative, at that point in time, to include other harmful substances into the regime, but it was hoped that the CLC could act as a model convention for other forms of pollution as well. The international community thus opted at a very early point in time for a strong fragmentation of responsibility regimes in the field of maritime casualties,[585] certainly in the search for the lowest

[582] Lucchini, '1970 "La pollution des mers"', 827.

[583] See the discussion on the scope of the Convention during the negotiations, *1973 "Official Records (CLC)"*, pp. 615 et seqq.

[584] Mensah, *2007 "Prevention"*, p. 43.

[585] Lucchini, '1970 "La pollution des mers"', 827.

common denominator between negotiating States, but giving rise to a system in which the risk of gaps would be imminent.

With regard to this, there have been cases in which victims have consciously chosen not to submit their claims to the CLC regime, and others in which national courts have found ways to circumvent the CLC rules. These cases are comparatively few in number, but they highlight some of the system's shortcomings, and there may be a lesson learned from this when it comes to an analysis of other civil responsibility regimes.

Chapter 4
Civil Responsibility Regimes for Other Forms of Pollution

Civil liability for oil spills by ships other than tankers had been consciously left out of the responsibility regime when the CLC was negotiated.[1] According to the delegates' perception, widening the scope of application to all ships was likely to present problems that might hinder or even render impossible a timely solution.[2] The same was apparently felt with regard to the inclusion of other pollutants than oil. However, it was always understood that there would be additional liability regimes in the future, probably with the CLC as a model, or that the CLC regime would be extended to encompass other forms of pollution than tanker oil spills.

Nevertheless, both the Bunker Oil Pollution Convention (BOPC) and the HNS Convention took a very long time to negotiate. In the case of the BOPC, a considerable time had to pass before there was even the first initiative to proceed to the draft of an international instrument. Originally, the liability for bunker oil pollution was meant to be included in a wider instrument dealing with responsibility for pollution by hazardous and noxious cargoes, but the delegations to the 1996 HNS conference decided against it, so that the issue was left for yet another instrument.[3] Even though a number of bunker oil spills, especially in Australia[4] and the United States[5] occurred in the meantime, highlighting the fact that ships other than tankers may also present a severe threat to the marine environment, the need for an international instrument was seemingly not felt as pressing—even though the governments of Australia and the United Kingdom struggled in more than one case with the problem of ships that were uninsured for this kind of

[1] Jacobsson, '2009 "Bunkers Convention in force"', 22.

[2] P. Boisson, 'L'OMI adopte une nouvelle convention pour indemniser les dommages dus à la pollution par les soutes' (2001) 53, Droit Maritime Français, 659–67 at 660.

[3] Ibid., p. 660; Gaskell and Forrest, '2008 "Marine Pollution Damage"', 126 et seq.

[4] See Ibid., pp. 108 et seqq. for a comprehensive overview.

[5] de la Rue and Anderson, *2009 "Shipping and the environment"*, pp. 255 and seqq.

S.F. Gahlen, *Civil Liability for Accidents at Sea*, Hamburg Studies on Maritime Affairs 32, DOI 10.1007/978-3-662-45555-5_4

damage.[6] The spills in the waters of the United States were however compensated under the 1990 Oil Pollution Act.[7] It was only on the initiative of the Australian government in 1994 and 1995[8] that the negotiation of a convention on bunker oil pollution was envisaged,[9] and the International Convention on Civil Liability for Bunker Oil Pollution Damage (BOPC)[10] was finally adopted on 23 March 2001. Once the Convention was open for ratification, however, it did come into force surprisingly quickly[11] in 2008 and must be termed a very successful convention in this respect.

A. The Bunker Oil Pollution Convention

The central provision of the BOPC is the shipowner's strict liability for bunker oil pollution and the compulsory insurance requirement, providing victims with an easy footing to obtain compensation that is comparable to the CLC regime. Most of the wording of the Convention is clearly modelled on the CLC, too, setting out, *inter alia*, the same rules concerning the geographical scope of application,[12] concerning jurisdiction and enforcement (arts. IX and X BOPC) and the exoneration of the shipowner from strict liability.

There are however some marked differences between the BOPC regime and the CLC regime. The most obvious one, which is arguably also very likely to cause many problems in the practical application, is the lack of a separate limitation regime under the BOPC. Unlike the CLC, the BOPC does not set out liability limits of its own, providing for a separate limitation fund for the pollution damage, but refers to other instruments allowing the shipowner to limit his liability. The Convention thus gives up the idea of internationally harmonised liability limits and a global fund for oil pollution damage. A closer analysis will be given at a later point in this study (see *infra* under IV.), but it should already be stressed that the Convention thus creates a split regime, and that the funds available for a given bunker oil spill may thus depend on various factors of the case—a result that has

---

[6] Gaskell and Forrest, '2008 "Marine Pollution Damage"', 127.

[7] See de la Rue and Anderson, *2009 "Shipping and the environment"*, p. 266, for a short summary of the civil responsibility for bunker oil spills under the OPA 1990.

[8] See C. Wu, 'Liability and Compensation for Bunker Pollution' (2002) 33, Journal of Maritime Law and Commerce, 553–67 at 554 et seq. for details.

[9] See Gaskell and Forrest, '2008 "Marine Pollution Damage"', 127 et seq. on the negotiation history of the Convention.

[10] 40 ILM 1493; BGBl. II 2006, S. 578.

[11] N. Gaskell, 'The Bunker Pollution Convention 2001 and limitation of liability' (2009) 15, Journal of International Maritime Law, 477–94 at 477; Gunasekera, *2010 "Civil liability"*, p. 273: "much earlier than anticipated".

[12] de la Rue and Anderson, *2009 "Shipping and the environment"*, p. 259.

already been termed arbitrary.[13] Furthermore, the system thus created is actually likely to produce problems of conflicts between the BOPC and other regimes.

Another obvious difference between the BOPC and the CLC system is to be found in the BOPC's provisions determining the liable party. The BOPC does not adopt channelling provisions similar to the CLC, but sets out the liability of the "shipowner" in a wide sense, which means that more than one entity may be found strictly liable for the pollution damage. Additionally, the Convention does not prohibit civil liability claims against other parties that may have caused the damage or contributed to it, which offers a wide range of possible defendants to the victims.

Unlike oil that is carried as cargo, fuel oil is used for propulsion and thus directly in the operation of the ship. There is therefore no other industry than the shipowning industry who could contribute to compensation payments. The civil liability regime for bunker oil pollution has to do without a second tier that would, comparable to the Fund and the Supplementary Fund in the CLC system, step in for pollution compensation beyond the limited liability of the shipowner. This fact accounts for some of the main differences between the CLC and the BOPC.

I. Scope of Application

The geographical application of the BOPC corresponds to the one of the CLC—the norms apply to damages occurring within the territory, including the territorial sea, of a contracting State, and in the EEZ if the State in question has declared it in accordance with international law (art. II (a) BOPC) and to all preventive measures, regardless of where they have been taken (art. II (b) BOPC).

1. "Ships"

Under the BOPC, art. VII para. 1, every ship over a tonnage of 1,000 gross tons needs to have compulsory insurance for bunker oil pollution. According to art. VII BOPC, ships flying the flag of a contracting State need to carry a certificate proving that they comply with this requirement. Port States can verify the existence of this certificate and only admit insured ships to their internal waters, even if they fly the flag of a non-contracting State. Contracting States can thus issue certificates even for ships that are not registered in them. Considering that under the BOPC, virtually every ship, except rather small craft, needs insurance and the corresponding certificate, issuing the certificate was a huge administrative effort for the States parties when the convention entered into force,[14] not only with regard to the

[13] Wetterstein, *2009 "Complete Freedom"*, p. 141.

[14] Gaskell, *2013 "Compensation"*, p. 72.

assessment of the ships that need to carry insurance, but also when assessing whether the insurance is issued by a financially sound entity.[15]

The requirement of mandatory insurance poses some problems with regard to the question of what a "ship" is. This is an eternal conundrum of maritime law, to which there is no comprehensive and uniform answer, and the ongoing developments in the shipbuilding and offshore industry do not facilitate the definition.[16] According to art. I para. 1 of the Convention, "ship" is defined as "any seagoing vessel and seaborne craft, of any type whatsoever". In order to draw the line between the CLC and the BOPC, art. IV para. 1 of the latter provides that the BOPC shall not apply to pollution damage as defined in the CLC, i.e. to oil spills by tankers.

It is fairly obvious that the BOPC was drafted in order to cover the biggest number of ships possible, with a view to close the gap in the existing liability regime for oil pollution, by setting out a regime applicable to all seagoing vessels other than tankers. There are of course some uncertainties with regard to the criterion of "seagoing"—it is not overly clear whether this applies to the abstract ability of the ship to sail on the open seas or to the actual set of facts, the question whether it effectively does sail the ocean.[17] This gives rise to problems with regard to harbour ferries and other craft that operate on the border between inland waters and the sea, and with jet-ski and other small vehicles that are essentially built to be operated for recreational purposes and near the shore. There is abundant jurisprudence under domestic law whether a craft of this kind can be considered a "ship",[18] but it is very unclear whether there can be a uniform understanding at all, as courts tend to interpret it with regard to the purposes of the domestic legislation. With regard to the aims of the BOPC to fill the gap in the existing liability regime on an international level, it seems that a vessel should be considered as "seagoing" if it effectively leaves the internal waters of a state on a regular basis, and this, regardless of whether it is built for this purpose or can be considered as "seaworthy" for voyages of this kind.[19] The BOPC could thus apply to a very wide extent without intruding on the sovereign right of States to deal with oil pollution in their inland waters, rivers and estuaries on a domestic level.

---

[15] L. Zhu, 'Can the Bunkers Convention Ensure Adequate Compensation for Pollution Victims?' (2009) 40, Journal of Maritime Law and Commerce, 203–19 at 209 et seqq.

[16] R. Shaw, 'What is a ship in maritime law?' (2005) 11, Journal of International Maritime Law, 247–9 at 247.

[17] Gaskell and Forrest, '2008 "Marine Pollution Damage"', 132.

[18] See M. Ashford, 'A Jet Ski: Vessel, Boat or Ship?' (2006) 20, Australian and New Zealand Maritime Law Journal, 64–8, considering this problem under English and Australian law and Shaw, '2005 "What is a ship"', 249, referring to United States law.

[19] In this sense: Gaskell and Forrest, '2008 "Marine Pollution Damage"', 132.

2. Bunker Oil

Considering that the BOPC is meant to apply to all oil spills to which the CLC does not apply, the drafters chose a very wide definition of "bunker oil". According to art. I para. 5 BOPC, "'bunker oil' means any hydrocarbon mineral oil, including lubricating oil, used or intended to be used for the operation or propulsion of the ship, and any residues of such oil". The Convention thus omits any detailed description of the substance, unlike the CLC that requires oil to be "persistent". The notion in the BOPC encompasses any kind of oily substance used in the operation of the ship, as opposed to persistent cargo oil covered by the CLC. There can however be difficulties in submitting a set of facts to either the CLC or the BOPC if there is an incident in which both a tanker and another ship are involved. This can for example be the case in accidents in which a tanker supplies another ship with fuel[20]—the applicability of the civil responsibility regime will then depend on the ship from which the oil escaped. Another problem may arise with regard to so-called "hybrid spills". Both the CLC and the BOPC contain provisions according to which shipowners shall be jointly and separately liable in incidents involving one or more ships if the damage is not reasonably separable (art. IV CLC, art. V BOPC). These provisions are however made for homogeneous spills, involving either two or more oil tankers or two or more ships other than tankers. Neither Convention gives a hint to the regime applicable to a spill which involves a tanker and another ship—possibly even a vessel transporting chemical substances[21]—and in which it is impossible to tell what part of the damage was caused by which ship. As has rightly been pointed out, the decision will in these cases be left with the courts[22] that have jurisdiction on the matter, i.e. the courts of the place where the pollution damage occurred. For policy reasons they may be more inclined to apply the CLC to a spill of this kind, considering that this Convention offers clear liability limits and a second tier compensation beyond the shipowner's liability limits. It is possible that courts may proceed to a very summary calculation of the damage caused by the tanker and leave the costs in excess of this to the other regime—the BOPC or the general maritime law.[23] This is of course very problematic if the other ship's cargo has caused the pollution to be more dangerous.[24] Furthermore, if the litigation thus takes place in two or more different legal systems, it is not inconceivable that defendants are exposed to double recovery.[25] On the other hand, the lack of channelling provisions under the BOPC

[20] Envisaged by L. Zhu, *Compulsory insurance and compensation for bunker oil pollution damage*, (Berlin, New York: Springer, 2006), p. 23.

[21] The scenario is considered by Tsimplis, '2005 "The Bunker Pollution Convention"', 89, referring to a possible liability under the HNS Convention which is not yet in force.

[22] Gaskell and Forrest, '2008 "Marine Pollution Damage"', 131.

[23] Suggested by N. Gaskell, 'Lessons of the Mont Louis: Part Two: Compensation for Hybrid Accidents' (1986) 1, International Journal of Estuarine and Coastal Law, 269–307 at 281 under the CLC 1969 regime.

[24] Ibid., pp. 281 et seq.

[25] Tsimplis, '2005 "The Bunker Pollution Convention"', 91.

(see *infra* under III.) may make it easier for claimants to sue parties other than the shipowner if the BOPC is applied.[26]

Similar to the problem that arises under the CLC, it is questionable whether the BOPC should apply to bunker oil spills that have been caused intentionally. The question seems more important under the BOPC than under the CLC, as the illicit spill of waste oil and oily mixtures from merchant ships is unfortunately a common occurrence, and happens considerably more frequent than illicit spills of this kind from tankers. The application of the BOPC to intentional spills, if they cause "pollution damage", is sometimes simply assumed.[27] On the whole, it seems that there is a lot less discussion about the question than under the CLC (see *supra*, Chap. 3, A. II. 2.). Under the CLC however, intentional or reckless conduct are mentioned as reasons to deny the limitation of liability, whereas the BOPC, not providing for liability limits of its own, does not mention the consequences of intent. Having regard to the consequences that intentional or reckless conduct may have under the applicable international limitation regime (see *infra*, under IV. 3.), it seems admissible to submit intentional bunker oil spills to the BOPC, especially since the Convention, which is entirely mute as to this question, is designed to apply to the widest possible extent.

II. Jurisdiction

The wording of the BOPC's rules on jurisdiction, recognition and enforcement corresponds to the provisions of the CLC. This modelling of the Convention on the earlier instrument suggests that the solutions discussed under the CLC should also be adopted for the BOPC, but these interpretations may have different consequences with regard to the differences between the two Conventions. Moreover, the BOPC's interplay with the general rules on jurisdiction, recognition and enforcement within the EU causes more difficulties than the corresponding rules of the CLC.

1. Choice of Forum for Damages Caused to More than One State Party

According to art. IX BOPC, actions for compensation of oil pollution damage within the scope of application of the BOPC "against the shipowner, insurer or other person providing security for the shipowner's liability may be brought only in the courts of any such States parties". Under the CLC, the formulation has given rise to a debate about cases in which cargo oil pollution damage is caused in more than one State Party, the question being whether claimants can exercise a choice of

[26] Ibid., p. 89.

[27] Gaskell and Forrest, '2008 "Marine Pollution Damage"', 131, 133.

venue in these cases, or whether claims have to be brought in different States for damage caused in the territory of different States respectively. With regard to the wording of art. IX CLC, the vast majority of authors recognise the claimant's right to pursue their claim in any of the jurisdictions where the pollution damage occurred (see *supra*, Chap. 3, B. I. 1.). Correspondingly, the same solution can be advocated for the BOPC.

It must however be kept in mind that this choice granted to the claimants and the danger of forum-shopping will be considerably bigger under the BOPC, considering that the BOPC does not have independent liability limits. The BOPC recognises the right of limitation of liability as applied "under the applicable national or international regime" (art. VI BOPC), i.e. the liability limits that apply in the courts where the action is pursued.

Considering that the BOPC does not have liability limits of its own, it is of course mute as to the shipowner's right to exercise a choice between forums. As the bunker oil pollution claims are only another kind of claim within the general limitation of liability for maritime claims, the fund that will answer for the bunker oil pollution claims may be established in any of the countries in which the shipowner wishes to do so. It is however obvious that in order to obtain the limitation of liability for a bunker oil claim under the BOPC, the limitation forum must be recognised in the State where the damage was suffered. The questions of recognition and enforcement of claims from other countries as against the general limitation fund are then a matter of the general limitation law applicable, and so is the question whether the limitation is granted in respect of one or of several incidents, possibly giving rise to the establishment of more than just one limitation fund.[28]

2. Interplay with the Law of the European Union

The BOPC was adopted in 2001, in the same year in which the EU adopted Council Regulation 44/2001 (the Brussels Regulation) on jurisdiction, recognition and enforcement of judgments in civil and commercial matters. The Brussels Regulation, based on the provisions of the earlier Brussels Convention, rendered the rules of jurisdiction, recognition and enforcement a matter of European Union law. Uniform rules on jurisdiction and on the recognition and enforcement from courts in other contracting States are however an important part of many international agreements, especially those on international civil liability. The value of uniform material law is certainly diminished if the judgments rendered under this uniform law cannot circulate freely between the contracting States to the international agreement. Art. 71 of the Brussels Regulation preserves the applicability of already existing international agreements on other matters that also comprised rules on jurisdiction, recognition and enforcement, the so-called "conventions on particular

[28] Gaskell, *2013 "Compensation"*, p. 73.

matters", such as, e.g., the CLC. It however makes it impossible for the EU member States to become a party to new conventions on particular matters, considering that member States are generally prevented from concluding agreements on matters in which there are already provisions of EU law.

During the negotiations to the BOPC, the problem was pointed out by the Swedish delegation, but it was not dealt with after all, in order not to prolong the negotiations,[29] and it does not seem that the problems of the interplay of European and international law were then already fully understood.[30] The delegates, namely those of the EU member States apparently trusted in the EU institutions to find a solution to the problem.

It however turned out to be rather difficult for the EU to adapt its legal instruments to allow the member States to ratify the BOPC. The main problem consisted in the fact that art. XII BOPC only understands States to become parties, so that the EU itself could not ratify it and therefore could, in principle, not allow its member States to adopt rules on jurisdiction, recognition and enforcement in a particular matter to which the EU itself could not adhere.[31]

The difficulty was eventually solved by rather political means, Council Decision 2002/762/EC allowing and encouraging member States to sign the BOPC, if they undertake to make a special reservation towards the other States parties regarding the continuing application of the rules on recognition and enforcement of the Brussels Regulation. All member States have effectively done so.[32] The Council Decision also calls upon the member States to undertake "best endeavours" to achieve an amendment to the BOPC that will allow the EU to eventually become a party. It can in this respect be considered that the member States concluded the agreement in the interest and on behalf of the EU,[33] in order to give effect to a desirable improvement in regulation.[34]

It must be stressed that the EU member States' reservation regarding the continuing application of the rules of the Brussels Regulation only concerns

---

[29] Tsimplis, '2005 "The Bunker Pollution Convention"', 98; E. Røsæg, 'The Athens Convention on Passenger Liability and the EU', in J. Basedow, U. Magnus and R. Wolfrum (eds.), *The Hamburg Lectures on Maritime Affairs 2007 & 2008* (Berlin, Heidelberg: Springer, 2010), p. 60.

[30] H. Ringbom, 'EU Regulation 44/2001 and its Implications for the International Maritime Liability Conventions' (2004) 35, Journal of Maritime Law and Commerce, 1–33 at 2, footnote 6; K. Ramming, 'Zur Zuständigkeit deutscher Gerichte für Ansprüche wegen Ölverschmutzungsschäden' (2007), Transportrecht, 13–8 at 15.

[31] See Ringbom, '2004 "EU Regulation 44/2001"', 9 et seqq. for more details.

[32] See the IMO's information on the States' reservations, http://www.imo.org/About/Conventions/StatusOfConventions/Documents/Status%20-%202014.pdf, p. 481 et seqq. (last visited 8 March 2014).

[33] H. Ringbom, *The EU Maritime Safety Policy and International Law* (Leiden, Boston: Martinus Nijhoff Publishers, 2008), pp. 93 et seq.; A. Rosas, 'EU Law and International Agreements Concluded by EU Member States, with Particular Emphasis on Maritime Law', in H. Rak and P. Wetterstein (eds.), *Environmental Liabilities in Ports and Coastal Areas - Focus on Public Authorities and Other Actors* (Åbo/Turku: Åbo Akademi University, 2011), pp. 35 et seq.

[34] Ringbom, *2008 "The EU Maritime Safety Policy"*, p. 93.

recognition and enforcement of judgments, not jurisdiction. This means that art. IX BOPC, allocating jurisdiction exclusively in the courts of contracting States of the BOPC in which there has been bunker oil pollution damage, applies without restrictions. Recognition and enforcement of judgments as between EU member States is however governed by the rules of the Brussels Regulation. Since the Regulation only applies within the EU, this reservation will only be relevant where both the State of origin of the judgment and the State in which it needs to be recognised are EU member States. Where one of the countries in question is not an EU member State, the rules of the BOPC, i.e. art. X, will prevail.

It is true that the text of the BOPC does not give the contracting States the opportunity to declare a reservation concerning the application of art. X BOPC, so that in principle, EU member States violate their obligations under the BOPC.[35] The rules on recognition and enforcement of the Brussels Regulation are in some respects less generous than art. X BOPC, e.g. in that art. 34 of the Regulation allows a denial of recognition for a violation of the recognising State's *ordre public* or where the judgment is irreconcilable with an earlier judgment taking effect in the recognising State. In other respects, art. 34 of the Brussels Regulation is more generous than art. X BOPC, e.g. where it allows the recognition of judgments to which there can still be an appeal, whereas art. X BOPC only applies to decisions that are "no longer subject to ordinary forms of review". There is thus a certain risk that an EU member State will recognise judgments concerning one and the same incident to different extents, depending on whether the contracting State to the BOPC they originate from is an EU member State or not. Since the decisive factor is the State of origin of the judgment to be recognised, and not the nationality of the claimant, claimants from non-EU States having obtained a judgment in an EU member State could still be at a disadvantage.[36] In view of the fact that claims under the BOPC will generally be subject to limitation, the fund being distributed between all claimants *pro rata*, recognition or denial of recognition of certain claims will have an impact on the amounts other claimants receive. Even if the deviation from art. X BOPC only takes effect as towards judgments from other EU member States, who have in turn declared the reservation, and even if the differences between the two regimes are not too striking,[37] parties from non-EU contracting States may still be affected by it.

Considering that the BOPC however does not provide limitation amounts of its own and does not give claimants a wide choice of jurisdictions for filing claims, the consequences of the reservation do not seem to be too important. The State in which claimants ask for recognition and enforcement will be the State in which the shipowner has constituted the limitation fund after all, and it is not the BOPC that governs his choice of jurisdiction, but the "applicable national or international

[35] K. Ramming, 'Das Bunkeröl-Übereinkommen' (2007), Versicherungsrecht, 306–22 at 321.

[36] Ringbom, '2004 "EU Regulation 44/2001"', 23.

[37] P. Egler, *Seeprivatrechtliche Streitigkeiten unter der EuGVVO* (Berlin: Springer-Verlag, 2011), p. 282.

regime" art. VI BOPC refers to. As the limitation of liability generally bars the enforcement of claims against other assets of the shipowner than the limitation fund, it is thus in all likelihood the shipowner who chooses the recognising jurisdiction. The concern that claimants may consciously choose a recognising jurisdiction that is particularly generous in the recognition of their claims[38] is thus unfounded.

The shipowner is rather unlikely to exercise his choice of jurisdiction with a view to the chances that certain claims may be recognised or not. From his point of view, the question of how the limitation amount is distributed matters less than procedural issues or the question of whether this jurisdiction is hostile to a breaking of the liability limits or not (see *supra*, Chap. 3, B. I. 1. on the corresponding question under the CLC). This can be considered as unfair, but the problem is directly linked to the fact that the BOPC only refers to the general limitation regimes applicable, and it is thus inherent in the Convention.

The EU member States' reservation concerning art. X BOPC may thus entail some unfairness in the recognition and enforcement of judgments, but the system that the Convention establishes is already much less homogeneous than the one under the CLC. Claimants may see their shares in the limitation fund reduced by other claims for other heads of damage than bunker oil pollution, and the shipowner will establish his limitation fund as a general limitation fund under the jurisdictional rules of this general limitation regime. In view of that, it can first be argued that the BOPC does not give claimants a secure right to a given limitation fund in the first place. Secondly, it does not seem that the differences in the recognition according to the Brussels Regulation and the BOPC are overly likely to take effect in bunker oil claims. A claim for bunker oil pollution damage is rather unlikely to trigger the *ordre public* exception, which is, as an exception, to be understood narrowly. With a view to the fact that bunker oil claims can only be brought in States where there has been pollution damage (art. IX BOPC) and that the shipowner's responsibility is in any case strict if there has been pollution damage, it is not too likely that there will be contradictory judgments in other EU member States on the same cases of bunker oil pollution damage.

III. Parties Liable and Lack of Channelling

The BOPC's provisions on the party liable differ considerably from the corresponding provisions in the CLC. According to art. III para. 1 BOPC, it is the "shipowner" who is strictly liable, and art. III para. 2 BOPC clearly assumes that there may be more than one liable person and provides for liability between them to

[38] Expressed by F. Dörfelt, 'Gerichtsstand sowie Anerkennung und Vollstreckung nach dem Bunkeröl-Übereinkommen' (2009), Praxis des Internationalen Privat- und Verfahrensrechts, 470–4 at 474.

be joint and several. This is a big difference between the BOPC and the CLC. Under the CLC, parties other than the shipowner have to be sued under domestic law, which will almost always mean that the claimant has to prove their wrongful conduct, if the channelling provisions do not shield them from liability altogether. In sharp contrast to that, the BOPC creates a strict liability for all persons that may be considered as "shipowners".

Whereas liability under the BOPC is strict, the Convention also sets out exceptions in art. III paras. 3 and 4, which the shipowner can invoke to escape liability. The exceptions are the same as under the CLC: act of war or hostilities, natural phenomena, third party's intentional acts or omissions, wrongful acts of public authorities and the victim's contributory fault in art. III para. 4 BOPC. The wording exactly mirrors the corresponding provisions of the CLC, so the problems that may arise will be the same (see *supra*, Chap. 3, F.). It must however be kept in mind that the BOPC, unlike the CLC, does not offer a second tier of compensation under which funds will be available even if the shipowner is exonerated, which makes it less acceptable to victims that these clauses exist.[39] With regard to that, it is not unconceivable that courts may be inclined to interpret the provisions slightly differently under the BOPC than under the CLC.

1. "Shipowner" in the Sense of the BOPC

The term "shipowner" is defined in art. I para. 3 BOPC as "the owner, including the registered owner, bareboat charterer, manager and operator of the ship". The definition recalls the definition of "shipowner" under the LLMC, art. I para. 2 LLMC defining the "shipowner" as "the owner, charterer, manager and operator of a seagoing ship", and in fact the BOPC provision has been drafted with the LLMC definition in mind.[40]

Even though there is no clear definition to the terms of "owner, charterer, manager and operator" in the BOPC,[41] the definition obviously hints to all the persons typically having control over the operation of the ship, except for the notion of "registered owner". The "registered owner", in art. I para. 3 BOPC, is defined as the person formally registered as the owner, without the requirement that he actually needs to have control of the ship.[42] According to art. VII BOPC, the obligation to carry insurance is only on the registered owner, who will for this reason probably be the first target of oil pollution victims.

Apart from him, however, victims can sue all other persons involved in the operation of the ship and invoke their strict liability. This will most probably lead to

---

[39] E. Røsæg, 'The impact of insurance practices on liability conventions', in, *Legislative approaches in maritime law*, MarIus (Oslo: Sjørettsfondet, 2000), vol. 283, pt. 2.2.3.

[40] Zhu, *2006 "Compulsory insurance and compensation"*, pp. 26 et seq.

[41] See Ibid., pp. 137 et seq. for a closer analysis of these terms.

[42] Tsimplis, '2005 "The Bunker Pollution Convention"', 88.

double or even multiple insurance obligations. Even if, according to art. VII BOPC, only the registered shipowner is obliged to take out insurance, charterers and managers and other persons involved in the operation of the ship may be wise to equally protect themselves by taking out insurance, in view of the strict liability they incur once there is an incident.[43]

Considering the fact that it will not be necessary to prove a person's or entity's fault as long as they can be considered to be a "shipowner" in the sense of art. I para. 3 BOPC, it may be in the interest of some parties remotely concerned with the operation of the ship to put forward that they are not "owners" in the sense of the provision. On the other hand, the parties considered as "owners" can limit their liability according to art. VI BOPC and are protected from claims outside the Convention's regime by art. III para. 5 BOPC.

It must be stressed that under the BOPC, there is no second tier of compensation. The only funds available will be the limited sums under the applicable instruments themselves. By providing a vast number of defendants to possible victims, the negotiating parties apparently hoped to mitigate this shortcoming in material compensation.[44]

This may however not be the case, at least not to the expected extent. According to art. VI BOPC, every party targeted under the strict liability of the BOPC can limit its liability under the applicable regime. The limitation of liability under the BOPC will be analysed subsequently in this study (*infra* under IV.), but it must be retained that under the LLMC, in which the liability limits are to be found if the respective State is a party, the limitation fund established by one defendant is deemed to have been established by all persons (art. XI para. 3 LLMC). Under the LLMC, the limitation fund is thus generally not constituted per person, but per incident, so that claimants can only recover their share of the general fund even if there are several defendants, as long as all defendants can limit their liability under the LLMC.[45] As a rule, the limitation fund will already be provided by the registered owner's compulsory insurance to this amount, so that, from the point of view of the claimants, it is rather irrelevant if there are further defendants.

The multitude of defendants is thus only advantageous to claimants if one party not carrying compulsory insurance is insolvent,[46] or where different responsibility regimes are applicable (see *infra* under IV.), or where one of the parties cannot invoke the limitation of liability due to reckless or intentional conduct.[47] The possibility to claim against more than one defendant has however also

[43] Gaskell and Forrest, '2008 "Marine Pollution Damage"', 139.

[44] Zhu, *2006 "Compulsory insurance and compensation"*, p. 29; Gaskell, '2009 "The Bunker Pollution Convention"', 492.

[45] Tsimplis, '2005 "The Bunker Pollution Convention"', 92 et seq.; de la Rue and Anderson, *2009 "Shipping and the environment"*, p. 262; Gaskell and Forrest, '2008 "Marine Pollution Damage"', 139, referring to the situation in Australia where the LLMC 1996 applies.

[46] Zhu, *2006 "Compulsory insurance and compensation"*, p. 30.

[47] Gaskell and Forrest, '2008 "Marine Pollution Damage"', 139.

disadvantages. Claimants need to identify the potential defendants and to decide which of them to sue, and the potentially necessary apportionment of the claim between defendants may in some cases delay and complicate compensation payments.[48]

2. The Problem of Responder Immunity

The BOPC, in art. III para. 1, sets out the strict liability of all parties that can be considered as "shipowners", and in art. III para. 5 ensures that compensation claims for bunker oil pollution will only be dealt with under the BOPC. Following the lack of a channelling provision comparable to art. III para. 4 CLC, however, all parties that cannot be considered as "shipowners" in the sense of art. I para. 3 BOPC, face claims based on other legal bases, such as general tort claims, without protection by the BOPC. Unlike the CLC in art. III para. 4, the BOPC does not create a class of persons that are completely exempted from liability.

The BOPC is therefore not concerned with the responsibility of parties that may be responsible for the bunker oil pollution without being a "shipowner", such as salvors, pilots or independent contractors that are involved in salvage operations or any other service to the ship. The BOPC does not oppose a claim in tort or any domestic provision against them. It goes without saying that under the BOPC, submitting bunker oil pollution claims to the general limitation fund that all claimants need to share, bunker oil pollution victims will especially be interested in defendants who cannot limit their liability under this general limitation regime, such as shipyards, classification societies and independent contractors. This choice not to grant protection to a certain class of persons was also taken with a view to the fact that responsibility is dealt with on a one-tier basis and that victims may have an interest in finding as many defendants as possible.[49]

The lack of protection is very often deplored with regard to salvors,[50] and it has rightly been pointed out that even in the cases in which the salvors have a good defence to a claim against them, the cost of possibly defending their cause in different jurisdictions[51] may already be a strong disincentive to provide services in a case of environmental salvage. Considering the fact that salvors generally only intervene once a ship experiences serious difficulties and that their task entails a higher risk and greater uncertainties than the normal operation of a ship (also see

[48] Wu, '2002 "Liability and Compensation"', 559; Tsimplis, '2005 "The Bunker Pollution Convention"', 88; Jacobsson, '2009 "Bunkers Convention in force"', 36.

[49] Zhu, *2006 "Compulsory insurance and compensation"*, p. 30; N. A. Martínez Gutiérrez, 'The Bunkers Convention and the Shipowner's Right to Limit Liability' (2012) 43, Journal of Maritime Law and Commerce, 235–54 at 240.

[50] Wu, '2002 "Liability and Compensation"', 560; Zhu, *2006 "Compulsory insurance and compensation"*, p. 104; Gaskell and Forrest, '2008 "Marine Pollution Damage"', 140; Jacobsson, '2009 "Bunkers Convention in force"', 36; Bishop, *2010 "Places of refuge"*, p. 355.

[51] Ibid., p. 355.

supra, Chap. 3, E. I. 6.), it is often considered unfair and, in the long term, unwise, to expose them to a possible responsibility under a domestic tort law or another provision, as this may deter them from intervening on behalf of a stricken ship.

The negotiating parties did however adopt a resolution in which it was strongly recommended to States parties to provide for domestic legislation that ensures responder immunity.[52] Some,[53] but by no means all States have followed that recommendation.[54] It has also been pointed out that salvors may have other ways of ensuring that they will be free from third party liability. Considering that salvage is a business featuring very few professional experts operating on a global market, it may be in the interest of States to incorporate provisions on responder immunity in their domestic legislation. Salvors may bluntly refuse to work for States in which there is no responder immunity under domestic law[55] or a comparable mechanism lowering the risk of a successful third party claim against them.

IV. The Limitation of Liability and the Relationship to Other Instruments

The fact that the BOPC liability regime does not create liability limits of its own is probably the biggest difference between the BOPC and the CLC and FUND regime. According to art. VI BOPC, the shipowner's or the insurer's right to "limit liability under any applicable national or international regime, such as the Convention on Limitation of Liability for Maritime Claims, 1976, as amended" remains unaffected by the BOPC. Art. VII BOPC provides for compulsory insurance or financial security to be taken out by the registered shipowner to an "amount equal to the limits of liability under the applicable national or international limitation regime, but in all cases, not exceeding an amount calculated in accordance with the Convention on Limitation of Liability for Maritime Claims, 1976, as amended."

Considering the wording of art. VI BOPC, the Convention does not even clearly set out that the responsible persons' liability should be limited, but only refers to this right under different instruments. The fact that the Convention's preamble recognises the importance of strict liability for all forms of oil pollution "linked to an appropriate limitation of the level of that liability" hints however at the fact that the limitation of liability, and not only of insurance cover, is regarded as the rule.

---

[52] "Resolution on protection of persons taking measures to prevent or minimize the effects of oil pollution", LEG/CONF. 12/18 of 27 March 2001, p. 3, available from the IMO Documents Service, http://www.imo.org/OurWork/Conferences/Pages/Documents.aspx (last visited 8 March 2014).

[53] Martínez Gutiérrez, '2012 "The Bunkers Convention"', 241, footnote 36, giving examples.

[54] Jacobsson, '2009 "Bunkers Convention in force"', 28, giving some examples.

[55] Gaskell and Forrest, '2008 "Marine Pollution Damage"', 140, referring to the *Tasman Spirit* incident in Pakistan.

The precise arrangement of this limitation is however left to the States parties and their decision to either ratify an international convention or to implement the limitation of liability into their national law.

The reasons for this approach are rather pragmatic. At the time that the BOPC was negotiated, the LLMC 1976 had just been revised and the limits had been amended by the 1996 Protocol. The new version of the LLMC did not contain a proviso as to the possible exclusion of bunker oil pollution claims from the LLMC, in contrast to the claims under the CLC and the HNS (art. III (b), art. XVIII para. 1 b) LLMC as amended). It was considered that the introduction of a regime under which another separate liability fund would be established would conflict with the just amended LLMC.[56] Furthermore, there was some resistance against the introduction of yet another separate fund for another class of damages, which was perceived as a waste of insurance capacity and resources.[57] In this situation, the linkage of the bunker oil pollution regime to the most recent instrument on liability limits was considered the easiest solution.[58] There was however the problem that ratification of the 1996 Protocol and even of the original 1976 version of the LLMC was not general, and the negotiating parties to the BOPC did not want to impose the LLMC limits on States that had not ratified the Conventions. The compromise found was the reference to the LLMC—or any other international or domestic instrument for the limitation of liability—in art. VI BOPC, along with a recommendation in a Conference resolution that meant to encourage States ratifying the BOPC to also ratify the LLMC 1996 Protocol.[59]

The question of how claims for bunker oil pollution may be treated under the LLMC will be addressed *infra* under 1. It must however be mentioned first that art. VI BOPC thus creates a very split regime, in which liability limits are not uniform, but differ according to the legal system in which the claim is pursued, and in which the limitation fund is set up.[60] If claimants sue more than one defendant, there may be different limitation amounts in different legislations, and it is highly unclear whether there will be some balancing between the payments made by these two or even more limitation funds.[61] It is easily conceivable that problems of double compensation may arise if victims put forward the same claim against the two funds.

This effectively means that States, by ratifying the LLMC, its 1996 Protocol or by implementing domestic legislation, can to some extent choose which regime will

[56] Gaskell and Forrest, '2008 "Marine Pollution Damage"', 150.

[57] Gunasekera, *2010 "Civil liability"*, p. 223.

[58] In this sense: Gaskell and Forrest, '2008 "Marine Pollution Damage"', 151.

[59] "Resolution on limitation of liability", LEG/CONF. 12/18 of 27 March 2001, p. 1, available from the IMO Documents Service, http://www.imo.org/OurWork/Conferences/Pages/Documents.aspx (last visited 8 March 2014).

[60] Zhu, '2009 "Can the Bunkers Convention"', 215. See Gaskell, '2009 "The Bunker Pollution Convention"', 493, for further considerations.

[61] Wu, '2002 "Liability and Compensation"', 559.

be applied in bunker oil pollution claims subject to limitation in their courts.[62] Shipowners and insurers will hardly ever know with certainty which liability limits they will be exposed to for a bunker oil spill in foreign waters,[63] since they will need to make sure that the BOPC claims can be easily recognised in the jurisdiction in which they set up the fund.

Theoretically, if the State in question has not ratified the LLMC at all, nor any earlier instrument such as the 1957 International Convention Relating to the Limitation of Liability of Owners of Seagoing Ships,[64] and does not provide for limitation of liability for maritime claims under its domestic law, liability for bunker oil pollution could, following the wording of art. VI BOPC, actually be unlimited. This has been perceived as a threat very early after the adoption of the Convention.[65] It is sometimes argued that the BOPC, not least by reference to the limitation of liability in the preamble, assumes that there will be limits to liability and thus opposes unlimited claims.[66] It is however very unclear how a court could handle this situation in the absence of provisions on limited liability. It seems rather likely that without existing liability limits to apply, a court would treat the claim as unlimited in amount, especially as a limitation is still regarded as an exception to the general concept of full compensation by most legal systems.[67]

1. Bunker Oil Pollution Claims Under the LLMC

The LLMC was drafted before the BOPC was even envisaged, and although the LLMC is a very general instrument applicable to third-party claims against shipowners in a wide sense, it does not encompass provisions dealing with the particularities of environmental claims. By virtue of art. III (b) LLMC, all claims falling under the CLC/FUND regime are expressly withdrawn from the LLMC regime (see *supra*, Chap. 3, E.), and the same applies to claims under the existing regimes for civil liability for nuclear damage (art. III (c) and (d) LLMC). It is therefore sometimes considered unfortunate that the BOPC explicitly refers to this instrument

---

[62] Tsimplis, '2005 "The Bunker Pollution Convention"', 92; Gaskell and Forrest, '2008 "Marine Pollution Damage"', 151.

[63] Wu, '2002 "Liability and Compensation"', 562; Zhu, *2006 "Compulsory insurance and compensation"*, p. 159.

[64] UKTS 52 (1986); BGBl. 1972 II, p. 672.

[65] Wu, '2002 "Liability and Compensation"', 562.

[66] Tsimplis, '2005 "The Bunker Pollution Convention"', 92; Gaskell, '2009 "The Bunker Pollution Convention"', 478; Jacobsson, '2009 "Bunkers Convention in force"', 29, but see D. R. Thomas, 'Bunker oil pollution - a new liability convention' (2008) 14, Journal of International Maritime Law, 375–6 at 375: "an unexpected circumstance, but, nonetheless, not necessarily one inconsistent with the provisions of the convention."

[67] Gaskell and Forrest, '2008 "Marine Pollution Damage"', 156, footnote 347, referring to Australian law.

for questions of environmental damage for bunker oil pollution.[68] There are in fact some difficulties with regard to the limitation of BOPC claims under the LLMC.

The LLMC provides for the constitution of two different limitation funds, one for "passenger claims" in art. VII, for loss of life or personal injury to passengers of a ship, and the other for all other claims. Liability limits in the first fund are double the amount of the second fund. Considering that the BOPC only applies to "pollution damage", resulting from the escape or discharge of bunker oil, death or injury claims under the BOPC will only arise if they are caused by the pollution rather than by the maritime accident itself, which is rather unlikely. With regard to damage done to reefs or other underwater structures, the damage caused by mere physical contact will not be covered by the BOPC, as this can hardly be termed "pollution damage".[69]

The LLMC, in its art. II para. 1, sets out specific classes of claims that can be subject to limitation. The provision reads:

> Art. II
> Claims subject to limitation
> 1. Subject to Articles III and IV the following claims, whatever the basis of liability may be, shall be subject to the limitation of liability:
> (a) claims in respect of loss or life or injury and loss or damage to property (including damage to harbour works, basins and waterways and aids to navigation), occurring on board or in direct connection with the operation of the ship or with salvage operations, and consequential loss resulting therefrom;
> (b) claims in respect of loss resulting from delay in the carriage by sea of cargo, passengers or their luggage;
> (c) claims in respect of other loss resulting from infringement of rights other than contractual rights, occurring in direct connection with the operation of the ship or salvage operations;
> (d) claims in respect of the raising, removal, destruction or the rendering harmless of a ship which is sunk, wrecked, stranded or abandoned, including anything that is or has been on board such ship;
> (e) claims in respect of the removal, destruction or the rendering harmless of the cargo of the ship;
> (f) ...

The claims that can occur in relation to bunker oil pollution are the cost of cleaning up the spilt oil and other removal measures, damage to private property and loss incurred because of this very damage, and pure economic loss such as a diminution of income to the fishing and tourism industries. Comparable to the questions that arise under the CLC, there can also be the conundrum of abstract compensation for damage to the environment.

The most common claims that can be expected are however the costs of cleaning operations on beaches and reefs. It is noteworthy that it is only with difficulties that

---

[68] Griggs, *2001 "International Convention on Civil Liability for Bunker Oil Pollution Damage"*, p. 6; Zhu, *2006 "Compulsory insurance and compensation"*, p. 157; Gaskell and Forrest, '2008 "Marine Pollution Damage"', 151; Martínez Gutiérrez, '2012 "The Bunkers Convention"', 252.

[69] Gaskell, *2013 "Compensation"*, p. 78.

they can be subjected to the LLMC provisions. If beaches and other areas—or the sea itself[70]—can be considered as "property", the contamination can fall under art. II para. 1 (a) LLMC.[71] The application of this provision however somehow stretches its wording.[72] By reference to "harbour works, basins and waterways and aids to navigations" as included in "property", the provision clearly assumes that there might be a doubt if pollution of shoreside installations and navigational waterways is covered as property damage. If the doubt is explicitly addressed and done away with as to these installations, but not with regard to any other parts of the environment, it seems questionable whether beaches, coastal landscapes and the sea itself should be covered as well.

Cost of clean-up and removal of oil off beaches could also fall under art. II para. 1 (d) LLMC as the cost of "removing anything that [. . .] has been on the ship", but this, somehow inconsistently with the purposes of the BOPC,[73] presupposes that the ship is sunk, wrecked, stranded or abandoned.[74] A comprehensive compensation of bunker oil pollution in all cases could probably only be achieved if this condition was interpreted very extensively. Still, for cases in which there is no bigger physical damage to the ship,[75] art. I para. 1 (d) cannot be applied without corrupting the wording of the provision.[76] Art. II para. 1 (e) LLMC cannot apply because bunker oil can of course not be considered as cargo.

Under English law, clean-up operations on beaches, reefs and other State-owned environment are deemed to fall under art. II para. 1 (a) LLMC, by implementation of a clause to this end into the *Merchant Shipping* Act.[77] This was apparently done because the English legislator effectively feared that claims of this kind could be considered as unlimited in liability by the courts, if the LLMC provisions did not cover them.

Claims for damaged or lost property such as damaged fishing equipment or shoreside installations and the consequential loss of income will be covered by art. II para. 1 (a) LLMC. Claims for economic loss suffered by the fishing and tourism industry that are caused by the deterioration of the environment rather than by the damage of equipment or property would only be covered if art. II para. 1 (a) LLMC was interpreted to mean "consequential loss" caused to persons other than those

[70] Gaskell, '2009 "The Bunker Pollution Convention"', 484.

[71] Martínez Gutiérrez, '2012 "The Bunkers Convention"', 249.

[72] Zhu, *2006 "Compulsory insurance and compensation"*, p. 155.

[73] Ibid., p. 156.

[74] Baatz, *2011 "Maritime Law"*, p. 458.

[75] See Gaskell, '2009 "The Bunker Pollution Convention"', 487, considering the *Pacific Adventurer* accident where an oil spill was caused by falling containers damaging the hull of the ship but the vessel remained fully under control.

[76] Tsimplis, '2005 "The Bunker Pollution Convention"': "remains unclear"; Gaskell and Forrest, '2008 "Marine Pollution Damage"', 158.

[77] See Ibid., p. 156.

whose property is damaged—always under the assumption that beaches and shores can be considered as "property".[78]

Claims for pure economic loss such as the loss of income for the fishing and tourism industry that is only indirectly suffering from the oil spill could also be compensated under art. II para. 1 (c) LLMC[79] as "claims in respects of other losses", but here again, an extensive interpretation of "occurring in direct connection with the operation of the ship" would be needed.[80] This could effectively render the scope of application of the provision so wide that one could doubt whether sub-paragraphs (d) and (e) are necessary at all[81] and thus does not seem to be the interpretation intended by the drafters of the LLMC.[82]

The problems occurring with regard to compensation for pure economic loss are the same under the BOPC and the CLC/FUND regime. Indirect losses that are only incurred as a consequence of the environmental pollution, such as loss of income in the tourism industry, may be considered as too remote by some jurisdictions to be attributed to the shipowner.[83] National courts will probably have regard to the link of causation between the oil pollution and the loss of earnings of the business in question, but there is the difference that the handling of claims by the Fund creates a certain, even if not officially recognised, authority in the interpretation of the terms of the FUND convention and thus also the CLC (see *supra*, Chap. 3, B. I. 2.). In the BOPC system, there is no comparable neutral authority[84] and if courts do not resort to the principles established under the CLC and FUND,[85] they may be left without guidelines. Art. I para. 9 BOPC, borrowing the definition of oil pollution from the CLC is just as ambiguous as the CLC definition.

The application of the LLMC to cargo oil pollution claims that are for some reason outside the CLC has been considered *supra*, Chap. 3, G. II. as very likely, but there is a difference between cargo and bunker oil under the LLMC. At least the removal of spilled cargo oil is clearly covered by art. II para. 1 (e) LLMC.

The uncertainty as to this question for bunker oil claims has serious consequences under the BOPC. The main problem in this respect is that a claim for bunker oil pollution may be treated as a strict liability claim under the BOPC, but is then left unlimited in amount because of the inapplicability of the LLMC.[86]

---

[78] Ibid., p. 157; Martínez Gutiérrez, *2011 "Limitation of Liability"*, p. 192.

[79] Martínez Gutiérrez, '2012 "The Bunkers Convention"', 250.

[80] Gaskell, '2009 "The Bunker Pollution Convention"', 486.

[81] Gaskell and Forrest, '2008 "Marine Pollution Damage"', 157.

[82] Gaskell, '2009 "The Bunker Pollution Convention"', 487, referring to the change of the provision from the 1957 Limitation Convention's wording.

[83] Martínez Gutiérrez, *2011 "Limitation of Liability"*, p. 160.

[84] Gunasekera, *2010 "Civil liability"*, p. 225.

[85] Advocated by Martínez Gutiérrez, *2011 "Limitation of Liability"*, p. 162.

[86] Tsimplis, '2005 "The Bunker Pollution Convention"', 99; Mandaraka-Sheppard, *2009 "Modern Maritime Law"*, p. 979; Martínez Gutiérrez, '2012 "The Bunkers Convention"', 248; Baatz, *2011 "Maritime Law"*, p. 458, considering this as "likely" for claims of consequential losses by the tourism and fishery industry.

This concern has especially been expressed with regard to claims for mere damage to the environment, without an actual value in money for reinstatement or restoration of affected resources.[87] If claims of this kind were treated by national courts as strict liability claims under the BOPC but not subject to the LLMC, the outcome could be quite alarming to the shipping industry, considering the uncertainty of the value of these claims. Under the CLC, claims of this kind are arguably considered to be wholly outside the CLC system (see *supra*, Chap. 3, G.) and can only be pursued under domestic tort law, at least not entailing strict liability of the responsible persons. The absence of any authority comparable to the Fund who could foster a uniform understanding of the BOPC and LLMC or at least a harmonised treatment of these claims adds to the problem.

Therefore, an interpretation of the LLMC's terms in a way that ensures their applicability to bunker oil claims covered by the BOPC is often strongly advocated.[88] With regard to the fact that the limitation of liability is generally seen as the counterweight for strict liability, this view can only be supported. It is true that the BOPC thus demands a certain interpretation or implementation of its provisions from the ratifying States.

2. Funds for Bunker Oil Pollution Available

Irrespective of whether it is the LLMC or another limitation regime that applies, there will be no separate fund for bunker oil pollution. Claims will always need to be directed against a general limitation fund and thus compete with other claims for the same incident. It is of course possible that the only claims arising after an incident are claims for bunker oil pollution, so that the entire limitation fund can be distributed amongst the victims, but it is submitted that this is not a very likely scenario.

Under the LLMC, one fund will be established for death and personal injury claims and a second one for the others, especially property claims, this general fund providing for half the amount of the first one. Generally, there will therefore be only one third of the limitation amounts under the LLMC available for bunker oil pollution victims. Furthermore, claims for death and personal injury exceeding the limitation amount available under the respective fund will be considered under the general fund, too. This means that the claims of bunker oil pollution victims possibly compete with a multitude of other claims and that their chances to obtain full compensation can be extremely reduced. As long as the HNS Convention is not in force, claims for pollution by hazardous and noxious substances are also subject

[87] Griggs, *2001 "International Convention on Civil Liability for Bunker Oil Pollution Damage"*, p. 6.

[88] Martínez Gutiérrez, *2011 "Limitation of Liability"*, p. 196.

to limitation under the general regimes, and will also be directed against the LLMC general fund—or another kind of limitation fund under the applicable regime.[89]

The liability limits have been considerably raised under the 1996 Protocol to the LLMC as opposed to the 1976 original version, but victims in States that still adhere to the 1976 LLMC and then ratify the BOPC may find the liability limits derisory to deal with all upcoming claims after a given incident.[90] Even under the 1996 LLMC limits, the available funds can be rather restricted.[91] Some years ago, the limits in the 1996 LLMC were not expected to be increased in the near future.[92] It remains to be seen whether the recent changes provide for more adequate compensation amounts.

The fact that the bunker oil pollution claims then are dealt with on a strict liability basis can actually also reduce the funds available to other claimants.[93] It is thus true to state that the BOPC introduced strict liability for the shipowner without creating additional funds for victims.[94] It still remains to be seen whether this causes tensions in the aftermath of maritime accidents where there are claims for different types of damage.

3. Breaking the Limits

As the BOPC does not provide for a limitation regime of its own, it does not set out provisions on when the limitation can be overcome either. By referring to the applicable liability regime, the BOPC also takes over the provisions on the loss of the limitation benefit. If one considers the respective provision in art. IV LLMC, which remained unchanged by the 1996 Protocol, it reads:

> Art. IV
> Conduct barring limitation
> A person liable shall not be entitled to limit his liability if it is proved that the loss resulted from his personal act or omission, committed with the intent to cause such loss, or recklessly and with the knowledge that such loss would probably result.

The wording is almost identical to the terms of art. V para. 2 CLC as amended, but for the reference to the wider notion of "loss" instead of "damage", and which is to be explained by the wider variety of claims covered under the LLMC as opposed to the "oil pollution damage" claims under the CLC.

With regard to the possibility of a multitude of defendants under the BOPC, the question for the consequences of a conduct barring limitation with regard to the

---

[89] Gaskell and Forrest, '2008 "Marine Pollution Damage"', 154; also see Gaskell, '2009 "The Bunker Pollution Convention"', 481, giving the example of the *MSC Napoli* incident where there could have been considerable claims for HNS pollution.

[90] See, as an example B. Marten, 'The "Rena" and liability' (2011), New Zealand Law Journal, 341–4 on the *Rena* accident in New Zealand waters, New Zealand not being a party to the BOPC, but to the LLMC 1976, the respective national legislation covering bunker oil pollution spills.

[91] Gaskell, '2009 "The Bunker Pollution Convention"', 479 et seq. giving numerous examples.

[92] Gaskell and Forrest, '2008 "Marine Pollution Damage"', 154.

[93] Tsimplis, '2008 "Marine pollution"', 126.

[94] Ibid., p. 126.

other defendants arises. It is argued and it actually also appears fair that only the party guilty of this conduct will be exposed to unlimited liability, whereas all other parties can still put forward the liability limits.[95] In the cases of the applicability of the LLMC or other liability regimes under which there is an aggregated fund by all defendants for one incident, breaking the limits will therefore only benefit the victims if the respective party has independent funds beyond the liability limits. It is however likely that in a situation where there is more than one entity controlling the ship that can be considered as a "shipowner" in the sense of art. I para. 3 BOPC, the wrongful conduct can be reproached to several of them.[96] Victims can then enforce their unlimited claims against all of them.

A claim for an intentional bunker or waste oil spill covered by the BOPC (see *supra* under I. 2.), would therefore, under the LLMC, be unlimited, as the limitation of liability is denied for intentional conduct. This seems a desirable outcome, so that there is no need to restrict the scope of application of the BOPC to accidental spills only.

4. Problems with the Referral

Apart from the problems with submitting bunker oil pollution claims to the LLMC that have been pointed out *supra* under 1., there may be numerous other problems with a referral to limited liability regimes that have been in force before the adoption of the BOPC. It must be kept in mind that at least the LLMC contains provisions on the limitation of liability, the constitution and the distribution of the limitation fund only and does not provide for bases of liability as such. This must not necessarily be the case for domestic limitation regimes, and it is thus difficult that the BOPC refers to national provisions whose scope is not entirely clear. The vague reference to the "applicable national regime" in art. VII para. 1 BOPC in order to calculate the amounts of compulsory insurance creates the "curious"[97] situation that a ship may be insured according to the national rules of the State in which it has been registered, and which issued the certificate, but that there may be a limitation action in the courts of the State affected by the pollution damage in which liability limits are very different. Without a wide-spread applicability of uniform limitation rules, for instance the provisions of the 1996 LLMC, there will be anything but harmonisation in bunker oil claims.[98]

---

[95] Zhu, *2006 "Compulsory insurance and compensation"*, p. 159; Jacobsson, '2009 "Bunkers Convention in force"', 30.

[96] Ibid., p. 30.

[97] See Martínez Gutiérrez, '2012 "The Bunkers Convention"', 243, speculating on whether the reference points to the State affected by the pollution damage, which is problematic as this State cannot be known before the accident.

[98] Ibid., p. 247.

Even in the interplay between the BOPC and the LLMC, a few gaps have already been detected. For instance, mobile offshore drilling unions that are self-propelled and thus carry bunker oil can easily be considered as sea-going ships for the purposes of the BOPC, but the LLMC, according to art. XV para. 5 (b), does not apply to them.[99] A similar problem could arise with regard to bunker oil spill from a wreck—to the extent that the wreck is still considered a "ship"—in the countries which have excluded wreck removal claims from the LLMC.[100] Again, the problem emerges that the BOPC generally sets out strict liability of the shipowner, but following the non-applicability of the LLMC, it is not clear whether the claim is then unlimited in amount or whether the BOPC does not apply, either.

Furthermore, there can be considerable difficulties with regard to the compulsory insurance that the shipowner needs to take out according to art. VII para. 1 BOPC. The amount has to equal the limit of liability under the applicable regime, but there is an absolute limit calculated in accordance with the LLMC as amended. This means that the shipowner's actual limit of liability can exceed the funds available under the compulsory insurance, if the limitation regime applicable provides for higher limits than the LLMC. This creates problems in direct actions against the insurer that are possible according to art. VII para. 10 BOPC.[101]

Not least, it is debatable whether in the case of the adoption of another civil liability convention, e.g. for pollution caused by offshore installations, the BOPC can be regarded as referring to another, and possibly more specific limitation regime than the LLMC and comparable general regimes for maritime claims.[102] Given the wide scope of application of the BOPC for "sea-going ships", it seems that the drafters of future liability conventions will need to be very careful in order to avoid clashes and contradictions between the BOPC and any new draft.

V. Conclusion: Changes Brought About by the BOPC

The BOPC has been acclaimed as being a necessary complement to the existing conventions on oil pollution and as filling the "last gap" in the regime on maritime pollution. Following a closer analysis, it seems very questionable that this enthusiasm is justified.

There are effectively two major changes brought about by the BOPC: it sets out the strict liability of every entity that can be defined as a "shipowner", and it obliges the registered shipowner to take out insurance for this responsibility. The strict liability provision is certainly advantageous to victims, and the insurance obligation

---

[99] Jacobsson, '2009 "Bunkers Convention in force"', 34.

[100] Zhu, '2009 "Can the Bunkers Convention"', 215; Martínez Gutiérrez, *2011 "Limitation of Liability"*, p. 191.

[101] See Gaskell, '2009 "The Bunker Pollution Convention"', 491, for more details.

[102] Gunasekera, *2010 "Civil liability"*, p. 243.

guarantees that there will be funds up to a certain limit available. It must be stressed that even under the regime of States parties that explicitly do not provide for a limitation of liability, the mandatory insurance under art. VII para. 1 BOPC is always limited to the amounts under the LLMC.[103]

The shipowner's liability limits vary according to the law of the court in which the claim is pursued, and equal the limitation amounts for maritime claims applicable there. The question of whether limited liability for bunker oil claims was "introduced" by the BOPC therefore depends on whether oil pollution claims have been covered by the general regime on limitation for maritime claims beforehand. Considering the difficulties to submit bunker oil pollution claims to art. II para. 1 LLMC, it is actually perceivable that it was in some contracting States actually the BOPC—subject to the interpretation of its interplay with the LLMC in this way[104]—that introduced limitation of liability. It must however be referred to the fact that quite a number of States had submitted bunker oil pollution claims to a national regime of limited liability comparable to the CLC at an earlier point in time, such as England with the *Merchant Shipping Act*. The entry into force of the BOPC and its ratification by States however triggered some legislative efforts to create certainty at least on a national scale and hopefully will continue to do so.[105]

Very generally, the coming into force of the BOPC has not provided bunker oil pollution victims with additional funds. An advantage on a financial level is only provided to them if more than one party can be held liable as the "shipowner" in the sense of art. I para. 3 BOPC, and if there is at least one of them subject to a limitation regime that does not set up funds per incident or does not recognise the fund set up in a different jurisdiction. Even then, the claimant must still sue in more than one jurisdiction and possibly make up his mind as to the most solvent or most available defendant.

One of the biggest shortcomings of the BOPC is the lack of unity with regard to the liability limits and its failure to provide mechanisms by which payments of limitation funds in different legal systems can be balanced or reconciled.

There does not seem to be a lot of precedent in bunker oil pollution claims since the coming into force of the BOPC. However, following the *Pacific Adventurer* incident off Brisbane in 2009, rather shortly before the BOPC entered into force in Australia, Australian authorities already referred to the fact that the liability limits for the incident would remain the same after the coming into force of the Convention and were in any case insufficient to cover the expected costs.[106] It is therefore not too speculative to maintain that there will in the future be cases in which bunker oil pollution victims will be faced with insufficient limits of liability[107] and perhaps

[103] Gaskell, *2013 "Compensation"*, p. 76.

[104] Martínez Gutiérrez, '2012 "The Bunkers Convention"', 252.

[105] Martínez Gutiérrez, *2011 "The Bunkers Convention 2001"*, pp. 13 et seq.

[106] See Martínez Gutiérrez, '2012 "The Bunkers Convention"', 252; Gaskell, *2013 "Compensation"*, pp. 74 et seqq.

[107] See Gaskell, '2009 "The Bunker Pollution Convention"', 482, for more statistical data as provided within the IMO.

make use of the shortcomings and systemic lacunae of the BOPC. Given the similarities between the BOPC and the CLC, most strategies that have already been discussed for the CLC in Chap. 3 of the study, are also open to claimants under the BOPC. Claimants may sue in a non-contracting State such as the US and may try to overcome the limitation of liability by establishing the defendants' reckless conduct. Considering that the BOPC's channelling provisions do not provide for a class of persons entirely exempt from liability, it will also be easier to invoke the tortious responsibility of persons other than the "shipowners", subject to the respective provisions of domestic law. Furthermore, following the respective decisions of, especially, European courts (see Chap. 3, H.), criminal and administrative law provisions may also help bunker oil pollution victims.

Not least, claimants may decide to put the BOPC's referral to separate limitation regimes to the test. The weaknesses of the LLMC with regard to its applicability to bunker oil pollution claims have already been pointed out (*supra* under IV.1.). If the referral is read literally, it means that bunker oil claims that cannot be considered to be covered by the LLMC or a domestic regime may effectively be subject to strict liability without limitation. This applies to claims for the mere impairment of the environment without impact on human activities, the so-called "purely economic damage", possibly to claims for damages that cannot easily be subsumed under art. II para. 1 LLMC, and to claims expressly excluded from the LLMC such as compensation for damage caused by mobile offshore drilling units. It is true that this problem has been identified by a number of legal systems and mitigated by domestic provisions, but the BOPC remains a Convention that needs a bigger effort in being implemented than the other elements of the civil liability system for oil pollution.

B. The Convention on Liability for Damage by Hazardous and Noxious Cargoes

It has been pointed out at a very early point in time that not only cargo and bunker oil, but also dangerous substances transported on board ships have the potential to cause considerable damage to the environment, economic losses[108] and, not least, to the health of victims and property at sea or on shore.[109] Significant efforts have been made to submit the civil liability of this kind to a responsibility regime similar to the oil pollution regime, but these efforts must be considered as failures up to now. The International Convention on Liability and Compensation for Damage in

[108] Gaskell, *2013 "Compensation"*, p. 77, also referring to the consequences of large scale evacuations.

[109] N. Gaskell, 'The Draft Convention on Liability and Compensation for Damage resulting from the Carriage of Hazardous and Noxious Substances', in P. Wetterstein and A. Beijer (eds.), *Essays in Honour of Hugo Tiberg* (Stockholm: Juristförlaget, 1996), p. 225.

Connection with the Carriage of Hazardous and Noxious Substances by Sea (HNSC)[110] which was adopted under the auspices of IMO in 1996, has not entered into force. A Protocol of 2010 has brought about substantial amendments to the rules of the HNSC, but it still remains unclear if it will ever take effect. It is a sad fact that the need for an international regime is apparently not felt as long as incidents involving chemicals and other dangerous substances do not reach the catastrophic scale of oil pollution incidents.[111]

Compensation for damages caused by hazardous and noxious substances (HNS) on board ships thus are currently still dealt with under domestic law and will continue to be a matter of national provisions until there is an international instrument to this end. Considering that HNS claims are considered to fall under the general regime of the LLMC or the applicable national regime, liability for these damages will most probably be limited in the LLMC contracting States and countries with comparable national regimes.[112]

Questions of conflict of laws and private international law with regard to maritime torts are the subject of the third part of this study and will thus be dealt with there. However, even if the HNS regime is not yet in force, it will be interesting to give an overview over the provisions that have been adopted and highlight the differences between the oil pollution regimes and the HNSC in order to inquire into the difficulties that have hindered the HNSC regime from entering into force.

I. The Provisions of the HNS Convention

The development of a civil responsibility regime for damage caused by hazardous and noxious cargoes has been an ambitious project, because the incidents that can give rise to damages are manifold. Unlike oil, that mainly causes damage by its mere presence in the environment after a spill and will almost exclusively create costs relating to its removal,[113] chemicals and other dangerous cargoes can cause damage in many different ways. Hazardous substances may be toxic and thus present a direct threat to humans or the fauna and flora in the marine environment, or they can be explosive or easily inflammable and thus rather cause damage indirectly. Furthermore, chemicals and other hazardous substances may be transported in many different forms and will possibly not even be recognisable as

[110] The Convention text, the text of the Protocol of 2010 and a consolidated version can be found at http://www.hnsconvention.org/Pages/TheConvention.aspx (last visited 8 March 2014).

[111] Bederman, '1986 "Dead in the Water"', 512 et seq. referring to delegates' opinions after the failure of the 1984 conference in respect to HNS pollution; J. Lopusky, 'El Convenio sobre sustancias peligrosas y nocivas (H.N.S.): ¿Estamos en el buen camino?' (1993) 10, Anuario de Derecho Maritimo, 235–56 at 238; Gaskell, *2013 "Compensation"*, p. 79.

[112] Gaskell, *1996 "The Draft Convention"*, p. 226.

[113] Lopusky, '1993 "El Convenio"', 243 et seq.

dangerous from the outside, or only become dangerous under certain circumstances. Furthermore, hazardous and noxious substances are produced, shipped and used by many different entities all over the world, and for various purposes. Whereas the market for crude oil production consists of a limited number of actors, who are in the public eye[114] and tend to take direct influence on the transport of the product as well,[115] the market for the production and trade of dangerous substances is almost unlimited,[116] and in many cases completely disconnected from the transport. Chemicals and other dangerous substances, especially in packaged form, are effectively carried by ships of almost all kinds and sizes. This lack of integration may also cause insurance capacity to be smaller than in the oil industry.[117] Moreover, the damages that are caused by an incident with HNS may be very different, too. In cases where packaged cargo is lost at sea the main cost may be incurred for the search and retrieval of the lost units, without any actual damage to the environment at all. On the other hand, the spill of a comparatively small amount of a highly toxic substance could give rise to very important damages, by far exceeding the cost of a comparable oil spill.[118] Fire and explosions may also lead to death and personal injury claims, a head of damage that is extremely rare in oil pollution incidents and generally excluded from the CLC and the BOPC regime, both only applying to "pollution damage" (arts. I para. 5, II CLC; art. I para. 9, II BOPC).

Given these differences, it almost seems surprising that the IMO, in the long-lasting drafting process of the HNS Convention,[119] modelled the liability regime for hazardous and noxious substances after the CLC and FUND Conventions.[120] The HNSC provides, in one single instrument, for the shipowner's strict liability and compulsory insurance as a first-tier compensation, and for additional funds available in a second-tier compensation system, which is a fund paid for by the industry. As both tiers are to be found in one instrument, States, in contrast to their position under the CLC and FUND, do not have the possibility to put the burden on the

---

[114] Ibid., p. 241.

[115] Ibid., p. 243; K. Le Couviour, 'La Convention SNPD (1): quelques réflexions sur la dernière pièce du dispositif' (1999) 51, Droit Maritime Français, 1075–94 at 1087.

[116] Baatz, *2011 "Maritime Law"*, p. 468.

[117] R. Balkin, 'The Rise and Rise of IMO's Liability and Compensation Regimes', in T. M. Ndiaye and R. Wolfrum (eds.), *Law of the sea, environmental law and settlement of disputes* (Leiden, Boston: M. Nijhoff, 2007), p. 698.

[118] G. Little, 'The Hazardous and Noxious Substances Convention: a new horizon in the regulation of marine pollution' (1998), Lloyd's Maritime and Commercial Law Quarterly, 554–67 at 562.

[119] A description of the drafting history of the HNS is outside the scope of this study. Extensive overviews can be found in Gaskell, *1996 "The Draft Convention"*, 225 et seqq.; A. F. M. de Bièvre, 'Liability and Compensation for Damage in Connection With the Carriage of Hazardous and Noxious Substances by Sea' (1986) 17, Journal of Maritime Law and Commerce, 61–88, and Bederman, '1986 "Dead in the Water"', 486 et seqq., focussing on the failure of the 1984 Conference.

[120] E. Røsæg, 'The Rebirth of the HNS Convention', in G. Berlingieri, A. Boglione, S. M. Carbone and F. Siccardi (eds.), *Scritti in onore di Francesco Berlingieri. Volume II* (Genova: Il Diritto Marittimo, 2010), p. 853.

shipowner alone. The decision was taken with a view to the fact that the compensation scheme needed the industry's contribution to be workable.[121]

It has been pointed out by some authors that the HNS Convention thus had an inherent flaw from the start,[122] as the principles used for oil pollution compensation are simply not well adapted to the problem of pollution by other substances. It is however true that the carriage of dangerous cargoes creates very complicated dynamics between the parties involved in the concerned industries, and it is understandable that the IMO focussed on existing models to find solutions. There are however some differences that have evidently been enacted with regard to the differences between the carriage of oil and of other dangerous cargoes.

1. Geographical Scope of Application and Jurisdiction

Between the geographical scope of application of the HNSC and the CLC, there is an important difference. Unlike the CLC, the HNSC possibly also applies to damages on the high seas. According to art. III HNSC, the Convention applies firstly, to all damages on the territory and in the territorial sea of a State Party, secondly, to damage by contamination of the environment of a State Party's EEZ or a corresponding zone, thirdly, to any damage other than environmental impairment outside the territory and territorial sea of any State, if it has been caused by a ship flying the flag or being registered in a State Party. Fourthly, it applies to the cost of preventive measures, irrespective of where they have been taken.

The differences in the geographical scope of application between the CLC and the HNSC are due to the fact that the HNSC is also meant to cover death and personal injury claims, especially for people on board the ship experiencing the accident. These damages, as long as there is a link to a contracting State by the flag the ship is flying, are covered, no matter where they occur. The same applies to damage to property that is located on the high seas, such as fishing gear or offshore installations. The gap that can be found in the CLC in this respect (see *supra*, Chap. 3, A. II.) thus does not exist under the HNSC. Environmental damage, on the other hand, is only covered if it occurs on the territory, including the territorial sea, of a contracting State, and in its EEZ. This is easily explainable by the fact that there would hardly be a party having standing to claim damages for an impairment to the natural resources of the high seas, which are not subject to a State's sovereignty.

The geographical scope of application of the HNSC is thus considerably more comprehensive than the CLC in this respect, and carefully considers the extent of a coastal State's sovereignty in the underlying law of the seas.

As to jurisdiction, the HNS borrows the provisions from the CLC and the FUND, allocating jurisdiction in the courts of all the countries in which damage has been suffered, and also providing for the jurisdiction of these courts for claims against

---

[121] Balkin, *2007 "The Rise and Rise"*, p. 697.

[122] Tsimplis, '2008 "Marine pollution"', 133.

the HNS Fund. For EU member States, there will be the problem that the HNSC is a convention posterior to the Brussels Regulation, so that member States are, on the face of it, prevented from signing the convention containing provisions on jurisdiction, recognition and enforcement of judgments. EU Council Decision 2002/971/EC allows and encourages the ratification of the HNSC, mirroring the solution for the BOPC (see *supra* under A. II. 2.), but the Decision still refers to the 1996 Convention, omitting the 2010 Protocol. It however appears likely that there will be an amendment of the Decision at some point in time.[123]

2. Substances Covered

Following the drafting history of the HNSC, it was obvious that the negotiating parties' aim was to cover the biggest number of possibly harmful substances, irrespective of whether they were dangerous for being toxic, corrosive, easily inflammable, explosive or for another reason.[124] This however raised the very practical problem of how to define "dangerous", given that basically harmless and biodegradable substances can become dangerous if stored under unfavourable conditions or transported in big amounts.[125] It proved very quickly that the development of a free-standing list of substances that would clearly set out the substances covered, demanded more working capacities and technical skills than the negotiating parties were willing to invest. The eventual solution was the reference to lists of substances that were drafted for other instruments. Art. 5 (a) HNSC refers to the substances, materials and articles listed in MARPOL (annex I for oil, annex II for noxious liquid substances), the International Code for the Construction and Equipment of Ships Carrying Dangerous Chemicals in Bulk, 1983, the IMO International Maritime Dangerous Goods Code (IMDGC), the International Code for the Construction and Equipment of Ships Carrying Liquefied Gases in Bulk and the Code of Safe Practice for Solid Bulk Cargoes. Additionally, all liquid substances with a flashpoint of less than 60 °C are covered (art. I para. 5 (a) (vi) HNSC).

A major problem arose regarding the extension of the Convention to packaged substances, i.e. mainly containerised goods. Almost all vessels can carry packaged goods, and there is a certain likelihood that in the cargo of an average containership, some of the substances will be found. Thus, including packaged cargo would have rendered the Convention almost universally applicable. If packaged, these cargoes may furthermore not be correctly labelled or declared so that the shipowner would not even be aware of the applicability of the Convention. However, the fact that hazardous cargoes are packaged normally does not influence their potential to cause

---

[123] Jacobsson, '2012 "Perspective of the Global Compensation Regimes"', 72.

[124] P. Wetterstein, 'Carriage of Hazardous Cargoes by Sea - The HNS Convention' (1997) 26, Georgia Journal of International and Comparative Law, 595–614 at 599, especially referring to the Nordic countries.

[125] Le Couviour, '1999 "La Convention SNPD"', 1085.

danger and damage,[126] so a solution had to be found to deal with the amount of packaged cargo. According to art. I para. 5 (a) (vii) HNSC, packaged substances are covered if they are subject to the 1996 version of the IMDG Code. In contrast to the references to the other instruments, the reference to the IMDG Code in art. I para. 5 (a) (vii) HNSC refers to the version of 1996 and disregards any later amendments. The reason for this is that the reference, in this way, allows the exclusion of coal from the scope of the HNSC. Coal, even though it is mentioned in annex B of the Code of Safe Practice for Solid Bulk Cargoes, it is not covered by the 1996 version of the IMDG if carried in packaged form and thus does not fall under art. I para. 5 (a) (vii) HNSC.

Even if the provision narrows down the number of transports of packaged goods covered, the IMDG Code still subjects more than 5,000 products to special safety provisions.[127]

The provision remains unchanged under the 2010 Protocol to the HNSC, but alterations have been made with regard to the contribution of the receivers of packaged cargo to the second tier of compensation under the HNSC,[128] the HNS Fund.

The definition of HNS by reference to existing lists of course means that all parties involved in the maritime transport of any matter need to check multiple instruments in order to decide whether the HNSC provisions are applicable or not.[129] The substances in the mentioned instruments cover by no means only chemicals. It is often criticized that the inclusion or exclusion of substances into the scope of application of the HNSC is in some cases policy-driven[130] or, through the reference to other instruments made for other purposes, arbitrary.[131] The most prominent example for this critique is the exclusion of coal, the transport of which can actually present a major hazard.[132] Coal is not only easily inflammable and

[126] Bederman, '1986 "Dead in the Water"', 495.

[127] P. Boisson, 'La Convention SNDP de 1996 et l'indemnisation des dommages causés par le transport maritime de marchandises dangereuses' (1996) 48, Droit Maritime Français, 979–85 at 980.

[128] See J. Schelin, 'The HNS Protocol - The Resurrection of a Convention', in H. Rak and P. Wetterstein (eds.), *Environmental Liabilities in Ports and Coastal Areas - Focus on Public Authorities and Other Actors* (Åbo/Turku: Åbo Akademi University, 2011), p. 289; R. Shaw, 'Pollution of the Sea by Hazardous and Noxious Substances - Is a Workable International Convention on Compensation an Impossible Dream?', in M. Clarke (ed.), *Maritime Law Evolving: Thirty Years at Southampton* (Oxford, Portland: Hart Publishing, 2013), p. 50.

[129] P. Griggs, 'Extending the frontiers of liability - the proposed Hazardous Noxious Substances Convention and its effect on ship, cargo and insurance interests' (1996), Lloyd's Maritime and Commercial Law Quarterly, 145–55 at 149.

[130] Wetterstein, '1997 "Carriage of Hazardous Cargoes"', 601, also hinting at the negotiation history.

[131] Bederman, '1986 "Dead in the Water"', 505; Le Couviour, '1999 "La Convention SNPD"', 1085.

[132] Ibid., p. 1085; Wetterstein, *2006 "Shipowners' Environmental Impairment Liability"*, p. 1245, footnote 49; Tsimplis, '2008 "Marine pollution"', 134; Baatz, *2011 "Maritime Law"*, p. 470.

causes slow-burning fires that are extremely hard to extinguish, but can also emit methane gas that may result in major explosions. The worldwide amount of coal transported by ship is however very great, and an insurance obligation on these transport would significantly increase the cost[133] and cross-subsidise the transport of other chemicals,[134] thus rendering an important source of energy more expensive.

The reference system also creates the constitutional problem that a State party to the HNSC may become factually bound by a list from another international instrument to which it may not be a party.[135] The system has however the advantage that it can be very easily updated, as it takes into account all the amendments that are made to the lists and appendices of the instruments referred to.[136] Changes that are independent from these instruments are however impossible to undertake under this system.[137]

3. Definition of "Damage"

It has already been mentioned that the definition of "damage" in the HNSC is different from the CLC and the FUND Convention. Art. I para. 6 HNSC reads:

> 6.
> "Damage" means:
> (a) loss of life or personal injury on board or outside the ship carrying the hazardous and noxious substances caused by those substances;
> (b) loss of life or damage to property outside the ship carrying the hazardous and noxious substances caused by those substances;
> (c) loss or damage by contamination of the environment caused by the hazardous and noxious substances, provided that compensation for impairment of the environment other than loss of profit from such impairment shall be limited to costs of reasonable measures of reinstatement actually undertaken or to be undertaken, and
> (d) the costs of preventive measures and further loss caused by preventive measures.

It is thus clear that claims for personal injury and death were a real concern to the drafters of the Convention, as explosions and fire are more likely to occur in an accident with hazardous substances than in an incident involving oil.[138] The HNSC is in fact the first Convention to combine strict liability and compulsory insurance for claims for personal injury and loss of life.[139] It must however be stressed that the

[133] M. Göransson, 'The HNS Convention' (1997) 2, Uniform Law Review, 249–70 at 254.

[134] Balkin, *2007 "The Rise and Rise"*, p. 703.

[135] Göransson, '1997 "The HNS Convention"', 252.

[136] Ibid., p. 253; Le Couviour, '1999 "La Convention SNPD"', 1086.

[137] A. Rengifo, 'The International Convention on Liability and Compensation for Damage in Connection with the Carriage of Hazardous and Noxious Substances by Sea, 1996' (1997) 6, Review of European Community and International Environmental Law, 191–7 at 193.

[138] de Bièvre, '1986 "Liability and Compensation"', 71.

[139] Ibid., p. 79.

HNSC excludes claims arising out of contracts for the carriage of goods and passengers from its scope of application, just as claims from crewmembers to the extent that they are incompatible with the applicable labour or social security law (art. IV para. 1 and 2 HNSC). The Convention's preoccupation with death and personal injury claims therefore only concerns tort victims, who are not contractually bound to the shipowner and could not foresee that they could be harmed by the HNS transport.

Considering that the HNSC applies to oil as a hazardous substance (art. I para. 5 (a) (i) HNSC), and only excludes claims subject to the CLC from its scope of applicability (art. IV para. 3 (a) HNSC), claims involving oil that are not for "oil pollution damage" in the sense of art. II CLC are effectively covered by the HNSC. This means that claims for death or personal injury following a fire or explosions on board of an oil tanker will be covered by the HNSC,[140] and the same applies to damage done by non-persistent oils.[141] This could render the Convention a very important default regime and do away with some of the injustices that are caused by the definition of "pollution damage" under the CLC.

The HNSC's provisions as to the compensation for environmental claims however closely mirror the corresponding CLC and FUND provisions, and guidance can be obtained from the IOPC's experience and claims regulation manual[142] (see *supra*, Chap. 3, B. I. 2.).

4. Strict Liability of the Shipowner

Mirroring the provisions of the CLC and FUND Conventions, the HNSC provides in art. VII para. 1 for strict, but limited liability of the registered shipowner, covered by compulsory insurance to the liability limits. There are however some possibilities for the owner to obtain exoneration, for acts of war or hostilities, acts or omissions of a third party done with the intent to cause damage, the wrongful acts of public authorities (art. VII para. 2 (a)–(c) HNSC) and the victim's contributory fault (art. VII para. 3 HNSC). Following art. XIV para. 3 HNSC which sets out the exoneration clauses for the HNS Fund, the Fund however remains liable for damages caused by natural phenomena. The problems that may arise in connection with these exoneration clauses are the same that have emerged in the context of the CLC (see *supra*, Chap. 3, F.). It is perhaps to be regretted that the HNSC simply repeats the CLC in this respect, instead of addressing some of the problems, especially piracy and war risks (see *supra*, Chap. 3, F. I. and III.) that have become

[140] Gaskell, *1996 "The Draft Convention"*, p. 251; Göransson, '1997 "The HNS Convention"', 257; Tsimplis, '2008 "Marine pollution"', 135; Mandaraka-Sheppard, *2009 "Modern Maritime Law"*, p. 981.

[141] Tsimplis, '2008 "Marine pollution"', 135; Mandaraka-Sheppard, *2009 "Modern Maritime Law"*, p. 981.

[142] Göransson, '1997 "The HNS Convention"', 259; Wetterstein, '1997 "Carriage of Hazardous Cargoes"', 602.

more pressing since the negotiation of the CLC in the late 1960s. On the other hand, the scope of the shipowner's responsibility thus uniform for both regimes. It would be hard to justify why war risks should be treated differently, depending on whether it is in the context of an oil pollution claim or an HNS claim, just because the HNSC was negotiated at a later point in time. The use of the CLC and FUND as a model for further conventions here clearly causes some of their shortcomings to be perpetuated.

The HNSC however provides for one further cause of exoneration for the shipowner that reflects the peculiarities of the carriage of hazardous and noxious substances other than oil. According to art. VII para. 2 (d) HNSC, the shipowner is exonerated if he can prove that the damage was, wholly or partly, caused by the shipper's or any other person's omission to inform about the nature of the cargo, or if the failure to inform led the shipowner not to obtain insurance for the cargo. The shipowner can however only invoke the exception if neither he nor any of his servants or agents knew or ought to have known about the dangerous nature of the cargo.

With regard to the fact that packaged, especially containerised, dangerous cargo will as a rule not be recognisable as such if it is not correctly labelled or declared, the provision seems only to be fair towards shipowners.[143] There is however an uncertainty in the wording of art. VII para. 2 (d) HNSC, referring to the "failure of the shipper or any other person to furnish information concerning the hazardous and noxious nature of the substances shipped". This of course covers situations in which there was no information at all regarding the nature of the cargo, but it might be more difficult to draw the line in cases in which information was insufficient. As the exoneration does not apply where the shipowner or his agents knew or ought to have known about the characteristics of the cargo and as, furthermore, the burden of proof is on the shipowner, it can be considered that the shipowner has a certain obligation to inquire whether the HNSC applies[144] and can possibly only invoke art. VII para. 2 (d) HNSC where the cargo was completely inconspicuous. This is however, again, a very subjective question. As the HNS Fund remains liable for the damages that have been caused by the failure to inform the shipowner about the nature of the cargo (art. XIV HNSC not setting out this exoneration for the Fund), claims of this kind may cause considerable litigation against the Fund,[145] even for small amounts.

According to art. VII para. 5 HNSC, liability is channelled to the shipowner. As the wording of the provision is an almost exact copy of art. III para. 4 of the CLC as

[143] Ibid., p. 607, however expressing some doubts with regard to the lack of homogeneity; Little, '1998 "The Hazardous and Noxious Substances Convention"', 556; Le Couviour, '1999 "La Convention SNPD"', 1080 et seq.

[144] Little, '1998 "The Hazardous and Noxious Substances Convention"', 557.

[145] Göransson, '1997 "The HNS Convention"', 261; Wetterstein, '1997 "Carriage of Hazardous Cargoes"', 607.

amended by the 1992 Protocol, the problems that will arise in this context will be the same as under the CLC (see *supra*, Chap. 3, E.).

There is however a difference with regard to the scope of application of the two Conventions. The CLC does only apply to "pollution damage", so that claimants can be tempted to maintain that their claim is for another head of damage than "pollution damage" (see *supra*, Chap. 3, G.). Given the fact that the HNSC applies not only to pollution damage, but also to other heads of damages that arise in context with the carriage of hazardous and noxious cargoes, it will be harder to overcome the channelling and limitation provisions of the HNSC. This is reflected in the more general wording of art. VII para. 5 HNSC: "... no claim for compensation for damage under this Convention or otherwise may be made against...", as opposed to the wording of art. III para. 4 CLC, referring to "pollution damage".

5. Limitation in Time

The HNSC's time bars for bringing actions against the shipowner and the HNS Fund are considerably longer than the prescription periods under the CLC and the FUND. Art. XXXVII para. 1 and 2 HNSC provide for a limitation in time of three years for claims both against the shipowner and the HNS Fund, calculated from the time when the damage was discovered, with an absolute time bar of ten years from the date of the incident. It is submitted that the relationship between the two time limits is the same as under the CLC and the FUND (see *supra*, Chap. 3, I.).

The absolute limit of ten years is however considerably longer under the HNSC, most probably because of the fact that long-term damage caused by chemicals is significantly more likely than the long-term risks of oil. There is very little experience with the impact of pollution by chemicals and hazardous substances other than oil, and the long-term damage may be more important than the damage immediately perceived.[146] This is a big difference to oil pollution, as oil is biodegradable in the long run, so that the biggest damage is caused by the mere presence of the substance in the environment shortly after the incident. It can thus of course be argued that with regard to this even the ten years' time limit in the HNSC seems arbitrary and unjust.

6. The Limits of Liability

Unlike the BOPC, the HNSC sets out independent liability limits for both the shipowner and the HNS Fund. This means that there is a separate amount of money available for the victims of HNS damage. There is however very little

[146] Ibid., pp. 609 et seq.; Le Couviour, '1999 "La Convention SNPD"', 1080, also giving examples.

experience with regard to the costs of an HNS incident in recent times,[147] catastrophic incidents involving explosions on ships, such as the *Mont Louis* or the *Grandcamp* disaster dating back to times before the negotiation of the first maritime civil liability conventions. Therefore setting the limits of liability for the HNSC at the time of negotiation was not an easy task and caused a lot of debate.[148] It has been pointed out by several critics that the limits of liability applicable that were finally agreed on were much more based on the amounts that the insurers were willing to guarantee than on the expected cost of an HNS incident.[149] To the extent that the limits are higher for a presumably smaller circle of claimants than the general provisions of the LLMC, the HNSC limits of liability are however advantageous to victims.

It can still be criticised that the limits are calculated according to the tonnage of the vessel in question (art. IX HNSC). Again, there is an important difference between the carriage of oil and the carriage of other hazardous cargoes in this respect that is likely to render the provision inadequate. Oil is generally transported by specialised tankers, which only rarely carry other cargoes in addition. Even if the tankers may not always be fully loaded, it is rather unlikely that a very large vessel only carries a very small amount of oil. As the damage caused by oil generally correlates with the amount of oil released, the calculation of liability limits in accordance with the tonnage of the ship is reasonable. This does not hold true for the carriage of other hazardous cargoes. Limited cargoes carried by rather small ships can cause considerable damage,[150] depending on their characteristics, and in a very large container vessel, there may only be a very small amount of dangerous substances whereas the rest of the cargo is completely harmless.[151] In effect, there are studies according to which the most dangerous cargoes tend to be transported in small packages on board of vessels with a very limited tonnage.[152]

Still, it does not seem that there is a better way of setting the limits of liability. The calculation in accordance with the tonnage of cargo actually carried in a given voyage would be completely impractical given the fluctuation of numbers, and would probably be equally unjust considering that the risks inherent in the carriage of dangerous cargo vary very much depending on the actual substance.

---

[147] Boisson, '1996 "La Convention SNDP"', 984.

[148] Lopusky, '1993 "El Convenio"', 245. See Gaskell, *1996 "The Draft Convention"*, pp. 266 et seqq. for more details.

[149] Wetterstein, '1997 "Carriage of Hazardous Cargoes"', 608; Le Couviour, '1999 "La Convention SNPD"', 1083.

[150] N. Gaskell, 'New limits for passengers and others in the United Kingdom' (1998), Lloyd's Maritime and Commercial Law Quarterly, 312–36 at 313.

[151] Le Couviour, '1999 "La Convention SNPD"', 1087.

[152] Gaskell, *1996 "The Draft Convention"*, p. 269; Griggs, '1996 "Extending the frontiers"', 150, both pointing to a study carried out in the UK and presented to the IMO Legal Committee in October 1994 (LEG 71/3/11), available from the IMO Documents Service http://www.imo.org/OurWork/Conferences/Pages/Documents.aspx; Le Couviour, '1999 "La Convention SNPD"', 1081.

Following the adoption of the 2010 Protocol to the HNS, there are however now slightly higher liability limits for packaged HNS than for HNS carried in bulk. This is due to the alterations that have been made with regard to the HNS Fund and the calculation of contributions to it. It was considered one of the major shortcomings of the original 1996 HNSC to make receivers of packaged HNSC liable to contribute to the HNS Fund (also see *infra* under 7.). This created a huge administrative effort for comparatively small amounts of contribution,[153] considering the limited amounts of HNS that are carried in package. Under the provisions of the 2010 Protocol, receivers of packaged HNS therefore do not contribute to the Fund any longer. In order to make up for this shortening of contributions to the Fund, the liability limits in the first tier of compensation have been increased (art. IX, para. 1 HNSC as amended). Whereas this is clearly a practical solution to the problem that has been identified under the 1996 HNSC, it is also a shifting of risks from the receivers of packaged cargoes to shipowners and their insurers.

Claims against the shipowner's limitation fund, according to art. XI HNSC, are prioritised according to the type of claim. Claims for death and personal injury are first paid for by two thirds of the limitation fund, and if the total sum claimed exceeds the two thirds, only the remaining third is distributed pro-rata between claimants for property and other damage. The prioritisation of course only becomes important once the claims exceed the total liability limits of the HNSC.

7. Contributions to the HNS Fund

The provisions that have clearly caused the most important problems in the negotiation of the 1996 HNSC and the 2010 Protocol are the provisions on the HNS Fund, representing the second tier of compensation.[154] The idea that the industry profiting from the carriage of chemicals and other dangerous substances should contribute to the inherent risks, is clearly a wise one, yet it is very difficult to put into practice.

In the original 1996 HNS Convention, contributions to the Fund were levied by charging the "receivers" of hazardous and noxious substances. "Receiver", according to the definition in art. I para. 4 HNSC, means the person who physically receives the cargo in the port of a State Party, or, the person who is designated to be the receiver according to the domestic law of a State Party. Whereas the definition creates legal certainty as to the person liable[155] to pay the contribution to the Fund and is, theoretically, easy to control through customs clearance,[156] the provision

[153] Tsimplis, '2008 "Marine pollution"', 146; Jacobsson, '2009 "Bunkers Convention in force"', 364; R. Shaw, 'The 1996 HNS Convention - An Impossible Dream?', in G. Berlingieri, A. Boglione, S. M. Carbone and F. Siccardi (eds.), *Scritti in onore di Francesco Berlingieri. Volume II* (Genova: Il Diritto Marittimo, 2010), p. 908.

[154] Boisson, '1996 "La Convention SNDP"', 982; Tsimplis, '2008 "Marine pollution"', 133.

[155] Le Couviour, '1999 "La Convention SNPD"', 1088.

[156] Boisson, '1996 "La Convention SNDP"', 983.

may create arbitrary results. Just like any other goods, chemicals and other dangerous substances may be sold in transit a number of times and between persons whose contractual relationships may be manifold.[157] It is by no means sure that the person receiving them is economically benefitting from the shipment. This is different from the oil trade where there is only a limited number of market actors, which will in the long run all benefit from the carriage of oil by sea. Furthermore, the horizontal structure of the oil industry in which market actors take direct influence on the transport, justifies holding them liable for possible incidents, whereas this is clearly not the case with the very diversified chemical industry, especially if the "receiver" only is liable.[158] In this question, again, the particularities and the variety of situations in the carriage of hazardous and noxious substances made it difficult for the negotiating parties to find a just solution adapted to every case. Still, the targeting of the "receiver" remained the obvious solution. In the original 1996 HNSC, the only exception consisted in the contributions payable for LNG cargoes, which were due by the "title holders". At that time, the transport of LNG was predominantly in the hands of public organisations and major energy companies, because the investments necessary for the transport were still too important for other players.[159] It was thus considered most practicable to levy contributions to the HNS Fund from the title holder to the gas, immediately prior to discharge. The gas industry, the amounts of LNG transported and the corresponding contractual arrangements however changed considerably over time, and during the negotiations to the 2010 Protocol, the provision was changed, now requiring the "receiver" to provide the contributions, aligning the provision for LNG with that of the other cargoes.

In order to reflect the very different security accounts of different substances covered by the Convention,[160] the HNS Fund will be separated into one general account, sub-divided into sectors, and three special accounts (art. XVI HNSC). Contributions will thus be made to four different accounts, first, into the general account (that is again sub-divided), second, for oil, third, for liquefied natural gases (LNG) and fourth, for liquefied petroleum gases (LPG). The contributions by a given sector will generally only be used to compensate damage caused by the substances of this sector. The division between the sectors is meant to prevent cross-subsidisation between the sectors[161] and could, notwithstanding the opaque method of calculating it, create responsibility in each branch of the chemical industry.[162]

[157] Bederman, '1986 "Dead in the Water"', 500; Tsimplis, '2008 "Marine pollution"', 145.

[158] Bederman, '1986 "Dead in the Water"', 508.

[159] M. Jacobsson, 'The HNS Convention - Prospects for its entry into force' (2009), CMI Yearbook, 418–30 at 422; Shaw, *2013 "Pollution of the Sea"*, p. 57.

[160] Wetterstein, '1997 "Carriage of Hazardous Cargoes"', 609; Tsimplis, '2008 "Marine pollution"', 133.

[161] Göransson, '1997 "The HNS Convention"', 265; Wetterstein, '1997 "Carriage of Hazardous Cargoes"', 609; Tsimplis, '2008 "Marine pollution"', 144; Jacobsson, '2009 "Bunkers Convention in force"', 361.

[162] Le Couviour, '1999 "La Convention SNPD"', 1089.

With regard to the administrative effort to levy contributions on very small shipments of HNS which may not pay off, contributions are only due for received shipments beyond certain thresholds (art. XIX para. 1 HNSC).

II. Problems that May Arise with the Ratification of the HNSC

The negotiation and drafting of the original 1996 HNSC and of the 2010 Protocol have taken considerable time and effort.[163] However, the Convention has still not attracted enough ratifications, even though there seems to be some optimism that with the adoption of the 2010 Protocol, and with the European Union supporting the ratification by its member States,[164] the HNSC could eventually enter into force.[165] It remains to be seen whether States can eventually overcome their reluctance to implement the civil liability regime for HNS, and some of the main problems that possibly keep States from ratifying the Convention will be outlined at this point.

1. The Problem of the Relationship with Other Instruments

It has often been maintained that the problems that were identified in the 1996 HNSC had exclusively to do with the practical and legal difficulties of calculating and levying the contributions to the HNS Fund.[166] Whereas this was—and possibly is—certainly still a major problem which will be discussed *infra* under 2., there are also problems with the implementation of another civil liability regime and the relationship to the other existing instruments.

The HNSC is a very comprehensive Convention, covering a wide variety of possibly dangerous cargoes and different damages for which claims might arise—death and personal injury, property, damage to the environment and the costs of preventive measures. The problem that it might interfere with other conventions is therefore bigger than with regard to, e.g., the CLC. Furthermore, many States already have unilateral legislation to deal with civil liability for pollution by chemicals and other dangerous substances in their territorial seas.[167] Following a ratification of the HNSC, these States may have to adapt their national legislation, especially with regard to possible domestic liability limits.

Until the coming into force of the HNSC, claims for damages caused by dangerous cargoes will be covered by the LLMC. In the States that have ratified the LLMC, victims of damage done by HNS therefore share the LLMC limitation

---

[163] See Røsæg, *2010 "The Rebirth"*, p. 852 et seqq. for an overview.

[164] EU Decision 2002/971/EC.

[165] Ibid., p. 859; Schelin, *2011 "The HNS Protocol"*, p. 290.

[166] Tsimplis, '2008 "Marine pollution"', 133; Schelin, *2011 "The HNS Protocol"*, p. 288.

[167] Gaskell, *1996 "The Draft Convention"*, p. 227, referring to the UK.

fund with other claimants, and there does not arise any problem to attribute a given property damage or injury to a certain cause—it does not matter whether it has been brought about by the spill of chemicals or by any other cause, such as a mechanical strain. This may be of some importance in collision cases between a ship carrying HNS and another vessel. The HNSC, in art. I para. 6, does state that damage that cannot be reasonably separated from the damage caused by the HNSC will be deemed to be HNS damage and is therefore covered by the Convention, but the provision in effect only shifts the uncertainty from one definition to the other.[168] Instead of establishing that the damage has been caused by the HNS, claimants will now need to establish that it is not separable from the HNS damage. Cases in which this will be difficult are easily conceivable, such as a fire on a non-HNS ship after a collision, that may, but does not necessarily need to have been ignited or intensified by inflammable HNS, or a fatality with an uncertain cause.

With regard to the fact that the liability limits available under the HNSC are higher than under the LLMC, and do not need to be shared with other claimants, the application of the HNSC could cause a big difference.[169] Especially in cases of fatalities, it may prove to be very cynical and inadequate to award different amounts of compensations to claimants, according to whether the death of the person has been caused by the presence of hazardous cargoes or not. The problem may of course be mitigated by a very generous interpretation of the term "separable". It can also always be argued that from the victims' point of view, applying higher limits of liability to some claimants is preferable to applying uniform, but low, limits of liability to everyone, but the resulting risk of unfairness and the additional cost to insurers[170] not fully justified is an inherent weakness of the regime.

Furthermore, the existence of other liability regimes with significantly lower limits of liability, such as the 1976 LLMC may cause problems[171] in cases between two States that are party to the same liability regime. In a case between two shipowners from these two countries, either of them could invoke the 1976 LLMC limits, even if one of the States has ratified the HNSC.[172] The other State not being bound by the instrument with the higher liability limits, the mutually accepted lower liability limits will apply. Under the 1996 LLMC, States can however exclude HNS claims from the scope of applicability of the Convention (art. XVIII para. 1 (b) LLMC 1996). Still, this means that States ratifying the HNSC will need to review their legislation in force and possibly adapt it in order to give full effect to the HNSC provisions.

---

[168] Ibid., p. 258; Tsimplis, '2008 "Marine pollution"', 135, hinting to this problem: "unless it can be proved that the damage or part of it was caused by the non-hazardous materials, the 1996 HNS will be applicable."

[169] Gaskell, *1996 "The Draft Convention"*, p. 257.

[170] Bederman, '1986 "Dead in the Water"', 503.

[171] Griggs, '1996 "Extending the frontiers"', 152.

[172] Example given by ibid., pp. 152 et seq.

2. The Practical Problems with Regard to the HNS Fund Contributions

With regard to the levying of contributions to the HNS Fund, problems arise both with calculating the amounts and with the practical procedure of actually collecting them. According to art. XVII HNSC, the calculation of annual contributions shall be made by the HNS Fund's Assembly, on the basis of contributing cargo received in that year. It has obviously been a constant fear of the potential and actual State Parties that large amounts of HNS remain unreported and that the system is thus financed by only a few receivers to the benefit of the whole industry, and this perhaps rightly so. In order to calculate the amount of contributing cargo, the HNS Fund depends on the information furnished by the contracting States. In the past, the record of contracting States' reports has been rather disappointing.[173] According to art. XDVI para. 1 HNSC, the annual receipt of at least 40 million tonnes of contributing cargo in the contracting States is a requirement for the entry into force of the convention, and irrespective of the number of ratifying States, it will not be fulfilled as long as they do not report. The reasons for the omissions in the reporting obligation can only be speculated on, probably consisting in the lack of the necessary administrative procedures[174] or capacities to carry them out. Theoretically, the control could be made through the mechanisms of customs clearance, but the underlying procedures may need to be adapted in some States.[175] The situation in the trade of dangerous substances is apparently very different from the oil transport industry, where, due to the existence of only a few market players and the use of specialised ships, the export or import of major amounts will not escape the attention of competitors and public authorities.

In the 2010 Protocol, the problem of omissions in reporting has been addressed by the introduction of an art. XXIbis which entitles the HNS Fund to claim compensation for loss from non-reporting States and to temporary deny the payment of compensation other than for death and personal injury claims to non-reporting States until they fulfil their obligation. Whereas there is certainly a need to create sanctions for non-reporting contracting States, the pressure that is thus created may render States even more reluctant to submit to the regime.

3. Uncertainties in Costs

Another major problem that may make the concerned industries—shipping, insurance[176] and chemical industries[177]—lobby for abstention from ratification consists

[173] See the chart in Røsæg, *2010 "The Rebirth"*, p. 858.

[174] Ibid., p. 858.

[175] Boisson, '1996 "La Convention SNDP"', 983.

[176] Gaskell, *1996 "The Draft Convention"*, p. 268, and Le Couviour, '1999 "La Convention SNPD"', 1084, apply the failure of the 1984 conference largely to the insurance industry's resistance.

[177] Røsæg, *2010 "The Rebirth"*, p. 853.

in the uncertainty about the economic impact that the introduction of the system may have. Difficulties start with the problems of reporting contributing cargoes, even if the technical means for this are available[178] and make it impossible to calculate even the dimension in which a given shipment would be charged by the HNS Fund.[179]

There is very little experience with the cost and the possible long-term consequences of damages caused by dangerous cargoes other than oil,[180] and incidents can be very different in their gravity. Therefore, it is difficult for all industries concerned to estimate the cost that a ratification of the convention with its limits of liability would save them, in comparison to the liability they currently incur for the same set of facts. Their reluctance to accept a new liability regime is certainly aggravated by the fact that a tort claim for pollution damage will entail a number of uncertainties for the claimants, especially with regard to the proof of causation,[181] and depending on the legal system in question, even uncertainties in law.[182]

If shipowners are subjected to the additional cost for compulsory insurance, it may become difficult for them to pass it on to shippers or cargo owners. Whereas the transport of crude oil is generally closely controlled by the oil industry and any additional cost will thus be consolidated in the industry as a whole and passed on to end customers, non-specialised container and bulk shipping companies may not be able to increase freight rates and may therefore consider abstaining from the transport of HNS.[183] The carriage of substances covered by the HNSC would then presumably be left to the most irresponsible shipowners that try to save costs in other, possibly safety-related, ways.[184]

III. Conclusion

Considering these uncertainties in the impact on industries that may be vital for many States, it is certainly not surprising that there has been a reluctance in ratifying the instrument and that States have "watched each other"[185] for a very long time, waiting for the others to make the first step.

An uncoordinated ratification by only a few States would be very detrimental to their industries, as they may incur a marked competition disadvantage as opposed to

---

[178] See Jacobsson, '2009 "The HNS Convention"', 424.

[179] Boisson, '1996 "La Convention SNDP"', 984.

[180] Ibid., p. 984.

[181] Le Couviour, '1999 "La Convention SNPD"', 1079 et seq.

[182] Griggs, '1996 "Extending the frontiers"', 147, referring to the problems that claimants experienced under English law in 1996.

[183] Little, '1998 "The Hazardous and Noxious Substances Convention"', 559; Røsæg, *2010 "The Rebirth"*, p. 853.

[184] Little, '1998 "The Hazardous and Noxious Substances Convention"', 559.

[185] Wetterstein, '1997 "Carriage of Hazardous Cargoes"', 610.

businesses in non-contracting States.[186] Shipping is a global industry, and customers may choose vessels flying the flag of a non-party to escape liability under the HNSC. Parts of the industry may also decide to ship their goods to ports in non-contracting States and rather transport them on land routes to the final destination.[187] Both occurrences would only damage the industry in contracting States without creating a benefit, neither for the environment nor for potential victims. Given that incidents with HNS cargoes have up to now been by no means as catastrophic as oil spills, the expected cost of such an accident may not be enough of an incentive for a given State to ratify the Convention. It is submitted that a marked political commitment will be necessary to achieve enough ratifications and the fulfilment of the reporting obligations to make the HNSC enter into force. Given the support that the EU has already shown in the adoption of the 2010 Protocol, it is likely that such an initiative will begin among its member States. It however remains to be seen whether the story of the HNSC will turn out to be a success in the end.

C. Conclusion: Civil Responsibility for Environmental Pollution

The existing framework of international agreements for the compensation of environmental pollution has been the result of a long development, in which gaps and shortcomings in the system have been addressed when they became apparent. The civil responsibility regimes that have been discussed in this part of the study are global regimes, open to ratification by all States of the world, and thus, the very instruments art. 235 para. 3 UNCLOS calls for. It is of course always open to debate whether conventional regimes can provide just and reasonable solutions in every case and whether some problems may not be tackled more easily on a domestic or regional level. There are many problems with regard to the establishment of the right liability limits in a world where there are severe differences in the cost of living and the valuation of national resources,[188] and there are practical problems such as currency conversion.[189] It is also true that the binding character of international agreements may keep States from unilaterally adopting higher standards of compensation, the denunciation of an existing regime being a very difficult step.[190]

186 Boisson, '1996 "La Convention SNDP"', 984; Le Couviour, '1999 "La Convention SNPD"', 1093.

187 Tsimplis, '2008 "Marine pollution"', 145.

188 P. Griggs, 'Obstacles to Uniformity of Maritime Law: The Nicholas J. Healy Lecture' (2003) 34, Journal of Maritime Law and Commerce, 191–208 at 201.

189 Gaskell, *2013 "Compensation"*, p. 76.

190 H. Ringbom, 'The "Erika" Accident and Its Effects on EU Maritime Regulation', in M. H. Nordquist and J. N. Moore (eds.), *Current Marine Environmental Issues and the International Tribunal for the Law of the Sea* (The Hague, London: Martinus Nijhoff Publishers, 2001), p. 277, referring to considerations within the EU following the *Erika* accident.

Still, the negotiation and ratification of international conventions is the solution encouraged by UNCLOS, and the existing framework does provide for some unification in this respect.

The current system is, to a very wide extent, a result of path dependency.[191] The dangers linked to different pollutants have been tackled one by one, in different political climates and with the important changes in the shipping industry since the 1960s in the background. With a view to that, the history of the IMO's civil liability conventions must certainly, despite all problems that there might be, be termed a success. For an international convention to take effect, the adoption of a cleverly drafted instrument is not enough; it also needs to be accepted, ratified and implemented by States.[192] It is therefore a good sign that almost all of the instruments discussed here, apart from the HNSC, entered into force and have sometimes even achieved impressive numbers of ratification. The importance of environmental protection through the establishment of civil liability mechanisms on a global level can obviously no longer be neglected by States. In the field of civil liability for passenger claims, this story of success seems to continue with regard to the 2010 Protocol to the Athens Convention (see *infra*, Chap. 5, B.), and that, irrespective of the serious scepticism whether the Convention would ever enter into force.

The fragmentation of the system certainly begs questions of possible overlaps or clashes between the instruments,[193] but at least the interplay between the conventions that are currently in force seems to work surprisingly well. The respective fields of application of the CLC and the BOPC are quite clearly regulated.[194] There is a certain danger of overlap between the BOPC and the HNSC[195] for accidents where there has been bunker oil pollution and HNS damage (see *supra* under B. I.), but the respective provisions allow a line to be drawn. In comparison to that, more important problems seem to arise regarding the interplay between the civil liability regimes and global limitation regimes such as the LLMC. The LLMC tries to rule out oil pollution claims to the widest possible degree, possibly even exaggerating so as to withdraw claims that are not covered by the CLC after all (see *supra*, Chap. 3, E. and G. II.). As shown, the interplay between the two Conventions can still cause some legal uncertainty.

Notwithstanding the fact that the environmental pollution Conventions generally stand alone, their provisions tend to closely mirror each other. The CLC, as the

---

[191] Gaskell, *2013 "Compensation"*, p. 65.

[192] Balkin, *2007 "The Rise and Rise"*, p. 712.

[193] Wetterstein, *2006 "Shipowners' Environmental Impairment Liability"*, p. 1263, footnote 160.

[194] Balkin, *2007 "The Rise and Rise"*, p. 702.

[195] Martínez Gutiérrez, *2011 "Limitation of Liability"*, pp. 162, 188, 189.

oldest civil liability Convention in this field, has thus been a very important model, and the CLC and FUND precedents, in terms of national courts' judgments and of the Fund's compensation practice, will certainly serve as precedents for the other conventions. This modelling is both a chance for and a danger to the international civil liability system. On the one hand, the uniform approach to many legal concepts will create a certain uniformity in the application of the conventions, and the existence of precedents in the application of other conventions will hopefully also foster uniformity of application between different States, this chapter having identified converging interpretations as one of the main dangers to the fairness of the system. On the other hand, the repeated use of provisions whose origins date from the 1960s makes it very difficult to adapt the system to the realities of modern shipping, for instance, with regard to the comparatively new risks of terrorism and piracy. Even if the CLC's provisions concerning the exoneration for risks of war and malicious actions by third parties are not very well adapted to the modern forms of terror and piracy, they have still been used in the BOPC and HNSC. Only when the compulsory insurance for war risks created considerable difficulties in the regime of passenger claims (see *infra*, Chap. 5, B. IV.), it was envisaged to adapt the established rules, but the IMO still preferred to find a solution through the possibility of reservations rather than to depart from the established provisions.[196]

Notwithstanding the overall success of the conventions, there are quite naturally limits in what they can provide for.

Considering the time that it takes for an international agreement to be negotiated and to become effective,[197] it is not overly surprising that there have been initiatives, in the form of court decisions, to overcome some of its effects, namely the limitation of liability, in the short run.[198] Providing for limited liability always means that there will be cases in which not all of the claims are satisfied, and it is rather obvious that claimants will under these circumstances pursue alternative strategies, e.g. by creating claims outside the oil pollution regime, or against persons who are arguably protected by the channelling provisions, or in criminal or administrative law. However, the domestic and European decisions that can be considered to favour alternative legal bases, have in common that the CLC regime has very rarely been directly violated. Most courts, especially the French courts in the *Erika* decisions, considered rather carefully to which extent the CLC would allow the parallel application of domestic law to cases that are arguably outside the scope of the Convention. Whereas it does not seem that the negotiating parties to the Convention foresaw the extent to which that could be driven, the CLC does not pretend to cover all possible claims in relation to an oil tanker accident. Concerning

---

196 Balkin, *2007 "The Rise and Rise"*, p. 709.

197 See the examples given by P. Griggs, 'Making Maritime Law - Do Conventions Work?', in G. Berlingieri, A. Boglione, S. M. Carbone and F. Siccardi (eds.), *Scritti in onore di Francesco Berlingieri. Volume I*, (Genova: Il Diritto Marittimo, 2010), pp. 542 et seqq.

198 Gaskell, *2013 "Compensation"*, pp. 63 et seq.

the claims against persons other than the shipowner, the CLC and the HNS clearly assume that there can be claims against some persons that are not covered by the channelling provisions.

Apart from the criticisms with regard to the liability limits for catastrophic incidents, there are two main consequences that are repeatedly quoted as shortcomings of the liability Conventions. One is related to the fact that the strict liability system in connection with the compulsory insurance fails to render shipowners officially responsible and thus rather encourages careless behaviour. This concept is however at the core of maritime law. Whereas it is submitted that a slightly more liberal understanding of conduct barring limitation could penalise shipowners at least in the worst examples of careless conduct, the idea to create responsibility in shipowners by wide-ranging amendments to the limitation provisions has been vigorously refused.[199] After all, a limitation of liability that is difficult to overcome and taken as granted from the start also serves to shorten proceedings, does away with many problems in enforcement and thus helps to provide quick compensation for victims.

The second criticism, which has been increasingly quoted in recent times, concerns the inapplicability of the CLC and the other civil liability conventions to compensation for purely ecological damage.[200] It is certainly one of the main characteristics of these conventions to only compensate damages to which there is a certain economic value, and in relation to which losses or costs have effectively been incurred. It should be mentioned that the definitions of "pollution damage" as formulated by the conventions already cover damages that could be considered as too remote in certain jurisdictions. An example is the compensation for pure economic loss that is only possible under very strict conditions in England, the common law being clearly less generous than the Fund's practice.[201] Since the system is financed by the shipping and other industries, compensation has to have its limits somewhere, and it does not seem surprising that the contracting States are only willing to make their industries pay for ascertained claims that can be immediately traced back to an incident.

The consequences of purely ecological damage, i.e. the consequential costs that the impact on an ecosystem may have in the long run, are almost invariably borne by the States the coastline of which is affected by the pollutants, and thus, in the very end, by the taxpayer in these countries. As undesirable as this development is, it is already reality for other, more immediate, damages, too. Whereas the civil liability conventions start from the idea that all creditors rank equally and that all claims will suffer a reduction *pro rata* if the total of damages exceeds the limitation fund available, coastal States have in many catastrophic oil spills stepped back from submitting claims, in order to ensure that in any case, private parties' claims would

[199] Balkin, *2007 "The Rise and Rise"*, p. 692.

[200] Meilán Gil, *2005 "Problemas Jurídico-Administrativos"*, p. 180; J. Radovic and A. Brandani, 'Protecting the aquatic environment: Are all costs and benefits accounted for?' (2013) 65, Droit Maritime Français, 839–61 at 850.

[201] Gaskell, *2000 "Pollution, Limitation and Carriage"*, p. 78.

be satisfied to the widest possible extent.[202] This conduct of States may be criticised, with a view to the fact that in this way the taxpayer subsidises on the one hand the coastal communities and businesses, and on the other hand the shipping and the oil industries that would otherwise need to justify themselves if the liability limits are indeed not high enough to cover all claims. Considering that the coastal States' reasons for stepping back in favour of private parties are highly political, it does not seem that the problem can be mitigated by changes in the conventions or the administration of the liability funds. Here again, it must be accepted that civil liability conventions can only provide for compensation as regarded under private law; the distribution of responsibility and the financial burden within States or as between States cannot be addressed by these instruments.

This does not mean that there should not be any civil liability for environmental damage beyond the scope of the civil liability system. It is merely submitted that the conventions discussed may not be the right instrument to regulate this. The analysis of the corresponding obligations under public international law are beyond the scope of this study, but it is to be deplored that States do not assume more responsibility for the ships flying their flag, and that States in general tend to turn a blind eye to pollution that slowly accumulates in the ocean without being caused by a single event.

It must however be stressed that there is, by now, a certain tendency especially of the Fund to compensate so-called "phantom spills", caused by unknown ships, as long as it can be established that the spill was caused by a ship as opposed to any other source.[203] The question has also been a matter of debate in the negotiation of the HNSC, considering that chemical pollution is actually more often than not caused by land-based sources.[204] It is now considered that there must be a reasonable probability that it was caused by a ship for a phantom contamination to be compensated by the HNS Fund. Even if these developments may be rather modest for now, they prove that the civil liability regimes could be a starting point for private actors to assume even more responsibility for environmental protection.

It is however understandable that States will only require these private actors to cover the responsibility for matters that can be attributed to their activity at sea. In this respect, the fragmentation of the system creates a rather long list of activities and matters that will, in the very end, not be covered by any regime. The CLC's and the BOPC's definition of "ship" for instance have the consequence that damages caused by the storage of oil at sea will in many cases be left out of the regime,[205] and the same applies to loading, unloading[206] or bunkering operations at sea where the oil is spilled from a pipe or hose that cannot be considered to form part of the

---

[202] Jacobsson, '2011 "An Ideal International Scheme"', 268; Gaskell, *2013 "Compensation"*, p. 75.

[203] Jacobsson, '2011 "An Ideal International Scheme"', 267.

[204] Göransson, '1997 "The HNS Convention"', 264; Balkin, *2007 "The Rise and Rise"*, p. 699.

[205] Jacobsson, '2011 "An Ideal International Scheme"', 266.

[206] Wetterstein, *2006 "Shipowners' Environmental Impairment Liability"*, p. 1258.

ship.[207] With a view to the growing importance of liquefied natural gas as a source of energy, it is often deplored that the HNSC has still not entered into force[208] and that victims of a LNG accident would only rank as normal creditors within the general maritime limitation regime. It is however submitted that the HNSC itself already creates a systemic lacunae in this respect, as according to art. I para. 5, it applies only to chemicals carried as cargo, so that there is a gap where the damage was caused by a ship using LNG as fuel.[209]

A major blank spot in the system of liability regimes is of course the liability for pollution resulting from the exploration and exploitation of the marine subsoil, namely the liability for accidents on oil platforms, as highlighted by the *Deepwater Horizon* accident in 2010. Without wanting to open the discussion whether a workable civil liability regime would be possible and desirable[210] and could take effect in the near future,[211] it must however be stressed that the balance of interests of the parties involved in the offshore oil industry is different from the situation of shipping. Drilling operations still only take place in waters where there States can put forward sovereign rights to this end (see *infra*, Chaps. 6 and 7), and the operation as such is stationary, so that there is the theoretical possibility of domestic regulation to this end.

These problems regarding the scope of application of the civil liability conventions could arguably only be overcome if there was a unique liability regime for all kinds of maritime pollution[212] or even a unique regime for all kinds of shipowner liability. The possibility has been envisaged in theory[213] more than once, but seems almost possible to put into practice. In the mid-1990s, the UK suggested to the IMO to establish a uniform international regime that would require shipowners to carry mandatory insurance for all possible third-party claims arising out of the operation of a ship. This initiative, an attempt to harmonise at least one feature that is truly common to all civil liability regimes, however failed at an early point in time.[214]

---

207 Radovic and Brandani, '2013 "Protecting the aquatic environment"', 844.

208 Gauci, '1999 "Protection of the Marine Environment"', 31.

209 Gaskell, *2013 "Compensation"*, p. 71, footnote 44.

210 See J. Allen, 'A Global Oil Stain - Cleaning Up International Conventions for Liability and Compensation for Oil Exploration/Production' (2011) 25, Australian and New Zealand Maritime Law Journal, 90–107; M. Smith, 'The Deepwater Horizon Disaster: An Examination of the Spill's Impact on the Gap in International Regulation of Oil Pollution from Fixed Platforms' (2011) 25, Emory International Law review, 1477–516; S. Rares, 'An international Convention on offshore hydrocarbon leaks?' (2011), Lloyd's Maritime and Commercial Law Quarterly, 361–71; Gaskell, *2013 "Compensation"*, p. 86.

211 See ibid., pp. 80 et seqq. for an overview on the more recent efforts in this matter.

212 See Jacobsson, '2011 "An Ideal International Scheme"', 263; Radovic and Brandani, '2013 "Protecting the aquatic environment"', 858 et seq.

213 Gaskell, *2013 "Compensation"*, p. 65.

214 See B. Soyer, 'Sundry Considerations on the Draft Protocol to the Athens Convention Relating to the Carriage of Passengers and Their Luggage by Sea 1974' (2002) 33, Journal of Maritime Law and Commerce, 519–41 at 520; Griggs, Williams and Farr, *2005 "Limitation of Liability"*, p. 111.

Instead, the IMO focussed on an amendment of the civil liability regimes for passengers and also pursued the adoption of an international instrument on wreck removal. It is revealing that a proposal to a more integrated regime resulted in the adoption of two more international agreements on specific matters, continuing the fragmentation of the system.

Chapter 5
Civil Responsibility Regimes for Passenger Claims

The civil responsibility for the claims of passengers involved in shipping accidents is likely to create a multitude of problems, both in law and in practice. Difficulties already arise with regard to the competent court and the law applicable to a given claim, since the passenger and the entity whose responsibility is invoked will very likely have different nationalities and places of business, and there may also be the legal system of the ship's flag and a possible port or coastal State coming into play. Beyond this, the contract of transport is very likely to contain provisions itself.

Furthermore, the principle of limitation of liability for maritime claims may give rise to surprising outcomes for passengers or their bereaved. These problems and the need for an international instrument addressing them have been recognised as early as the nineteenth century, when migration to the New World caused an unprecedented volume of transatlantic passenger travel.[1] A number of tragic accidents, of which the sinking of the *Titanic* in 1912 was the most famous, however resulted first and foremost in the adoption of regulations for the safety of ships, rather than in regulation of liability issues. Solutions for the corresponding problems in air traffic have been found as early as 1929 with the Convention for the Unification of Certain Rules relating to International Transportation by Air (Warsaw Convention). It may thus seem surprising that the first attempts to create international solutions for maritime passenger claims only date from the 1950s and 1960s, and that the first convention to take effect was the 1974 Athens Convention relating to the Carriage of Passengers and their Luggage by Sea (PAL),[2] which only came into force in 1987. Due to the limited number of States

[1] C. Legendre, 'La Convention d'Athènes relative au transport par mer de passagers et de leurs bagages' (1976) 28, Droit Maritime Français, 451–61 at 451.

[2] 1463 UNTS 19.

S.F. Gahlen, *Civil Liability for Accidents at Sea*, Hamburg Studies on Maritime Affairs 32, DOI 10.1007/978-3-662-45555-5_5

party to it and the practical problems with its application, the responsibility regime for sea passengers is still far from being satisfactorily harmonised.[3]

The Convention establishes a liability regime under which the "carrier" of a contract of passage is responsible for fault whereas the burden of proof is reversed to the benefit of the victim in certain cases. The carrier's liability is limited, but there is no compulsory insurance obligation.

The PAL was adopted with a view to facilitate the compensation of passengers[4] by doing away with the difficulties of the conflict of laws and jurisdiction problems. It also aims at protecting passengers from extensive exemption clauses that may be incorporated into their contracts of passage.[5] However, it was not very successful in terms of ratification. A Protocol bringing about substantial amendments was adopted in 2002 and has entered into force on 23 April 2014. The Protocol has already taken some effect even before its entry into force, since EU Regulation (EC) 392/2009, incorporating most of its provisions into EU law, has entered into force on 31 December 2012.

The reasons for the difficulties in harmonising civil liability for passenger trade are manifold. Passenger shipping can be undertaken in very many ways. Not only are there different standards in comfort and, unfortunately, in safety,[6] in different countries, but there are already marked differences between ferries, transport of passengers on cargo ships,[7] short day cruises, and cruise ships even in one country,[8] which also influence the attitude of passengers towards the voyage.[9] Cultural differences and differences in cost of living in a given country also create problems in setting appropriate liability limits—or in deciding whether liability should be limited at all. Uniformisation regarding claims for death and personal injury is very

[3] G. Thomas-Ciora, 'Le transport des passagers par voie maritime' (2006) 58, Droit Maritime Français, 977–84 at 978; P. Delebecque, 'L'évolution du transport de passagers', in J. Muriel Fabre-Magnan, J.-S. Borghetti, O. Deshayes and C. Péres (eds.), *Études offertes à Geneviève Viney* (Paris: Librairie Générale de Droit, 2008), p. 308.

[4] Legendre, '1976 "La Convention d'Athènes"', 455.

[5] H. Honka, 'Nordic Reactions to Maritime Disasters: The Estonia Case', in H. Koch and A. Willingmann (eds.), *Großschäden - Complex Damages*(Baden-Baden: Nomos, 1998), p. 80; W. Müller, 'Passengers Carried by Sea. Should the Athens Convention 1974 be modified and adapted to the liability regime in air-law?' (2000), CMI Yearbook, 665–8 at 667.

[6] W. Müller, 'Faut-il réviser la Convention d'Athènes de 1974 sur le transport de passagers par mer?' (1999) 51, Droit Maritime Français, 5–8 at 8; Soyer, '2002 "Sundry Considerations"', 539, with regard to ferry services in the Far East; K. Lewins, 'Australian cruise passengers travel in legal equivalent of steerage - considering the merits of a passenger liability regime for Australia' (2010) 38, Australian Business Law Review, 127–42 at 132, footnote 42.

[7] B. Kröger, 'Passengers Carried by Sea - Should they be Granted the Same Rights as Airline Passengers?' (2001), CMI Yearbook, 244–52 at 247.

[8] Lewins, '2010 "Australian cruise passengers"', 135, stressing that limits of liability will apply irrespective of the price of the voyage.

[9] N. Gaskell, 'The Zeebrugge Disaster: Application of the Athens Convention 1974' (1987), New Law Journal, 285–8 at 286; Kröger, '2001 "Passengers Carried by Sea"', 246, with regard to self-insurance that may be taken out for a longer cruise, but not for a ferry crossing.

difficult where standards are different, and may not be beneficial to any party if the limits are set in the middle between extremes.[10] Consequentially, the limits of liability in the PAL were criticised as too high and too low by different States at the same time,[11] and this is widely considered to be the major reason for the limited success of the convention.

Unlike the victims of environmental harm, who generally do not have a contractual relationship with one of the actors of the maritime trade, passengers have consciously entered into an agreement by which they exposed themselves to the risks of the maritime transport. Ironically, this fact seems to improve their situation with regard to the fact that in most cases, when taking the ship for recreational purposes on a cruise, or in order to reach their holiday destination, they will be considered as consumers under the applicable domestic law.[12] Especially if EU law applies, there may be an enhanced protection available that can create some conflicts with the principles of maritime law,[13] in which liability is generally limited for the sake of the viability of the industry. This is one of the big differences between the liability regimes for passenger claims and for environmental harm. Victims of oil or other pollution can generally not invoke consumer rights, as the interest in an unharmed environment can hardly be termed a consumer good. Businesses in the tourism or fishing industry may be injured in their economic expectations, but being entrepreneurs themselves, cannot put forward consumer rights.

The special situation of passengers can therefore create many new rights, not only as against the shipowner or "carrier" or the entity controlling the ship, but possibly also as against the shipyard or other parties for delivering or operating a faulty product. At this point in the study, the analysis centres on claims brought under or according to the principles of the existing liability regime for passenger claims, but it will allude to possible claims outside the system. Further considerations can be found in Part III of the study (Chaps. 6 and 7), dealing with questions of jurisdiction and the conflict of laws.

In comparison to incidents that have caused environmental pollution, the number of maritime accidents that have given rise to passenger claims, e.g. involving cruise ships or ferries, is comparatively small, even if accounting for the tragic *Estonia* catastrophe in 1994. It must however be mentioned that accidents like the *Costa Concordia* grounding or the three days' power failure on board the *Costa Allegra* in

---

[10] Honka, *1998 "Nordic Reactions"*, p. 80.

[11] F. L. Wiswall, '2002 Protocol to the Athens Convention' (2003) 1, Benedict's Maritime Bulletin, 17–9 at 17; B. Soyer, 'Boundaries of the Athens Convention: What you see is not always what you get!', in D. R. Thomas (ed.), *Liability regimes in contemporary maritime law* (London: Informa, 2007), para. 11.1, footnote 6.

[12] Delebecque, *2008 "L'évolution du transport de passagers"*, p. 311.

[13] See E. Røsæg, 'News under the Athens Sun - New Principles and Lost Opportunities of the Athens Convention 2002' (2004) 46, Scandinavian Studies in Law, 153–73 at 165; C. Scapel, 'Le droit positif de la croisière maritime en France' (2012) 64, Droit Maritime Français, 306–13 at 306: "a compromise".

early 2012 have apparently alerted the maritime industry for the physical and legal consequences that may arise following an accident with a cruise ship[14] or a big ferry.

A. The 1974 Athens Convention (PAL)

The PAL is not a very successful Convention in terms of ratification. Having been adopted in 1974, there are only 35 States parties to it as of February 2014, only representing 46.53 % of worldwide maritime tonnage, and only 24 States parties to a Protocol of 1976 increasing the liability limits, representing about 45 % of worldwide maritime tonnage.[15] The main reason for the limited number of ratification is said to be the difficulty to find adequate liability limits for passenger claims, not only in the PAL itself, but also in its interplay with other conventions or domestic provisions on the general limitation of maritime claims.

The PAL differs from general limitation regimes such as the LLMC in that it applies limits to each passenger claim separately, on a *per capita* basis, whereas the LLMC 1976 and 1996 respectively set out limits for all maritime claims to a given incident, on a *per vessel* basis. Under the PAL, it therefore does not make any difference for a claimant if other passengers have also been injured, as there is a limitation fund for each passenger, whereas the LLMC sets out a single limitation fund for a given incident on one ship that all claimants would have to share. Art. XIX PAL however clarifies that all the passenger claims arising under its provisions may again be subject to a general limitation of liability under a regime such as the LLMC. This means that the shipowner will be able to limit his liability on a *per vessel* basis in a disastrous accident, where the total of limited passenger claims under the PAL exceeds the limitation fund under the LLMC.

The PAL has, however, significantly influenced domestic legislation on passenger claims. Many States that did not sign the Convention have adopted national laws that are rather close to the provisions of the Convention, only diverting from it with regard to the liability limits.[16] The PAL has, in this respect, become an instrument to safeguard passengers from abuses of the right to contractual

---

[14] Ibid., p. 306.

[15] See http://www.imo.org/About/Conventions/StatusOfConventions/Documents/Summary%20of%20Status%20of%20Conventions.xls (last visited 7 March 2014).

[16] Müller, '1999 "Faut-il réviser la Convention d'Athènes ?"', 5; Lewins, '2010 "Australian cruise passengers"', 130; F.-X. Pierronnet and A. Luquiau, 'L'indemnisation des dommages subis par les passagers transportés par mer: de nouvelles voies possibles?' (2000) 18, Annuaire de droit maritime et océanique, 265–96 at 265, referring to France before the coming into force of the provisions of the 2002 Protocol to the PAL by way of the European Regulation. Also see the results of the 1997 IMO questionnaire on national legislation, P. K. Sotiropoulos, 'Carriage of Passengers and their Luggage by Sea - The Athens Convention 1974 and its Protocols. Synopsis of the Replies to the Questionnaire' (1998), CMI Yearbook, 129–44.

freedom.[17] Furthermore, at least many cruise operators try to incorporate parts of the PAL into their contracts of carriage[18] and thus submit passengers to the Convention's legal framework through contractual provisions. It is, and seemingly rightly, argued that their main interest to do so are the limits of liability set out in the 1974 PAL that are, to modern standards, ridiculous.[19] Moreover, the contractual provisions often contain jurisdiction clauses that make it difficult for passengers to bring a claim in a legal system that is hostile to the incorporation of these clauses. Contractual incorporation will however not be focused on at this point, as the focus here is on the statutory provisions of the PAL. With the UK, Greece, Liberia and the Bahamas, the latter two being popular flag States for cruise vessels,[20] the Convention was in its 1974 version ratified by important maritime nations, even if its overall number of contracting States is modest.

I. Scope of Application

The PAL's scope of application raises questions that are very similar to the problems arising under the liability regimes for environmental pollution, especially with regard the applicability of the Convention in the courts of non-contracting States. Still, the Convention essentially deals with claims that arise in a contractual relationship, between the passenger and the carrier, and therefore also determines its applicability through the existence of a contract of carriage.

Within this contractual relationship, there are however only certain claims covered by the Convention. It must be stressed at this point already that the PAL only applies to claims of passengers against the carrier for death and personal injuries and the loss of damage. The Convention does not apply to the multitude of other claims that could arise in this contractual relationship,[21] e.g. for delays or other shortcomings in the performance of the contract. These claims outside the scope of the PAL will be governed by the provisions of the contract and the domestic law applicable to the claim.

---

[17] Martínez Gutiérrez, *2011 "Limitation of Liability"*, p. 117.

[18] See Lewins, '2010 "Australian cruise passengers"', 140, in the appendix to the article setting out the terms and conditions of certain cruise providers.

[19] Ibid., p. 135.

[20] Kröger, '2001 "Passengers Carried by Sea"', 246.

[21] I. Volner, 'Forum Selection Clauses: Different Regulations from the Perspective of Cruise Ship Passengers' (2006) 8, European Journal of Law Reform, 439–76 at 475.

1. Geographical Scope of Application

According to art. II PAL, the Convention applies to any international carriage if, firstly, the ship in question is flying the flag of or is registered in a State party, or secondly, if the contract of carriage has been made in a State party, or, thirdly, if the place of departure or destination according to the contract of carriage, is found in a State party. Art. XVII PAL further sets out a list of competent courts, which the claimant can choose from, "provided that the court is located in a State Party to this Convention".

This however begs the question for the law applicable to claims arising in contracts of carriage that fulfil the conditions of art. II PAL, but are brought in non-contracting States. This problem has repeatedly emerged in France, which is not a party to the PAL. Unfortunately, it does not seem that there is a uniform answer to the question as treated by the French courts. A rather recent decision of the *Cour d'Appel* of Nancy[22] concerned the *Sea Diamond* incident in 2007, in which a cruise ship grounded off the island of Santorin, Greece, while taking a usual sea route in calm weather. Almost all of the crew and passengers on board the *Sea Diamond* could be rescued by help from shore, but two passengers died when they were trapped in the hull of the listing ship. Surviving passengers sued the tour operator of the cruise for the loss of personal belongings and for immaterial damages, both for mental pain and suffering during the accident and for the loss of enjoyment in their holidays. The tour operator argued that he was the "carrier" under the broad understanding of the PAL, and invoked the limits of liability of the Convention, the provisions of which had been incorporated into the tourist travel contract with the claimants. The *Cour d'Appel* however rejected this argument. Following the courts' reasoning, the contract was governed by French law, containing consumer protection provisions in favour of travellers, and which, at the time, did not allow for the implementation of international conventions on the limitation of liability,[23] not even by way of a contract clause.

The decision can of course be criticised for disregarding the principle of party autonomy on the one hand and the existence of an international agreement that is meant to protect passengers' rights on the other hand. The *Cour d'Appel* has however decided it in favour of consumer protection beyond the level provided for by the PAL, relying on the mandatory provisions of the French law as it then was.

The exact opposite was held in an earlier decision of the *Cour d'Appel* of Paris in 2000,[24] and this even in the situation in which the PAL applied by virtue of its own scope of application rule in art. II PAL, the vessel's destination being a port in the United Kingdom. The court expressly stated that by virtue of art. II para. 1 (c) PAL,

[22] "*Sea Diamond*", (2011) 64 Droit Maritime Français 604 (Cour d'Appel de Nancy, 20 January 2011).

[23] Ibid., p. 608.

[24] "*Prince Laurent*", (2000) 52 Droit Maritime Français 26 (Cour d'Appel de Paris, 19 May 1999).

every maritime transport of passengers to and from the United Kingdom was governed by the dispositions of the PAL, even if France was not a party to the Convention.[25] The decision however concerned a passenger's accident on board the vessel and was mainly concerned with the application of the PAL's prescription period which was significantly shorter than the corresponding time frame under domestic law. It is not inconceivable that the decision was rendered for policy considerations, allowing the court to dismiss the action without inquiring into the facts of the case that dated back to 1990.[26] In its assertiveness to apply the PAL to every case that fulfils the conditions of art. II PAL, the decision is however interesting. If it was recognised on an international scale that the PAL is applicable to all cases taking place during voyages to and from a State party, on a ship flying the flag of a State party and to all contracts of carriage that have been made in a State party, the PAL could become a very important international regime notwithstanding its limited number of ratifications. Especially with regard to the criterion of art. II para. 1 (b) PAL, the contract made in a State party, the scope of application could become very wide and possibly unclear in the times of internet bookings.[27]

In the decision rendered by the *Cour d'Appel* of Paris, it is furthermore noteworthy that the court completely omitted to proceed with a proper conflict-of-law analysis. It never inquired into the law governing the injured passenger's contract of carriage or into the qualification of the prescription period that could possibly either have been seen as a procedural question—governed by the law of the forum—or as a question of a contractual term, governed by the law of the contract.

Concerning France, both decisions are however outdated, as the underlying domestic law on tourist travel contracts has been amended as of 1 December 2010, now allowing for the implementation of international conventions for the matter of limitation of liability.[28] The amendment has been effectuated with regard to the EU's adherence to the 2002 Protocol to the PAL. By way of a European directive, some of the Protocol's provisions will be given effect in the EU member States, even before the entry into force of the Protocol on an international scale (see *infra* under B.).

Similar to the question that arose in context with the application of the CLC by the courts of a non-contracting State (see *supra*, Chap. 3, A. I. 1.), there is thus a lot of uncertainty whether a given domestic court will choose to apply the PAL. As a rule, the application of the PAL by the courts of a State that is not a party may be rejected,[29] but the questions of competence and conflict of laws that arise in this context are however likely to create surprising outcomes. With regard to the fact

---

[25] Ibid., p. 28.

[26] J.-S. Rohart, 'Observations CA Paris 19.05.1999' (2000) 52, Droit Maritime Français, 29–31 at 31.

[27] Lewins, '2010 "Australian cruise passengers"', 134, footnote 61.

[28] Scapel, '2012 "Le droit positif de la croisière"', 312.

[29] Very explicitly for the United States: R. D. Peltz, 'The Athens Convention Revisited' (2012) 43, Journal of Maritime Law and Commerce, 491–520 at 517.

that there are significantly less State parties to the PAL than to the CLC, the problem may be more pressing. Furthermore, claims for death and personal injury to passengers may, possibly even more than environmental claims, give rise to policy considerations by courts.

In this context, it is interesting to mention that the possibility of a "national exception", the existence of which is assumed by certain authors under the CLC (see *supra*, Chap. 3, A. III.), is effectively codified under the PAL. According to art. XXII PAL, every State party to the PAL may declare at the time of ratification that it intends not to apply the PAL to cases where both the carrier and the passenger are nationals or subjects of that State. The possibility to provide for the application of domestic law in domestic cases has to some extent been enacted with regard to the situation of States such as Greece or Norway, which rely on ferry services as a part of their infrastructure and have created suitable, well-adapted national legislation[30] on this. This of course means that a certain group of passengers that are nationals of the State they sue in may receive a more favourable compensation than other passengers under the PAL, even if they are all victims of the same accident.[31] Furthermore, it should also been mentioned that there are countries in which domestic legislation is entirely insufficient to even guarantee the most basic safety standards,[32] let alone a compensation system for accidents. Countries of this kind may make use of the opt-out possibility to protect their domestic passenger services from additional cost.

The national exception under art. XXII PAL can be criticised with the same arguments as the unwritten national exception under the CLC. Of course, the need to declare the non-application of the PAL to domestic cases in written form is preferable to the legal uncertainty created by the mere assumption of the exception under the CLC in certain countries. It however seems that cases with unjust outcomes for different groups of passengers are more likely under the PAL than under the CLC. The oil pollution damage covered by the CLC will, as a rule, be perceived on land, and under the CLC, victims can bring a claim in the courts of the State where the damage took place. Oil pollution victims will therefore generally be able to sue in their place of residence. Even if they are treated differently, at least every group of victims will be subject to their home jurisdiction. Claims covered by the PAL, on the other hand, will take place on board ships. If claimants want to bring their actions under the PAL, they may only theoretically have the choice of venue as provided for by art. XVII para. 1 PAL, but be restricted in it by the fact that not all of these countries are contracting States to the PAL. If, out of the remaining venues in which the PAL is effectively in force, none is their home jurisdiction, they will not be able to invoke the benefits of the national exception after all.

[30] Kröger, '2001 "Passengers Carried by Sea"', 245; R. Shaw, 'Carriage of Passengers', in Y. Baatz (ed.), *Southampton on shipping law* (London: Informa, 2008), p. 147.

[31] Legendre, '1976 "La Convention d'Athènes"', 460.

[32] P. Boisson, 'Navigabilité du navire et responsabilité des États', in A. Knops (ed.), *Liber Amicorum Roger Roland* (Gent: Larcier, 2003), p. 63.

If the national exception in art. XXII PAL has been codified in order to induce States with a more generous domestic liability regime to ratify the convention,[33] it is thus still of a rather reduced importance.

2. Claims and Contracts Covered

Art. III PAL sets out the liability of the carrier for damages suffered as a result of death or injury by a passenger, or by the loss or damage of passengers' luggage items, provided that the damage was sustained during the "carriage" and is the result of the fault or neglect of the carrier. The notion of "carrier" and the circumstances under which he can be held liable are discussed *infra* under II., but even the question of the persons who can invoke the PAL and the claims they can put forward gives rise to some discussions.

Following the wording of art. III para. 1 PAL and the definitions under art. I PAL, the question for the understanding of "passengers" and "carriage" arises. Whereas this may seem an easy inquiry at first sight, it can create some problems. According to art. I para. 4 PAL, the understanding of "passenger" presupposes that there is a contract of carriage for the voyage of this very passenger, or an agreement under which the person has the right to accompany a vehicle or animals. This can be difficult with regard to different understandings of what a "contract of carriage" is constituted of and how it is concluded in different legal systems. Art. I para. 2 PAL defines "contract of carriage" as "a contract made by or on behalf of a carrier for the carriage by sea of a passenger and his luggage, as the case may be". This is a rather wide definition, especially as it does not set out that the contract must necessarily be concluded between the carrier and the passenger. Problems have been detected under English law with regard to voyages that are provided free of charge,[34] which means that for a lack of consideration between the carrier and the passenger, there may not be properly speaking a contract. It however seems possible to still subject these cases to the convention regime, by arguing that the contract is made by the person offering the free tickets on behalf of the passenger. Furthermore, the English legislation implementing the PAL addresses the problem by including gratuitous passages into the regime.[35]

This seems possible as long as there is a commercial dimension to the voyage. Considerable problems will arise with regard to sea carriage that is offered for

[33] In this sense, Legendre, '1976 "La Convention d'Athènes"', 460.

[34] Gaskell, '1987 "The Zeebrugge Disaster"', 285; Soyer, *2007 "Boundaries of the Athens Convention"*, para. 11.17, footnote 46, both referring to the *Herald of Free Enterprise* disaster where certain persons on board had obtained tickets for a nominal fee following an advertising promotion; Griggs, Williams and Farr, *2005 "Limitation of Liability"*, p. 98, however pointing out that the nominal fare, though not adequate, would have been enough to establish existence of a contract of carriage.

[35] See Soyer, *2007 "Boundaries of the Athens Convention"*, para. 11.17.

private reasons, so as a gratuitous voyage on a friend's yacht.[36] It seems on the one hand right to rule out voyages that are clearly not made for commercial reasons and not to submit the hospitable owner of a small pleasure craft to a mandatory liability regime.[37] The distinction may in some cases however be hard to draw if the trip is offered free of charge or only for a contribution to the cost of the operation of the craft and as a half-private favour, but with a commercial purpose in mind, such as improving a relationship that is properly speaking a business relationship. In domestic legal systems that are generous in recognising a contract in situations even where there is no direct exchange of performances, such as Germany,[38] the PAL would possibly be applicable to a wide range of rather non-commercial voyages. This is possibly another reason that renders States reluctant to ratify the Convention.

Following the wording of art. I para. 4 PAL, the definition clearly rules out claims of blind passengers or stowaways,[39] and of persons that have been admitted on board of ships for reasons other than a contract of carriage, e.g. persons rescued at sea. Whereas this seems fair with regard to the fact that the "carrier" never consciously agreed to take these people on board and may not be prepared to guarantee their safe passage, the complete omission of crew claims from the scope of application of the PAL can certainly be criticised, especially as there may be persons on board a vessel that cannot be easily classified as either crew members or passengers.[40] Claims of crew members, even if they arose under the same circumstances as the passenger claims, may thus be completely differently treated, which appears hardly fair at first sight.[41] Of course, the underlying labour law may lead to different considerations for these claims.[42] It seems likely that crew claims have been omitted from the scope of application of the PAL in order not to confound the law of maritime labour with the passenger regime governed by considerations of consumer protection.

---

[36] Ibid., para. 11.17, referring to the *Marchioness* accident in which the boat was chartered by an individual for celebrating a party with guests.

[37] In this sense Gaskell, '1987 "The Zeebrugge Disaster"', 285.

[38] German law recognises contracts even in situations where performances are provided free of charge, but also knows the mere favour, not considered as a contract and not entailing obligations to the other party. The distinction is drawn according to the facts of the case, namely the economic importance of the performance and the purpose of the parties in granting and accepting it. See O. von Palandt and H. Sprau, *Bürgerliches Gesetzbuch*, 73rd edn. (München: Beck, 2014), Einf. v. § 662, para. 4.

[39] I. Arroyo Martínez, 'The Contract of Maritime Passage - A Commentary on the Athens Convention Relating to the Carriage of Passengers and their Luggage by Sea, 1974', in P. Wetterstein and A. Beijer (eds.), *Essays in Honour of Hugo Tiberg* (Stockholm: Juristförlaget, 1996), p. 16.

[40] Gaskell, '1987 "The Zeebrugge Disaster"', 286, referring to staff employed by outside companies rather than the carrier himself.

[41] Ibid., p. 287; Gaskell, '1998 "New limits for passengers"', 331.

[42] See Gaskell, '1987 "The Zeebrugge Disaster"', 287, on considerations under English law for crew members victims of the *Herald of Free Enterprise* incident.

This distinction in the PAL is different from the HNSC regime that applies the same provisions to passenger and crew claims[43] if the damage has been suffered by hazardous and noxious cargo (see *supra*, Chap. 4, B. I. 3.).

As to the understanding of "carriage", the PAL gives a rather detailed definition in art. I para. 8. In short, the period between the embarkation and the disembarkation of the passenger is covered, including the tender from the shore to the ship if this service is provided by the carrier. With regard to luggage, the transport is generally covered as soon as the luggage has been taken over by the carrier. The provision may be of utter importance in the rather frequent cases of accidents during embarkation and disembarkation, both on gangways and on smaller craft used for the tender to the ship.[44]

The PAL only covers claims for death or personal injury of passengers or for loss or damage of their luggage. Many claims that may arise in connection with sea voyages, such as claims for delay in ferry passages or claims for insufficient quality of services in cruise contracts,[45] are therefore not subject to the PAL. Furthermore, it is highly debatable to what extent consequential damage caused by injury or the loss or damage to luggage is covered by the PAL.[46] The answer to this question will depend on the jurisdiction in which the claim is pursued. Actions for very remote damages may be ruled out in some jurisdictions, but admissible in others.

With regard to the cases of "death and personal injury", it has however to be defined to what extent "personal injury" can comprise fear, anguish or mental distress that has been experienced during the sea voyage, when there was no corporeal harm. It has very adroitly been argued that the notion of "personal injury" differs from the term "bodily injury" in other international conventions, which can be taken as a hint that the drafters of the PAL intended at least not to bluntly rule out pure mental distress from the convention's scope of application.[47] In the first Spanish decision concerning the *Costa Concordia* accident,[48] decided with a view to the PAL as contained in EU Regulation 392/2009 (see *infra* under B.), the claimants also filed suit for mental anguish suffered during the accident and the evacuation of the vessel. The claim was put forward together with the claim for physical injuries suffered. Even if the court did not explicitly state so, it seems that it treated the psychological and immaterial damages under one head of damage with the physical injuries, and thus, as a PAL claim. In the proceedings, the amounts claimed by the passengers for mental suffering were not high enough to trigger the

[43] Gaskell, *1996 "The Draft Convention"*, p. 270.

[44] Arroyo Martínez, *1996 "The Contract of Maritime Passage"*, p. 39, footnote 26.

[45] See Volner, '2006 "Forum Selection Clauses"', 443 et seq.

[46] See the discussion in Soyer, *2007 "Boundaries of the Athens Convention"*, pp. para. 11.26.

[47] Kröger, '2001 "Passengers Carried by Sea"', 249; also see Soyer, *2007 "Boundaries of the Athens Convention"*, para. 11.27; Martínez Gutiérrez, *2011 "Limitation of Liability"*, p. 131, footnote 137, both referring to the identically worded provision under the 2002 Protocol.

[48] Juzgado de Primera Instancia N° 63 de Madrid, 18.04.2013, unreported. The case is presented in detail by C. Lubach, 'Zur Havarie der "Costa Concordia": Erste Gerichtsentscheidung in Spanien' (2013), Transportrecht, 405–7.

limitations of liability, the application of which would have been a clear indication of the claim being a PAL claim. The conclusion is however open to debate as the court also stated that the claim was mainly based on a bad performance of the package travel contract underlying the cruise, even if Regulation 392/2009 was also referred to in order to limit liability for the claims for lost luggage, and in order to justify the amounts rewarded for physical and psychological injury. The case illustrates very well the difficulties to decide whether claims are covered by the PAL or not.

Especially with regard to psychological injury, it may thus be hard to draw a distinction. Whereas it only seems fair that extreme anguish on board a sinking ship that may result in a further medical condition even after the end of the voyage, should be treated in the same way than a physical injury, mental distress can take a wide variety of symptoms and be triggered by very minor events. It should not be forgotten either that a claim brought under the PAL for a maritime accident may at first glimpse be beneficial to claimants because of the rules on liability (see *infra* under II.), but that claimants may, in a second step, then be faced with the limitation of liability.

3. Jurisdiction

In contrast to the civil liability regimes for environmental pollution that have been discussed *supra* in Chaps. 3 and 4, the PAL provides for rules on responsibility in a legal relationship that is governed by a contract. A passenger and the carrier know that they will be dealing with each other, and they have the possibility to anticipate a judicial venue for conflicts arising out of the contract. In the absence of contractual provisions, virtually all legal systems have refined procedural law on the allocation of jurisdiction between parties to a contract, very often linked to their respective domiciles or places of business. In both cases, the judicial venue will be obvious, and it will be irrelevant that the events leading to the claim took place at sea. It is therefore not for providing claimants with an ascertainable judicial venue if the PAL contains provisions on jurisdiction.

Art. XVII para. 1 PAL allocates jurisdiction, at the choice of the claimant, in the home jurisdiction of the defendant, at the place of destination or departure of the voyage. The claim can also be brought in the home jurisdiction of the claimant or in the jurisdiction where the contract was concluded, but only if these jurisdictions are also the home jurisdiction of the defendant. The provision allocates jurisdiction according to rather well-established factors for contractual relationships. The more important characteristic concerning the allocation of jurisdiction is however to be found in art. XVIII PAL. The provision declares any contractual term by which the passenger's rights as set out by the PAL are curtailed, namely a restriction of the jurisdictions presented in art. XVII PAL, is void. It is obvious that the PAL means, in this respect, to protect the passenger as the weaker party to the contract, ensuring that notwithstanding the superior bargaining power of carriers, passengers will still enjoy the rights they have been granted under the PAL.

The PAL being anterior to the coming into force of the Brussels Regulation within the EU, its provisions on jurisdiction, recognition and enforcement remained in force and superseded the Regulation's rules. The adoption of EU Regulation 329/2009 for the EU member States and the coming into force of the 2002 Protocol to the PAL in 2014 however create serious difficulties in the interplay of the legal frameworks (see *infra* under B. I.).

II. The Carrier's Liability

According to art. III para. 1 PAL, "the carrier shall be liable for the damage suffered as a result of the death of or personal injury to a passenger and the loss of or damage to luggage if the incident which caused the damage so suffered occurred in the course of the carriage and was due to the fault or neglect of the carrier or of his servants or agents acting within the scope of their employment".

The liability provision for passenger claims is in a number of respects very different from the liability for environmental pollution. First of all, under the 1974 PAL, the liability for passenger claims is still based on fault, and it is only with regard to the burden of proof for certain incidents that the claimants obtain advantages when bringing their claim. Secondly, the notion of the "carrier" requires some interpretation, especially as the PAL also defines and mentions the notion of "performing carrier" (arts. I para. 1 (b), IV PAL) and the relationship between these two.

1. The Notion of "Carrier"

According to art. I para. 1 (a) PAL, "carrier" means "a person by or on behalf of whom a contract of carriage has been concluded, whether the carriage is actually performed by him or by a performing carrier". The PAL thus clearly sets out the responsibility of the carrier by virtue of the conclusion of the contract of carriage, irrespective of whether the carrier is actually involved in the maritime business. This means that travel agents or tour operators merely organizing cruises or sea crossings, but notwithstanding concluding contracts for these voyages, can be liable under the PAL.[49] The possible conflict with tourist travel law is therefore inherent in the Convention.[50] The understanding of the term "carrier" under the PAL is extremely broad and can cover parties that may not have expected to be liable under a maritime convention.

---

[49] Griggs, Williams and Farr, *2005 "Limitation of Liability"*, p. 97; Martínez Gutiérrez, *2011 "Limitation of Liability"*, p. 117.

[50] See the extensive discussion in Soyer, *2007 "Boundaries of the Athens Convention"*, para. 11.35 et seqq.

The term of "performing carrier" is defined in art. I para. 1 (b) PAL as "a person other than the carrier, being the owner, charterer or operator of a ship, who actually performs the whole or a part of the carriage". It is therefore here that the actual maritime players may be concerned under the PAL. Following the provisions in art. IV para. 1 and 2 PAL, the carrier as defined by art. I para. 1 (a) PAL however remains liable for the whole of the carriage, even if the incident giving rise to the damage occurred during the carriage as carried out by the performing carrier. In addition, the performing carrier is bound by the provisions of the PAL for his performance. According to art. IV para. 4 PAL, the liability of the carrier and the performing carrier is joint and severable to the extent that they are both responsible under the PAL. The recourse between the carrier and the performing carrier of course remains possible, art. IV para. 5 PAL.

Without alleviating the responsibility of the carrier, art. IV PAL therefore adds the liability of the performing carrier, giving the claimant passenger under the PAL the choice between two—or even more—possible defendants.[51] With regard to the fact that the performing carrier is only liable for the part of the carriage performed by him, claimants may however be more inclined to address the carrier who is responsible for the whole of the contract of carriage.

2. Liability and Burden of Proof

As has been mentioned above, the PAL 1974 still sets out that the carrier's liability is based on fault. It is only with regard to the burden of proof for typical shipping incidents that claimants obtain advantages to set out their claim.[52] Art. III PAL concerning the carrier's liability reads:

Art. III
Liability of the carrier

1. The carrier shall be liable for the damage suffered as a result of the death or the personal injury to a passenger and the loss of or damage to luggage if the incident which caused the damage so suffered occurred in the course of the carriage and was due to the fault or neglect of the carrier or of his servants or agents acting within the scope of their employment.
2. The burden of proving that the incident which caused the loss or damage occurred in the course of the carriage, and the extent of the loss or damage, shall lie with the claimant.
3. Fault or neglect of the carrier or of his servants or agents acting within the scope of their employment shall be presumed, unless the contrary is proved, if the death of or personal injury to the passenger or the loss of or damage to cabin luggage arose from or in connexion with the shipwreck, collision, stranding, explosion or fire, or defect in the ship. In respect of loss of or damage to other luggage, such fault or neglect shall be presumed, unless the contrary is proved, irrespective of the nature of the incident which caused the loss or damage. In all other cases the burden of proving fault or neglect shall lie with the claimant.

[51] Arroyo Martínez, *1996 "The Contract of Maritime Passage"*, pp. 26 et seq.

[52] Legendre, '1976 "La Convention d'Athènes"', 454.

With regard to claims for death and personal injury and damage to cabin luggage, the provision draws a marked difference between damages that are caused by "shipwreck, collision, stranding, explosion or fire, or defect in the ship" and other claims. The reason for this is of course the difference between damages that are caused by a typical maritime peril and other damages that could also have happened on shore, such as a passenger slipping and falling on one of the decks.[53] In claims that have been caused by a danger that is genuine to the transport by sea or that are directly linked to the function of the ship as a means of transport, such as a foundering or collision, it will be very hard for the passenger to prove any fault on behalf of the carrier or the performing carrier.[54] Therefore, the burden of proof is reversed according to art. III para. 5 PAL and the carrier's fault is presumed, however with the possibility for him to rebut the presumption.

With regard to all other claims, the passenger bears the burden of proof. This rule acknowledges very clearly the fact that a passenger on board a vessel—during a cruise of several days or even weeks, but also during a short sea crossing on a ferry—will move freely, possibly embark on sports and other activities, in the same way that he would do so during a stay in a hotel.[55] It is furthermore argued, and just as rightly, that the situation is neither comparable to air traffic or road traffic, where passengers are more immobile in the vehicles and are under better surveillance, which can justify strict liability schemes for passenger air transport.[56] At a very early point in time, it has been perceived as unfair towards carriers and performing carriers to render them liable with a reversed burden of proof for dangers that are not inherent to the sea passage, but present very general, everyday life risks.[57] The liability rule under the PAL, by which the carrier's responsibility is presumed for typically maritime risks only, is therefore equitable.[58]

If the cases in which the carrier's liability is presumed are given a closer look, it becomes obvious that the PAL's drafters were concerned with claims in which the danger of the sea transport as such materialises. This is obvious for the cases of "shipwreck, collision, stranding" as they will not occur on land. With regard to "explosions or fire", they can of course also—perhaps with less likelihood—take place on land, but will not be as dangerous there as in the secluded environment of a ship. The last case given of a "defect in the ship" is both rather narrow, as it would rather not encompass shortcomings such as inadequate manning,[59] and very wide, as it does not refer to the cause by which the defect arose. Certainly, a "defect in the ship" may occur due to sabotage or act of God and it is hardly fair to presume the

53 Lewins, '2010 "Australian cruise passengers"', 130.

54 Gaskell, '1987 "The Zeebrugge Disaster"', 286; Tsimplis, '2009 "Liability in respect of passenger claims"', 126; Scapel, '2012 "Le droit positif de la croisière"', 311.

55 Müller, '1999 "Faut-il réviser la Convention d'Athènes?"', 7.

56 See Soyer, '2002 "Sundry Considerations"', 522, especially with regard to air traffic.

57 Legendre, '1976 "La Convention d'Athènes"', 455.

58 In this sense, Müller, '1999 "Faut-il réviser la Convention d'Athènes ?"', 8.

59 Griggs, Williams and Farr, *2005 "Limitation of Liability"*, p. 102.

carrier's responsibility for that.[60] It is however submitted that in these cases it will be possible for the carrier to rebut the presumption. In the 2002 Protocol, a definition has been introduced in art. III para. 5 (c), which essentially refers to a shortcoming in equipment necessary for the safe operation, but it does not do away with the difficulties mentioned and has been criticised as needlessly complicated.[61]

The rules on liability are slightly different for luggage claims, where a difference is drawn between cabin luggage that is still under the control of the passenger (art. I para. 6 PAL) and other luggage that is in the custody of the carrier or that the passenger cannot access during the voyage. With regard to loss or damage to the latter, the carrier's liability is presumed, as it will be very hard for a passenger to establish a claim for fault or negligence if the luggage is not under his control.[62] With regard to cabin luggage, the same system as for death and personal injury claims applies, the carrier's liability only being presumed for the damages caused by typically maritime perils.

The contractual relationship between the carrier and the passenger is stressed in the provision on the liability for valuables. According to art. V PAL, the liability for valuables such as monies and jewellery is excluded unless the parties agreed that the carrier would keep them safe.

The PAL does not answer the question of how "fault or neglect" are to be understood and leaves this issue to the domestic law as applied by domestic courts.[63] This can naturally be a source of conflicting interpretations by different courts. A strict liability, comparable to the shipowner's strict liability under the environmental liability regimes, has been discussed in the past,[64] and the 2002 Protocol to the Athens Convention does set out the carrier's strict liability for claims caused by a shipping incident up to a certain sum. The provision will be discussed *infra* under B. II.

Under the original PAL 1974, there is no insurance obligation to the carrier. It has however been observed that the total absence of insurance for a passenger ship would hardly be conceivable and that most claims are currently covered by P & I insurance,[65] so that an introduction of compulsory insurance, as in the 2002 Protocol, actually has the effect of doing away with unfair competition by uninsured carriers.[66] The matter may therefore rather be one of adequate limits than of sheer existence of insurance.

---

[60] Legendre, '1976 "La Convention d'Athènes"', 454.

[61] Griggs, *2010 "Making Maritime Law"*, p. 544.

[62] Tsimplis, '2009 "Liability in respect of passenger claims"', 127.

[63] Martínez Gutiérrez, *2011 "Limitation of Liability"*, p. 118.

[64] P. Godin and A. Vialard, 'Cinquième journée Ripert, deuxième intervention: Vers une révision de la Convention d'Athènes de 1974 sur le transport de passagers par mer' (1998) 50, Droit Maritime Français, 733–4 at 734.

[65] Gaskell, '1998 "New limits for passengers"', 331.

[66] Müller, '2000 "Passengers Carried by Sea"', 666 et seq.

III. Liability Limits

The PAL sets out liability limits of its own for passenger claims in its arts. VII and VIII. The limits of liability differ for claims for death and personal injury on the one hand and claims for loss or damage to luggage on the other hand. In this respect they resemble the provisions of the LLMC that set out two different limitation funds, for damages following death or personal injury and for damage to property.

There is however a marked difference in the PAL's way to calculate the limits. According to arts. VII para. 1 and VIII para. 1, the limits are calculated *per capita* and per voyage, so that there is the limit of liability available for every injured or otherwise aggrieved passenger. To the individual voyager, it thus does not make any difference under the PAL whether other travellers have been injured in the same incident: the limit of liability applies to each passenger and does not need to be shared.

The PAL however preserves in art. XIX the general right of the carrier to limit his liability in his function as an owner, charterer or other maritime player. This general right to limit, is, in the LLMC and in most domestic legislations, calculated according to the passenger capacity of the ship (art. VII LLMC) and applies per vessel. The accumulated amount of all passenger claims under the PAL may therefore be limited by virtue of the LLMC. If the total amount of passenger claims exceeds the LLMC limits, passengers' claims may be capped under the LLMC, when the funds available under the general limitation Convention are distributed *pro rata* among the claimants. It must be kept in mind that under the LLMC, passenger claims may furthermore be competing with other LLMC claims, such as crew claims, claims for bunker oil pollution or HNS damage, as long as the HNSC is not yet in force.

It is therefore important whether the state in which limitation funds are set up is a party to the PAL 1974 or the Protocol of 2002 to the PAL, comprising higher limits, and whether the LLMC 1976 or 1996 applies to a given claim. Very generally, and given the extremely low limits under the PAL 1974,[67] the LLMC limits are high enough to only cap the PAL's limits for passenger claims in the event of a major incident giving rise to claims from a big number of passengers on board the vessel.[68] Further considerations as to the interplay between the PAL and the LLMC will be treated *infra* under V.

According to art. VII para. 2 PAL, States can however set higher limits of liability for death and personal injury claims for carriers that are their nationals. The provision gives States who wish for higher liability limits some leeway, but the significance of the norm is doubtful. By only ordering their own nationals to grant higher limits of liability to passengers, States do not necessarily obtain a better protection of passengers who are their nationals, and furthermore create a business

---

[67] Tsimplis, '2009 "Liability in respect of passenger claims"', 133.

[68] Lewins, '2010 "Australian cruise passengers"', 130.

disadvantage for their national sea travel providers. Certain States such as the United Kingdom have however made use of the provision.[69]

According to art. XII para. 1 PAL, the limits of liability apply to the aggregate of the amounts recoverable in all claims concerning a given incident. This means that even if a passenger sues more than one party under the PAL, or one and the same party in more than one jurisdiction, he will not recover amounts in excess of the limit.[70] The provision acts as a safeguard to the parties liable under the Convention, but will of course not take effect in claims against persons that are not brought under the PAL or in a non-contracting State.

Given the fact that the liability based on the PAL is exclusively for fault, there is no need to grant the carrier exoneration clauses for intentional acts of third parties, act of God or similar occurrences. Art. VI PAL however allows a national court, in accordance with domestic provisions, to wholly or partly exonerate the carrier if he can prove that the damage was caused or contributed to by the neglect or fault of the passenger.

The carrier's right to limit liability is lost where "it is proved that the damage resulted from an act or omission of the carrier done with the intent to cause such damage, or recklessly and with knowledge that such damage would probably result", art. XIII para. 1 PAL. By virtue of para. 2, the provision is extended to the servants or agents of the carrier, whose right to limited liability is also lost if it resulted from an intentional or reckless act "of that servant or agent". The wording therefore implies that the carrier on the one hand and the servants or agents on the other hand are always only denied the benefit for their own acts, and that there is no attribution of the agents' or servants' misconduct to the carrier or *vice versa*.[71]

It has, especially in English policy[72] and legal literature,[73] been argued that sea passengers should, at least in this matter of the carrier's loss of the right to limitation, be treated equally with air passengers.[74] The wording of art. XIII para. 1 PAL does however not reflect the wording of the respective liability conventions on air traffic exactly,[75] and there are multiple other exceptions in the air traffic conventions that make it very difficult to align the two liability regimes for passenger claims.[76]

---

[69] Tsimplis, '2009 "Liability in respect of passenger claims"', 134; Baatz, *2011 "Maritime Law"*, p. 233.

[70] Tsimplis, '2009 "Liability in respect of passenger claims"', 134.

[71] Griggs, Williams and Farr, *2005 "Limitation of Liability"*, p. 106; Tsimplis, '2009 "Liability in respect of passenger claims"', 140.

[72] Gaskell, '1998 "New limits for passengers"', 325, referring to the discussions among the UK government in 1997.

[73] Tsimplis, '2009 "Liability in respect of passenger claims"', 141, giving a comparison between art. XIII PAL and the corresponding provisions of the 1929 Warsaw Convention.

[74] Haddon-Cave, '2000 "Limitation against passenger claims"', 236 et seq.; Tsimplis, '2009 "Liability in respect of passenger claims"', 141, 142.

[75] See the differences pointed out by Gaskell, '1998 "New limits for passengers"', 325, and Tsimplis, '2009 "Liability in respect of passenger claims"', 142.

[76] Kröger, '2001 "Passengers Carried by Sea"', 252.

The wording of the provision recalls the corresponding terms of the other maritime civil liability conventions, and much of what has been said there, especially for the CLC and the LLMC *supra*, Chap. 3, D., will also apply to the PAL. Under this Convention, too, the interpretation of the provision, at least as considered by English courts, raises the threshold to find a defendant having acted "recklessly and with knowledge that such damage would probably result" so high that it is virtually impossible to break the liability limits in this way.[77] There are some small differences in wording but which are generally considered not to cause the provisions to be applied differently.[78] Whereas art. IV LLMC refers to the "personal" act or omission of the carrier, the PAL omits that word. It is in fact not conceivable in which way this would alter the meaning of the provision. The reference to the "damage" that will occur, as opposed to "loss" in the LLMC is perhaps best explained by the fact that the PAL only refers to certain heads of damage, namely death and personal injury and lost luggage, whereas the LLMC covers all possible maritime claims.

It is submitted that the breaking of the liability limits may be slightly less unlikely if the provision is interpreted by the courts of other countries than England, such as France. In this context it is perhaps noteworthy that during the negotiations to the PAL, the French delegation advocated a wording of the provision that would have allowed denying the right of limitation more easily, and, specifically made the servants' or agents' acts or omissions attributable to the carrier.[79] With regard to the fact that this could act as an incentive to carriers to act more responsibly, a more generous interpretation of the understanding of "recklessly and with knowledge that such damage would probably result" is also advocated by others.[80]

Notwithstanding the difficulties that arise with regard to the interpretation of this provision, it is maintained in the 2002 Protocol to the PAL, which otherwise very markedly changes the Convention's liability regime.[81]

IV. Channelling and the Relationship to Consumer Protection Law

The PAL also contains a channelling provision in art. XIV, setting out that "no action for damages for the death of or personal injury to a passenger, or for the loss of or damage to luggage, shall be brought against a carrier or performing carrier otherwise than in accordance with this convention". First of all, it has to be noted

77 Martínez Gutiérrez, *2011 "Limitation of Liability"*, p. 127.

78 Ibid., p. 126.

79 Legendre, '1976 "La Convention d'Athènes"', 457.

80 Arroyo Martínez, *1996 "The Contract of Maritime Passage"*, p. 49.

81 P. Bonassies, 'Conventions internationales. Protocole à la Convention d'Athènes de 1974 sur le transport de passagers' (2003) 55, Droit Maritime Français (Hors série n° 7), 7–9 at 8.

that it is only the carrier and the performing carrier who will benefit from the channelling, as by its wording, art. XIV PAL does not prohibit a claim against their servants or agents. These can, according to art. XI PAL, however invoke the Convention's limits of liability. Other than that, there is no exempted class of maritime players or service providers, which means that nothing in the PAL keeps claimants from bringing actions outside the Convention against classification societies and, as long as they cannot be considered as "carriers" under the PAL, charterers or shipowners.

Even with this rather narrow scope of art. XIV PAL, the provision's importance is yet more reduced by domestic courts. Especially cruise voyages are very often sold as a package travel, comprising the travel to the port of destination, the sea voyage and diverse services on board and the travel from the port of destination. Thus, the question arises whether the applicability of the PAL rules out parallel actions under tourist travel law. At first glance, if the cruise line or the travel agent are liable as "carriers" under the Convention, art. XIV PAL should rule out any further claims against them.

In the English courts, the UK being a party to the PAL, actions for damage based on the legislation adopted under the 1992 European Package Holiday directive[82] have however been allowed. In the case of *Lee v. Airtours*,[83] passengers of the cruise ship *Sun Vista* that had caught fire and sunk, brought a claim for mental stress and anguish, the loss of personal belongings and loss of enjoyment in their holidays under the package travel direction, whereas the carrier sought limitation of liability under the Athens Convention, arguing that it applied by virtue of the English legislation implementing the PAL. The judge however held that the package holiday directive created a compensation regime independent from the PAL,[84] prevailing over the English legislation implementing it because of the primacy of European over domestic legislation. According to the provision in the package travel law, the PAL could only have taken priority if it had been incorporated as a term of the contract of carriage, which was not the case in the instance. This effectively means that in this case, European Law was held to prevail over the international regime meant to establish uniform compensation rules,[85] to the extent in which the package travel directive conflicted with the channelling benefit accorded to the carrier by the PAL.

This finding is however in conflict with a strong assertion as to the importance of treaty law in an earlier decision of the High Court of England,[86] even at a point in

[82] European Directive 90/314/EEC.

[83] "*Lee v. Airtours*" 2004 1 Lloyd's Law Reports 683 (Central London County Court (Business List), 26 April 2002).

[84] Ibid., p. 688.

[85] A. Saggerson, 'That Sinking Feeling - Psychiatric Injury, Lost Valuables and the Demise of the Cruise Ship "Sun Vista"' (2002), International Travel Law Journal, 198–202 at 202 et seq.

[86] "*The Lion*" 1990 2 Lloyd's Law Reports 144 (High Court (England), Queen's Bench Division, 20 March 1990).

time in which the PAL was not yet in force. The case concerned the carriage of a coach on a car ferry. The contract of carriage was subject to conditions that simply consisted in the provisions of the 1974 PAL in their entirety. Even though the Convention was not yet in force at the time of the carriage, its provisions had already been implemented into English law. The respective legislation however also required the carrier to explicitly draw the passenger's attention to a number of facts, namely to state that the PAL limited the carrier's liability. A mere repetition of the PAL's provisions was not enough, and the failure to comply with this information requirement was even an offense punishable by a fine. When the claimants' coach was damaged on the ferry and the defendant ferry line tried to limit liability under the provisions of the PAL, the claimants opposed this, invoking the fact that the defendants had failed to duly refer to the PAL's consequences in their terms and conditions, and that these could therefore not apply. The High Court however stated explicitly that the provisions of the PAL applied by law, and even if the defendants were guilty of an offense for their omission and liable to pay the corresponding fine, the provisions of the PAL and the limitation of liability could not be set aside. Notwithstanding the fact that the PAL was not yet in force, the court pointed out that the United Kingdom was under an international obligation to give effect to the Convention's provisions in the contracts of carriage covered.[87] The application of the PAL was therefore here not held to be a matter of contractual incorporation, but of granting effect to international law.

Similarly, the case of *Norfolk v. MyTravel PLC*,[88] decided by Plymouth County Court two years after *Lee v. Airtours*, the judges held up the PAL's two years' time bar and ruled out the action as barred, even though there was also responsibility under the package holiday legislation, which contained no time bar. However, the domestic package holiday legislation allowed for contractual clauses for "compensation to be limited in accordance with international conventions". It made no mention of limitations in time. The plaintiffs therefore argued that the PAL's time bar could not be invoked by the defendants, as this restriction of their rights under the tourist travel legislation was, unlike the limitation in amount, not explicitly allowed. The judge however found the two regimes did not apply alternatively, but rather concertedly, the PAL's time bar being also applicable to the action under domestic law transposing the EU directive.[89] In the same spirit as the judges argued this in the case of *The Lion*, the PAL's provisions were held to be applicable by virtue of law, irrespective of whether there was a contractual clause to this end and of whether tourist travel law allowed for such a clause.[90] The decisions cited are

[87] Ibid., p. 152.

[88] "*Norfolk v. MyTravel Group PLC*" 2004 1 Lloyd's Law Reports 106 (Plymouth County Court, 21 August 2003).

[89] Ibid., p. 109.

[90] A. Saggerson, 'Case Notes: A Tale of Two Conventions' (2004), International Travel Law Journal, 14–7 at 15.

hard to reconcile,[91] so that there is considerable uncertainty as to the scope of art. XIV PAL, especially if the alternative claim is based on EU legislation.[92]

The findings of the English court in *Lee v. Airtours* are effectively mirrored by the French *Cour d'Appel* of Nancy in the *Sea Diamond* decision discussed *supra* under I. 1. By stating that the cruise on the *Sea Diamond* was governed by the mandatory provisions of the French tourist travel legislation that did not allow for a limitation of liability to be implemented at that point in time, the court effectively stressed the primacy of these provisions over the PAL. It is however true that the convention was only implemented into the contract by agreement of the parties, and not as an international convention, since France is not a party.

The impact of consumer protection provisions on the PAL's liability regime however creates problems in the recourse between a tour operator liable under the tourist travel legislation and the PAL carrier. This is highlighted by the *Sea Diamond* case, in which the ship's foundering was clearly caused by the negligence of the crew and thus possibly attributable to the shipowner. The tour operator, being bound by the tourist travel legislation, is liable without limitation towards the passengers even though he is not directly responsible for the accident. A negligent shipowner can however, in the tour operator's recourse action against him, possibly invoke limitations of liability.[93]

With the coming into force of the provisions of the 2002 Protocol to the Athens Convention and the adaptation of the European tourist travel legislation to allow for the limitation of liability, this problem will most likely disappear.[94]

The problem of alternative claims available to victims under consumer protection law may however not only emerge in connection with tourist travel law, but possibly also with more general legislation, e.g. on compensation for injury claims. Especially under domestic liability regimes providing for strict liability where a consumer was injured, it may be hard to determine whether the PAL—still providing for fault-based liability—should, by virtue of art. XIV, prevail over the domestic consumer protection legislation.[95] It should be kept in mind that the wording of art. XIV PAL is actually not very open to interpretations allowing competing

[91] Shaw, *2008 "Carriage of Passengers"*, p. 152; Mandaraka-Sheppard, *2009 "Modern Maritime Law"*, p. 931; Tsimplis, '2009 "Liability in respect of passenger claims"', 131 et seq.; see the comprehensive discussion of the problem under English law in Soyer, *2007 "Boundaries of the Athens Convention"*, para. 11.42 et seqq.

[92] See Baatz, *2011 "Maritime Law"*, p. 239, submitting that the prioritisation of EU law amounts to a breach of the State's obligations under the PAL.

[93] Soyer, *2007 "Boundaries of the Athens Convention"*, para. 11.50; S. Schweitzer, 'Observations CA Nancy 20.01.2011' (2012) 64, Droit Maritime Français, 609–12 at 610 et seq.

[94] Soyer, *2007 "Boundaries of the Athens Convention"*, para. 11.6.

[95] See K. Lewins, 'The Cruise Ship Industry - Liabilities to Passengers for Breach of s52 and s74 "Trade Practices Act 1974" (Cth)' (2004) 18, Australian and New Zealand Maritime Law Journal, 30–54 at 49 et seq. and B. Marten, 'Limitation of Liability for Personal Injury in New Zealand: ACC Meets the Sea' (2006) 16, Australian and New Zealand Maritime Law Journal, 16–24 at 20 et seq., on a discussion of the relationship between the PAL and Australian and New Zealand legislation for injury compensation respectively, should the countries sign the Convention.

claims, and reflects the fact that one of the convention's goals is to provide certainty for carriers as to their liabilities. Deciding on the scope of the channelling mechanism for passenger claims of course means striking a balance between consumer protection and business interests and should be one of the concerns of the State Parties. The impact of consumer protection law on the PAL's channelling provisions or vice versa is worth considering in an important number of cases.

V. Relationship with Other Limitation Instruments

It has already been mentioned *supra* under III. that the liability limits in the PAL may, by virtue of art. XIX PAL, be subject to the general limits of liability "in international conventions relating to the limitation of liability of owners of seagoing ships". In major disasters in which the shipowner is faced with a multitude of passenger claims, the question of whether there is a superposing international liability regime that applies may be of considerable interest to the claimants.

Following the wording of art. XIX PAL, however, global limits of liability to passenger claims are only admissible if they are found in international conventions. The PAL therefore allows the interplay with the LLMC where it is in force, but not with domestic legislation.

This study will consider the impact of the 1976 and the 1996 LLMC respectively, but they of course do not apply universally. Some countries have enacted their own domestic legislation on limitation of liability for maritime claims, which may be inspired by the LLMC provisions but also have very different provisions. This can give rise to considerable problems of forum shopping and conflict of laws[96] if defendants in maritime claims are eager to set up limitation funds in a jurisdiction featuring very low levels of limitation or a very slow litigation process[97] whereas claimants may want to bring their claim in jurisdictions more favourable to them.

For an incident that is dealt with under the 1976 LLMC, there are different limitation amounts for claims for death and personal injury and all other claims. The limits for loss of life or personal injury are double the amount of the general claims, and according to art. VI para. 2 LLMC there are in fact two limitation funds: claims for death and personal injury will be paid with priority, and if their total amount exceeds the funds available for these claims, the unpaid balance will be made valid against the general fund and rank rateably with the property and other claims.

The limits for passenger claims under the LLMC are calculated according to the number of persons that the vessel is allowed to carry, but there is an overall limit of

[96] Lewins, '2004 "The Cruise Ship Industry"', 52.

[97] Soyer, '2002 "Sundry Considerations"', 533 with regard to the 2002 Protocol to the PAL, possibly encouraging forum shopping.

25 million SDR. This means that under the LLMC, very big passenger ships are under a considerable advantage, since the claims may be reduced to marginal amounts if all of the passengers are claimants.[98]

Under the 1996 LLMC, art. VII para. 2 LLMC was amended and the overall limit for death and personal injury claims was abandoned. Furthermore, the liability limits were considerably increased, by about 250 %.[99] Moreover, the 1996 LLMC introduced a new art. XV para. 3bis allowing States to increase liability limits for loss of life or injury even further under their national law, up to unlimited liability.[100] It was apparently mostly politic pressure that led to these changes, in order not to oppose limitations of liability to sea passengers whereas victims of road or train accident would generally be fully compensated.[101] The LLMC 1996 thus now only creates a minimum limit.[102] Victims bringing their claim under the PAL 1974 however still face the PAL's limit.[103] This theoretically creates a contradiction for States having exercised the possibility of unlimited liability for death and personal injury claims but remain party to the 1974 PAL[104] which limits liability in a more perceivable way for the individual passenger. The UK, as a State party to the 1974 Convention and as one of the States having made use of the possibility in art. XV para. 3bis have substituted different limits under the PAL,[105] which is possible under art. VII (2) PAL 1974. Furthermore, under the 2002 Protocol to the PAL, there is equally the possibility to remove the limitation of liability for death and personal injury claims (see *infra* under B. III. 1.).

With regard to claims to which there is an alternative basis in consumer protection legislation, defendants may again try to oppose the liability limits—now the global limits under the LLMC. As has been mentioned earlier, these will only start to play a role in catastrophic incidents, where the accumulated amount of separate passenger claims exceeds the LLMC limits. It does not seem that there has as of yet already been a case in which a shipowner or other beneficiary of the LLMC limitation would have invoked the global limits. It is however maintained by some authors that the global limitations under the LLMC have better chances to be successfully invoked by defendants than the PAL limits, even if the claim arose under a consumer protection legislation.[106]

Still, the claims following catastrophic incidents are most often settled in agreements out of court.[107] As the *Costa Concordia* incident has recently shown,

---

[98] Gaskell, '1998 "New limits for passengers"', 324.

[99] Ibid., p. 315.

[100] Tsimplis, '2009 "Liability in respect of passenger claims"', 134.

[101] Gaskell, '1998 "New limits for passengers"', 324.

[102] Honka, *1998 "Nordic Reactions"*, p. 83; Gaskell, '1998 "New limits for passengers"', 329.

[103] Ibid., p. 326.

[104] Ibid., p. 329.

[105] Griggs, Williams and Farr, *2005 "Limitation of Liability"*, p. 104; Mandaraka-Sheppard, *2009 "Modern Maritime Law"*, p. 927.

[106] Tsimplis, '2009 "Liability in respect of passenger claims"', 135.

[107] Honka, *1998 "Nordic Reactions"*, p. 84.

courts, at least in France, significantly encourage settlements of this kind. The *Tribunal de Grande Instance* of Nanterre,[108] and, in the second instance, the *Cour d'Appel* of Versailles[109] decided on the form of the shipowning company's amicable settlement offer, finally judging that the time allowed to the victims to accept or reject it should be considerably prolonged.

VI. Limitations in Time

The PAL contains, in art. XVI, provisions on the limitation of claims in time. Art. XVI PAL sets out a general time bar of two years for both death and personal injury and luggage claims. In general, it begins with the disembarkation of the passenger and his luggage or the time he would have disembarked in cases of death claims. Art. XVI para. 3 PAL sets out that the procedural rules of the court deciding on a case shall govern possible suspensions and interruptions of the period, but also states an absolute limit of three years, after which all claims are barred.

The prescription period of two years and even three years after disembarkation is rather short in comparison to the prescription periods for physical injury in many countries.[110] Even if art. XVI para. 3 PAL in its first part refers to the procedural law of the court in question and therewith allows claimants to trust the procedural rules of their home forum to some extent, the absolute bar after three years can be a real obstacle, especially if the procedural law of the court in question is generally very generous. This was also the matter in the *Prince Laurent* decision[111] cited above where the claimant was still a minor at the time of his accident on board the ship. Under French procedural law he would have had the opportunity of bringing his claim only after his coming of age, the respective provisions under certain conditions suspending the prescription until the claimant's majority. In the *Norfolk v. MyTravel* decision,[112] the claimant tried to overcome the short prescription period under the PAL by suing under tourist travel law which did not provide for any special prescription period at all.

Considering that liability under the PAL 1974 is still based on fault and that claimants may need to carry out enquiries or wait for an official investigation to be completed, the short prescription period can appear unfair towards them. The difference between the claims under the PAL 1974 and tourist travel law based

---

[108] "*Costa Concordia*", (2012) 64 Droit Maritime Français 750 (Tribunal de Grande Instance de Nanterre (Référés), 13 February 2012).

[109] "*Costa Concordia*", (2012) 64 Droit Maritime Français 598 (Cour d'Appel de Versailles, 09 May 2012).

[110] Soyer, '2002 "Sundry Considerations"', 534; Lewins, '2010 "Australian cruise passengers"', 136, comparing it to Australian law.

[111] *Cour d'Appel de Paris, 19.05.1999, "Prince Laurent"*, 28.

[112] *Plymouth County Court, 21.08.2003, "Norfolk v. MyTravel Group PLC"*, 108.

on the 1992 Package Holiday directive thus appears even more marked. It is not surprising that the 2002 Protocol to the PAL provides for a slightly longer period in the absolute time bar, with more refined provisions having regard as to whether the claimant knew or should have known of the damage caused.

B. The 2002 Athens Protocol

It is a sad fact that the liability limits of the 1974 PAL were widely considered as being already out of date by the time the Convention entered into force in 1987.[113] This was seen as one of the reasons for States' reluctance to ratify the PAL, and the Convention was at times described as being "perpetually under construction".[114] Especially within Europe, several ferry disasters, first and foremost the sinking of the *Estonia*, triggered concern for the legal difficulties that might arise in connection with passenger ship accidents.[115] Among other difficulties, the absence of compulsory insurance was perceived as a considerable shortcoming in the settling of passenger claims.[116] Furthermore, the limitation of liability for death and personal injury claims has met with a growing resistance since the adoption of the 1974 PAL.[117] Passenger claims in Japan,[118] and in the United States,[119] for instance, and, to a certain extent in Scandinavia[120] are not subject to global statutory limitation. Furthermore, Art. XV para. 3bis LLMC 1996 offers its contracting States the possibility to enact unlimited liability for loss of life and injury.

The initiative to review the PAL 1974 and make substantial amendments to its liability regime was a remnant of an original proposal to oblige shipowners and other maritime players to obtain insurance for any possible claim by third parties that may arise against the ship, but that was eventually abandoned as being too ambitious.[121] It is therefore not surprising that the introduction of compulsory

[113] Gaskell, '1998 "New limits for passengers"', 330; Soyer, '2002 "Sundry Considerations"', 531; Mandaraka-Sheppard, *2009 "Modern Maritime Law"*, p. 923.

[114] Delebecque, *2008 "L'évolution du transport de passagers"*, p. 308.

[115] H. D. Tebbens, 'The European Union and the Athens Convention on Maritime Carriers' Liability for Passengers in Case of Accidents: An Incorporation Adventure' (2008) 61, Revue Héllenique de Droit International, 653–67 at 653.

[116] Honka, *1998 "Nordic Reactions"*, p. 86, referring to the *Scandinavian Star* accident in the waters off Norway and Denmark in 1990.

[117] Delebecque, *2008 "L'évolution du transport de passagers"*, p. 313.

[118] Shaw, *2008 "Carriage of Passengers"*, p. 150; Baatz, *2011 "Maritime Law"*, p. 235.

[119] Røsæg, *2010 "The Athens Convention"*, p. 57; also see Peltz, '2012 "The Athens Convention Revisited"', 504 et seq. on a possible contravention of US public policy by contractual damage caps.

[120] Baatz, *2011 "Maritime Law"*, p. 235.

[121] Soyer, '2002 "Sundry Considerations"', 520; Griggs, Williams and Farr, *2005 "Limitation of Liability"*, p. 111.

insurance for passenger claims, still omitted in the 1974 PAL, was one of the main changes brought about by the 2002 Protocol.

The Protocol has been modelled on the 1999 Convention for the Unification of Certain Rules for International Carriage by Air (Montréal Convention) but it also comprises elements that are typical for the maritime civil liability conventions for environmental pollution[122] discussed *supra* in Chaps. 3 and 4. It was adopted on 1 November 2002, and it has entered into force on 23 April 2014, following the ratification by Belgium as its tenth contracting State. This can be seen as a considerable success for the negotiating parties, since the Convention has met with considerable scepticism in the past.[123] Even if the 2002 version of the PAL is technically only a protocol to the 1974 original Convention, the amendments made by the Protocol are considerable and thoroughly change the 1974 PAL's liability regime. States ratifying the Protocol have to denounce the original convention (art. XVII PAL 2002), in order to do away with the so-called "grandfather problem", the difficulty that the old rules have to apply as between contracting States of the old Convention, even if one of them is party to the Protocol. The Protocol therefore establishes a liability regime that is widely independent from the original 1974 PAL. The EU and the IMO have provided a consolidated version of the Convention. For the sake of simplicity, this study will refer to the Protocol as the PAL 2002 and quote the articles of the consolidated text.

Even before its coming into force, the Protocol's provisions have already gained importance by their implementation into the law of the European Union. By Regulation (EC) No. 392/2009, the EU has redrafted the majority of the Protocol's rules as a part of European Law and summoned its member States to ratify the 2002 Protocol. Failing this, the regulation's provisions applied as European Law within the member States "from no later than 31 December 2012". This means that since 1 January 2013, the PAL 2002—albeit with minor amendments especially as to competence, and recognition and enforcement of judgments—has been in force in the EU member States, not as an international convention, but as part of the legislation of the EU. This will lead to a situation in which the PAL's provisions will be subject to interpretation by the ECJ.[124]

The EU, having ratified the 2002 Protocol, has however negotiated certain clauses, by virtue of which it can apply slightly different rules, e.g. on recognition and enforcement of judgments and on the geographical scope of application. Even if Regulation 392/2009 has the clear aim to introduce the liability regime of the PAL 2002 into European law, the Regulation effectively does not transpose the entire Convention into EU law. It is effectively only the arts. I, Ibis, II para. 2, III to

---

[122] Balkin, *2007 "The Rise and Rise"*, p. 707.

[123] Soyer, '2002 "Sundry Considerations"', 539; Lewins, '2010 "Australian cruise passengers"', 132; Shaw, *2008 "Carriage of Passengers"*, p. 147.

[124] Legros, *2012 "L'intégration"*, p. 368; Jacobsson, '2012 "Perspective of the Global Compensation Regimes"', 72.

XVI, XVIII, XX and XXI that the Regulation refers to.[125] This allows the EU to adapt the PAL's regime to its internal legislation.

The EU's regulation is, however, still meant to be supported by the ratification of the 2002 Protocol by all member States, and considering that the Protocol has now been signed by some of them, it seems that its coming into force is to some extent due to the EU's action in this matter.

I. The PAL 2002 and the EU Regulation

The years of 2013 and 2014 represented a period of transition with regard to the application of the PAL 2002 in the EU member States. Within the EU, most of the rules of Regulation 392/2009 have applied from 1 January 2013 on, and have already taken effect in at least one Spanish decision concerning passenger claims in the *Costa Concordia* accident (see *supra* under A. I. 2.). However, out of the 28 EU member States, only Denmark, the Netherlands, Latvia, Croatia, Malta and Belgium have ratified the PAL 2002 so far.[126]

Since 23 April 2014, the PAL 2002 is in force and applies as between its contracting States, of which there are currently 14, representing a little more than 8 % of the world merchant shipping tonnage. For the EU member States having ratified the Convention, the provisions of the 2002 PAL will apply both as a part of EU law between them and other EU member States, including those that have not ratified, and as an international convention between them and other States parties to the PAL.

During the negotiations of the 2002 Protocol to the PAL, the EU took more interest in the Protocol's provisions and in the way in which the regime could be applied within the EU. The protection of passengers in different means of transport had been on the EU's agenda for some time already.[127] By the introduction of harmonised rules within Europe, the EU aimed for a uniform standard of passenger protection to the level granted to other consumers, and thus for a fair competition between European carriers, all having to comply with the same rules.[128]

---

[125] J. Karsten, 'Im Fahrwasser der Athener Verordnung zu Seereisenden: Neuere Entwicklungen des europäischen Passagierrechts' (2009), Verbraucher und Recht, 213–25 at 214, footnote 10.

[126] See http://www.imo.org/About/Conventions/StatusOfConventions/Documents/status-x%20temporary.xls (last visited 7 February 2014).

[127] B. Czerwenka, 'Neue Haftungs- und Entschädigungsregeln in der Personenschiffahrt - Harmonisierung durch Europarecht' (2010), Transportrecht, 165–74 at 165; M. Piras, 'International Recent Developments: European Union - Maritime Passenger Transport' (2012) 36, Tulane Maritime Law Journal, 627–39 at 634.

[128] Ibid., pp. 627 et seq.

1. The Rules on Jurisdiction, Recognition and Enforcement

The EU's plan to give effect to the PAL 2002 in all of its member States met with the same difficulties the EU had already experienced with regard to the BOPC (see *supra*, Chap. 4, A. II. 2.). EU member States were prevented from ratifying the PAL 2002 due to the fact that it contained rules on jurisdiction, recognition and enforcement in civil and commercial matters, an area in which there were already general rules by virtue of the Brussels Regulation. In the negotiations to the PAL 2002, the EU however negotiated art. XIX, by virtue of which "regional economic integration organisations" can become a party to the Convention. This allowed the EU to ratify the PAL 2002 itself, so that its provisions could become a part of EU law and the member States would no longer be faced with a situation in which an international convention they wanted to become a party of conflicted with the EU law they were bound by. The EU has effectively been a party to the PAL since 15 December 2011.

It is however one of the EU's policies to maintain the application of the rules of the Brussels Regulation on recognition and enforcement as between the member States by all means.[129] It therefore negotiated a so-called "disconnection clause" to the PAL 2002. According to art. XVIIbis para. 3 PAL 2002, contracting States "may apply other rules for the recognition and enforcement of judgments, provided that their effect is to ensure that judgments are recognised and enforced at least to the same extent as under paragraphs 1 and 2." The PAL 2002 therefore allows a deviation from its rules on recognition and enforcement, in contrast to the situation under the BOPC (see *supra*, Chap. 4, A. II. 2.), where the EU member States declared a reservation without actually being allowed to do so by the Convention's terms.

The disconnection clause however presupposes that judgments are recognised and enforced under the applicable alternative rules to the same extent as they are under art. XVIIbis para. 1 and 2 PAL 2002. The PAL's provisions on recognition and enforcement having the same wording as the provision in the IMO's other civil liability conventions, judgments under the PAL need to be generally recognised and enforced in all contracting States, without re-opening of the merits of the case. The only exceptions are judgments obtained by fraud (art. XVIIbis para. 1 (a)) and judgments in which the defendant did not have a fair opportunity to answer the claim filed against him (art. XVIIbis para. 1 (b)). Whereas the EU authorities consider the rules of the Brussels Regulation to be at least as generous[130] as art. XVIIbis para. 1 and 2 PAL 2002, it has already been pointed out that this is not the case.[131] As the PAL's rules on recognition and enforcement are exactly the same as the BOPC's, the points of conflict are the same (see *supra*, Chap. 4, A. II. 2.). Art.

[129] K. Takahashi, 'External Competence Implications of the EC Regulation of Jurisdiction and Judgments' (2003) 52, International and Comparative Law Quarterly, 529–34 at 533.

[130] Ibid., p. 531, quoting a communication of the EU Commission; Røsæg, *2010 "The Athens Convention"*, pp. 61 et seq.

[131] Ibid., pp. 62 et seq., explaining several examples in detail.

34 of the Brussels Regulation allows the denial of enforcement if the judgment violates the recognising State's *ordre public* or if there is a conflicting judgment that takes effect or could take effect in the recognising State.[132] Under the BOPC, these differences did not appear too important, as claims for the clean-up or other costs in connection with bunker oil pollution are not overly likely to touch matters that can be considered as public policy in an EU member State. It is submitted that with regard to claims under the PAL, namely claims for death and personal injury, this could be different. Following art. 5 of Regulation 392/2009, EU member States can provide for global limitation of passenger claims, as long as the national legislation in this matter is based on the 1996 LLMC. The 1996 LLMC however allows removing the limitation of liability for death and personal injury claims altogether.

It has already been mentioned that many States are today very hostile towards the concept of limited liability for personal injury or death claims. It is not entirely inconceivable that a judgment from an EU member State which declares such a claim to be limited in amount could be considered as contravening public policy in another EU member State where recognition is sought, and where such a claim would not be limited. The UK is in this context an example of an EU member State providing for unlimited liability for personal injury and death claims, both under the PAL 1974 and the LLMC 1996.

Different limits of liability could possibly also trigger the other exception of general recognition under art. 34 of the Brussels Regulation that does not exist under the PAL: the irreconcilability with an earlier judgment between the same parties in the recognising State, or even with an earlier judgment from a third State if it could be recognised in the EU member State addressed. Considering that a carrier could always bring a claim for a declaratory judgment stating that a certain limitation under the PAL 2002 applies, there is a danger that a court in a State providing for unlimited liability for this claim may deliver a conflicting judgment. The risk of such a direct conflict is also greater than under the BOPC because the PAL provides for individual limits for each contract of carriage. As long as the global limits are not triggered in a second step, which will only occur in very catastrophic accidents aggrieving an important number of passengers on board (see *supra*, under A. V.), a passenger under the PAL is entitled to his personal limitation amount under his individual claim. This situation contrasts with the situation under the BOPC, where the amount claimants receive will only be determined during the distribution of the global fund.

The risk of irreconcilable judgments under the PAL is also greater than under the BOPC because the PAL, in art. XVII, provides for a variety of possible forums, whereas under the BOPC, jurisdiction is restricted to the courts of the countries in which there has been bunker oil pollution damage. The different judicial venues to the PAL 1974 (see *supra*, A. I. 3.) remain unchanged by the 2002 Protocol, so that

[132] Example pointed out by Tebbens, '2008 "The European Union and the Athens Convention"', 658, and Røsæg, *2010 "The Athens Convention"*, p. 67.

the claimant can exercise a choice between a number of different jurisdictions if the contract of carriage has links with different countries. A claim brought by the carrier against the defendant, such as a negative declaratory action or a claim for contributory negligence may be brought in a completely different forum, in all likelihood in a venue that the carrier has designated in a contractual term. The multitude of possible forums increase the chance of multiple proceedings, and thus also of conflicting decisions.

Following the wording of the disconnection clause, this would actually mean that in these cases, where recognition and enforcement is unconditionally possible under art. XVIIbis para. 1 and 2 PAL 2002, but impossible due to art. 34 of the Brussels Regulation, recognition and enforcement would again need to be effectuated according to the PAL 2002. The provisions of the Convention can of course only bind States parties to it, and art. XVIIbis PAL 2002 is not part of the provisions taken over by EU Regulation 392/2009. EU member States that are also party to the PAL 2002 may however see themselves forced to either violate the conditions of the disconnection clause if they disregard the Protocol's provisions in this respect or to accept that the coming into force of the PAL 2002 significantly alters the scope of applicability of the Brussels Regulation.[133] This would be admissible by virtue of art. 67 of the Regulation, giving primacy to the enforcement and recognition provisions in special agreements, and there is also an Opinion of the ECJ stating that a disconnection clause in an international agreement can affect community legislation.[134] It however remains to be seen whether EU member States parties to the PAL 2002 will interpret the disconnection clause in this way, or whether the problem will be disregarded or even pass unnoticed, as an internal issue between the EU member States.

2. Geographical Scope of Application Under the EU Regulation

The rules of Regulation (EC) No. 392/2009 on geographical applicability slightly differ from the PAL 2002. Art. 2 of the Regulation sets out that the Regulation will apply to voyages where first, the ship is flying the flag of a member State, or second, where the contract of carriage has been made in a member State, or third, where the place of either departure or destination is located in a member State. The Regulation thus mirrors the scope of application of the PAL—which remained unchanged in the 2002 Protocol—but replaces the reference to the contracting State by a reference to the member State.[135] This means that as long as there are EU member States which have not ratified the PAL, there are effectively two groups of States in which

[133] Tebbens, '2008 "The European Union and the Athens Convention"', 658; Røsæg, *2010 "The Athens Convention"*, p. 67.

[134] Opinion 1/03, pt. 154.

[135] Czerwenka, '2010 "Neue Haftungs- und Entschädigungsregeln"', 171.

the provisions apply.[136] Cases fulfilling the conditions of art. II PAL, by having a connection with a State party to the PAL which is not a member State of the EU, but being tried in a member State will therefore not be subject to the provisions of either the Regulation or the PAL.[137]

Furthermore, concerning the geographic scope of application, art. 2 of Regulation 392/2009 renders the liability regime mandatory for any international carriage as defined by the PAL, meaning, in between ports of two different States or of one State if the ship calls in the port of another State in between. Member States can decide to submit any domestic voyages to the regime, but following the reference to internal EU legislation, voyages within one and the same member State are always covered as soon as the vessel in question operates in more than 15 miles distance from a place of refuge, or five miles from the coast.[138]

This means that the liability regime may eventually be applicable to several coastal ferry services that are provided at sub-marginal cost, e.g. to link islands with the mainland, and may render these even more unprofitable.[139] It is very debatable if the mandatory application of the PAL to domestic voyages is appropriate, considering that the PAL, following the wording of art. II para. 1 in both its 1974 and 2002 versions only applies to "international" carriages and art. XXII para. 1 PAL 1974 even offered States parties the possibility to withdraw cases only involving nationals from the international regime.

II. Liability Regime

The 2002 Protocol to the PAL has changed the Convention's liability regime to a considerable extent. It establishes a two-tier regime in which the liability for damages for death and personal injury following "shipping incidents" up to a certain liability limit is strict. For any claims exceeding these liability limits, the carrier's responsibility is fault-based, but responsibility is presumed. The PAL 2002 thus maintains the differentiation between damages following "shipping incidents" that can be traced back to a typical risk of maritime transport, and other claims where the damaging event could also have occurred on land. The establishment of the two-tier regime has therefore refined, and in some authors'

---

[136] Furthermore, in the one group, the provisions will be subject to interpretation by the ECJ, while this is not the case in the other group, see Legros, *2012 "L'intégration"*, p. 386.

[137] F. Berlingieri, 'The Athens Convention on the Carriage of Passengers and Their Luggage by Sea and the European Parliament Regulation (EC) No. 392/2009' (2011) 113, Il Diritto Marittimo, 1125–40 at 1127.

[138] See Ibid., p. 1128; N. A. Martínez Gutiérrez, 'New European rules on the liability of carriers of passengers by sea in the event of accidents' (2012) 18, Journal of International Maritime Law, 293–305 at 296.

[139] Thomas-Ciora, '2006 "Le transport des passagers"', 983.

opinion, complicated[140] the liability regime for death and personal injury claims. Except for a significant rise in the liability limits, the rules on claims for loss of or damage to luggage have not been subject to changes.

1. Strict Liability for Shipping Incidents

Notwithstanding the differences between air traffic and the carriage of passengers by sea as to their freedom to move and embark on activities on board, the aim to align the liability regime for sea voyages with the existing liability regimes for the air passenger trade was one of the reasons to introduce a first tier of strict liability[141] for death and personal injury claims up to a limit of 250,000 SDR (art. III para. 1 PAL 2002). These limits, that apply per passenger per incident, are actually quite high, even in comparison with the amended limits of the 1996 LLMC, where the global limitation per vessel per incident is 175,000 SDR, multiplied with the number of passengers. The PAL 2002 is thus the first Convention to couple strict liability with rather high liability limits.[142]

The understanding of the notion of "shipping incident", which has been defined in art. III para. 5 (a) PAL 2002 is therefore even more important than under the PAL 1974, where the understanding of "shipwreck, collision, stranding, explosion or fire, or defect in the ship" in art. III para. 1 only triggered the reversal of the burden of proof in favour of the claimant. Art. III para. 5 (a) PAL 2002 reads:

> 'shipping incident' means shipwreck, capsizing, collision or stranding of the ship, explosion or fire in the ship, or defect in the ship.

The definition thus draws heavily on the understanding of "typical maritime perils" reversing the burden of proof in the PAL 1974. The term "defect in the ship" is now defined in art. III para. 5 (b) PAL 2002:

> 'defect in the ship' now means any malfunction, failure or non-compliance with applicable safety regulations in respect of any part of the ship or its equipment when used for the escape, evacuation, embarkation and disembarkation of passengers; or when used for the propulsion, steering, safe navigation, mooring, anchoring, arriving at or leaving berth or anchorage, or damage control after flooding; or when used for the launching of life saving appliances.

The term "defect in the ship" under the 1974 PAL has been criticised for offering claimants the possibility to bring the numerous cases of injuries following slipping or falling on board as a consequence of a "defect in the ship", whereas accidents of this kind, at least where they happen indoors and on installations that could also be

---

[140] P. Griggs, 'Le protocole d'Athènes' (2002) 54, Droit Maritime Français, 291–305 at 296; R. D. Peltz and V. J. Warger, 'Amendments to Athens Convention Threaten US Maritime Law' (2001), International Travel Law Journal, 170–91 at 181.

[141] Soyer, '2002 "Sundry Considerations"', 522.

[142] Røsæg, '2004 "News under the Athens Sun"', 157.

found in a holiday resort, are rather not a typically maritime peril.[143] Following the wording of the definition in art. III para. 5 (b) PAL 2002, these accidents are not covered by the strict liability regime, as the referral to "equipment when used for escape, evacuation, embarkation and disembarkation" clearly points at facilities only used when in emergency, or in embarkation or disembarkation, the two latter processes clearly only covering the short periods of time when the passenger comes on board and leaves the ship again. Given the referral to "the applicable safety regulations", the determination of whether there was a defect or not will depend on the compliance with technical safety provisions such as are found in SOLAS and in the ISM Code,[144] which is a rather objective standard.

Even if it appears detailed at first sight,[145] the definition is not free from uncertainties[146] and arguably quite narrow.[147] It is noteworthy that the problems that have already been identified with the term (see *supra*, A. II. 2.) have not been solved by the definition. Problems arise if the defect in the ship, or, for that matter, the explosion or fire,[148] has been caused by an event for which the carrier is only partly responsible, but which does not enter in one of the exoneration exemptions under art. III para. 1 (a) and (b) PAL 2002. Given the strict liability in the PAL 2002, this problem has become considerably more pressing than under the PAL 1974, where the shipowner could prove that he did not act negligently with regard to the incident.[149] Furthermore, the definition does not address shortcomings in the ship's condition such as yielding parts that may be very dangerous to passengers by their sheer physical impact.[150] It can however still be argued that collapsing building structures are a danger that may also occur in a building on land and is therefore not, properly speaking, a typically maritime peril.

It is debatable to which extent the introduction of strict liability is more favourable to passengers than the fault-based liability with the burden of proof reversed that did work without problems in practice.[151] It remains to be seen whether the funds that are thus easily available under the first tier of the PAL 2002 will encourage frivolous claims,[152] especially given the existence of a right of direct action against the insurer.

---

[143] Peltz and Warger, '2001 "Amendments to Athens Convention"', 179 et seq.

[144] Mandaraka-Sheppard, *2009 "Modern Maritime Law"*, p. 935.

[145] Berlingieri, '2011 "The Athens Convention"', 1130.

[146] Griggs, Williams and Farr, *2005 "Limitation of Liability"*, p. 115; Shaw, *2008 "Carriage of Passengers"*, p. 148, footnote 148: "remarkably cumbersome".

[147] Griggs, '2002 "Le protocole d'Athènes"', 295; Soyer, '2002 "Sundry Considerations"', 525.

[148] Peltz and Warger, '2001 "Amendments to Athens Convention"', 180.

[149] Griggs, Williams and Farr, *2005 "Limitation of Liability"*, p. 115.

[150] Soyer, '2002 "Sundry Considerations"', 525.

[151] Griggs, '2002 "Le protocole d'Athènes"', 296; Shaw, *2008 "Carriage of Passengers"*, p. 147.

[152] This is a general concern raised by Soyer, '2002 "Sundry Considerations"', 523 with regard to strict liability.

Above the limit set out in art. III para. 1 PAL 2002, the carrier is further liable for fault or negligence, but his responsibility is presumed. For death and personal injury claims that are not caused by a "shipping incident" as defined by art. III para. 5 (a) PAL 2002, the carrier is only liable if and to the extent that his fault or neglect are proved by the claimant (art. III para. 2 PAL 2002).

There is no strict liability for luggage claims, the carrier's liability is based on fault. Considering cabin luggage that is under the control of the passenger during the voyage, the carrier's fault or neglect is presumed if the damage was caused by a shipping incident. In all other cases, the burden of proof is on the claimant. For the loss of or damage to luggage that is handed over to the custody of the carrier, the carrier's fault or neglect is always presumed, considering that it will be difficult for the passenger to prove the causes for the damage if he could not access the luggage during the voyage.

Apart from the introduction of strict liability for death and personal injury claims in a first tier, there have thus only been very few changes to the 1974 PAL's liability regime.[153] It is in any case to be kept in mind that the 2002 PAL still presents a protocol to the 1974 PAL, and that this begs the question for the extent to which the rules of the 2002 PAL can be interpreted in the light of the 1974 PAL. One of these issues arises with regard to valuables. The provision may be of some importance on cruise ships if passengers' jewellery and other valuables can be stored in the ship's safe rather than in an individual safe in the cabins.[154] According to art. V of the 1974 PAL, which has remained unchanged by the 2002 Protocol, the carrier's responsibility for valuables such as monies and jewellery is excluded, unless the items have been deposited with the carrier for the agreed purpose of safe-keeping. In this case, the carrier is liable up to the amounts set out for luggage claims, but the basis of liability is not set out. It has thus been argued that the carrier's liability for valuables should thus, in the absence of a closer designation, be strict.[155] Even if some assent to this view could be drawn from the thought that the valuables have been deposited with the carrier "for safe-keeping" and that he may therefore assume a greater responsibility as to their transport, the concept of strict liability was entirely inexistent in the 1974 PAL, the carrier being only responsible for fault or negligence. It therefore seems that the loss of or damage to valuables should rather be submitted to the liability regime for luggage other than cabin luggage, where the responsibility is based on fault, but presumed in any case.

[153] Bonassies, '2003 "Conventions internationales"', 8; Berlingieri, '2011 "The Athens Convention"', 1131.

[154] See *Central London County Court (Business List), 26.04.2002, "Lee v. Airtours"*, p. 686, but where the problem was not decided as the applicability of the 1974 PAL was ruled out, also see Saggerson, '2002 "That Sinking Feeling"', 200 et seq.

[155] Berlingieri, '2011 "The Athens Convention"', 1134.

2. Exoneration

With regard to the introduction of strict liability for death and personal injury claims up to a certain amount, the carrier's possibilities of exoneration become significantly more important than under the 1974 PAL. Furthermore, after the World Trade Center attacks in 2001, the problem of hostile and terrorist attacks especially on cruise ships became a matter of concern for the shipping interests. Given the fact that there is generally no P & I cover for terrorism and war risks, insurers were concerned about the strict liability for passenger claims, especially in view of the high limitation amounts per claim.[156] By providing for some exoneration clauses by which the carrier may escape strict liability, the problem of the lack of insurability is partly mitigated, but of course, the overwhelming question of who should be liable for the damage done to innocent victims[157] (also see the discussion with regard to the shipowner's exoneration in oil pollution cases, *supra*, Chap. 3, F.) is thus not solved. It however seems that the clauses allowing the carrier's exoneration are not very far-reaching, and the PAL 2002 and the IMO's guidelines on implementation of insurance (see *infra* under III.) assume the existence of war risk cover for the passenger trade.

The exoneration clause in art. III para. 1 (a) and (b) read:

> 1. For the loss suffered as a result of the death or personal injury to a passenger caused by a shipping incident, the carrier shall be liable to the extent that such loss in respect of that passenger on each distinct occasion does not exceed 250,000 units of account, unless the carrier proves that the incident:
> (a) resulted from an act of war, hostilities, civil war, insurrection or a natural phenomenon of an exceptional, inevitable and irresistible character; or
> (b) was wholly caused by an act or omission done with the intent to cause the incident by a third party.
> [...]

It is noteworthy that the wording of the provision mirrors exactly the exoneration clauses of art. III para. 2 (a) and (b) CLC except for the referral to the "intent to cause the incident" in the PAL instead of the "intent to cause the damage". Considering that art. III para. 1 (a) and (b) PAL 2002 is evidently modelled on the corresponding provisions in the CLC, the LLMC and the HNS Convention, it should be interpreted in the light of the solutions found under these maritime conventions.[158] It is maintained that in cases covered by insurance other than for war and terrorism, the PAL's exemption clauses are less important than, for instance, under the oil pollution regime.[159] This is because of the P & I clubs' general policy to settle cases amicably outside of court, and this tendency may even

---

[156] Thomas-Ciora, '2006 "Le transport des passagers"', 977, 979; Røsæg, *2010 "The Athens Convention"*, p. 57.

[157] Thomas-Ciora, '2006 "Le transport des passagers"', 981, advocating ways to provide for public funds to this end.

[158] Berlingieri, '2011 "The Athens Convention"', 1137.

[159] Røsæg, *2010 "The Athens Convention"*, p. 58.

be stronger in passenger claims where the decisions are more likely to have an impact on an individual's personal fate.[160] There have however been concerns on the mere capacity of P & I Clubs to cover liability for passenger claims under the high limits of the 2002 PAL, and it is not inconceivable that insurers and shipowners need to rely on exoneration clauses at some point.

This is heavily debated with regard to terrorist and war risks to which cruise ships may be exposed. Considering the values at stake, the exoneration clause in art. III para. 1 (b) for intentional acts by third parties is of some importance for carriers and their insurers. A terrorist attack aiming at the damaging or destruction of a passenger ship will of course be considered as an "act with intent to cause the incident".[161] It even seems that the wording of art. III para. 1 (b) PAL 2002 is broader than the corresponding provision under the CLC referring to the "intent to cause the damage". Under the CLC, the wrongdoing third party must foresee and intend to cause the oil pollution damage, thus aiming at damage to the ship that will invariably cause the escape of oil from the hull. Under the PAL, the intent to cause the "incident" appears to be less determined, as it seems to encompass mere unforeseen events on the ship. It is submitted that this would cover a pirate attack in which the ship was forced astray from its planned route, with the pirates not necessarily wanting to cause damage to the ship's structure or equipment[162] (also see the discussion of the same problem under the oil pollution regime, *supra*, Chap. 3, F. III.). The wrongdoing third party must clearly not aim at the "damage" in shape of the death or personal injury to a passenger, it will be enough if they premeditated the event that caused this.

The damage must however be "wholly" caused by the third party's act or omission, which creates the same problems as the corresponding art. III para. 2 (b) CLC. Even if the death or personal injury was mainly caused by a terrorist attack, a carrier may still be strictly liable if he contributed to the attack taking place, for instance by not having carried out sufficient controls to detect terrorists among the passengers[163] or crew, by having negligently drawn attention to the ship,[164] by unnecessarily navigating in war zones[165] and in the most extreme case simply by having failed to foresee and prevent a hostile attack.[166] Furthermore, the

[160] Gaskell, '1998 "New limits for passengers"', 331, referring to "informal indications" of P & I Clubs not to leave passenger claims unsettled even where there was a legally valid defence; Shaw, *2008 "Carriage of Passengers"*, pp. 149 et seq.

[161] Soyer, '2002 "Sundry Considerations"', 523; Røsæg, '2004 "News under the Athens Sun"', 161; Mandaraka-Sheppard, *2009 "Modern Maritime Law"*, p. 935; Tsimplis, '2009 "Liability in respect of passenger claims"', 145.

[162] Probably also endorsed by Soyer, '2002 "Sundry Considerations"', 523, giving the example of a pirate attack.

[163] E. Røsæg, 'Passenger liabilities and insurance: Terrorism and war risks', in D. R. Thomas (ed.), *Liability regimes in contemporary maritime law* (London: Informa, 2007), para. 12.18.

[164] Example given by Soyer, '2002 "Sundry Considerations"', 523, but referring to art. III para. 1 (a) PAL 2002.

[165] Røsæg, *2007 "Passenger liabilities"*, para. 12.18.

[166] Røsæg, '2004 "News under the Athens Sun"', 161.

contributory negligence may not even be imputable to the carrier, but to other persons such as the harbour authority[167] checking the staff in the port.

In an exact parallelism to the corresponding provision of the CLC, carriers may therefore have an interest to invoke the exoneration clause of art. III para. 1 (a) PAL 2002, as the link of causation between the "act of war or hostilities" and the incident appears to be weaker, the incident only needs to "result" from it.[168] It is however conceivable that the carrier may be denied the exoneration if he contributed substantively to the situation in which the damage occurred, i.e. if the incident was not only caused by the attack.[169] Furthermore, there is the question whether a pirate attack can be considered as an "act of war or hostilities". At least under English law, there is a certain tendency to only consider as "war" situations in which there has been a proper act of war by a governmental authority, following the understanding of the term in public international law. According to the view of the English courts, even "acts of hostilities" presuppose acts from agents benefitting from a certain support from a government or government-like entity, assuming authority over these actions.[170] It is therefore argued that a terrorist attack cannot be covered by the exception.[171] Neither would it apply to most pirate attacks committed by individuals from failed States such as Somalia for their own profit.

Another possibility of exoneration for the carrier is, according to art. VI PAL 2002, the contributory negligence of the passenger. The corresponding provision is however rather vague, providing for exoneration wholly or partly, in accordance with the provisions of the domestic court seized.

III. Liability Limits and the Relationship with the LLMC

With a view to the fact that the 1974 PAL's liability limits were, by most industrial States, already considered as too low when the Convention came into force in 1987, the limits have been significantly increased in the 2002 Protocol. Even from an objective point of view and in comparison with other limitation instruments such as the LLMC 1996, the limits are extremely high,[172] also—to the extent that the analogy is possible—for cargo ships of a comparable size.[173] For the first time,

[167] Røsæg, *2007 "Passenger liabilities"*, para. 12.19.

[168] Soyer, '2002 "Sundry Considerations"', 523.

[169] Ibid., p. 523.

[170] See Gunasekera, *2010 "Civil liability"*, pp. 193 et seq., giving a wide overview over the English understanding of the identical provisions in the BOPC.

[171] Tsimplis, '2009 "Liability in respect of passenger claims"', 145.

[172] Røsæg, *2007 "Passenger liabilities"*, para. 12.6.

[173] Ibid., para. 12.10.

the advice of the insurance industry as to the limits that were insurable on the market has been disregarded.[174]

1. Liability Limits and the Possibility of Unlimited Liability for Loss of Life and Personal Injury

The liability for claims regarding death and personal injury as set out in art. VII para. 1 PAL is limited to 400,000 SDR per passenger per incident. It should be repeated that the carrier's liability for claims caused by shipping incidents up to 250,000 SDR per passenger per incident is strict. The limits for luggage claims have equally been significantly increased. The PAL 2002 is clearly an instrument for Western and developed States,[175] as it is hardly conceivable that carriers from less wealthy parts of the world can provide the necessary funds for their strict liability obligation and compulsory insurance to these amounts.

The Convention even goes further by allowing its State Parties, in its art. VII para. 2, to provide unilaterally for higher limits, and even to remove the limitation altogether. In this respect, the PAL 2002 mirrors the development in the 1996 LLMC, equally allowing States in its art. X para. 3bis to provide for higher limits of liability. This leeway given to States rather than obliging them to adhere to fixed limits will very probably become more and more common.[176] The possibility for States to legislate higher liability amounts unilaterally however creates some problems in the application of the limitation as between States. If the higher limits of liability are enforced in the courts of the legislating State in question, but against vessels and individuals having a different nationality, the State may actually fail in his obligation under the Convention to apply the limits as set out in the international agreement.[177] Furthermore, it will of course encourage forum shopping.[178] Within the European Union, during the drafting of the Regulation No. 392/2009, European authorities expressed their wish to that there should be common limits for all member States,[179] in the interest of a fair competition between carriers.[180] Art. 5 para. 1 of the Regulation however now provides that the Regulation's provisions are without prejudice to the global limitation as set out by the LLMC in its 1996 version. Whereas a limitation by virtue of any other global limitation regime is forbidden to EU member States, those having ratified the 1996 LLMC can make use

---

[174] Soyer, '2002 "Sundry Considerations"', 539; W. R. Dorsey, 'The 2002 Protocol to the Athens Convention Relating to the Carriage of Passengers and Their Luggage by Sea' (2003) 1, Benedict's Maritime Bulletin, 12–6 at 15; Røsæg, '2004 "News under the Athens Sun"', 160 et seq.

[175] Ibid., p. 159.

[176] Ibid., p. 159.

[177] Griggs, '2002 "Le protocole d'Athènes"', 299.

[178] Soyer, '2002 "Sundry Considerations"', 533.

[179] Røsæg, *2010 "The Athens Convention"*, p. 70.

[180] Karsten, '2009 "Im Fahrwasser"', 215.

of its provisions. This rules out the application of the lower limits of the 1976 LLMC and even earlier instruments[181] on the one hand, but on the other hand, the 1996 LLMC allows for higher limitations and even unlimited liability.

The procedure for adapting the liability limits in the PAL 2002 is, according to art. XXIII, very similar to the one adapted for the 1996 LLMC. The amendment can be triggered by half of the State Parties and can, after due consideration and revision, be adopted by a two-thirds majority of the State Parties (art. XXIII paras. 2, 3, 5 PAL 2002). Out of the context of the other provisions of the PAL 2002, it must however be understood that the amendment procedure cannot apply to the limits for claims for death and personal injury that have been set unilaterally by States parties to the PAL.[182] It however seems possible to amend the minimum limits applicable to those States that have not legislated for limits of their own, but it is then very questionable whether all States should have a vote in the amendment procedure or rather only those who will effectively apply them.

The rules on the calculation of damages and the question of what can constitute as damage are not harmonised[183] and will therefore be subject to the appreciation by judges in domestic courts, doubtlessly influenced by the point of view of their home jurisdictions. The PAL 2002 however sets a limit in art. III para. 5 (d), by explicitly stating that punitive or exemplary damages cannot be claimed under the Convention. The provision reveals a concern of the negotiating parties that the Convention could become an instrument to enforce the abnormally high punitive or exemplary damages on an international scale, and its existence is acclaimed by certain authors.[184] It has however been pointed out that even in the US, there have not yet been many examples of punitive damages in admiralty cases and that the likelihood of this ever to happen, at least in death and personal injury cases, was rather small.[185] Furthermore, even under US conflict-of-law rules, it is very unlikely that punitive damages for torts that have been committed outside US territory will be enforced.[186] However, there have been more recent developments with regard to the availability of punitive damages for pollution claims.[187] It is thus

[181] Czerwenka, '2010 "Neue Haftungs- und Entschädigungsregeln"', 169.

[182] Berlingieri, '2011 "The Athens Convention"', 1139.

[183] Røsæg, '2004 "News under the Athens Sun"', 159.

[184] Ibid., p. 159.

[185] Shaw, *2008 "Carriage of Passengers"*, p. 151, also see J. P. Jones, 'The Sky Has Not Fallen Yet on Punitive Damages in Admiralty Cases' (2009) 83, Tulane Law Review, 1289–302 at 1295, on claims of a non-dependent relative for damages for a seaman's death, precluded by specialised provisions of domestic law.

[186] See S. Symeonides, 'Territoriality and Personality in Tort Conflicts', in T. Einhorn and K. Siehr (eds.), *International Cooperation Through Private International Law* (The Hague: T. M. C. Asser Press, 2004), p. 409, on the qualification of punitive damages as a conduct-regulating rule, that can only be applied within the territory where it is in force.

[187] Jones, '2009 "The Sky Has Not Fallen"', 1291 et seq., but dealing mainly with punitive damages for environmental pollution; L. E. Hume, 'Are We Sailing in Occupied Waters?: Rethinking the Availability of Punitive Damages under the Oil Pollution Act 1990' (2011) 86, New York University Law Review, 1444–81 at 1451 et seq. on a new interpretation of certain principles following recent decisions.

perhaps right that an instrument dealing with passenger claims should have a provision ruling out the possibility of punitive damages, if only for the sake of clarity.

2. Relationship with Other Limitation Instruments

It has been pointed out earlier in this study (see *supra*, Chap. 4, A. IV.) that the introduction of new instruments with higher limits of liability may be without effect, if the number of States having ratified the new instrument is still low. In all situations involving States that have only ratified the former instrument, the States parties to the new instrument must apply the liability limits of the old instruments, considering that they are still party to a mutual instrument with low limits. This "grandfather problem"[188] can hamper the effectiveness of a new convention or a Protocol considerably. It can be addressed by the States unilaterally denouncing the old instrument, and this is why the IMO, along with the promotion of the ratification of the BOPC, strongly advised that States ratifying the new instrument would also denounce the 1976 LLMC and ratify the 1996 LLMC in order to render it the general limitation regime for bunker oil claims (see *supra*, Chap. 4, A. IV.).

Before the 2002 PAL, the problem was however never directly dealt with in a new convention. Art. XVII para. 5 PAL 2002 requires every State ratifying the PAL 2002 to denounce the PAL 1974 and the 1976 and 1990 Protocols that have not entered into force.

Still, the Convention however does not address the more pressing problem of low limits in global limitation by the LLMC. Art. XIX of the PAL 1974, declaring that the Convention does not restrict the right to global limitation, remains unchanged by the 2002 PAL. Thus, the application of the low global limits of the 1976 LLMC, that apply per vessel and per incident, could seriously restrict the claims that passengers have under the PAL 2002.[189] It is only under the higher limits of the 1996 LLMC and the possibility given to States to legislate for yet higher limits that the provisions of the PAL 2002 can really take effect.

3. Overcoming the Liability Limits

Even if the liability is unlimited in theory, it will often be limited to the amount of compulsory insurance in practice.[190] It is often argued that in merchant shipping, vessels are most often operated by single-ship companies, not possessing assets above the value of the vessel and the funds available under compulsory insurance,

[188] Røsæg, '2004 "News under the Athens Sun"', 157.

[189] Bonassies, '2003 "Conventions internationales"', 8 et seq.

[190] Soyer, '2002 "Sundry Considerations"', 534.

so that claimants will not recover unlimited amounts even if it is legally possible to do so.[191] This can also influence their decision to invoke the rules on the loss of the right to limit liability (see, for the problem under the CLC, *supra*, Chap. 3, D. IV.). However, it does not seem that this holds entirely true for the cruise market. Big cruise lines are very often quite wealthy entities, possessing several ships[192] that are not necessarily run by entirely independent single-ship companies. It does seem that claimants may in very catastrophic events have some interest to challenge the limits of liability under the PAL. Furthermore, as the rules on the conduct barring the right to limit liability are exactly the same, art. IV LLMC 1976 and art. IV LLMC 1996 and art. XIII para. 1 and 2 PAL having the same wording in this respect, it is conceivable that both the limitation *per capita* under the PAL and the global limitation under the LLMC will be denied to the defendant.

This means that the problems as to the interpretation of "recklessly, and with knowledge that such damage would probably result", especially in England and France will occur again (see *supra*, Chap. 3, D., and Chap. 4, A. IV. 3.). Under the EU regulation, this may become a stumbling block for the harmonisation of passenger claims throughout Europe. Of course, given the ECJ's competence to interpret provisions of EU law, the understanding under European law of this frequently-used concept may finally be decided on, but it will of course only be binding on EU member States, with a risk of causing disparities between contracting States that are EU member States and others that are not.[193] It remains to be seen whether there will be such a judgment and whether it will can become a persuasive authority for maritime law on a global level.

IV. Compulsory Insurance and Special Rules for War and Terrorism Risks

It was perceived as one of the most important concerns of the PAL 2002 to introduce compulsory insurance for passenger claims, so that victims' losses could be compensated irrespective of the carrier's ability to pay.[194] Certainly, compulsory insurance protects the interests of passengers and may also be an incentive to shipping providers to enhance safety measures on board their ships to obtain cheaper premiums for the reduced risk.[195]

---

[191] Martínez Gutiérrez, '2012 "New European rules"', 301.

[192] See *Cour d'Appel de Versailles, 09.05.2012*, "*Costa Concordia*", where the proceedings on the settlement of claims for passengers of the *Costa Concordia* were conducted by the cruise line, *Costa Crociere SPA*.

[193] Legros, *2012 "L'intégration"*, p. 386.

[194] Gaskell, '1998 "New limits for passengers"', 331; Griggs, '2002 "Le protocole d'Athènes"', 297.

[195] Soyer, '2002 "Sundry Considerations"', 527.

Art. IVbis para. 1 of the 2002 PAL requires carriers to provide insurance for the amount of 250,000 SDR per passenger per incident at least. This means that the funds available under the insurance contract may be lower than the actual liability incurred by the carrier, whose absolute limit of liability is of 400,000 SDR (art. VII para. 1 PAL 2002). Even where a State party has made use of the right to legislate unilaterally for higher limits or even unlimited liability, the insurance cover does not need to exceed the 250,000 SDR. The consequence of this is that the claimant is guaranteed to obtain the 250,000 SDR for which the carrier is strictly liable. In claims exceeding this amount, the claimant may be exposed to the carrier disputing the claim, which is then, in addition, not backed up by compulsory insurance. Claimants may therefore be very reluctant to put forward claims for more than 250,000 SDR unless they expect the carrier to have more assets than just the insurance fund at their disposal.

1. Insurance and Channelling

During the first discussions of a Protocol to the 1974 PAL, it was considered whether the introduction of compulsory insurance would be of avail without a channelling provision clearly designating one liable party that would need to take out insurance.[196] Considering that there is no channelling provision in the PAL 2002 either, the problem of double insurance needed to be addressed in its drafting, especially as the high limits of liability and the strict liability for claims up to 250,000 SDR made insurance generally costly. It has been suggested that parties that could be potentially liable should take out insurance together, under the same insurance policy.[197] This of course presupposes that they are both eligible for the same insurance terms. Considering that most third party risks in the passenger trade are still covered by P & I cover, which is a mutual shipowner's insurance, this may not be the case. In order to obtain P & I cover, the insured needs to be member of a P & I club, which is not possible for companies that are not shipowners. Depending on the business the "carrier" as opposed to the "performing carrier" does, it may be a part of a completely different insurance market.[198]

Without introducing any channelling provision, the problem of double insurance has actually been addressed in the PAL 2002 by the wording of art. IVbis para. 1, and a slight change to art. I para. 1.[199] In art. I para. 1 (c) PAL 2002, a new definition has been added. Following the provision, a "carrier who actually performs the whole or a part of the carriage" is either the carrier or the performing carrier, depending on who actually provides the sea transport. This is the person

[196] Gaskell, '1998 "New limits for passengers"', 332.

[197] Berlingieri, '2011 "The Athens Convention"', 1132.

[198] Kröger, '2001 "Passengers Carried by Sea"', 250.

[199] Griggs, '2002 "Le protocole d'Athènes"', 293; Griggs, Williams and Farr, *2005 "Limitation of Liability"*, p. 112.

which is, according to art. IVbis para. 1 PAL 2002, the addressee of the insurance obligation. Even if there are both a carrier, such as a travel agency, and a performing carrier, such as a ferry or cruise line, both liable under the PAL, the insurance obligation only lies with the entity that actually performs the sea trip, and thus, most likely with a shipping line. This means that the insurance offered will almost invariably be a maritime insurance, part of a market that is a lot more likely to be able to provide for the insurance cover at reasonable cost than the accident insurance market in which travel agencies buy their cover.

2. Insurance for War Risks

Notwithstanding the limitation of liability for individual claims, the burden of compulsory insurance for a ship of 3,000 or more passengers is extremely high and will have a significant impact on the profitability of the maritime passenger trade once the 2002 Protocol comes into force on a global level. The P & I Clubs that have hitherto provided insurance cover for third party indemnity of this kind have been reluctant to offer insurance on this level. Furthermore, there remains the question of the war and terrorism risk that is generally excluded from P & I cover, and that is certainly not easily insurable for a passenger ship the size of an average cruise ship. As passenger trade is only a very small part of the market on which the Clubs operate, a real risk of P & I Clubs simply refusing to provide cover for passenger ships in the future was spotted.[200] Insurance capacity is always hard to measure,[201] but it is fairly obvious that the burden on carriers under the 2002 PAL is considerably bigger than the burden on any other maritime player in the framework of another civil liability convention. It had thus been realised at an early point in time that the compulsory insurance obligation as set out in the 2002 PAL would be a serious impediment to the convention coming into force. Still, the respective provisions were kept in the instruments, partly because they corresponded to the rules on compulsory insurance in the IMO's other civil liability conventions, and it was felt that it would be better to find an "informal" solution to the problem than officially abandoning the established rules of the CLC and BOPC.[202]

The IMO's Legal Committee therefore, after considerable debate, issued a "reservation" in 2006 that is meant to be used as a "standard" reservation by ratifying States.[203] The IMO also issued "guidelines for the implementation of

[200] Røsæg, *2000 "The impact of insurance practices"*, pt. 4.5.4; Tebbens, '2008 "The European Union and the Athens Convention"', 662, footnote 38; Berlingieri, '2011 "The Athens Convention"', 1125.

[201] Røsæg, *2000 "The impact of insurance practices"*, pt. 2.3.2 and 3.1.

[202] Balkin, *2007 "The Rise and Rise"*, p. 708.

[203] See the statement on the IMO's website, http://www.imo.org/About/Conventions/ListOfConventions/Pages/Athens-Convention-relating-to-the-Carriage-of-Passengers-and-their-Luggage-by-Sea-%28PAL%29.aspx (last visited 8 March 2014).

the Athens Convention"[204] (the Guidelines). Following its terms, the States can reserve the right to limit both the carrier's liability under art. III para. 1 and 2 PAL 2002 and the insurer's liability under the art. IVbis PAL 2002 for war and terrorism risks to 250,000 SDR per passenger per incident or 340 million SDR per ship per incident, whichever is the lower. This means that the IMO for terrorism and war risks effectively on the one hand limits the liability to 250,000 SDR (ruling out the higher limit of 400,000 SDR in art. VII PAL) and, on the other hand, introduces a "global" cap per vessel again,[205] in order to meet the carriers' and insurers' concerns considering war and terrorism risks.

This strategy presents a very pragmatic solution[206] to deal with the insurance industry's resistance to cover war risks in full. It means of course that victims of war and terrorism will find that there is a cap to their claims that is not opposed to victims of other risks. With regard to this, it is sometimes maintained that in the absence of available insurance cover for a danger that has its roots in conflicts between States and is essentially a problem of public international law, States rather than private entities should assume responsibility to compensate victims of war and terrorism risks.[207]

Following art. 2 of the Guidelines, the insurance provided for a given vessel can be split between one insurer covering war risks and another insurer covering all other risks. The provision aims at giving carriers the opportunity to maintain P & I cover for general risks and buy war-insurance with another insurer additionally, for the risks that the P & I insurance does not cover.[208] The requirement in art. IVbis para. 2 PAL 2002 to provide for one single document of insurance is therewith altered. The Guidelines furthermore set out some exoneration clauses in the war insurance as acceptable, both concerning special risks such as radioactive contamination and biological weapons and a termination and exclusion clause (cl. 2.1 and 2.2. 1 to 2.2.3 of the Guidelines).

The definition of "war and terrorism risk", i.e. the risks covered by war insurance are defined in cl. 2.2 of the guidelines and are very broad, comprising almost every possible risk caused by armed conflict, irrespective of the motives or aims of the armed party. With regard to terrorism, the war cover however only comes into play if the terrorists acted with a political motive.[209] The provision on war cover is therefore clearly meant to ensure that the carrier can obtain insurance for all the war and terrorism risks for which he is not exonerated under art. III para. 1 (a) and

[204] IMO Circular letter No. 2758, Ref. A1/P/5.01. The document can be found through the IMO Documents Service, http://www.imo.org/OurWork/Conferences/Pages/Documents.aspx (last visited 8 March 2014).

[205] Berlingieri, '2011 "The Athens Convention"', 1135.

[206] Thomas-Ciora, '2006 "Le transport des passagers"', 979; Røsæg, *2007 "Passenger liabilities"*, para. 12.27.

[207] Thomas-Ciora, '2006 "Le transport des passagers"', 981.

[208] Tsimplis, '2009 "Liability in respect of passenger claims"', 148.

[209] Røsæg, *2007 "Passenger liabilities"*, para. 12.54.

(b) PAL 2002 (see *supra*, II. 2.).[210] War cover as opposed to terrorism cover should, for instance, also cover events like capture, seizure or arrest of the ship and possible consequences, acts of terrorists or other persons acting maliciously, irrespective of their motives and also the risks created by derelict weapons of war.

The IMO's Guidelines, except for the technical provisions on the supervisory competence of the IMO in cl. 4 and 5, have been incorporated into the EU Regulation (EC) No. 392/2009 by virtue of art. 1 (b)—therefore, the member States can apply the rules on insurance cover as a part of EU law. The EU, being a party to the PAL 2002 has also declared the 2006 "reservation",[211] and it is to be expected that all member States ratifying the Convention subsequently will enter it, too. In the long run, the Guideline's provisions, namely on the cap of liability for war risks at 340 million SDR per vessel per incident, will therefore become an integral part of the 2002 PAL.

The existence of a second set on rules on insurance for terrorism and war risks does however complicate[212] the application of the Convention's provisions. Even where all or almost all of the States parties have entered the reservation, the existence of another instrument significantly altering the 2002 PAL can create problems. One of them relates to the amendment of the Guidelines, in practice especially with regard to the amendment of the separate liability limits set out in them. The Guidelines, in their cl. 1.13, provide for an amendment procedure by the IMO, on the initiative of the Legal Committee. The unprejudiced right of a State in public international law to withdraw or amend its reservation is specifically referred to. It has been pointed out that any amendment effectuated by the IMO has however either to be accepted or abolished altogether,[213] so that States are not free to maintain a reservation that has been changed by the IMO.

There is thus some uncertainty as to the interplay between the PAL and the Guidelines.[214] It remains however to be considered that the Guidelines were introduced in the interest of the global insurance market and that any subsequent modification will, for the same reason, take effect on a global level as well.[215] The amendment for the Guidelines is, in any case, entirely independent from the amendment procedure of the limits in the PAL as such.[216] It remains to be seen whether any amendments can be aligned or whether the rating of risks on the insurance market will trigger a completely independent development.

---

[210] Ibid., para. 12.50.

[211] See the IMO Document http://www.imo.org/About/Conventions/StatusOfConventions/Documents/Status%20-%202014.pdf, p. 328 (last visited 8 March 2014).

[212] Røsæg, *2010 "The Athens Convention"*, p. 71: "The entire arrangement is, of course, very messy".

[213] Ibid., p. 71.

[214] Tsimplis, '2009 "Liability in respect of passenger claims"', 149.

[215] Røsæg, *2010 "The Athens Convention"*, p. 71.

[216] Røsæg, *2007 "Passenger liabilities"*, para. 12.33; Berlingieri, '2011 "The Athens Convention"', 1139.

The existence of a "global" cap, per vessel, on war risks, secondly creates a certain conflict with the global levels under the LLMC.[217] For disastrous events where the sum of all passenger claims actually exceeds the 340 million SDR cap under the Guidelines, there will be yet another cap under the LLMC 1976 or 1996, which will be most likely lower than the Guidelines limit. According to art. XIX PAL recognising global limitation, the LLMC limit then prevails. If the LLMC limit exceeds the Guidelines limit, it is not overly clear if the PAL Guidelines allow to cap the LLMC limit for the war and terrorism claims, but it has been argued that this sub-limit is acceptable under the wording of the LLMC and the Guidelines.[218] This problem will of course only arise where States have not made use of the possibility in the 1996 LLMC to abolish the limitation of liability for death and personal injury claims.

The introduction of the reservation and the implementation Guidelines complicate the legal framework of the PAL 2002 even more and are likely to create problems in its application. Claimants may furthermore be surprised to find that the actual limits of liability and of insurance for damages that have been caused by acts of war terrorism and related events, are considerably lower than for any other, probably less traumatic, risk. However, the IMO has made a conscious choice by drafting the reservation and the Guidelines, considering that the PAL 2002 did not have good chances to enter into force without them, and that it would be better for passengers to obtain the limited protection under the PAL 2002 and the Guidelines than not getting adequate protection at all because the Convention would never enter into force. It is however regrettable that the insurance issue was not considered earlier and inserted into the 2002 Protocol itself.

V. Time-Bar for Actions

The 1974 PAL's provisions on the limitation of actions in time have been altered in the 2002 Protocol, but the general limit of a two years' time bar in art. XVI has been maintained. It is still the domestic law of the court in question that shall govern any possible suspension or interruption of the time limit, but the absolute limit in art. XVI para. 3 PAL has been slightly prolonged. The general limit of three years after the actual or originally planned disembarkation is replaced by either a five years' limit or a three years' limit starting with the moment the claimant knew or should have known of the injury, whichever is the earlier. The provision, which has already been criticised for being needlessly complicated[219] has been enacted due to the growing importance of transport by high speed craft. Accidents in these vehicles

---

[217] Røsæg, *2007 "Passenger liabilities"*, para. 12.89; Berlingieri, '2011 "The Athens Convention"', 1136.

[218] Røsæg, *2007 "Passenger liabilities"*, para. 12.90 et seq.

[219] Griggs, '2002 "Le protocole d'Athènes"', 300.

can cause whiplash injuries[220] that are often only be diagnosed after considerable time. The understanding of "knew or should have known of the injury" is of course open to interpretation. On the whole, with regard to the generally[221] very long limitation periods for personal injury in domestic law (see *supra*, A. VI.) the prolongation, even if the time frame is still short, should be acclaimed.

It has however, and rightly, pointed out that stretching the time bar will make it very hard to reconcile the liability regime of the PAL with global limitation regimes[222] under the LLMC, or even under the global cap on war risk claims set out in the Guidelines. A global fund that is meant to cover all claims arising under a given incident can only be definitely distributed once all possible claims have been assessed, and thus, only after the expiration of the limitation period. A longer limitation period under the PAL will therefore mean that the global limitation can only be calculated after years, and claimants will be uncertain as to the definite amount that they can recover. This practical problem can of course be solved by practical solutions such as interim and *pro rata* payments until the number and the amount of claims is ascertained. It is however another circumstance that shows that the interplay between the PAL, especially as amended by the 2002 Protocol, and existing global limitation regimes is not very well established. It remains to be seen whether upcoming issues can be solved in future claims.

C. Conclusion

There are still very many unresolved issues in the existing liability regime for passenger claims, and with the application of the 2002 PAL as a part of European law, big changes have already realised for the EU countries. Several difficulties that may arise with the international regime put into an EU Regulation have been pointed out. It will be interesting to see whether the EU member States that are not parties to the PAL 2002 can, when applying the Convention's provisions, handle these difficulties. It stands to expect that more EU member States, being bound by the Convention's provisions by EU law already, may ratify the PAL 2002 and thus make it applicable as between parties from contracting States outside the EU, too. Widespread ratification in this way would of course mean that it is a successful, practical regime and that its advantages outweigh its shortcomings.

Still, its chances of taking effect in less developed countries where business entities will hardly be able to provide insurance or security for the liability limits,

---

220 Soyer, '2002 "Sundry Considerations"', 534; Røsæg, '2004 "News under the Athens Sun"', 154.

221 See however Peltz and Warger, '2001 "Amendments to Athens Convention"', 189, on the rather short limitation periods in US maritime law, consequentially criticising a prolongation such as the one under the Protocol.

222 Røsæg, '2004 "News under the Athens Sun"', 160.

are almost nil,[223] and, sadly, it is in any case unlikely that personal injury claims in these countries would reach the dimension of the liability limits in the PAL.[224] The PAL effectively accepts that it will only be valid for Western and industrialised countries as the price for an effective protection of the sea passengers in these parts of the world. Even within these countries, it has been very difficult to find at least common minimum limits that could be agreed on, which is of course a consequence of marked differences in the standard of living, the social security provided and even the development of costs in medical treatment.[225]

Even under the civil liability regime as established by the PAL 2002, there are however some issues that may be criticised. The existence of strict liability coupled with compulsory insurance up to rather high limits may considerably reduce the carrier's interest to avoid risks, at least on a small scale towards individual passengers. The values at stake in the passenger transport are in this respect quite different from the protection of the environment in the other civil liability conventions, where this concern was also raised.

Still, passenger claims are very different from the environmental pollution claims that have been dealt with in the Chaps. 3 and 4 of this study. It has already been pointed out that in most catastrophic ferry and cruise ship accidents claims were not settled according to the liability and limitation regimes in place, but in private agreements between the associated victims and the shipowner and his insurance.[226] The limits of liability as provided by liability regimes in place can however serve as a negotiation guideline in these settlements.[227] The limits that will be provided by the PAL 2002 and already are fixed by the corresponding EU Regulation will thus influence victims' views on what they can reasonably expect.

---

[223] Soyer, '2002 "Sundry Considerations"', 539.

[224] Røsæg, *2007 "Passenger liabilities"*, para. 12.8.

[225] Røsæg, '2004 "News under the Athens Sun"', 163, giving the development of these costs in Norway in comparison to the general cost of living as an example.

[226] Honka, *1998 "Nordic Reactions"*, p. 84, referring to the *Estonia* disaster, *Cour d'Appel de Versailles, 09.05.2012, "Costa Concordia"*, with regard to the settlement offer of the *Costa Concordia's* cruise line.

[227] Shaw, *2008 "Carriage of Passengers"*, pp. 149 et seq.

Part III
Jurisdiction and Conflict of Laws

Chapter 6
Jurisdiction

The civil liability regimes that have been given a closer look in Part II (Chaps. 3–5) are essentially international regimes in that they set out matter-of-fact rules, applying directly to all cases that come before the courts of State parties. Even if interpretation and sometimes definition of some terms is left to the national courts, the rules on civil liability in these conventions are in themselves international.

The UNCLOS calls for the establishment of regimes of this kind, at least concerning the compensation for pollution damage, in its art. 235 para. 3, and also sets out a safeguard provision for existing conventions in art. 304.[1] The negotiation, ratification and the enforcement of civil liability conventions is thus generally in accordance with the law of the sea.

Art. 229 UNCLOS clarifies that none of the provisions of the Convention affects the institution of civil proceedings for loss or damage resulting from the pollution of the marine environment. Whereas the Convention sets out many procedural and other safeguards as to the institution of criminal proceedings for marine pollution, the law of the sea is considerably more generous regarding civil proceedings. Still, art. 229 UNCLOS does not set out a ground of jurisdiction in itself; it merely insists that a civil jurisdiction otherwise established remains unaffected by the provisions of the Convention.[2] In art. 235 para. 2 UNCLOS, the Convention calls upon States to ensure that their national legal systems provide for mechanisms to obtain prompt and adequate compensation for pollution damage, without however setting out these mechanisms in more detail.

Art. 235 para. 2, as a provision in Part XII of the UNCLOS on the marine environment, is restricted to pollution damage, and it is debatable whether there is a corresponding obligation of States to provide for legal mechanisms to obtain

[1] M. H. Nordquist, S. Rosenne and A. Yankov, *United Nations Convention on the Law of the Sea, 1982. A Commentary*, Volume IV (Dordrecht, Boston: Martinus Nijhoff Publishers, 1991), para. 235.10(f).

[2] Ibid., para. 229.5.

S.F. Gahlen, *Civil Liability for Accidents at Sea*, Hamburg Studies on Maritime Affairs 32, DOI 10.1007/978-3-662-45555-5_6

compensation for other heads of damage, such as death or personal injury. Some considerations will follow *infra,* analysing the grounds on which States generally assume jurisdiction for events that take place at sea. It must however be kept in mind that the mere existence of State power does not mean that the State will exercise it in any event. Even if there is an uncontested ground of jurisdiction, e.g. where the State could exercise its jurisdiction without infringing the rights of any other sovereign nation, a State may decide not to act on it after all.[3] This voluntary waiver of jurisdiction is particularly important in the context of events at sea[4] where there may be conflicting jurisdictional grounds. It will however be seen that States within the EU may be bound by European law. The question of whether, outside of clear treaty provisions, customary international law generally binds States to exercise jurisdiction in certain cases, is outside the scope of this study.

In the absence of international rules, national courts wishing to deal with sets of facts and to establish liability for a given behaviour must have recourse to domestic rules. With regard to the fact that the law of the sea only accords sovereignty over parts of the oceans, functional rights over others and entirely withdraws the high seas from claims of appropriation by States, it is by no means clear that domestic legal systems can apply their own rules to events that occurred on the oceans.[5] In answering the questions whether States have competence to adjudicate a given set of facts under their jurisdiction, and, in a second step, which rules should be applied to the case, it becomes necessary to consider the scope of jurisdiction and conflict-of-law provisions in a maritime context.

The following two chapters of the study deal with the application of domestic conflict-of-law rules to maritime incidents, and will centre on the principles that are employed to designate a domestic court and a national law that can govern the relationship.

The question of the extent to which States can assume competence for events that happen at sea is at the root of the law of the sea and the UNCLOS. The oceans are, according to customary public international law and according to the Convention, an area in which sovereign rights cannot be exercised by States to the extent States can do so on their land territory. Even in the territorial waters of a coastal State, foreign-flagged ships enjoy innocent passage and can only be subject to intervention by the coastal State under certain circumstances, namely in all cases in which their passage has an impact on the State in question.

Thus, events that take place on board ships or in relation to ships may only with difficulty be submitted to a given sphere of sovereignty. As a moving entity, the ship enters and leaves maritime zones over which coastal and port States have

[3] See A. Mills, *The Confluence of Public and Private International Law* (Cambridge, New York: Cambridge University Press, 2009), pp. 6 et seq.

[4] Churchill and Lowe, *1999 "The Law of the Sea"*, p. 66; A. Pozdnakova, *Criminal jurisdiction over perpetrators of ship-source pollution* (Leiden, Boston: Martinus Nijhoff, 2013), p. 130.

[5] See George, '2007 "Choice of Law"', 137: "the application of choice of law principles, however, is anathema to the operation of an ancient body of overarching sea rules".

different degrees of sovereignty, having to consider the interests of other States to varying degrees. Furthermore, their interest in the events that happen on board may be very different, depending on the consequences of the events and the damages that arise. Jurisdiction for an accident on board the ship in which a crewmember or passenger is injured may be considered by coastal State authorities in a very different way from jurisdiction for an accident in which there is a marked risk of pollution or where other vital interests of the coastal State are at stake.

There thus may be a different jurisdictional regime for different events. Furthermore, regimes will, under the law of the sea, differ according to the maritime zone in which they apply. Moreover, it is submitted that State practice in jurisdictional matters by no means always exactly follows the provisions of the UNCLOS,[6] so that there may be precedents blurring the distribution of competences as provided for by the Convention.

Given the many different ways in which the terms "jurisdiction" and "competence" are used in both public and private international law, it is important to briefly point out the meaning of these terms in the context of this study.

Jurisdiction can be assumed on different grounds, territoriality being one of the most important, but by no means the only one. It is of course debatable to which extent territorial jurisdiction can be assumed in areas of reduced state sovereignty, so that coastal State jurisdiction over ships at sea is one of the most problematic issues in that context.

The competence of a State to submit persons, nationals or aliens, to the expression of its sovereignty, can be divided into three types of power: first, the legislative jurisdiction, i.e. the competence to create laws and rules that will be valid within the scope of sovereignty of the State; second, enforcement jurisdiction, i.e., the power of State authorities to effectively take measures against persons or property, either in preparation for a trial or in execution of a judgment already delivered. Third, there is judicial jurisdiction, meaning a State's power to try cases in its courts and to deliver judgments that will have further effects.

UNCLOS contains provisions on the first two powers for the respective maritime zones, but almost none on judicial competence. Even with regard to legislative and enforcement jurisdiction, the regimes are fragmented[7] and leave certain questions open. It is often argued that the regime as created by legislative and enforcement jurisdiction should be considered to have an impact on judicial jurisdiction,[8] especially as in many cases, enforcement measures by a coastal State can trigger the proceedings, and sometimes, proceedings are impossible without previous gathering of evidence or arrest of persons that can only be achieved by enforcement

---

[6] See R. R. Churchill, 'The Impact of State Practice on the Jurisdictional Framework Contained in the LOS Convention', in A. G. Oude Elferink (ed.), *Stability and change in the law of the sea: The role of the LOS convention* (Leiden, Boston: Martinus Nijhoff, 2005), pp. 125 et seqq.

[7] V. Labrot, 'Compétences des États et Droit de la mer: regard sur les compétences pénales', in A. Cudennec (ed.), *Le droit penal et la mer* (Rennes: Presses universitaires de Rennes, 2006), p. 19.

[8] Pozdnakova, *2013 "Criminal jurisdiction"*, p. 130, with regard to criminal judicial jurisdiction.

measures. Reference in this chapter will therefore be made to legislative and enforcement powers where they have an impact on judicial jurisdiction, which is the subject matter of this part of the study.

A. Judicial Jurisdiction in Public International Law

In criminal cases, judicial jurisdiction is generally based on the principles of territoriality and nationality.[9] States usually claim competence to adjudicate infractions that have taken place on their territory, even if committed by foreigners, and competence to adjudicate infractions committed by their nationals, even if committed abroad, at least if the act is also punishable under the law of the place where it was committed. Furthermore, there is the protective or security principle, under which States claim competence for acts threatening their security, even if committed abroad and by foreigners. The limits of this ground of jurisdiction are however very much debated, and it is only well-established in rather obvious cases, such as the counterfeiting of the State's currency abroad. For some crimes which have a truly international character and are rather hard to link to the territorial or national jurisdiction of any State, universal jurisdiction is assumed by most States as a matter of public policy. However, the only examples for which this universal jurisdiction is, on a global scale, rather unanimously accepted are piracy, and, already with considerably less certainty, the unlawful seizure of aircraft and possibly, drug trafficking.[10] Outside these customary sources of criminal jurisdictional competence, jurisdiction can of course be assumed for a certain number of crimes that are crimes under international law. Most often, this is done under an agreement to this end as between States, by which they are both obliged and entitled to persecute individuals who have committed these deeds, irrespective of their nationality and the place where the crime took place.

In criminal cases at sea, the essential difficulties with judicial jurisdiction thus always circle around the question whether the alleged criminal or whether the ship that was involved in the crime can be subject to the competence of the adjudicating State.[11] The UNCLOS contains a number of provisions concerning States' power to assume criminal jurisdiction over persons for conduct at sea, such as art. 27 on the coastal State's criminal jurisdiction over a ship in the territorial sea, or art. 97 conferring criminal jurisdiction for collisions on the high seas exclusively to the flag

---

[9] See I. Brownlie, *Principles of Public International Law*, 6th edn. (Oxford, New York: Oxford University Press, 2003), pp. 297 et seq. for a comprehensive overview on criminal jurisdictional competence.

[10] Ibid., p. 303.

[11] See I. Papanicolopulu, 'A Missing Part of the Law of the Sea Convention: Addressing Issues of State Jurisdiction over Persons at Sea', in C. H. Schofield (ed.), *The Limits of Maritime Jurisdiction* (Leiden: Martinus Nijhoff, 2014) and Hakapää, *1981 "Marine pollution"*, pp. 151 et seq. for an overview of the grounds of criminal judicial jurisdiction at sea.

State or to the State the concerned individual is a national of. The detailed provisions on the enforcement of pollution prevention regulation in Part XII of the Convention arguably only concern criminal jurisdiction, considering that art. 229 UNCLOS explicitly upholds the institution of civil proceedings for compensation of marine pollution damage.[12] Criminal cases are brought by the public prosecution of a State, so that there is a public authority involved that will most probably address these questions beforehand.

It has been maintained that the legitimacy of a State's assumption of judicial jurisdiction in civil trials has in general followed the same principles, especially with regard to the ensuing enforcement in which repressive measures, very similar to the enforcement of criminal sanctions, are used.[13] The matter is however by no means as easy as this. Civil law claims are brought by one of the parties, so that the jurisdiction that is seized does not generally have the opportunity to withdraw from the case before it first becomes pending with a court. In civil law claims, jurisdiction can follow from many factors that may not necessarily have a link with the case as such, such as the domicile of either party or the existence of assets in the jurisdiction. In choosing a venue, the claimant will most likely already consider the chances of enforcing the judgment afterwards. In order to complicate things more, many civil proceedings entail claims of both parties against each other. Last but by no means least, the court in a civil law claim does not necessarily apply its own law. Following a conflict-of-law analysis, the case might be decided under a law that is not the law of the country in which the suit is brought.[14] In the law of the sea, there are hardly any provisions restricting or guiding the exercise of civil jurisdiction.[15]

It is obvious, from a mere consideration of the bases on which States effectively have assumed civil jurisdiction in the past[16] and today, that States are considerably less restrictive in this assumption of civil judicial jurisdiction and that State practice with regard to civil jurisdiction is very different from the one regarding criminal jurisdiction. As has rightly been pointed out,[17] decisions in civil trials are significantly less likely to give rise to diplomatic protests against the adjudicating State than criminal trials. The consequences of a criminal trial are more perceivable to the accused, so that any other State having an interest in the case is more likely to act on behalf of the interested persons. In a civil trial concerning a relationship between private entities, it is less likely, yet not unconceivable, that a State may consider its

[12] P. Bonassies, 'Après la décision "Erika": Observations sur la compétence des juridictions pénales françaises après abordage en haute mer' (2013) 65, Droit Maritime Français, 195–200 at 197.

[13] Brownlie, *2003 "Principles of Public International Law"*, p. 298.

[14] See Papanicolopulu, *2014 "Addressing Issues"*, p. 391.

[15] Bonassies, '2013 "Après la décision "Erika""', 197.

[16] D. W. Bowett, 'Jurisdiction: Changing Patterns of Authority over Activities and Resources' (1983) 53, British Yearbook of International Law, 1–26 at 4.

[17] M. Akehurst, 'Jurisdiction in International Law' (1972) 46, British Yearbook of International Law, 145–257 at 170.

interests aggrieved.[18] There have been a few cases in which certain civil judgments have given rise to diplomatic protests, but these protests mostly did not concern the assumption of jurisdiction, but rather the assumption of sovereign rights on which the adjudicating State based its claim of jurisdiction.[19] It can be argued that the reason for this lack of interest on an international basis may be due to the fact that the mere delivery of a judgment generally does not yet injure another State's national's rights. It is only by enforcing this decision that the defeated party will feel the effect of the civil trial,[20] and enforcement in another country generally presupposes recognition of the judgment in the enforcing State. Leaving aside the stage of enforcement, it is thus not very likely that the mere delivery of a judgment in one State concerning facts that have a strong link with another State will be resented by this latter State. Still, the judgment may become enforceable at some later point in time or in another State, so that an assumption of unlimited civil jurisdiction will ultimately endanger the interests of all other States.[21]

It is however noteworthy that some of the older cases that have given rise to diplomatic intervention following civil decisions concern incidents involving vessels,[22] namely, collisions. One example is the case of the *City of Meccah*,[23] in which a British ship collided with a Portuguese ship off the Portuguese coast. The collision was subject to proceedings in Lisbon when the *City of Meccah* arrived there after the accident. The Portuguese courts instituted proceedings *in rem* against the vessel in order to obtain it as security, as well as *in personam* against the British officers who were held liable for having caused the collision. The British diplomatic representative in Lisbon protested against the latter, maintaining that the behaviour that was adjudicated by the court took place outside the waters that were subject to Portuguese sovereignty.[24] Interestingly, the protest was later dropped on the ground that Portuguese courts had assumed jurisdiction not as a matter of territorial sovereignty over the waters in question but, following the passive personality principle, as a competence to adjudicate torts that were caused outside Portugal to

---

[18] T. B. de Maekelt, 'Relaciones entre el derecho internacional privado y el derecho internacional público', in D. P. Fernández Arroyo and C. L. Marques (eds.), *Derecho internacional privado y derecho internacional público* (Asunción: CEDEP, 2011), p. 28.

[19] Akehurst, '1972 "Jurisdiction"', 177.

[20] See C. Parry, 'The Theory of Restrictive Practices' (1958) 44, Transactions of the Grotius Society, 109–23 at 116.

[21] F. A. Mann, 'The Doctrine of International Jurisdiction Revisited after Twenty Years' (1964) 111, Recueil des Cours, 19–115 at 33.

[22] See S. C. Symeonides, 'Accommodative Unilateralism as a Starting Premise in Choice of Law', in H.-E. Rasmussen-Bonne, R. Freer, W. Lüke and W. Weitnauer (eds.), *Balancing of Interests: Liber Amicorum Peter Hay zum 70. Geburtstag* (Frankfurt am Main: Verlag Recht und Wirtschaft, 2005), pp. 424 et seq. on the theory of State interests in the outcome of private law disputes, namely maritime cases.

[23] See the summary of the case as presented by Parry, '1958 "The Theory of Restrictive Practices"', 118.

[24] Ibid., p. 119.

the detriment of Portuguese nationals.[25] The British authorities apparently did not object to this footing of competence. Furthermore, the protest did not concern the proceedings *in rem*, where jurisdiction was based on the presence of the ship in a Portuguese port.[26] The Portuguese judgments to this end were later given effect in England,[27] the court considering that jurisdiction was rightly assumed by the Portuguese courts and that the English courts had to enforce the claim against the ship when it arrived in England and was subject to English sovereignty.[28] The case does show that civil jurisdiction needs to be based on a given link between the set of facts and the adjudicating State, and it has been cited as an example for the fact that there must be a limit to a State's assumption of civil jurisdiction over foreign nationals and sets of facts taking place outside their territory.[29] Another prominent example of the limits of States' civil jurisdiction is, interestingly, the civil jurisdiction over foreign-flagged ships in ports,[30] where it is in many States maintained that events happening exclusively on board the ship and not having an impact on the interests of the port State cannot be adjudicated in the latter's courts.[31] It appears that in the maritime context civil law may, especially in tort cases, be a way to regulate the behaviour of persons in an area where the adjudicating State's rights are reduced, which is clearly one of the situations in which the exercise of civil jurisdiction should be reconcilable with the standards of public international law.[32]

Rights and obligations of States under public international law are to some extent defined by the States themselves and by their practice in claiming them, and encroachments on the domain of other States will remain unrequested, and perhaps even unnoticed, if the aggrieved State does not protest. Given the fact that assertion of jurisdiction in civil claims has so far only rarely been the cause of diplomatic protest, it was for a long time put forward that, in theory, civil jurisdiction of States is unlimited and that it is restricted, if at all, only by considerations of effectiveness[33] or by its proper interests.

---

[25] Akehurst, '1972 "Jurisdiction"', 177.

[26] See A. V. Dicey, J. H. C. Morris and L. Collins, *Dicey, Morris and Collins on the Conflict of Laws*, 15th edn. (London: Sweet & Maxwell, 2012), para. 13-002 for an explanation of the concept of a claim *in rem*.

[27] *"The City of Meccah"*, 5 Law Reports, Probate Division 28 (High Court (England), Probate, Divorce and Admiralty Division, 25 November 1879).

[28] Ibid., pp. 32 et seq.

[29] Parry, '1958 "The Theory of Restrictive Practices"', 120.

[30] Mann, '1964 "The Doctrine of International Jurisdiction"', 22, but rather considering legislative jurisdiction.

[31] Churchill and Lowe, *1999 "The Law of the Sea"*, p. 350.

[32] See Bowett, '1983 "Jurisdiction"', 4.

[33] See A. V. Dicey and A. Berriedale Keith, *The Conflict of Laws*, 3rd edn. (London: Stevens, 1922), p. 40: "the sovereign of a country, acting through the courts thereof, has jurisdiction over [...] any matter with regard to which he can give an effective judgment".

The absence of protest and the uncertainty of the limits of legitimate jurisdiction in civil cases[34] can however not mean that there are no limits to jurisdiction. An excess in the exercise of civil jurisdiction can be an encroachment on the rights of other States, which may have an equal claim to deal with the matter,[35] and jurisdictional competence in civil cases is just as much subject to the restrictions of public international law[36] as criminal or administrative law in which the defendant party directly faces State authorities.[37]

It is submitted that maritime law, especially considering maritime torts, is one of the fields of law in which the legitimacy of applying conflict-of-law rules becomes most pertinent and in which the chances of a jurisdiction being considered as exorbitant are rather important. Similar problems in respect of the legitimacy of civil jurisdiction arise in connection with human rights law,[38] but the law of the sea is an area where the direct interest of States may be at stake.[39] The UNCLOS sets out a well-established balance in the rights and obligations of coastal States and flag States with regard to the competences they have. However, the Convention, as has been pointed out already, hardly ever addresses the problem of judicial jurisdiction, and it seems impossible to derive comprehensive rules from the few provisions that there are.

It is however submitted that the competences that flag States, coastal States, port States and possibly even other States enjoy in maritime contexts are inherent in the framework of the law of the sea as a part of public international law, and that these competences should be given regard in the consideration of jurisdiction and choice-of-law rules for maritime cases, namely maritime tort cases. Thus, the circumstances that normally trigger judicial jurisdiction of a State will be given regard, before asking the question whether they can be of use in a maritime environment.

---

[34] See Mann, '1964 "The Doctrine of International Jurisdiction"', 31, and D. Opertiti Badán, 'Derecho internacional público y derecho internacional privado. Hacia un diálogo renovado', in D. P. Fernández Arroyo and C. L. Marques (eds.), *Derecho internacional privado y derecho internacional público* (Asunción: CEDEP, 2011), p. 76 on the rarely discussed relationship between private and public international law.

[35] A. T. von Mehren, 'Adjudicatory Jurisdiction: General Theories Compared and Evaluated' (1983) 63, Boston University Law Review, 279–340 at 338.

[36] See the comprehensive discussion in A. L. Strauss, 'Beyond National Law: The Neglected Role of the International Law of Personal Jurisdiction in Domestic Courts' (1995) 36, Harvard International Law Journal, 373–424; de Maekelt, *2011 "Relaciones"*, pp. 36 et seq.

[37] Bowett, '1983 "Jurisdiction"', 32, footnote 44.

[38] See de Maekelt, *2011 "Relaciones"*, pp. 40 et seq.

[39] T. Smith, 'Fighting on the Ocean Blue: New Zealand's Extra-Territorial Jurisdiction and Maritime Protest' (2001) 32, Victoria University of Wellington Law Review, 499–520 at 501.

B. Legitimacy of Rules of Civil Jurisdiction for Maritime Torts

The purpose of this section on the legitimacy of judicial jurisdiction for maritime torts is an analysis of the interplay between common grounds of jurisdiction and the law of the sea, in order to highlight the situation in which an assumption of civil jurisdiction by a coastal State may, under public international law, be difficult. This will always be the case in sets of facts effectively taking place at sea, where there may be concurring jurisdictions of the flag State, the coastal State and possibly even a third State.

The rules on civil jurisdiction differ from State to State, and, especially in the past, some States assumed very far-reaching jurisdiction over cases in a maritime context that were considered as admiralty cases. To some extent, these ancient claims of civil jurisdiction for maritime cases still influence their modern rules on jurisdiction.[40]

In the EU member States, jurisdiction and enforcement of judgments in civil and commercial matters have been harmonised at a rather early point in time by the Brussels Convention, applicable to the member States of the then European Community, and the Lugano Convention, applicable to the States of the European Free Trade Association (EFTA).[41] The Brussels Convention was replaced by Regulation (EC) No. 44/2001 (Brussels Regulation), which now applies to all civil and commercial matters in all courts and tribunals of the European Union. The project of harmonising the rules on jurisdiction and enforcement of judgments throughout the member States of the then European Economic Community was certainly an important step in outlining the realm of civil jurisdiction that a State can claim towards other States.[42] In the United States however, a legal system that is rather antagonistic to the European system with regard to jurisdiction,[43] questions of jurisdiction have developed through case law, and there is generally no differentiation between jurisdiction for tort and for contract cases. This section, under a rather global approach, aims to consider the situations in which States generally assume civil jurisdiction over a case with foreign elements and to ask the question

---

[40] See J. J. Fawcett, J. M. Carruthers and P. M. North, *Cheshire, North & Fawcett - Private International Law*, 14th edn. (Oxford, New York: Oxford University Press, 2008), p. 860, still maintaining that the English admiralty jurisdiction has competence to hear cases concerning injurious actions on the high seas even if both parties are foreigners.

[41] See Dicey, Morris and Collins, *2012 "Conflict of Laws"*, pp. 11-004 et seq. for a detailed historic outline.

[42] Mann, '1964 "The Doctrine of International Jurisdiction"', 69; J. Halpern, '"Exorbitant Jurisdiction" and the Brussels Convention: Toward a Theory of Restraint' (1983) 9, Yale Journal of World Public Order, 369–87 at 369.

[43] See the comparative analysis in G. Tu, 'Finding A Proper Nexus for Constructing Specific (Special) Jurisdiction Regarding Commercial Contract and Tort Cases' (2009) 5, Journal of Private International Law, 243–79.

whether this results in an assumption of jurisdiction that is in line with the rules of the law of the sea.

The Brussels Regulation recognises the fact that there may already be conventions between member States regulating the mutual recognition of jurisdiction and the enforcement of judgments, either as an international agreement expressly to this end, or as a convention on other matters, but also containing rules on jurisdiction for these matters. Art. 71 and 72 of the Brussels Regulation provide that these conventions, as far as they have been enacted prior to the coming into force of the Regulation, will remain unaffected. This has already been pointed out *supra* in Part II, with regard to the civil liability regimes discussed there. The precedence of agreements on "particular matters" however also holds true for the 1952 Brussels Conventions on Civil Jurisdiction Regarding Collisions[44] that specifically sets out rules on jurisdiction for collision cases.

Historically, the far-reaching assumption of jurisdiction in maritime claims, especially by the courts of sea-faring nations,[45] was certainly due to the fact that jurisdiction was hard to allocate for cases taking place at sea, and that legal systems feared an assumption of exorbitant jurisdiction less than the legal vacuum which would result if they denied jurisdiction in a given case.[46] This difficulty to designate the appropriate jurisdiction, however, mostly arises in cases where the set of facts actually took place at sea. If a maritime accident is the direct result of a decision or an act that has taken place on shore, e.g. in the headquarters of a shipping firm, a classification society or in a shipyard, jurisdiction in most cases still lies with the State that has the territorial and, possibly, national sovereignty over the wrongdoers and their premises, just as if the case did not have a maritime dimension. On the other hand, and this also in view of historic developments, there may be additional judicial venues in maritime cases, such as the place of the arrest of the ship or the place where the ship is currently to be found, if claimants have tried to obtain the vessel as a security for their claims.

Some of the problems that arise when considering which court in which country will be competent to adjudicate a case appear again in the quest for the law that this court should then apply. This concerns especially the localisation of acts that take place on a ship on the high seas or in any other maritime zone, if the jurisdiction or the law applicable should be found according to the place where, geographically, the events took place. Questions of jurisdiction and venue certainly influence each other. It is generally considered desirable that a court should apply its own law,

---

[44] 439 UNTS 217; BGBl. 1972 II, p. 653, 663.

[45] See F. Berlingieri, 'Jurisdiction and the Choice of Law in Collision Cases and an Overview of the Concept of Fault and its Apportionment' (1977) 51, Tulane Law Review, 866–79 at 866 et seq., considering England before the coming into force of the Brussels Convention and the United States respectively.

[46] See C. F. Finlayson, 'Shipboard torts and the conflict of laws' (1986) 16, Victoria University of Wellington Law Review, 119–45 at 134.

i.e., that the rule on jurisdiction and the conflicts rule point to the same country.[47] In tort cases, this can be obtained if the court of the place where the tort occurred *(forum delicti)* applies the law of the place where the tort occurred (*lex loci delicti*). This is however not always possible, especially when one takes into account that most claimants may think it desirable to bring a claim in a jurisdiction where the defendant has his place of business and some assets, so that the judgment can easily be enforced. The principle that the claimant should come to the jurisdiction in which the defendant is to be found and bring his claim there is, however, a very old and well-established one.

It is submitted that in these cases, where the competence of the forum to decide on the set of facts is not established entirely without doubt under public international law, or where there is a concurring jurisdiction, the application of a foreign law under the choice-of-law rules can, at least to some extent, mend the doubtfulness of the jurisdiction.

Considering that this section scrutinises the reconcilability between the assumption of civil jurisdiction and the law of the sea, questions of allocation of justice within a given legal system are left out of the picture. In general, the distribution of competences between the courts of one and the same country is to be determined by the rules of the domestic legal system in question.

I. Domicile, Residence or Place of Business of the Defendant

The principle that a defendant may be sued in the courts of the country where he resides, is a national of or has his place of business is a very old and quite well-established one.[48] From the point of view of legal theory, it is generally considered that a person living or carrying out business in a given country is subject to the territorial sovereignty of that State, and has a close link to the authorities of that State, so that he must expect to be subject to their judicial powers,[49] under the aspects of both fairness[50] and foreseeability. Thirdly, it seems logical that under normal circumstances, namely if both parties are of similar economic and financial

[47] G. Wagner, 'Internationales Deliktsrecht, die Arbeiten an der Rom II-Verordnung und der Europäische Deliktsgerichtsstand' (2006), Praxis des Internationalen Privat- und Verfahrensrechts, 372–91 at 375.

[48] Mann, '1964 "The Doctrine of International Jurisdiction"', 69; D. P. Fernández Arroyo, 'Exorbitant and Exclusive Grounds of Jurisdiction in European Private International Law: Will They Ever Survive?', in H.-P. Mansel, T. Pfeiffer, H. Kronke, C. Kohler and R. Hausmann (eds.), *Festschrift für Erik Jayme* (München: Sellier, 2004), p. 169.

[49] H. G. Maier and T. R. McCoy, 'A Unifying Theory for Judicial Jurisdiction and Choice of Law' (1991) 39, American Journal of Comparative Law, 249–92 at 250, but generally criticising the attribution of jurisdiction for considerations of convenience.

[50] See however the strong critique in ibid., p. 272.

strength, the claimant should address the courts in the defendant's country rather than forcing the defendant into proceedings abroad.

In practice, claimants often choose to bring proceedings in the defendant's jurisdiction because enforcement will presumably, unless the defendant has important assets abroad, take place in the same country, so that the decision and its enforcement are subject to the same rules and will thus be quicker and easier. In most jurisdictions, the legal venue at the domicile, residence or place of business of the defendant is the rule, and any other admissible venue is rather regarded as a deviation from the well-established principle, undertaken for a certain reason. This is also the structure of the Brussels Regulation, naming in art. 2 (1) the domicile of the defendant as the pertinent venue and only setting out additional venues for "special jurisdiction" in art. 5.

Legal persons are generally sued in the place of their registration or their statutory seat. In the Brussels Regulation, art. 60 (1) sets out that a company or other legal person is deemed to be domiciled at the place of its statutory seat, central administration or principal place of business.

Following maritime incidents, the likelihood of proceedings in the defendant's home jurisdiction is therefore considerable. To the extent that maritime accidents can give rise to claims that could be considered as claims under "individual contracts of employment", arts. 18 to 21 of the Brussels Regulation are applicable, providing in art. 19 (2 b) for a subsidiary venue at the place of business of the employer. A seaman's claim arising out of an accident at sea can however also be brought as a claim in tort—which seems possible especially under English law, also see *infra* under II. Subject to the subsidiary venue coming into play under the Brussels Regulation, there is thus also a fair chance that a maritime employer, such as a bareboat charterer, may be sued in his home jurisdiction for a tort that has been committed and bore its consequences on the high seas, and where there is no other connection to the forum jurisdiction apart from the domicile, residence or place of business of the defendant. Especially in a maritime context, the persons on board the ship that were involved in the tort may not even be aware of the place of business or domicile of a future defendant such as the shipowner or charterer.

This is perhaps the setting which best explains that there can be an interest in having a venue in the defendant's home jurisdiction but with the court applying the law of the place where the tort actually took place[51] (see *infra*, Chap. 7 on the problems of choice of law for maritime torts). In tort cases, it does not seem acceptable to submit the persons involved in the tort to the *lex fori* of a forum State they could possibly not envisage. The situation is similar to criminal cases that are tried in a forum that is not the *forum delicti*, and where there are very good arguments to allow the accused to plead the *lex loci delicti*.[52] If one does not want to sacrifice the legal venue at the domicile, residence or place of business of the

[51] See ibid., p. 250, setting out this situation in a more general context, and also strongly criticising it.

[52] Brownlie, *2003 "Principles of Public International Law"*, p. 306.

defendant, which has however the most obvious advantages in terms of fairness and ease of enforcement, the defendant's home jurisdiction should at least apply—or take into account—the law of the place where the tort actually occurred. Even if the determination of this law can be extremely difficult in maritime tort cases, as will be shown *infra* in Chap. 7, it should however not be forgotten that the application of this law is to some extent mandatory to render the judicial venue reconcilable with the requirements of public international law. From this point of view, the generally accepted point of view (also see *infra* in Chap. 7) that the *lex loci delicti* should be applied to torts can be explained rather well, and it is submitted that "homing ambitions" by which courts ultimately try to submit the case to the *lex fori,*[53] are at least questionable if seen with the question of legitimacy of jurisdiction in mind.

II. Jurisdiction Based on the Subject Matter of the Claim

Given the importance that the place where the tort occurred has for the expectations of the persons involved in it, it is not surprising that the existence of a judicial venue at the place of the tort is a rather old phenomenon.[54] The Brussels Regulation codifies it in art. 5 para. 3, providing for a venue "in the courts for the place where the harmful event occurred or may occur", in matters "relating to tort, *delict* or *quasi-delict*". The jurisdiction presents an additional venue to the general jurisdiction in the home jurisdiction of the defendant under art. 2.[55] In general, the court designated under art. 5 para. 3 of the Brussels Regulation will, when applying the *lex loci delicti*, apply its own law, so that jurisdiction and the choice of law concur.

Another example of a venue rather based on the subject matter of the claim is the jurisdiction over individual contracts of employment in art. 19 of the Brussels Regulation. The employee may sue either in the courts of the employer's home jurisdiction or in another member State in the courts of the place where he habitually carries out his work (art. 19 para. 2 (a)).

From the point of view of legitimacy of jurisdiction under public international law, the choice of a venue in the courts of the place where the tort occurred seems quite logical—the parties were aware of the place in which they were at the time and could obviously expect to be subject to the adjudicating powers of that State. It is however needless to say that this territorial link creates important problems in maritime tort cases, where ships flying the flag of a given State are on the one hand

[53] See R. Mortensen, 'Homing Devices in Choice of Tort Law: Australian, British and Canadian Approaches' (2006) 55, International and Comparative Law Quarterly, 839–78.

[54] Mann, '1964 "The Doctrine of International Jurisdiction"', 71: "the traditional search for the *locus delicti*".

[55] A. Layton and H. Mercer, *European Civil Practice*, (London: Thomson/Sweet & Maxwell, 2004), vol. 1, para. 13.016.

subject to the powers of this State and may on the other hand also be subject to certain sovereign rights of the State through whose waters they are sailing.

1. Forum Delicti in Maritime Torts

Art. 5 para. 3 of the Brussels Regulation only refers to the jurisdiction of "the courts for the place where the harmful event occurred or may occur". The problem of torts taking place on a ship is clearly not envisaged, much less than the cross-border cases in which there is an action in one country that bears consequences in another, i.e. where the place where the tort was committed and the place where the damage occurred are in different jurisdictions. It is however well-established through the famous ECJ precedent *Mines de Potasse d'Alsace*[56] that the claimant has a choice between the two venues in these cases.[57] The Brussels Regulation however does not give an explanation as to the problem of localising "the place where the harmful event occurred or may occur" if the tort involves a ship or an installation at sea.

It has already been mentioned that in tort cases it is highly desirable, not least as a matter of public international law, that the rules on the *forum delicti* and the *lex loci delicti* should point to the same legal system, so that the court designated as the *forum delicti* can apply its own law as the *lex loci delicti*.[58] The problems of determining a court and an appropriate legal system are therefore, in the context of torts happening in a maritime environment, the same concerning jurisdiction on the one hand and the appropriate choice of law on the other hand. Both rules can, for torts that happened on the high seas, in some cases point to a non-jurisdiction, and if the case concerns a ship sailing the EEZ or territorial waters of a State that is not the flag State, to more than one possible jurisdiction. The outcome for both the question of jurisdiction and the choice of law should therefore be the same.

Within the European Union, where the rules on choice of law for non-contractual obligations have been harmonised under the Regulation (EC) No. 864/2007 (Rome II Regulation) as of January 2009, this desirable synchronisation between the *forum delicti* and the *lex loci delicti* has been recognised and put into place. It is widely accepted that the provisions of the Rome II Regulation and the Brussels Regulation should be interpreted in the same way for similar questions in the interest of a homogeneous interpretation of European Law.[59] In fact, the ECJ's decisions

---

[56] *Bier v. Mines de Potasse d'Alsace* Reports of Cases (ECJ) 1735 (European Court of Justice, 30 November 1976), pt. 19.

[57] Also see the critique of this approach in Tu, '2009 "Finding A Proper Nexus"', 263 et seq.

[58] Wagner, '2006 "Internationales Deliktsrecht"', 375; Tu, '2009 "Finding A Proper Nexus"', 277.

[59] Z. Tang, 'The Interrelationship of European Jurisdiction and Choice of Law in Contract' (2008) 4, Journal of Private International Law, 35–59 at 36, mainly on the interplay between the Brussels Regulation and the Rome I Regulation, but also referring to the Rome II Regulation; A. Briggs, *Civil Jurisdiction and Judgments*, 5th edn. (London: Informa, 2009), para. 2.148, concerning the Rome I Convention but referring to "what will one day soon be a private international legal code for the Member States"; Egler, *2011 "Seeprivatrechtliche Streitigkeiten"*, p. 292, albeit also expressing some doubts in footnote 119.

regarding art. 5 para. 3 of the Brussels Regulation are often discussed in the context of the corresponding provision in art. 4 of the Rome II Regulation.[60] It should however not be omitted that the provisions of the two regulations are not entirely identical. For instance, the Rome II Regulation offers escape clauses, allowing the initial choice of law to be set aside and the submission of the set of facts to a law that has, in accord with the circumstances of the individual case, a closer connection to the events. Such an escape clause would hardly be admissible in the determination of the venue, where it is crucial for the claimant to know that his case can be brought under the rules on jurisdiction of the forum. There may thus be cases in which the parallelism of the *forum delicti* and the law applicable is not put into effect.

a) Torts Occurring on the High Seas

It is needless to repeat that the high seas are not subject to the sovereignty of any State and that, therefore, a competent court cannot be determined based on a geographical claim to these waters. A ship sailing the high seas has hardly any link to a State other than its flag, by which it is submitted, in the scope set out by art. 92 UNCLOS, to the sovereignty of the State granting the flag. The allocation of jurisdiction under art. 5 para. 3 of the Brussels Regulation, and, more widely, under the well-established *forum delicti* principle, must therefore necessarily be found in the courts of the flag State, if the ship where the tort took place is found on the high seas.[61] Not least, since it is the flag State's duty to investigate the causes of an accident (art. 94 para. 7 UNCLOS), it therefore only seems logical also to confer the judicial competence to decide on the ensuing civil liability to the flag State.[62]

aa) Environmental Torts on the High Seas

There is however a conflict between the jurisdiction of the flag State and of coastal States if high seas pollution can possibly bear consequences for the coastline of a given State. Pollution of waters beyond national jurisdiction may, to a wider extent than pollution on land, subsequently cause effects in different coastal States. The coastal State will, as a rule, want to assume jurisdiction for this case itself rather than referring the matter to the courts of a possibly disinterested flag State. Under the regime as created by art. 5 para. 3 of the Brussels Regulation and the landmark case of *Mines de Potasse d'Alsace*, this remains possible. The claimant would then

[60] J. Basedow, 'Die private Haftung aus Seedelikten zwischen Völkerrecht und internationalem Privatrecht', in H. Hestermeyer, D. König, N. Matz-Lück, V. Röben, A. Seibert-Fohr, P.-T. Stoll and S. Vöner (eds.), *Coexistence, Cooperation and Solidarity: Liber Amicorum Rüdiger Wolfrum* (Leiden, Boston: Martinus Nijhoff Publishers, 2012), p. 1887.

[61] Egler, *2011 "Seeprivatrechtliche Streitigkeiten"*, p. 299.

[62] Ibid., p. 299.

have the choice between the courts of the flag State, representing the place where the tortious act was committed, and the courts of the coastal State where the damage was perceived. Considering that pollution victims in the coastal State will most likely prefer to bring a claim in their home jurisdiction, the judicial venue in the courts of the coastal State will generally be the rule. Whereas this seems fair in the cases of perceptible or actual pollution, there may however be cases in which it is questionable whether the alleged consequences are real or at least probable enough.

An illustration of this can be found in a rather exceptional case in the French and Italian courts. The so-called "Red Sludge Case" (*l'affaire des boues rouges*) (also see *infra,* Chap. 7, A. IV.) concerned a factory in Scarlino, south of Livorno in Tuscany, producing titan bioxyde. The waste products of this chemical production were, from the opening of the factory and onwards, meant to be dumped at sea. The Italian company running the factory had two ships especially constructed for this purpose, and obtained a corresponding authorisation from the Italian authorities. Both ships flew the Italian flag. From early May 1972 until late September 1973, when both ships were seized and immobilised following complaints and legal action, the ships dumped large amounts of waste chemicals into the high seas of the Mediterranean, about 70 km from the coast of Tuscany but only 40 km from the French island of Corsica.[63] The product dumped was in fact very toxic, but it was by no means established that there was a physical impact to the natural resources of either Corsica or Tuscany. Scientific evidence from Italy revealed that there was a measurable, albeit minor impact on the plankton off the Italian coastline, but its consequence for the ecosystem as a whole was not established.

The set of facts however gave rise to proceedings in both Italy and France. Notwithstanding the fact that the dumpings as such had been permitted by the Italian authorities, four directors of the company were criminally persecuted. The criminal proceedings in the court of Livorno resulted in suspended prison sentences.[64] The *Pretura*'s reasoning for establishing competence over the dumpings that had taken place on the high seas was based on the competence that the Italian legislation pronounced itself to have over these waters, according to, mainly, fisheries regulation.[65] The very detailed considerations of the court however do not mention the fact that the polluting ships flew the Italian flag; this circumstance was evidently not perceived as important in establishing jurisdiction. The *Pretura*'s reasoning of course only concerns the jurisdiction in criminal law on the high seas, and it is questionable whether conclusions concerning civil jurisdiction can be drawn from it.[66]

[63] See A. Kiss, 'Un cas de pollution internationale: l'affaire des boues rouges' (1975) 102, Journal du Droit International (Clunet), 207–48 at 207 et seq. for a comprehensive outline of the facts.

[64] *"Affaire Montecatini Edison" (French translation)*, 102 Journal du Droit International (Clunet) 237 (Pretura de Livorno, 27 April 1974), p. 237.

[65] Ibid., pp. 237 et seqq.

[66] Kiss, '1975 "Un cas de pollution internationale"', 211.

The French courts, which were subsequently considering the case in civil law, used a different reasoning to establish their competence. The claim in the French courts was brought by a Corsican fishermen's association, the *Prud'hommie de Bastia*, which claimed compensation for the damage that the dumping of the chemicals did to their habitual fishing grounds.

Having established that the Italian society actually did not even deny that the dumping of the chemicals was detrimental to the environment, the French *tribunal de grande instance* of Bastia, Corsica stated that "the waste that was disposed of at sea represents a ferment of damage that will, absorbed by the fauna and flora that are transported by the currents, diffuse as far as the territorial waters of Corsica, and other than only ruin a fisheries ground, also produce economic, ecological and touristic incidents that will be perceived in the circuit of the court of Bastia".[67] The competence to hear the case was thus based on the mere assumption that there would be an impact to be felt on the island of Corsica, and the court then used the ECJ's reasoning in *Mines de Potasse d'Alsace* to establish jurisdiction in the courts of the place where the damage was perceived, e.g., Corsica,[68] it being understood that there was no other, concurrent, jurisdiction, since the high seas are devoid of jurisdiction.[69] Again, no reference was made to a possible jurisdiction of the Italian courts as the courts of the polluting ships' flag States, which is perhaps illustrative of a certain reluctance to address the tortfeasor's home jurisdiction. It must however be repeated that the link between the pollution and the effect on natural resources in Corsican waters is extremely weak, as (also see *infra*, Chap. 7, A. IV.) there was effectively no evidence of actual damage in French waters. It must thus necessarily be questioned whether the realm of the provisions on subject matter jurisdiction cover cases like these.

It is furthermore submitted that a change in the wording in art. 5 para. 3 of the Brussels Regulation, as opposed to the earlier Brussels Convention, may give rise to even greater uncertainty in this matter. Whereas art. 5 para. 3 of the Brussels Convention, which was the provision interpreted by the ECJ in the *Mines de Potasse d'Alsace* decision, referred to jurisdiction in "the courts of the place where the harmful event occurred", this wording has been changed to "the courts of the place where the harmful event occurred or may occur" in art. 5 para. 3 of the Brussels Regulation. The change, evidently meant to cover cases in which the damage has not yet realised, causes however difficulties with regard to the degree of probability with which the damage is to be expected.[70] Under the altered wording, prospective damages such as the fishermen's claim in the Red Sludge

[67] *"La Prud'hommie des pêcheurs de Bastia v. Montedison"* 1977 Recueil Dalloz 427 (Tribunal de Grande Instance de Bastia, 08 December 1976), p. 428.

[68] Ballarino, '1990 "Questions de droit international privé"', 363, footnote 105; de Raulin, '1993 "L'épopée judiciaire"', 80.

[69] B. Audit, 'Note C. Cass. 03.04.1978' (1978), 1978 Recueil Dalloz (Informations Rapides), 367–8 at 367.

[70] Layton and Mercer, *2004 "European Civil Practice"*, vol. 1, para. 15.098.

case could be covered by the provision even more easily, generating a jurisdiction in the courts of any alleged victim.

bb) Collisions

A different problem of course arises for collisions on the high seas between ships flying different flags. The conundrum of finding a judicial venue without giving preference to either flag State is not easily solved. On an international level, the question of civil jurisdiction for ship collision has been addressed in the 1952 Brussels Convention on Civil Jurisdiction Regarding Collisions.[71] According to art. I para. 1 of this Convention, the judicial venue for a collision case is either the defendant's habitual residence or his place of business, or the place of the arrest of the vessel if either of the ships has been arrested following the collision (also see *infra* under III.). The place where the collision occurred is only of importance if the accident took place in "port or inland waters" (art. I para. 1 (c) of the Convention), i.e. in waters that are generally subject to the unrestricted territorial sovereignty of a State. The 1952 Civil Jurisdiction Convention simply omits the flag of the ships as a factor for determining civil jurisdiction.[72] The flags however serve to determine the scope of application of the Convention, the instrument being applicable to collisions in which both ships fly the flag of a contracting State (art. VIII). In this context, it is interesting to consider that the Convention does consider the law of the ships' flags for the initial, important question of whether there can be civil jurisdiction over a ship on the high seas at all—the question whether the flag State has, by ratifying the Convention, consented to submit the ships flying its flag to internationally uniform rules of civil procedure. That these uniform rules then omit the flag as a factor for determining civil jurisdiction rather appears as a consequence of the practical difficulty in giving preference to one of the flags of two colliding ships and not as a sign of the decreased importance of the law of the flag at sea.

b) Torts Occurring in the Exclusive Economic Zone

Whereas the general result that the *forum delicti* for high seas torts is the law of the flag does not seem to be debated, things are more difficult concerning the allocation of jurisdiction for torts taking place in the EEZ. The coastal State has certain sovereign rights over these areas, and the "place where the harmful event occurred" could thus be considered to be found in the coastal State. It is however questionable whether the sovereign rights the State can invoke for these waters under the law of the sea encompass the exercise of civil jurisdiction over ships sailing these waters and over installations that have been built there.

---

[71] 439 UNTS 217; BGBl. 1972 II, p. 653, 663.

[72] Egler, *2011 "Seeprivatrechtliche Streitigkeiten"*, p. 273.

In the EEZ, the coastal State exercises sovereign rights of a functional character, with regard to certain uses of these waters rather than with regard to the waters as such. Functional rights of the coastal State in the EEZ encompass, according to art. 56 UNCLOS, the right to explore and exploit the natural resources of the water column, the seabed and its subsoil, and the right to economic exploitation of the zone, for instance through the production of energy. According to art. 60 para. 2 UNCLOS, the law of the sea grants the coastal State civil jurisdiction over artificial islands and over installations that have been built to these ends. In contrast to that, with regard to shipping, foreign-flagged ships have the right to free navigation as described in art. 87 UNCLOS, which is the right of free navigation on the high seas. For the purposes of navigation, foreign-flagged ships therefore enjoy the same rights in the EEZ as on the high seas.

Regarding the functional rights as to exploration and exploitation of the resources, the coastal State can certainly assume civil jurisdiction for all activities by which private entities make use of these rights. Art. 60 para. 2 UNCLOS clearly allocates jurisdiction over installations for the exploration and exploitation of the EEZ's natural resources with, and exclusively with, the coastal State. The persons working on and in connection with the activities on these installations are aware of their geographical position and of the fact that the aim of the activity is the exploration and exploitation of resources that can only be considered to belong to the coastal State.[73] It therefore seems logical that the judicial venue for torts that occur in connection with the use of these sovereign rights should be considered to be in the courts of the coastal State.[74]

The tendency to consider the underlying law of the sea in the determination of jurisdiction has been confirmed by the ECJ in the decision of *Weber v. Ogden*,[75] albeit in the context of the provisions on individual employment contracts of the Brussels Regulation, in determining the place where "the employee habitually carries out his work". It can be inferred from that provision's wording and from its underlying aim—to determine a judicial venue with which the employee is familiar and that he can reasonably expect[76]—that the habitual working place should present some degree of stability and certainty. Arguably, this venue needs to be easier to determine than the *locus delicti* for torts, which may be, by definition, of a very ephemeral and arbitrary character. The Brussels Regulation's provisions on jurisdiction over individual contracts of employment will be given a closer look

[73] M. Rémond-Gouilloud, 'Pollution accidentelle et responsabilités' (2010) 62, Droit Maritime Français, 663–8 at 666.

[74] S. Pestke, *Offshore-Windfarmen in der Ausschliesslichen Wirtschaftszone: Im Zielkonflikt zwischen Klima- und Umweltschutz* (Baden-Baden: Nomos, 2008), p. 80; Egler, *2011 "Seeprivatrechtliche Streitigkeiten"*, pp. 296 et seq.

[75] *Weber v. Ogden* I Reports of Cases (ECJ) 2016 (European Court of Justice, 27 February 2002).

[76] T. Garber, 'Zum Begriff des gewöhnlichen Arbeitsortes i. s. d. Art. 19 Abs. 2 lit. a EuGVVO insbes. bei der Verrichtung der arbeitsvertraglichen Tätigkeit an Bord eines Schiffes', in R. Geimer and R. A. Schütze (eds.), *Recht ohne Grenzen: Festschrift für Athanassios Kaissis zum 65. Geburtstag* (München: Sellier European Law Publishers, 2012), p. 222.

infra under 2. However, in attaching the judicial venue for employment contracts to a territorial connecting factor, the ECJ's reasoning as to the place of habitual employment in *Weber v. Ogden* can easily be transferred to the determination of the *locus delicti* in art. 5 (3) of the Brussels Regulation, and with a view to the desirable concurrence between the judicial venue and the *lex loci delicti,* arguably even to choice-of-law considerations in the EEZ (see *infra*, Chap. 7, B.).

The case of *Weber v. Ogden* was brought by a German national, employed by a Scottish company for work carried out on board of vessels and installations operated in the Netherlands continental shelf area and subsequently on a floating crane operating in Danish territorial waters. The claimant brought the case in the Dutch courts that, according to Dutch internal legislation, assumed civil jurisdiction for employment contracts of workers on board any "mining installation" on or over the Netherlands' continental shelf. The defendant company disputed the Dutch courts' jurisdiction over the case and claimed that the case should be brought in Scotland.

The Netherlands court, in order to decide whether it could exercise jurisdiction under the provisions of the Brussels Regulation, as the court of the claimant's habitual workplace, eventually addressed the ECJ on that question. The case was decided by the ECJ with regard to the Continental Shelf Regime as established by the 1958 Geneva Convention on the Continental Shelf. The UNCLOS and its regime for the EEZ could not be directly taken into account as the set of facts occurred prior to the coming into force of the UNCLOS in the Netherlands. For the matter of the coastal State's functional rights over the Continental Shelf area, the continental shelf regime as considered by the ECJ and the EEZ regime in UNCLOS are however very similar, so that the case can be considered to allow for conclusions for the latter. The ECJ in fact put forward considerations of public international law, setting out that the exploitation of the seabed resources of the continental shelf is an activity for which the Netherlands can claim sovereign rights equal to those on its territory, so that the claimant's activity must be considered to have been carried out in the territory of the Netherlands.[77]

The ECJ however stressed that this result only holds true for cases in the context of the functional rights that the coastal State can assume in the continental shelf area, or for that matter, in the EEZ. The case is not concerned with sets of facts that arise in connection with navigation in the EEZ, bearing in mind that the coastal State does not have functional rights over this activity, and that for the matters of navigation it is essentially the high seas regime that applies to the EEZ. From the explicit mentioning of the realm of the coastal State's rights in this area, one can infer from *Weber v. Ogden* that the outcome may be different for matters of navigation.[78]

Apart from the differences in State sovereignty from the point of view of public international law, it is furthermore submitted that the subject matter of the case in

[77] *European Court of Justice, 27.02.2002, Weber v. Ogden*, p. 2040, pt. 25.

[78] This is also put forward by General Advocate Jacobs in his opinion on the case, ibid., p. 2023.

the EEZ is very different with regard to incidents in navigation. Due to the mobility of ships, these are likely to cross several zones of maritime jurisdiction on a given journey, and the persons present on board are not likely to attach considerable importance to the vessel's geographic position in the EEZ, given that the coastal State does not have any sovereign rights over the mere passage. There must however be a marked exception for cases which have a serious impact on the marine environment in the EEZ, considering that the coastal State enjoys functional sovereign rights with regard to the protection and preservation of the marine environment (art. 56 para. 1 (b) (iii) UNCLOS) within the respective provisions of the Convention. Art. 220 paras. 2 to 6 explicitly confers enforcement jurisdiction over pollution incidents to the coastal State, but clearly defined according to the gravity of the pollution and the degree of probability with which the incriminated ship is responsible for it. It is certainly remarkable that the question of whether a coastal State has jurisdiction over a ship in its EEZ should depend on the gravity of the effects for the coastal State,[79] and thus on a criterion ultimately begging for interpretation. Still, this represents a balance between ships' right to free navigation and the interest of the coastal State in a clean marine environment. Art. 220 UNCLOS however only concerns enforcement jurisdiction. As can be inferred from art. 229 UNCLOS, stating that nothing in the Convention affects the institution of civil proceedings for marine environmental pollution, the UNCLOS does not contain explicit provisions on civil judicial jurisdiction. This means that the assumption of civil jurisdiction must be considered to be governed by the general principles of public international law. With a view to the coastal State's functional rights regarding the preservation and protection of the marine environment, it seems that the assumption of civil judicial jurisdiction should follow the same principles. A tort in the EEZ that has a marked impact on the marine environment must be considered to be within the coastal State's functional rights as described in art. 56 para. 1 (b) (iii) UNCLOS, so that a grave incident of pollution caused by a ship would trigger the coastal State's judicial jurisdiction.

A shipboard tort in the EEZ or over the continental shelf that does not cause effects on the waters outside the ship has too little connection with the coastal State to create a jurisdictional venue in this State just on the subject matter of the case. Without doubt, the venue can be established according to one of the other factors, discussed *supra* under I. and *infra* under III.–V., but the mere fact that the subject matter of the case took place in the EEZ should not trigger jurisdiction under art. 5 (3) of the Brussels Regulation or similar *forum delicti* rules for torts that are uniquely linked to navigation.

Admittedly, the line between navigation and an activity for which the coastal State can put forward sovereign rights in its EEZ may be extremely hard to draw in some cases. Navigation in the EEZ may be undertaken as a mere transit through these waters, but it may also be linked to exploration and exploitation of the EEZ's

[79] P. Bonassies, 'L'arrêt "Erika" et la compétence des juridictions françaises' (2012) 64, Droit Maritime Français, 995–9 at 998.

resources, such as the transport of supplies to mining or other installations, or the transport of the products of the exploitation.[80] A very similar problem arises with torts involving fishing vessels, the exploitation of the living resources of the waters in the EEZ being one of the domains in which the coastal State can invoke sovereign rights (art. 56 para. 1 (a) UNCLOS). It may in a given case be very difficult to determine whether the tort occurred in connection with the vessel's function as a means of fishery or with its character as a means of transport. In general, it will however be possible to determine the main character of the activity as either part of the coastal State's sovereign rights or as a matter of the regime of free navigation. The distinction that the law of the seas draws between the different uses of the waters of the EEZ seems too important not to give it regard in the determination of a judicial venue for the case's subject matter.

c) Torts Occurring in Territorial Waters

In sharp contrast to the regime applicable on the high seas and in the EEZ, the territorial waters of a State extending 12 nautical miles from the baselines are, according to art. 2 UNCLOS, subject to the territorial sovereignty of the coastal State. This fact has driven a certain number of authors to submit that in any case, the *forum delicti* for torts that occur in the territorial waters of a State must be with the courts of that State.[81] It is in this context submitted that this would meet the expectations of the parties, e.g. on a ferry, to have a judicial venue in the courts of the coastal State in the waters of which the accident occurred.[82]

Art. 2 UNCLOS however immediately refers, in its para. 3, to the fact that the territorial sovereignty of the State can only be exercised subject to the provisions of UNCLOS and other rules of international law. In fact, one[83] of the most important restrictions[84] of the coastal State's competence is the right of innocent passage that foreign-flagged ships enjoy in the territorial sea (art. 17 UNCLOS). The coastal State cannot object to a ship's transit through the territorial sea, unless the passage is prejudicial to the peace, good order or security of the State. Examples of activities that fulfil this criterion are given in art. 19 para. 2 UNCLOS, and clearly show that the right of innocent passage is only lost where the ship carries out activities that objectively pose a threat to the coastal State.

---

[80] Egler, *2011 "Seeprivatrechtliche Streitigkeiten"*, p. 296.

[81] Ibid., pp. 294 et seq.

[82] Ibid., pp. 294 et seq.

[83] W. Graf Vizthum, 'Aquitoriale Souveränität', in P.-M. Dupuy, B. Fassbender, M. N. Shaw and K.-P. Sommermann (eds.), *Völkerrecht als Wertordnung: Festschrift für Christian Tomuschat* (Kehl: N.P. Engel, 2006).

[84] F. Ngantcha, *The Right of Innocent Passage and the Evolution of the International Law of the Sea* (London, New York: Pinter, 1990), p. 38.

In terms of legislative competence, art. 21 UNCLOS prevents the coastal State from adopting laws and regulations "relating to innocent passage" of foreign-flagged ships that go beyond the coastal State's legitimate interests in the protection of its sovereign rights. It is rather unclear whether and to which extent there is further legislative jurisdiction of the coastal State,[85] enabling it to enact laws not directly bearing on innocent passage that will be applicable in territorial waters. In any case, art. 24 UNCLOS prohibits coastal State legislation that indirectly bears on innocent passage and impairs or denies this right to foreign-flagged ships. With regard to criminal enforcement jurisdiction, art. 27 UNCLOS, dealing with physical measures against passing ships such as stopping the vessel or arresting persons on board, hints at the convention drawing a distinction between the cases in which the coastal State has an interest in the events and others in which there is none. According to art. 27 para. 1 UNCLOS, the coastal State "should not" exercise its criminal jurisdiction unless the consequences of the crime either extend to the coastal State or are of a nature to disturb the peace of the country or the good order of the territorial sea, or in cases where the coastal State's assistance has been requested by the master of the ship or the authorities of the flag State. However, the use of the formulation "should not" clarifies that there is no obligation under public international law for the coastal State to refrain from the exercise of jurisdiction; a State carrying out full-scale enforcement jurisdiction would not violate its obligations under the Convention,[86] even if UNCLOS discourages this. Art. 28 (1) UNCLOS, for enforcement jurisdiction as against persons, employs the same formulation, without actually distinguishing between claims to which there is a link with the coastal State and others. This distinction only re-emerges in art. 28 (2) UNCLOS considering enforcement jurisdiction against ships. This is absolutely forbidden, except if the civil claim concerns a liability that the ship has incurred during or for the purpose of its innocent passage (art. 28 para. 2 UNCLOS).

In enforcement jurisdiction, the Convention therefore strikes an important distinction between cases in which there is a coastal State interest, the incident affecting the coastal State, and others, that are purely internal to the ship and where there is a marked tendency in UNCLOS's provisions on legislative and enforcement jurisdiction to leave the matter to the authorities of the flag State. UNCLOS, in setting out that the coastal State "should not" exercise its enforcement jurisdiction, apparently starts out from the idea that a coastal State may exercise full-scale enforcement based on the sovereignty that it exercises over its territorial sea, but that it should voluntarily restrict itself as a matter of comity in all cases in which there is no coastal State interest in the matter.

It is debatable whether conclusions can be drawn from the enforcement regime for the regime of judicial competence, considering that UNCLOS is almost entirely

---

[85] See Churchill and Lowe, *1999 "The Law of the Sea"*, p. 95.

[86] I. A. Shearer, 'Problems of Jurisdiction and Law Enforcement Against Delinquent Vessels' (1986) 35, International and Comparative Law Quarterly, 320–43 at 327; Ngantcha, *1990 "The Right of Innocent Passage"*, p. 105; Labrot, *2006 "Compétences des États"*, p. 25.

mute on coastal States' competence to adjudicate over acts in the territorial sea. The only exception is art. 220 para. 2 in which it is mentioned that the coastal State can "institute proceedings" for violations of pollution prevention laws and regulations, pollution representing of course a marked coastal State interest both with regard to enforcement and to judicial competence. In all cases where the damage can be taken to have occurred in the waters of the coastal State, there will be a judicial venue in the coastal State anyway. This coincides with the ECJ's reasoning in *Mines de Potasse d'Alsace*. In all other cases, UNCLOS does not contain rules on judicial competence of coastal States over ships in their territorial waters.

It is evidently a widespread State practice not to assume subject matter jurisdiction for events that only happened on board a ship in innocent passage and where there is nothing other than the weak geographical link with the coastal State.[87] A voluntary waiver of an existing jurisdictional venue is however not easy to exercise for national courts under binding rules of jurisdiction, at least not with regard to the Brussels Regulation. In a landmark decision[88] on the concept of *forum non conveniens*, the ECJ explicitly stated that the courts of member States can generally not choose to decline jurisdiction if there is a jurisdictional venue set out under the Brussels Regulation, not even if the more appropriate forum is found in a country outside of the EU. It is not possible for courts to exercise discretion, deciding whether they represent an appropriate forum or not for a given case. This means that the question of how to interpret art. 5 para. 3 Brussels Regulation for torts that take place on a ship and are "internal" to the ship should be considered in view of the underlying law of the sea, as a discretionary decline of jurisdiction by the court itself is hardly possible.

Any tort whose harmful consequences are entirely restricted to the ship should therefore, with regard to the right of innocent passage, be considered to have occurred "on the ship", and thus, in the realm of the flag State. "The courts for the place where the harmful event occurred or may occur" should be the judicial authorities of the flag State. Examples for "internal" events of this kind are of course accidents on board the ship and other incidents between the crew or passengers that do not have an impact on the transit passage of the ship. Admittedly, the decision of whether an event is internal or external to the ship may be harder to strike in other cases. In all accidents in which persons drown, it may be difficult to decide whether a person drowned in the waters of the coastal State or inside the ship's hull, especially if the vessel otherwise remained afloat, and it seems difficult and inappropriate to draw distinctions of this kind, inquiring into the exact circumstances of the death. It is however submitted that almost any maritime accident in which there is water ingress into the ship's hull is severe enough to interest the

---

[87] Ngantcha, *1990 "The Right of Innocent Passage"*, p. 97; Churchill and Lowe, *1999 "The Law of the Sea"*, p. 66, holding this to be the practice even in internal waters where there is no right of innocent passage; also see Pozdnakova, *2013 "Criminal jurisdiction"*, p. 130, with regard to criminal judicial jurisdiction.

[88] *Owusu v. Jackson*, I Reports of Cases (ECJ) 1445 (European Court of Justice, 01 March 2005).

coastal State, as it would be rather unlikely that notwithstanding a damage of this kind, the ship could pursue its voyage completely undisturbed. Liabilities resulting from an accident of this kind, e.g. for search and rescue operations or other kinds of assistance would, under the enforcement regime of art. 28 para. 2 UNCLOS, even trigger the coastal State's right to arrest the ship. Judicial jurisdiction should be triggered under the same circumstances. Following the ECJ's jurisprudence on cross-border torts, the accident would therefore be considered to have originated in the ship but the damage to have taken place in the coastal waters, so that there would be a judicial venue in the courts of both the coastal State and the flag State.

Another difficult case in distinguishing between internal and external acts is considered in an ECJ case[89] that can be interpreted to rather point to the general applicability of the law of the flag for torts occurring on ships. The circumstances of the case are however rather special, and the findings of the court are anything but explicit. The case of *DFDS Torline v. SEKO Sjöfolk* concerned the *Tor Caledonia*, a cargo ship flying the Danish flag and serving on the route between Gothenburg and Harwich. The crew was Polish and worked on individual employment contracts that were subject to Danish law and based on a framework agreement between Danish unions and Danish associations of shipping companies. Following a request of one of the Danish unions on behalf of the crew for a collective agreement that the *Tor Caledonia's* shipping company rejected, the union served a notice of industrial action and called for sympathy action of other trade unions. This caused a Swedish trade union to give notice of sympathy action, which might have caused considerable difficulties in the loading and unloading operations on the *Tor Caledonia* in Gothenburg, in the course of her normal route. DFDS Torline therefore decided to withdraw the vessel from the liner service and replace her with another vessel, which had to be chartered. The shipowner, pleading that the threat of industrial action and the call for sympathy action was unlawful, brought a claim in tort against the Danish trade union for the cost of the *Tor Caledonia's* replacement. The Danish court which was seized addressed the ECJ, *inter alia*, with the question of whether the damage caused to the *Tor Caledonia's* owner could be considered, under art. 5 (3) of the Brussels Regulation, to have arisen in Denmark.

It should be pointed out that from the facts of the case, it is very debatable whether the damage arose on the *Tor Caledonia* or in the headquarters of the shipowning company in Denmark, and the ECJ did not decide on this but left the answer to this question of fact to be inquired into by the national court. The ECJ did however give a rather clear statement on the role of the law of the flag:

> ...the flag State, that is the State in which the ship is registered, must be regarded as only one factor, among others, assisting in the identification of the place where the harmful event took place. The nationality of the ship can play a decisive role only if the national court reaches the conclusion that the damage arose on board the *Tor Caledonia*. In that case, the flag State must necessarily be regarded as the place where the harmful event caused damage.[90]

---

[89] *DFDS Torline v. SEKO Sjöfolk* I Reports of Cases (ECJ) 1420 (European Court of Justice, 05 February 2004).

[90] Ibid., pt. 45.

As the court did not decide whether the damage, objectively, arose on the *Tor Caledonia*, this statement is necessarily *obiter dictum*. It is however very interesting in its general assertion that under the Brussels Regulation, the place where the harmful event occurred should be taken to be the flag State for torts on ships. The court does not consider whether there are cases in which a coastal State or a port State could be taken to be the venue. This is all the more noteworthy[91] as General Advocate Jacobs, giving his opinion in the case,[92] refrained from expressing himself as to this question and simply asserted that in this case, the *Tor Caledonia's* nationality did not matter.[93] Without doubt, the geographical position of the vessel during the call for sympathy action did not make a difference to the ship's owner. He would have incurred the cost for a replacement vessel in any case, and the tort is in this respect very likely an "internal" one in which the ship's position does not matter, neither to the persons involved nor to the coastal State or port State in which the ship is to be found.

The ECJ's statement can therefore be read as an interpretation of art. 5 para. 3 of the Brussels Regulation for torts internal to the ship, pointing out that in these cases, where the damage can be taken to have occurred on the ship only, the event should be adjudicated upon by the courts of the flag State.

With regard to the jurisdiction of the coastal State for torts occurring in its territorial waters, it must also be recalled that the 1952 Convention on Civil Jurisdiction for Collisions explicitly disregards the coastal State's territorial sovereignty for collisions that have taken place in territorial waters. According to art. I para. 1 (c) of the Convention, a venue only arises out of the collision having taken place in port or inland waters, where territorial sovereignty is virtually unrestricted.[94] The provision explicitly does not provide for the coastal State's judicial competence if the collision takes place in territorial waters. Unlike the 1952 Convention on Criminal Jurisdiction for Collisions, the Civil Jurisdiction Convention does not even provide a possibility for States to enter a reservation to this end and submit collisions that have occurred in their territorial waters to their civil jurisdiction.[95] The surprising consequence of this rule can be seen from a recent French case[96] concerning a collision that occurred close to shore, just outside a French port, between two pleasure crafts, one flying the French and the other the British flag. When the owners of the British vessel were sued in France, they contested the jurisdiction of the French courts. The case effectively hinged on

---

[91] Basedow, *2012 "Die private Haftung"*, p. 1886.

[92] *European Court of Justice, 05.02.2004, DFDS Torline v. SEKO Sjöfolk*, pp. 1420 et seq.

[93] Ibid., p. 1438, pt. 78.

[94] See Özçayir, *1998 "Liability for oil pollution"*, pp. 394 et seq. for an overview of the very limited exceptions.

[95] P. Bonassies, 'Observations CA Aix-en-Provence 04.10.2012' (2013) 65, Droit Maritime Français, 312–4 at 313.

[96] *"Triton II"*, 65 Droit Maritime Français 310 (Cour d'Appel d'Aix-en-Provence, 04 October 2012).

whether the collision had, geographically, taken place in French "inland waters", meaning internal waters landwards of the baseline, as opposed to the territorial waters beyond the baseline. As both France and the United Kingdom are parties to the 1952 Civil Jurisdiction Convention, jurisdiction of the French court was only established following an exact localisation of the accident in the French internal waters. The 1952 Convention effectively restricts the exercise of the coastal State's sovereignty in its territorial waters, ruling out an assumption of jurisdiction for collisions that have taken place there.

The Convention's rule confirms the view that the judicial venue for torts taking place in a State's territorial waters is not automatically to be found in the courts of that State, but it is however surprising that the Convention seems to consider that a collision is not an event triggering the interest of the coastal State. A serious collision between two ships may easily cause death or injury to passengers or crewmembers on either ship, but also pollution damage to the waters outside the ship. It seems strange that there is not even a judicial venue in the coastal State for these damages.

The Convention was however negotiated at a time when pollution damage was not yet a major concern for coastal States, and when there was hardly any surveillance of territorial waters, so that coastal States may in fact not have been too interested in death and personal injury cases. It is submitted that this consideration may be different today, but the 1952 Convention on Civil Jurisdiction for Collisions favours well-established judicial venues such as the *forum arresti* or defendant's domicile or place of business, based on an interest in restricting the possible number of venues in a collision case to a few clear and usual ones. The fact that the venue of the coastal States' courts is not among them is another hint at the fact that this venue is not an automatically granted one for torts in the territorial sea, and the hint is even stronger considering that the geographical venue does exist for torts in internal waters.

2. Place of Employment for Maritime Labour Cases

Within the Brussels Regulation, subject matter jurisdiction with regard to a venue based on territorial jurisdiction also exists in the context of labour law. Art. 19 of the Regulation, which has already been mentioned *supra* under 1. b) in the context of the *Weber v. Ogden* decision provides for a judicial venue in claims against an employer. These can be brought either in the courts of the State where the employer is domiciled or in another member State under the circumstances set out in art. 19 (2) of the Regulation. These are (a) in the courts for the place where the "employee habitually carries out his work" or for the place where he last did so or (b) "if the employee does not or did not habitually carry out his work in any one country, in the courts for the place where the business which engaged the employee is or was situated."

In seamen's claims for damages that arose out of maritime incidents, such as injury claims, the jurisdictional venue can thus depend on the place where they are considered to "habitually carry out their work". Arts. 19 to 21 of the Regulation are however concerned with claims that arise out of contracts of employment. As will be shown *infra* in Chap. 7, seamen's claims for death and injury can also be brought in tort, with, possibly, a different venue rather to be found in the *forum delicti*.

The question of when a working place can be called "habitual", especially in the cases where a seaman works on different installations or ships, is an important one,[97] but determining the "habitual" working place among several possible locations is rather a matter of labour law and a problem that also occurs between different working places on shore. In the context of this study, it is rather the problem of work carried out in areas beyond national jurisdictions that is of interest.

Under art. 19 of the Brussels Regulation, the quest for the place where a seaman "habitually carries out his work" entails problems that are very similar to the localisation of the "place where the tort occurred" and that have been presented *supra* under 1. It must however be considered that the term of the "habitual" workplace generally bears a connotation of stability and duration, and, to a certain extent, of foreseeability. Both the employer and the employee should know beforehand which will be the likely forum under art. 19 (2) of the Regulation.[98] On the other hand, the provision's wording clearly refers to matter-of-fact considerations in setting the judicial venue in the courts of the habitual working place and, failing the existence of this working place, in the courts of the engaging business.

With a view to this, from the very outset it seems difficult to establish a "habitual" working place according to changing geographic factors for employees on sea-going ships. An exception can probably be made for personnel on ships that only operate in territorial waters,[99] especially where the scope of employment according to contract is clearly confined to the coastal waters of a State, such as employment contracts for small pleasure craft, excursion boat crewmembers or coastal fishermen.[100] If all the facts of the case are only connected with the coastal State in question, it may actually be a purely domestic one and possibly not even subject to considerations of international jurisdiction.

Jurisdiction for labour disputes of employees working in the offshore industry has been considered in the ECJ's case of *Weber v. Ogden*, discussed *supra* under 1. b). The court clearly set out the coastal State's jurisdiction in the cases in which

---

[97] See P. Mankowski, 'Europäisches Internationales Arbeitsprozessrecht - Weiteres zum gewöhnlichen Arbeitsort' (2003), Praxis des Internationalen Privat- und Verfahrensrechts, 21–8 at 23 et seq.; Layton and Mercer, *2004 "European Civil Practice"*, vol. 1, para. 18.018; P. Mankowski, 'Das Internationale Arbeitsrecht und die See - Die Fortsetzung folgt' (2005), Praxis des Internationalen Privat- und Verfahrensrechts, 58–61 at 60; Egler, *2011 "Seeprivatrechtliche Streitigkeiten"*, pp. 211 et seqq. on this.

[98] Mankowski, '2003 "Europäisches Internationales Arbeitsprozessrecht"', 23.

[99] U. Magnus, 'Seearbeitsverhältnisse und die Rom I- und II-Verordnungen', in T. Borić, B. Lurger, P. Schwarzenegger and U. Terlitza (eds.), *Öffnung und Wandel - Die internationale Dimension des Rechts I: Festschrift für Willibald Posch zum 65. Geburtstag* (Wien: LexisNexis, 2011), p. 449; Garber, *2012 "Zum Begriff des gewöhnlichen Arbeitsortes"*, p. 233.

[100] Egler, *2011 "Seeprivatrechtliche Streitigkeiten"*, p. 149.

the employee's work is linked to an activity for which the coastal State can assume functional sovereign rights under the law of the sea. This is the case for employees in the offshore industry exploring and exploiting the resources of the marine subsoil, and for fishermen working in the EEZ. Mere navigation in the EEZ, which is explicitly not subject to the functional sovereign rights that the coastal State enjoys in this zone, can however not create a judicial venue. If it is submitted that the coastal State still has a link with these waters,[101] this view ignores the character of the zone under the law of the sea, UNCLOS submitting the navigation there to the high seas regime. Furthermore, such a judicial venue for seamen's pay contracts would create considerable practical difficulties. The EEZ stretches up to 200 nm from the coastal State's baselines, and the navigation in this area, unlike coastal shipping, is hardly to be differentiated from genuine high seas navigation. Whereas a difference between a seaman's pay contract for coastal shipping and high seas navigation may generally be easy to draw, a difference between EEZ shipping and high seas navigation appears rather ridiculous. It must be repeated that under the law of the sea, navigation in the EEZ is subject to the regime of freedom of navigation that also applies to the high seas.

For labour disputes of seamen working on seagoing ships, localisation of the habitual working place can in fact be difficult, and the determination of the judicial venue should be subject to some considerations that do not come into play in determining the *forum delicti* according to art. 5 (3) of the Brussels Regulation. A tort will normally be a matter of a clearly defined timespan. Even if there are also torts that take place over a longer time, the question of the *forum delicti* will in most cases be a matter of a localisation of the ship in a given moment. The habitual working place, on the other hand, should be determined with regard to a working relationship of some duration, during which the ship will possibly have passed a multitude of zones of maritime jurisdiction. It is, in this context, rather hard to imagine any appropriate forum to which art. 19 (1) could point other than the courts of the flag State.

With regard to the mobile character of ships, it is sometimes advocated[102] that seamen on seagoing ships do not have a "habitual working place" and that the determination of the competent forum should be done according to art. 19 (2 b) of the Regulation, granting jurisdiction to the courts for the place where the business which engaged the employee is or was situated. The same solution is often submitted for the reason that the flag State may, especially when the ship flies a flag of convenience, be very badly placed to adjudicate labour disputes of this

[101] Ibid., p. 209.

[102] P. Gottwald and H. Nagel, *Internationales Zivilprozessrecht*, 6th edn. (Köln: O. Schmidt, 2007), p. 120, but only considering staff on board of aircraft; O. Hartenstein, 'Rom I-Entwurf und Rom II-Verordnung: Zur Bedeutung zukünftiger Änderungen im Internationalen Privatrecht für das Seerecht' (2008), Transportrecht, 143–61 at 161.

kind.[103] The same, however, holds true for the place of business engaging the employee in the cases in which seamen are hired by manning agencies, which is a very common occurrence today. These agencies are equally likely to have their place of business in States with rather low social standards.[104] Furthermore, the place of business is a venue directly determined by the employer, which is at odds with the Regulation's aim to provide the employee, as the weaker party, with a venue other than the home jurisdiction of the employer.[105] The argument of a better protection of seamen by ruling out the jurisdiction of the flag State thus hardly always holds true. Moreover, from a pragmatic point of view, it must be submitted that the flag is at least a symbol of nationality that is certain, unique and identifiable to the crewmembers. In contrast to that, a manning agency can have branches of different sizes and legal character[106] in different countries working together, so that it may be difficult to identify the pertinent place of business.

Granting jurisdiction to the courts of the flag State in maritime labour disputes guarantees, with regard to the law of the sea, the legitimacy of decisions under public international law much better than the judicial venue in the courts of the engaging business. After all, seagoing ships tend to sail the high seas and EEZs for by far the majority of their time at sea,[107] and it should be kept in mind that according to arts. 92, 94 UNCLOS, the flag State has, during that time, the exclusive jurisdiction over the ship. The balance between the interests of flag States and coastal States that has been created by the law of the sea should not be disregarded in the attribution of jurisdiction.[108] It has rightly been submitted that the short periods in which the ship sails the internal or territorial waters of a State and would thus be subject to the jurisdiction of the coastal State should not alter the stable connection of the flag State jurisdiction for labour cases.[109] Especially in the field of maritime labour law, there may be quite a number of provisions on maritime safety and health that a State may, after all, consider to be provisions of immediate application, to which the ship is subjected without first proceeding on a conflict-of-law analysis (see *infra* in Chap. 7). Generally paralleling the judicial venue for maritime labour law with the law of the flag—even if, through the escape clauses of

---

[103] Concern pointed out by P. Lagarde, 'Le nouveau droit international privé des contrats après l'entrée en vigueur de la Convention de Rome du 19 juin 1980' (1991), Revue critique du droit international privé, 287–340 at 319, albeit in the context of the similar conflicts rule for employment contracts under the then Rome I Convention.

[104] Mankowski, '2003 "Europäisches Internationales Arbeitsprozessrecht"', 27.

[105] Garber, *2012 "Zum Begriff des gewöhnlichen Arbeitsortes"*, p. 227.

[106] Magnus, *2011 "Seearbeitsverhältnisse"*, p. 450, with regard to the corresponding problems in determining the law applicable to the seamen's pay contract under the Rome I Regulation.

[107] Hartenstein, '2008 "Rom I-Entwurf und Rom II-Verordnung"', 161.

[108] See Mankowski, '2005 "Das Internationale Arbeitsrecht"', 59, and Garber, *2012 "Zum Begriff des gewöhnlichen Arbeitsortes"*, p. 231, on the importance of the law of the sea.

[109] Ibid., pp. 233, 234.

the conflicts rules, parallelism may not always be the case[110]—thus leads to a concurrence of venue and applicable law that could otherwise become a problem if the judicial venue would be found in a State that is not the flag State.

This finding should by no means trivialise the problem of flags of convenience, potentially triggering maritime labour jurisdiction in countries that do not have a legal system that would be up to this task. Still, it must be kept in mind that maritime employees can in most cases bring their claim in the countries of the employer's place of business, under art. 19 (1) of the Brussels Regulation or under the general venue in the defendant's home jurisdiction (see *supra* under I.) if the subject matter jurisdiction rather points to an absurd judicial venue or in the absence of a link between the flag and the vessel.[111] Shipping and maritime labour are however international businesses in which it is very hard to establish jurisdiction that appears fair in the vast majority of cases. The rules considering the balance between flag State jurisdiction and coastal State jurisdiction that are established by the law of the sea should, for the sake of legal certainty not be disregarded.

III. The Venue of the Ship's Arrest

The particularities of jurisdiction in the maritime trade have created a special strategy to enforce maritime cases at a rather early point in time.[112] Under admiralty jurisdiction, claimants had, under certain circumstances, the possibility of bringing claims *in rem* that were virtually[113] directed against the ship itself rather than the owner or the persons on board. The underlying interest in most of these claims was to obtain the vessel as a security for the value of the claim, which generally triggered the arrest of the ship in a given port, so as to ensure that the vessel would be physically available as a security. In contrast to that, seizure of a ship in civil legal systems such as France[114] or Germany was rather a way of securing a valuable asset before the actual trial, seizure of ships not being subject to different rules than the seizure of any other of the defendant's assets.[115]

---

[110] Mankowski, '2005 "Das Internationale Arbeitsrecht"', 60 et seq.; Garber, *2012 "Zum Begriff des gewöhnlichen Arbeitsortes"*, p. 222.

[111] Ibid., p. 235.

[112] See V. Ruiz Abou-Nigm, *The Arrest of Ships in Private International Law* (Oxford: Oxford University Press, 2011), para. 2.01 et seq. on the history of the arrest of ships.

[113] See Dicey, Morris and Collins, *2012 "Conflict of Laws"*, para. 13-002.

[114] See P. Bonassies, 'La Cour de Cassation française et la Convention de 1952 sur la sasie conservatoire des navires de mer', in G. Berlingieri, A. Boglione, S. M. Carbone and F. Siccardi (eds.), *Scritti in onore di Francesco Berlingieri. Volume I*, (Genova: Il Diritto Marittimo, 2010), p. 216.

[115] N. Gaskell and R. Shaw, 'The Arrest Convention 1999' (1999), Lloyd's Maritime and Commercial Law Quarterly, 470–85 at 471.

Under the jurisdiction of the English courts, there was some interest in having a judicial venue in the courts of the country where the ship was arrested, so as to create a geographical proximity between the asset seized and the competent court. This allowed the immediate enforcement of the judgment if the claim was granted and was thus considered as a help to the foreign claimant.[116] In effect, under English common law, the possibility to physically arrest the ship depended on the prior service of an *in rem* claim form. This is actually the inverse solution to the procedure set out later in international agreements, where there is always a venue for seizure of the ship as an interim measure to obtain security,[117] which triggers jurisdiction on the merits of the claim under additional conditions.[118] Either way, given the difficulties of enforcement of cross-border claims against, possibly, single-ship companies,[119] which continues to be an issue today, the jurisdiction of the port where the ship was arrested has of course always been a very popular venue.

The possibility of a venue in the courts of the place where the a ship has been arrested under a maritime claim has been enacted in international conventions, namely the 1952 Convention on Civil Jurisdiction for Collisions,[120] as a more specified codification, and the 1952 International Convention for the Unification of Certain Rules Relating to the Arrest of Sea-Going Ships (Arrest Convention).[121] The former provides in its art. I para. 1 (b), for claims arising out of collisions, a venue at the place of the ship's arrest, or the place where it could have been arrested, in addition to the venues of the defendant's habitual residence or place of business and the *forum delicti* for inland waters or port collisions, two venues more familiar to civil law systems.[122] The Arrest Convention is a more general convention setting out the limits under which a ship can be arrested for a maritime claim (art. II). The Arrest Convention provides for a venue in the courts of the place of arrest, but only where the ship has effectively been arrested.[123] It creates, in addition to the venue for the mere seizure of the ship, also a venue for the merits of the claim, albeit only under certain additional circumstances (art. VII), based on particular links that the case may have with the jurisdiction,[124] but some of which, however, are also doubtful with regard to legitimacy.[125]

---

[116] Berlingieri, '1977 "Jurisdiction and the Choice of Law in Collision Cases"', 867.

[117] Gaskell and Shaw, '1999 "The Arrest Convention 1999"', 482.

[118] F. Berlingieri, *Arrest of ships: A commentary on the 1952 and 1999 Arrest Convention*, 5th edn. (London: Informa, 2011), para. 52.753; Ruiz Abou-Nigm, *2011 "The Arrest of Ships"*, para. 7.85.

[119] S. Brækhus, 'Choice of Law Problems in International Shipping (Recent Developments)' (1979) 164, Recueil des Cours, 259–330 at 265.

[120] 439 UNTS 217; BGBl. 1972 II 653, 663.

[121] 439 UNTS 193; BGBl. 1972 II 653, 655.

[122] Berlingieri, '1977 "Jurisdiction and the Choice of Law in Collision Cases"', 868.

[123] Berlingieri, *2011 "Arrest of ships"*, para. 52.760; Ruiz Abou-Nigm, *2011 "The Arrest of Ships"*, para. 7.88; Dicey, Morris and Collins, *2012 "Conflict of Laws"*, para. 13-032.

[124] See P. Yessiou-Faltsi, 'The 1952 Brussels International Convention on the arrest of ships for maritime claims and on jurisdiction' (2011), Aegean Review of the law of the sea and maritime law, 175–80 at 179.

[125] Ruiz Abou-Nigm, *2011 "The Arrest of Ships"*, para. 7.92.

With the 1999 Arrest Convention[126] that came into force in 2011, but has only ten States parties as of 2014,[127] these questionable factors conferring jurisdiction on the merits to the *forum arresti* have been done away with, among other important changes to the 1952 regime.[128] Under the 1999 Arrest Convention, the conferral of jurisdiction for the merits of the claim to the *forum arresti* is general, but the court may refuse to exercise it under the law of the forum, always provided that another court assumes jurisdiction. Whereas it is generally perceived that this alteration was rather a matter of pragmatism and simplicity,[129] abandoning conditions on jurisdiction that were considered as an obstacle to uniformity,[130] the change, from the point of view of public international law, creates the responsibility to consider the legitimacy of jurisdiction under the domestic law of the forum.[131] Due to the small number of ratifications, the 1999 Convention is however of limited importance in practice. Furthermore, there are serious doubts whether EU member States can ratify the 1999 Arrest Convention, considering that, as a convention posterior to the Brussels Regulation, the matter of jurisdiction and enforcement is withdrawn from the competence of the member States.[132]

Both 1952 conventions have been widely influenced by the English rules on admiralty procedure, and in both cases, the codification of the venue of the ship's arrest met with criticism at the time of the negotiations of the conventions.[133] At that time, the arguments against it were seemingly rather pragmatic than dogmatic, criticising the fact that it could create jurisdiction almost everywhere.[134] Today's doubts in the jurisdiction however rather concern the difficulties in reconciling it with public international law.

Under the regime as established by the Brussels Regulation, clearly aiming at doing away with exorbitant or "unreasonable" exercise of jurisdiction,[135] the *forum arresti* for the merits of claims as codified by the 1952 Arrest Convention and the 1952 Convention on Civil Jurisdiction for Collisions nevertheless survived, as a matter of art. 71 of the Regulation. Both international agreements present conventions prior to the Regulation, the provisions of which can be applied as between

---

[126] UN Document A/CONF.188/6 of 19 March 1999.

[127] See http://treaties.un.org/pages/ViewDetails.aspx?src=TREATY&mtdsg_no=XII-8&chapter=12&lang=en (last visited 23 February 2014).

[128] See F. Berlingieri, 'The 1952 and 1999 Arrest Conventions. A Comparison' (2012), Il Diritto Marittimo, 367–96 for a comprehensive summary.

[129] Ibid., p. 395.

[130] Ruiz Abou-Nigm, *2011 "The Arrest of Ships"*, para. 7.97.

[131] Gaskell and Shaw, '1999 "The Arrest Convention 1999"', 483. In this sense, Ruiz Abou-Nigm, *2011 "The Arrest of Ships"*, para. 7.98: "a far-reaching jurisdictional ground such as *forum arresti* has been combined with its natural correlative."

[132] Ibid., para. 7.99; Berlingieri, '2012 "The 1952 and 1999 Arrest Conventions"', 395.

[133] Berlingieri, *2011 "Arrest of ships"*, para. 99.227; also see Ruiz Abou-Nigm, *2011 "The Arrest of Ships"*, para. 7.05. for more details.

[134] Ibid., para. 7.06.

[135] See Fernández Arroyo, *2004 "Exclusive and Exorbitant Grounds"*.

States parties.[136] Even if the ECJ has limited the scope of art. 71 of the Regulation, or of the preceding art. 57 of the Brussels Convention in certain regards,[137] the judicial venue is still a valid one. Following more recent jurisdiction, art. 71 of the Regulation should be restricted to jurisdictional rules of specialised Conventions that are "highly predictable, facilitate the sound administration of justice, and enable the risk of concurrent proceedings to be minimised and that ... ensure, under conditions at least as favourable as those provided for under the Regulation, the free movement of judgments in civil commercial matters".[138] It is however easily arguable that the 1952 Arrest Convention fulfils these criteria.[139]

Whereas the judicial venue of the place of the arrest of the ship can be criticised as rather arbitrary, and devoid of a real connection with the court that would confer legitimacy to it under public international law, it is a venue that was in the first place driven by pragmatic considerations, giving shipowners a choice based on practical considerations for claims as between them.[140] It ended up being a matter of international consensus in the 1952 conventions. The contracting States of these agreements apparently sacrificed a sound dogmatic basis of jurisdiction to considerations of practicability, introducing the judicial venue into the realm of public international law, albeit, with regard to the ongoing criticisms, not into customary law. This ground for jurisdiction emerged with regard to—and still mainly concerns—claims that are restricted to the value of the ship as a security, an asset that is susceptible of constant movement in any case.[141] Furthermore, the venue only arises for maritime claims, matters that are closely connected with the operation of the ship[142] and thus in a special field of business relationships and where actors are most likely aware of the existence of this venue. In the case of the 1952 Convention on Civil Jurisdiction for Collisions, the venue was set out for cases in which it would be in all cases rather difficult to find a legitimate venue. It is

---

[136] Berlingieri, *2011 "Arrest of ships"*, para. 52.774; Ruiz Abou-Nigm, *2011 "The Arrest of Ships"*, para. 7.15; H. Yang, *Jurisdiction of the Coastal State over Foreign Merchant Ships in Internal Waters and the Territorial Sea* (Berlin, New York: Springer, 2006), p. 179; Dicey, Morris and Collins, *2012 "Conflict of Laws"*, para. 13.031.

[137] *"The Tatry"* I Reports of Cases (ECJ) 5442 (European Court of Justice, 06 December 1994). The case is mostly on the question whether the jurisdiction under the Arrest Convention should also take precedence over earlier proceedings in another member State, which the court denied. Also see the comment of A. Briggs, 'The Brussels Convention Tames the Arrest Convention' (1995), Lloyd's Maritime and Commercial Law Quarterly, 161–6 and, more generally, E. Blackburn, 'Lis alibi pendens and forum non conveniens in collision actions after the Civil Jurisdiction and Judgments Act 1982' (1988), Lloyd's Maritime and Commercial Law Quarterly, 91–102.

[138] *TNT Express Netherlands v. AXA* I Reports of Cases (ECJ) 4137 (European Court of Justice), pt. 56.

[139] Ruiz Abou-Nigm, *2011 "The Arrest of Ships"*, para. 7.49, para. 7.105 et seqq.

[140] Brækhus, '1979 "Choice of Law Problems"', 263 et seq.

[141] Ruiz Abou-Nigm, *2011 "The Arrest of Ships"*, para. 7.29.

[142] Ibid., para. 7.39, see Berlingieri, '2012 "The 1952 and 1999 Arrest Conventions"', 371 et seq. on the changes under the 1999 Convention.

certainly with a reason that the venue survived the changes that were brought about by the Brussels Regulation. Within the limited scope of its applicability, the *forum arresti* is certainly a useful jurisdictional base, even if its dogmatic foundation is clearly open to criticism.

IV. Domicile or Residence of the Claimant

Following considerations of fairness, the attribution of jurisdiction to the courts at the place of the domicile or residence of the claimant is a rather exceptional occurrence. In modern times, this judicial venue is however provided for in claims where the economic power of the parties is manifestly unequal, as an advantage to the weaker party.[143] This is the underlying rationale for the consumers' venues in the Brussels Regulation, arts. 15 to 17. In consumer matters where an individual brings a claim against a party that concludes contracts of this kind in the course of business, it is perceived as fair that entrepreneurs should expect and be ready to defend themselves in all jurisdictions in which they choose to contract.[144] It must in this context be stressed that the Regulation's section only applies to contracts, not to claims in tort. The legitimacy of jurisdiction for claims in contract is however not as considerable a problem as the legitimacy of jurisdiction for claims in tort, where the parties probably never envisaged that there would be a lawsuit between them.

With regard to maritime cases, consumer contracts are however unlikely to occur except in the case of passenger trade. Contracts of transport are however, by virtue of art. 15 (3) of the Regulation, withdrawn from its scope of application, except for consumer contracts that can considered to be package travel ("a contract which, for an inclusive price, provides for a combination of travel and accommodation"). As has been shown in an earlier part of this study on passenger claims (*supra*, Chap. 5), it is thus mostly cruises and similar contracts of travelling for the sake of travelling that can be considered as package travel and thus as subject to consumer jurisdiction. There may be borderline cases, e.g. for passenger voyages on cargo ships, but distinguishing between transport contracts and package travel contracts will in most cases be possible if the meaning of the voyage for the passenger is given regard to.[145] In any case, contractual relationships of this kind, in which the parties can foresee the probability of a lawsuit between them and provide for a venue, are unlikely to give rise to problems of legitimacy under public international law. If the consumer is given a rather exorbitant venue in his home jurisdiction, this is to be seen as a protection of the consumer and a compensation for the entrepreneur's superior bargaining power, allowing him to dictate the terms of the contract.

---

[143] Fernández Arroyo, *2004 "Exclusive and Exorbitant Grounds"*, p. 171.

[144] von Mehren, '1983 "Adjudicatory Jurisdiction"', 317.

[145] Egler, *2011 "Seeprivatrechtliche Streitigkeiten"*, p. 57.

V. Presence of the Defendant in the Adjudicating Country

This study of jurisdictional bases would not be complete without mentioning the ground of jurisdiction that has, originally, been the only one in the law of England as a major maritime nation. The approach to jurisdiction in England, and for that matter, in the United States, has been markedly different from the considerations on the continent. Originally, courts in England assumed competence following a service of the defendant in the territory of the country, irrespective of whether his presence there was merely temporarily or coincidental or of a lasting character.[146] This basis of jurisdiction however only existed for claims *in personam*, by which, generally speaking, the defendant was to be compelled to do something,[147] as opposed to the claims *in rem* for which admiralty claims "against" ships have been exemplified *supra* under III.

This doctrine of service originated in the historical use of physically ordering, and if necessary, bringing a defendant to the courts by the authorities of the country, which was a governmental act that was of course only possible within the territory of the governmental power providing the court. The act of physical presence was later replaced by the service of the writ, but for which, still, the defendant needed to be physically present in the country so that this act could be performed within the geographical limits of the sovereign power that the court represented.[148] The decisive moment was however the serving of the writ and not any other point in time, so that there are effectively cases in which defendants have been served during a stay in the country that was very momentary or even coincidental.[149] Under certain circumstances, the defendant could also be served outside the jurisdiction,[150] but this remained the exception.

This circumstance presents the difference between the ground of jurisdiction that has been presented *supra* under I., of domicile, allegiance or habitual residence in the adjudicating country, and mere presence. In the former case, the defendant has, by settling in the country, submitted to the jurisdiction of its courts, whereas under service of process jurisdiction there may be no link whatsoever between the defendant and the adjudicating court. The concept of due service of process within the geographical limits was however the general base of jurisdiction in the countries influenced by English legal doctrine, and continues to be so in the United States.[151] In the United Kingdom, with the advent of the Brussels Regime under European Law, this ground of jurisdiction has been abandoned for all matters which fall under

---

[146] See Maier and McCoy, '1991 "A Unifying Theory"', 260 et seq.

[147] Dicey, Morris and Collins, *2012 "Conflict of Laws"*, para. 11-002.

[148] Maier and McCoy, '1991 "A Unifying Theory"', 261.

[149] See von Mehren, '1983 "Adjudicatory Jurisdiction"', 291 et seq. for an early account of cases in the United States where defendants had been served in US airports while in transit or on foreign-flagged ships in US ports.

[150] Layton and Mercer, *2004 "European Civil Practice"*, vol. 1, para. 3.001.

[151] See Maier and McCoy, '1991 "A Unifying Theory"', 262.

the scope of what is now the Brussels Regulation. The jurisdiction for national processes, within England[152] (for this matter, the rules are clearly different in the other parts of the United Kingdom), and for matters that are expressly excluded from the scope of application of the Brussels Regulation[153] are however still based on due service of process.

On the other hand, the broad assumption of jurisdiction gave rise to a correcting factor. Originally a Scottish concept, the doctrine of *forum non conveniens* allowed a court to decline jurisdiction on the ground of the existence of another, more appropriate forum.[154] It was thus within the discretion of the courts whether they wanted to entertain a claim possibly not having any real connection with the forum or rather to refer the claimant to another, more closely connected jurisdiction. The grounds that were considered in taking this decision were, *inter alia*, cost-effectiveness and convenience for the parties, and thus, seemingly, originally more based on practical considerations than on the legitimacy of jurisdiction under public international law. This has changed to a certain degree. The doctrine of *forum non conveniens* has, especially with a view to the dogmatic legitimacy of jurisdiction, very much developed in the United States, where jurisdiction based on due service of process clearly remains the main concept, and, albeit to a considerably lesser extent, in England, but where the rules on jurisdiction in cross-border litigation are now subject to the provisions of the Brussels Regime.

It was debated for some time whether the doctrine of *forum non conveniens* should survive the advent of the Brussels Regime, at least for matters in which the courts of a non-member State could be considered the more appropriate forum. It was not until the ECJ's decision of 2005,[155] already discussed *supra* under II. 1. c), that the provisions of the Brussels Regime became universally accepted as absolute, not conferring any kind of discretion to the court.

Even if the assumption of jurisdiction following due service of process within the jurisdiction has been very much reduced in England by the Brussels Regime, this concept is still of an utter importance in the United States and in other jurisdictions that have been influenced by English legal concepts. In general, jurisdiction in these legal systems is subject to the discretion of the court, following the application of *forum non conveniens*, and it is by no means unlikely that defendants in maritime cases will see themselves faced with proceedings under this basis of jurisdiction. Depending on the circumstances of the case, bringing the case and contesting jurisdiction may entail considerable uncertainty for the parties. Reference will here be made again to the case of the *Prestige* in the US courts (see *supra*, Chap. 3, A. I. 1. b)), where it was heavily debated whether the United States

---

[152] See Layton and Mercer, *2004 "European Civil Practice"*, vol. 1, para. 3.001, 3.010 et seq. on the practical implications.

[153] See Dicey, Morris and Collins, *2012 "Conflict of Laws"*, para. 11-035 et seqq.

[154] See Maier and McCoy, '1991 "A Unifying Theory"', 273, on the role of *forum non conveniens* as a corrective of an, arguably, too general assumption of jurisdiction.

[155] *European Court of Justice, 01.03.2005, Owusu v. Jackson.*

was the appropriate forum to hear a case concerning the negligent classification of an oil tanker that had sunk off the coast of Spain, even if the defendant classification society was based in the United States. An even better example is probably the *Amoco Cadiz* case (see *supra,* Chap. 3, A. I. 1. a)), where it was discussed by observers whether the defendants should have pleaded *forum non conveniens* when the case was brought in the courts of the United States, considering that the only link with the adjudicating country was the nationality of the defendant mother company. In this context, it is interesting to hypothesise whether the court would have denied to grant *forum non conveniens* on the basis that there is a moral duty to adjudicate the behaviour of a US company according to American law.[156] In any case, the defendants' conscious decision not to plead *forum non conveniens* amounts at least to a tacit acceptance of the jurisdiction. As the defendant thus consents to being judged by the court, questions of legitimacy seem less pressing under these circumstances.

The advantages and disadvantages of the assumption of jurisdiction based on due service as opposed to the assumption of jurisdiction on domicile or allegiance of the defendant shall however not be discussed here. From the point of legitimacy of decisions under public international law in maritime tort cases, the competence of a court to adjudicate events that took part on a ship and in a maritime area under a foreign jurisdiction, or no jurisdiction at all, may be just as questionable under both systems. It even seems that the more flexible doctrine of *forum non conveniens* at least allows the seized court to reconsider the legitimacy of a future decision in this jurisdiction, whereas the fixed rules of the civil law systems and, today, the Brussels Regulation, do not allow for a correction of the jurisdictional rules *a posteriori*. The value of the doctrine of *forum non conveniens* however depends considerably on the way it is exercised, and on the considerations taken into account in assessing whether a venue is appropriate for the case or not.

VI. Conclusion

The question of jurisdiction for civil claims is a difficult matter and has, in comparison with the debate on criminal jurisdiction, received considerably less attention. This problem is seemingly often overlooked due to the fact that many connecting factors that trigger a court's competence can also be used to establish the link between the case and the adjudicating State—the parties are however likely to sue in the flag State (at least if they are not faced with a flag of convenience) or in the defendant's domicile or country of registry in order to easily enforce the claim, or the claimant may want to sue in his home jurisdiction. In all of these cases, there will be a link between the case and the country of adjudication somehow justifying the assumption of competence. Even if some of the earliest cases in which States

[156] Kloss, '1989 "Der Untergang der "Amoco Cadiz"'", 185.

have protested against another State's assumption of civil competence arose in a maritime context, the considerations of the role of public international law in adjudicating civil claims often pass unnoticed.

It is only when inquiring into the underlying principles of sovereignty, personality and territoriality that difficulties become obvious, and even then, the long-established principles of either suing the defendant in his home jurisdiction or establishing jurisdiction upon due service of proceedings are too well established to be seriously questioned. The same holds true for the *forum arresti*, which may be rather arbitrary in many cases, but which is, from a pragmatic point of view, a very convenient venue for claims aiming at or arising out of the arrest of the ship as a security. The Arrest Conventions that have been discussed in this part of the study however recognise this problem of consistency between a venue dictated by considerations of convenience and questions of legitimacy under public international law and only allow jurisdiction on the merits of the case if there are additional links between the case and the venue. Still, even then, the ties between the set of facts and the adjudicating court may be rather remote. It is submitted that in all of these venues the court must be particularly careful when considering the legal relationships between the parties that, especially in tort cases, may be governed by rules and principles that are very different from the *lex fori*.

With regard to that, the *forum delicti*, which allows the parties to litigate a tort in the courts of the place where it happened under the laws of this very place, becomes overly important in tort cases. Torts occurring on the high seas however give rise to the problem that the *forum delicti* may not be very easy to determine, that the rules properly speaking point to a non-jurisdiction or that there are two concurring jurisdictions with regard to the competence of a coastal State and the flag State. It is submitted that it is the law of the seas that decides these questions, even if UNCLOS refrains from explicitly determining the question of judicial competence for civil claims. Even under the Convention, however, the respective realm of the concurring jurisdictions of the coastal State and the flag State in the territorial waters is not entirely clear where the events only take place on the ship in question and do not have an impact on the interests of the coastal State.

In any case, it appears obvious that the requirements of public international law cannot be served by the choice of jurisdiction alone. It is both the questions of competence and the choice of law governing the set of facts, and their interplay that must be considered to obtain a decision that is reconcilable with the balance of interests in the law of the sea.

Chapter 7
Conflict of Laws

The questions of jurisdiction and the choice of law for a given claim in civil law should by no means be confused. Even if both matters are often decided on the same facts, the links with the jurisdiction in question often also being used as a connecting factor to determine the applicable law, conflict-of-law rules are domestic law. A court can only proceed on the application of a conflicts rule once it has determined its judicial jurisdiction. This does not mean that jurisdiction and choice of law are not interdependent. Especially in tort cases, where the parties never envisaged that they would have to deal with each other and where there cannot be previous agreements on jurisdiction and choice of law, jurisdiction and choice of law are often determined following the same circumstances.

However, the application of a conflicts rule, even if it does point to a law different from the *lex fori*, already represents the exercise of domestic judicial jurisdiction. The law that is designated by the conflicts rule of one country may not be the law designated by the conflicts rule of another country.[1] A certain harmonisation is obtained by mechanisms such as *renvoi*, but considering that conflicts rules are still a part of domestic law, that the courts of a State apply as an expression of their competence to adjudicate, the procedure of a conflict-of-law analysis is in itself a sovereign act,[2] which should be compatible with the State's jurisdiction under public international law.[3]

---

[1] See T. M. de Boer, 'Living Apart Together: The Relationship between Public and Private International Law' (2010) 62, Netherlands International Law Review, 183–207 at 195, addressing the interplay between public and private international law.

[2] In this sense, Maier and McCoy, '1991 "A Unifying Theory"', 253 et seqq; G. J. Simson, 'Rethinking Choice of Law: What Role for the Needs of the Interstate and International System?', in M. H. Arsanjani, J. K. Cogan, R. D. Sloane and S. Wiessner (eds.), *Looking to the Future: Essays on international law in honor of W. Michael Reisman* (Leiden, Boston: Martinus Nijhoff Publishers, 2011), p. 242.

[3] Bowett, '1983 "Jurisdiction"', 2.

S.F. Gahlen, *Civil Liability for Accidents at Sea*, Hamburg Studies on Maritime Affairs 32, DOI 10.1007/978-3-662-45555-5_7

After having discussed the questions of jurisdiction and the legitimacy of judicial competence *supra* in Chap. 6, the study now turns to the conflict-of-law rules that are applied or that have so far been applied to maritime incidents. As shown, the choice of law is sometimes regarded as a way to legitimise adjudicating power in cases where States assume civil jurisdiction for pragmatic or matter-of-fact reasons. Conflict-of-law rules differ from country to country, but there is a certain worldwide uniformisation which has created general principles. Furthermore, throughout the EU, the rules on the choice of the applicable law for contractual relationships and non-contractual relationships have been harmonised by the Rome I Regulation[4] and the Rome II Regulation,[5] respectively.

In tort claims, the principle of designating the law applicable according to the laws of the State in which the tort was committed has been of a very considerable importance all over the world[6] and has also been codified in art. 4 (1) of the Rome II Regulation. The so-called principle of the *lex loci delicti* can of course be debated, especially in cases in which the tort was committed in the territory of one jurisdiction and took effect in another, or in cases where there is another law manifestly better placed to deal with the case. Numerous codifications of conflict-of-law rules have accounted for these difficulties and allow for more flexibility in applying the rule.[7] In the United States, many State law jurisdictions have departed to a considerable extent from the principle of the *lex loci delicti*.[8]

For maritime torts, there is always the very obvious problem that a territorial connecting factor, by which an event or a chain of events is geographically localised, may be considered to point to conflicting jurisdictions, considering the ship as a moving entity, which is actually secluded from the waters that it sails. The law of the flag as a factor to determine the law applicable should never be underestimated, taking into account that it remains stable or is at least not likely to change without documentation. In contrast to that, the geographical connecting factor may be determined according to the waters in which the ship sailed when the accident occurred. For torts committed in areas beyond national jurisdiction or in regions that are not subject to territorial sovereignty, the *lex loci delicti* rule may then of course point to nothingness,[9] as there is no State in which the tort was committed. It is therefore necessary to analyse the regime that applies, according to the law of the sea, to the respective zone and to consider in which way it allows the application of private international law. When determining the legal system applicable to a given maritime zone following the *lex loci delicti* rule, it may also be interesting to consider questions of legislative competence over these waters. The

---

[4] Regulation (EC) No. 593/2008.

[5] Regulation (EC) No. 864/2007.

[6] G. A. L. Droz, 'Regards sur le droit international privé comparé' (1991) 229, Recueil des Cours, 25–421 at 282.

[7] Tomljenović, '1999 "Maritime Torts"', 252.

[8] See Symeonides, *2004 "Territoriality and Personality"*, for an overview.

[9] Tomljenović, '1999 "Maritime Torts"', 252.

interrelationship between the question of the State that would be entitled to enact laws and regulations for these waters—possibly also only with regard to certain matters—and the question of the law that will be applied to events in these waters in a court may be of considerable use in the often difficult quest for the applicable legal system.

Long before the coming into force of the Rome II Regulation harmonising the conflicts rules applicable to tort claims, legal systems in most maritime nations, especially in England, have found more or less consistent ways to deal with these problems in a more detailed way than the Regulation that does not contain special rules applicable to maritime torts and leaves quite a number of questions without answer. This study therefore presents these earlier domestic solutions before considering the rules of the Rome II Regulation and the changes they possibly brought about.

This study of the conflict-of-law rules will address the maritime zones separately, according to the maritime zone under the law of the seas in which the incident took place, and distinguish between events taking place only on the ship as opposed to events that have an impact on the surrounding waters. Even if the analysis centres on tort law, maritime casualties can give rise to claims under other legal bases, such as maritime labour law for crewmembers. Some regard will therefore also be given to these legal relationships, possibly entailing different connecting factors for the choice of law.

Apart from conflict-of-law rules, there may, in effect, be rules that are intended for application in a maritime context and to a certain extent withdraw the cases to which they are applicable from the scope of general private international law rules.[10] In most cases in which States embark upon legislations of this kind, there is a competence norm in international law giving them the right to do so. One example could be domestic legislation against vessel-source pollution in the Exclusive Economic Zone (EEZ) and in territorial waters,[11] or the domestic specific legislation on installations for the exploration and exploitation of the marine subsoil or the exploitation of other resources in the EEZ,[12] the corresponding competence being conferred to States by art. 60 UNCLOS. Most national legislation on licensing, operational issues and a possible civil liability will be specifically created for the application in the EEZ, and national courts will apply these provisions to the exploration and exploitation activities without proceeding on a conflict-of-law analysis beforehand, even if the incident occurred in an area where sovereign rights are restricted and even if there may be links to other jurisdictions.

Maritime law is however still a field of law presenting a considerable degree of harmonisation in many respects. It should not be forgotten that even if a claim's subject matter is to be governed by domestic law following a conflict-of-law

[10] Ibid., p. 288.

[11] Ibid., p. 288.

[12] See P. Ehlers, 'Meeresfreiheit und aquitoriale Ordnung. Zur Entwicklung des Seerechts' (2013) Verwaltungsarchiv, 406–428 at 411 et seqq. for a detailed discussion from a German point of view.

analysis, other aspects of the case such as the limitation of liability may still be subject to international agreements such as the LLMC,[13] or, as a procedural aspect of the case, by the *lex fori*.[14] Within the EU, the Rome II Regulation claims to cover cases in their entirety, including the limitation of liability (art. 15 para. (b) of the Regulation). It can be debated whether this has caused the law applicable to the limitation of liability to always coincide with the law applicable to the merits of the case.[15]

A. The High Seas

According to art. 87, 90 UNCLOS, the high seas are open to all States for the purposes of navigation, overflight and similar activities. Art. 89 UNCLOS codifies the well-established principle that the high seas are not subject to sovereignty claims of any State. No State can put forward claims based on a territorial link of any kind with these waters. No State can enact laws and regulations that would be applicable in these waters. Considering the status of ships and the criminal jurisdiction that applies to ships sailing the high seas, arts. 91 and 92 UNCLOS set out very clearly that it is only the vessel's flag State that can assume jurisdiction and competence over ships on the high seas. The rationale for this competence is, according to modern legal theory,[16] a voluntary grant of nationality by the flag State. The flag State's competence over a ship flying its flag is in fact created by the act of according the flag—and therewith the registry in the national register—to the ship, under the individual conditions and the procedure that the State sets in place. Following this well-established principle of the law of the sea, the only State that can, in general, assume competence over a ship on the high seas is the flag State.

Even if the law of the flag is thus an important connecting factor in all legal relationships involving the ship that may arise, there can be other laws called to govern given cases. The law of the flag may in effect have no particular link with the operation of the ship at all, especially in cases where the vessels fly flags of convenience that have only been chosen in order to avoid the applicability of a more complex and possibly more expensive flag regime. Furthermore, in bareboat charter contracts, the ship is generally exclusively run by the charterer, so that the law of the flag attributed according to the nationality of the owner or freely chosen by him

---

[13] See K. Ramming, 'Internationalprivatrechtliche Fragen der Haftung des Reeders' (2010), Transportrecht, 284–302 at 299 et seqq.; M. N. Tsimplis, 'The effect of European regulations on the jurisdiction and applicable law for limitation proceedings' (2011), Lloyd's Maritime and Commercial Law Quarterly, 307–21.

[14] See M. N. Tsimplis, 'Law and jurisdiction for English limitation of liability proceedings' (2010) 16, Journal of International Maritime Law, 289–316 at 306 et seqq.

[15] Ibid., p. 293; Tsimplis, '2011 "The effect of European regulations"', 316 et seq.

[16] For a short overview on historic theories, see W. Tetley, 'The Law of the Flag, "Flag Shopping", and Choice of Law' (1993) 17, Tulane Maritime Law Journal, 139–84 at 145.

are of reduced importance to the management. The crew and passengers of a ship will most likely have nationalities other than that of the flag State, and their presence on the vessel may be due to contractual relationships that are governed by yet another legal regime.

In considering which could be the decisive connecting factors and thus the law called to govern a given relationship, differences must be drawn between different kinds of damages, in the same way than it has been set out *supra*, Chap. 6, B. II. 1. for jurisdictional considerations. Torts involving man-made, immoveable structures at sea are very different from torts involving ships. Within the incidents involving ships, there is yet another distinction between torts that are "internal" to the ship, i.e., where the damage is only perceived among the people on board the ship, and torts that also have effects on third parties, outside the community of people on board.[17] Examples for the first category could be fatalities or injuries of passengers or crewmembers while the ship is en route. Torts that cause damages outside the ship often consist in pollution damage that then has consequences for third parties. The difference between these two is however not very easy to strike in some cases.

Thirdly, there will be some consideration given to the conflict-of-law rules that have been applied to purely environmental damage on the high seas, even if this entails the problems of jurisdiction and standing in international law that have been mentioned *supra*, Chap. 6, B. II. 1. a) aa). It will however be interesting to consider how domestic courts have solved the problem of the choice of law on the high seas when there is no law of the flag that they could refer to.

At last, there is the rather special case of collisions on the high seas. In all cases in which there is more than one ship involved, there will be, in all likelihood, more than one law of the flag called to govern the case. This of course causes a problem of choice of law that is, as will be seen, not easy to solve if regard is given to the legitimacy of the decision under the law of the sea.

I. Offshore Installations, Cables and Pipelines on the High Seas

When considering man-made structures at sea, the law of the sea distinguishes between, on the one hand, secluded installations resembling artificial islands that are actually operated in one spot, such as oil rigs, and installations that are running

[17] George, '2007 "Choice of Law"', 137; Basedow, *2012 "Die private Haftung"*, p. 1871. Also see the very detailed considerations in W. van der Velde, *De positie van het zeeschip in het Internationaal Privaatrecht* (Groningen, 2006), p. 186.

through the high seas, such as cables and pipelines,[18] and this, even where pipelines link immoveable offshore installations such as oil rigs.[19]

As of now, there are no offshore installations for the commercial exploitation of marine subsoil resources on the high seas, techniques of this kind being still in the stage of research and planning. The legal regime applicable to structures of this kind would, even under the existing law of the sea regime for the Area[20] be rather problematic to determine. It should briefly be mentioned that UNCLOS seems to presuppose that installations of this kind are registered in a given State or have another link with a State that renders them subject to the laws and regulations of that State—art. 109 para. 3 (b) UNCLOS refers to the "State of registry of the installation" when considering jurisdiction for unauthorised broadcasting on the high seas, and art. 209 UNCLOS, dealing with pollution from activities in the Area equates States' authority over "installations, structures and other devices" with the authority over vessels, "flying their flag or of their registry or operating under their authority, as the case may be".[21] In contrast to the registry of ships, that States must maintain under art. 94 para. 2 (a) UNCLOS, the Convention does not oblige States to have a registry of installations, which renders the attribution of nationality almost impossible in practice. It can however be expected that a solution to this problem will be found under the regime for the Area as soon as the construction of offshore installations on the high seas becomes reality.

In contrast to that, the importance of cables in waters beyond national jurisdiction was recognised at a much earlier point in time and the first domestic considerations on how to ensure that they would not be damaged date back to the nineteenth century.[22] With the advent of the oil industry and the growing importance of pipelines at sea, the way in which pipelines were considered resembled closely the regime for cables. The main concern at that time was the possible damage that ships could cause to cables, the typical accident being the weighing of an anchor that had become entangled in the cable.[23] Considering the involvement of a ship, subject to the laws of a flag State, incidents of this kind can however be considered as cases in which a wrongful action on a ship causes damages outside the ship. Consequentially, the existing and rather old precedents quoted have been decided according to the solution under English law.[24] This view is confirmed by

[18] S. Kaye, 'International Measures To Protect Oil Platforms, Pipelines, and Submarine Cables from Attack' (2007) 31, Tulane Maritime Law Journal, 377–423 at 379.

[19] Ibid., p. 403.

[20] H. Esmaeili, *The Legal Regime of Offshore Oil Rigs in International Law* (Aldershot, Burlington: Ashgate/Dartmouth, 2001), p. 99, with regard to criminal jurisdiction.

[21] Examples pointed out by Kaye, '2007 "International Measures"', 387 et seq.

[22] Ibid., pp. 396 et seq.

[23] See "*The Sub-Marine Telegraph Co. v. Dickson*", 143 English Reports 983 (Court of Common Pleas (England), 20 January 1864), where a submarine cable in the Channel was found to have been damaged on the high seas, the territorial waters then only stretching three miles from the shore.

[24] Also see Marsden, Gault, Hazelwood and Tettenborn, *2003 "Collisions at sea"*, para. 17-09.

art. 113 UNCLOS, obliging flag States or States otherwise assuming jurisdiction over the ship or the persons on board to render the wrongful breaking of a cable a punishable offense.

With regard to the problem of ships damaging submarine cables and pipelines that are placed in areas beyond national jurisdiction, it is however often advocated that the claims for compensation should be distinguished from the regime applicable to damages caused by ships. Considering that these devices have been built under the supervision and the authority of a given State and that their presence has been internationally announced by this State,[25] it is understandable that authors call for an application of the law of the State with which the damage has the "most real connection",[26] which would naturally be the State benefitting from and maintaining the installation.[27] However, laying cables and pipelines is one of the activities which are covered by the freedom of the high seas (art. 87 para. 1(c) UNCLOS) and which therefore can be carried out by any State.[28] The fact that a certain State fosters and supports the use of the high seas in this manner[29] should not be considered a connecting factor that triggers the application of this State's law. If considered in its most extreme extent, applying the law of the State beneficiary of a cable or pipeline to any accident with these devices would mean that a State's competence on the high seas is considerably extended by virtue of an act to which any State has the right, and that the parts of the seabed on which the cable or pipeline lies become to some extent entangled with the State that benefits from it. Furthermore, there are considerable practical difficulties with this approach, as it has become rather unlikely that a pipeline or cable is only benefitting one State. Cables and pipelines are today run by multi-national companies for the benefit of several States. In conclusion, it is the ship's flag State's law that should apply to accidents of this kind for choice-of-law purposes, rather than the law of a distant beneficiary State.

In rather the same spirit and with a view to the fact that there are international agreements on the principles applicable to collisions, it has moreover been argued that it might be desirable to extend the scope of application of these agreements and apply them, as a matter of customary law, to allisions.[30] It has also been pointed out

---

[25] S. M. Carbone, 'Conflits de lois en droit maritime' (2009) 340, Recueil des Cours, 73–269 at 206.

[26] Tetley, '1993 "The Law of the Flag"', 163.

[27] Carbone, '2009 "Conflits de lois"', 206; A. Hille, C. Schröder, M. Dettmer and M. Visser, 'Offshore-Windkraftanlagen - Haftung und Haftpflichtversicherung' (2010), Versicherungsrecht, 585–98 at 590, regarding the (still hypothetical case) of offshore wind energy devices.

[28] See R. Lagoni, 'Cable and Pipeline Surveys at Sea', in H. Hestermeyer, D. König, N. Matz-Lück, V. Röben, A. Seibert-Fohr, P.-T. Stoll and S. Vöner (eds.), *Coexistence, Cooperation and Solidarity: Liber Amicorum Rüdiger Wolfrum* (Leiden, Boston: Martinus Nijhoff Publishers, 2012), p. 952, on the problem of surveying.

[29] See R. Lagoni and A. Proelß, 'Festlandsockel und Ausschließliche Wirtschaftszone', in W. Graf Vizthum (ed.), *Handbuch des Seerechts* (München: Beck, 2006), para. 129.

[30] Tetley, '1993 "The Law of the Flag"', 166.

that the applicability of the conventions dealing with collisions of course depends on the understanding of what a "ship" is under these conventions, the term not being defined, and that certain cases, especially considering moveable oil rigs, could be covered[31] without making an analogy. Still, there are some structures such as offshore wind farms that will not be considered as a "ship" in any case, and the analogous application of the collision conventions seems hardly feasible with regard to their clearly defined scope of application as a matter of international agreement—it seems very doubtful whether there is a customary law on collision. Furthermore, the collision conventions generally presuppose that both parties involved in the collision operate ships and suffer a damage that is at least comparable in kind. It will be hard to imagine an apportionment of damages that consists on the one hand in physical damage to a ship and on the other hand in a breakdown of the power supply in a given country. Admittedly, the application of a convention *mutatis mutandis* may lead to a law that is well placed to govern a case of this kind, but if the same result cannot be obtained by the application of a conflict-of-law rule,[32] it should, for the sake of clarity and foreseeability, not be brought about by an analogy.

The problem of damages without any wrongful action of a ship must clearly be distinguished from this. For now, leaking pipelines or telecommunication cables introducing electric energy into the marine environment[33] will however in the first place cause environmental damage, which triggers the difficulties of standing and venue to claim the compensation for damages in an area beyond national sovereignty (see *infra* under IV.) As soon as there are plans for carbon dioxide pipelines, the discussion on the law applicable to claims for damages will take a very different turn, hopefully entailing an international civil liability regime for these cases.

II. Damages Outside the Ship

The well-established principle to determine the law applicable to a tort following the place in which it occurred[34] of course causes problems where the tortious action and the ensuing damage occur in different jurisdictions, or where the consequences of the tort are perceived in the territory of more than one country.[35] On shore, there is the question of what place will be decisive, but in an accident solely occurring on

---

[31] Esmaeili, *2001 "The Legal Regime"*, pp. 102 et seq.; Hille, Schröder, Dettmer and Visser, '2010 "Offshore-Windkraftanlagen"', 586 et seq., discussing the case of floating wind turbines.

[32] Solution favoured by Basedow, *2012 "Die private Haftung"*, p. 1876.

[33] Lagoni, *2012 "Cable and Pipeline Surveys"*, p. 947.

[34] Hartenstein, '2008 "Rom I-Entwurf und Rom II-Verordnung"', 151.

[35] F. K. Juenger, 'Mass Disasters and the Conflict of Laws' (1989), University of Illinois Law Review, 105–28 at 110.

the high seas, there is not even a place subject to sovereign jurisdiction,[36] and it is not clear which law would be called to govern.[37]

It should be kept in mind that the answer to the question might be different, depending on whether the tortious act and the consequences occur on one ship—e.g., the injury of a passenger or crew member—or whether the tortious behaviour has an impact on parties outside the ship community. The classical example for the latter case would be pollution damage on the high seas, caused by one vessel, that results in damage to a third person's assets, such as the damage of fishing equipment or another ship's hull or machinery by oil, or injury caused by the presence of chemical pollution on the high seas. Cases of this kind presuppose that assets of third parties have been brought to the high seas and sustained damage there. In former times, the importance of cases of this kind had been considered as extremely reduced, to the point that the delegates, during the negotiation of the CLC excluded compensation for oil pollution damage on the high seas from the field of applicability of the Convention (see *supra*, Chap. 3, A. II.). With the advent of new technologies aiming to use the oceans for carbon capture and storage (CCS), the problem of damage caused to third parties on the high seas may become more pressing. Most CCS techniques require that large amounts of carbon dioxide are brought to the open ocean. The gas is not toxic, but if large amounts are released, they will reduce the level of oxygen in the air and almost invariably cause humans and animals to suffocate. Accidents in the operation of CCS techniques could therefore bring about cases in which there would be a very pressing need to determine the law that is applicable to damages caused outside of ships on the high seas.

1. Solutions in English Law

England is one of the jurisdictions where the problem of torts involving a foreign element has been considered at an early point in time, and this also in a maritime context. For torts having taken place outside of English territory, the so-called rule of *Phillips v. Eyre*[38] applied, also known as the double actionability rule. The behaviour complained about needed to give rise to a tort claim under English law as the *lex fori*. In addition to that, under the original double actionability rule, there needed to be liability, but not necessarily in tort, under the law of the place where the tort had taken place.[39] The rule was generally accepted, but there are exceptions

---

[36] A. Dickinson, *The Rome II Regulation: The Law Applicable To Non-Contractual Obligations* (Oxford: Oxford University Press, 2008), para. 3.311.

[37] M. Anderson, 'Transnational Corporations and Environmental Damage: Is Tort Law the Answer?' (2002) 41, Washburn Law Journal, 399–425 at 416.

[38] "*Phillips v. Eyre*", 6 Law Reports, Queen's Bench Division 1 (Court of Exchequer Chamber (England), 23 July 1870).

[39] A. Briggs, *The conflict of laws*, 2nd edn. (Oxford, New York: Oxford University Press, 2008), p. 189.

to it, themselves not too clear in character and scope,[40] and this, especially in the field of maritime and aerial torts.[41]

When it came to the question of the law applicable to torts on ships, English law distinguished between torts within one and the same vessel, and collision cases, where two vessels, possibly flying different flags, would be involved.[42] Considering torts committed on board of a ship and not involving other vessels, thus being "internal" to the ship and its crew and passengers, regard would be given to the law of the ship's flag in the first place. The place where the tort occurred would, in most cases, taken to be the land of the ship's flag or registry.[43] Thus, the double actionability rule could be applied both to torts in the waters of a foreign State[44] and to torts taking place on the high seas, where there is, at first sight, no territorial jurisdiction offering a *lex loci delicti*.

The rule in *Phillips v. Eyre* was statutorily abolished[45] by the Private International Law (Miscellaneous Provisions) Act 1995, s. 10 (a), and according to s. 11 (1), the *lex loci delicti* alone is applicable.[46] With regard to torts occurring in one jurisdiction but causing damage in another, s. 11 (1) names the place where the events constituting the tort occurred (*lex loci delicti commissi*) as the decisive criterion. With regard to torts that take place on one ship only and where the consequences are confined to the ship, its crew and passengers, the statutory changes therefore enhanced the importance of the law of the ship's flag as the *lex loci delicti*.[47]

However, under English law, it has traditionally been held that the high seas were not devoid of jurisdiction, but that the law applicable to this area was the "general maritime law as administered in England".[48] The concept of this "general maritime law" first emerged in the nineteenth century in connection with maritime

[40] See the comprehensive discussion in The Law Commission, The Scottish Law Commission, *Working Paper No. 87* (London: Her Majesty's Stationery Office, 1984), para. 2.23 et seqq.

[41] C. G. J. Morse, 'Torts in Private International Law: A New Statutory Framework' (1996) 45, International and Comparative Law Quarterly, 888–902 at 889, footnote 16.

[42] See George, '2007 "Choice of Law"', for a detailed outline.

[43] *1984 "Working Paper No. 87"*, para. 2.110.

[44] Ibid., para. 2.112, but expressing uncertainty as to whether the *lex loci delicti* should be the law of the ship's flag or the law of the State in whose waters the ship was situated.

[45] See Morse, '1996 "Torts in Private International Law"' for an overview of the changes.

[46] It must be stressed that the 1995 Law Reform only abolishes "ordinary" common law conflicts rules in torts, not "special" tort choice of law rules (s. 14 (2) of the 1995 Act). The realm of the respective categories is however not easily determined, see Fawcett, Carruthers and North, *2008 "Private International Law"*, pp. 860 et seq.; Dicey, Morris and Collins, *2012 "Conflict of Laws"*, para. 35-017.

[47] *Private International Law: Choice of Law in Tort and Delict*, No. 193 (London: Her Majesty's Stationery office, 1990), para. 3.27; George, '2007 "Choice of Law"', 146.

[48] Fawcett, Carruthers and North, *2008 "Private International Law", p. 862.*

collisions[49] on the high seas (also see *infra* under V.). It seems that in the beginning, the provisions applicable were those as to the apportionment of fault rather than those on compensation and liability.[50] Out of the thought that the traffic rules that decided which ship was to blame for the collision were generally the same all over the world, English law could be applied as the "general maritime law".[51]

When the application of English law as the "general maritime law" began to encompass not only the generally accepted traffic rules and the apportionment of fault, but also the provisions on liability, it was then rightly considered[52] that it amounted in reality to "being in truth nothing more than English law".[53] It is interesting that the English courts thus effectively apply the *lex fori* to torts taking place on the high seas, but do not term it accordingly. The wording chosen by the courts seems to imply that English law in this matter is applied not in its capacity as the law of the court, but as the law universally applicable to the high seas,[54] which is presumably to be explained by the way in which this set of rules developed.

Even today, it is assumed in English doctrine[55] that the high seas are governed by the "general maritime law" and that the law applicable to damages on the high seas therefore is the maritime law, which in the English courts, happens to be the *lex fori*. The law of the ship's flag only applies if the causes of the tort are "wholly internal" to one ship.[56]

The case of harmful substances lost by a vessel on the high seas and causing harm outside the ship is not fitted into this category without difficulties—even though only one vessel is involved in the case, the actual harm takes place outside the ship. The same situation occurs in cases of people falling overboard and drowning on the high seas. It has been argued that in these accidents, the actual damage takes place on the high seas—to which, according to English law, the "general maritime law" applies.[57] In former times, it could easily be argued that the place where the damage occurred, i.e. where the victim died, should not be

---

[49] D. Winter, 'Maritime Torts: The Choice-of-Law Principles' (1954) 3, International and Comparative Law Quarterly, 115–25 at 121.

[50] D. P. O'Connell, 'The English Choice of Law Rule in Collision Suits and Shipboard Torts', in H. P. Ipsen and K.-H. Necker (eds.), *Recht über See*: *Festschrift Rolf Stödter zum 70. Geburtstag* (Hamburg, Heidelberg: R. v. Decker's Verlag, 1979), pp. 102 et seq.

[51] Ibid., p. 102.

[52] As cited by *1984 "Working Paper No. 87*", para. 2.107. Also see Marsden, Gault, Hazelwood and Tettenborn, *2003 "Collisions at sea*", pp. 17-03 on the concept of maritime law as a part of English law rather than a different set of rules.

[53] "*Lloyd v. Guibert*", 1 Law Reports, Queen's Bench Division 115 (Court of Exchequer Chamber (England), 27 November 1865), p. 123, by Willes, J.

[54] But see the criticism in Winter, '1954 "Maritime Torts"', 121: "the courts [. . .] have applied the *lex fori*, adopting the euphemism of our own court by saying that they are applying the general maritime law."

[55] George, '2007 "Choice of Law"', 148.

[56] *1984 "Working Paper No. 87*", para. 2.110.

[57] George, '2007 "Choice of Law"', 148, but terming this "arguably unwelcome".

confused with the place where the tort was committed, i.e. the ship where the victim was caused to fall overboard.[58] Since the introduction of the 1995 Private International Law (Miscellaneous Provisions) Act, s. 11 (2) (a) now provides that for causes of action in respect of personal injury or death the applicable law should be taken to be the law of the country where the injured individual was when sustaining the injury. Hence the argumentation that the damage was sustained on the high seas is justified and not easily argued against. The same holds true for damage to property by pollution on the high seas, as s. 11 (2) (b) of the 1995 Act calls for the application of the law of the country where the property was when it was damaged. It is therefore not the *lex loci delicti commissi* that is applied to these torts where the action and the cause take place in different jurisdictions, but the law of the place where the damage was sustained (*lex loci damni*). Under the English common law doctrine applying English law in the form of the "general maritime law" to the high seas, which was not changed by the 1995 Act, it is therefore argued by certain authors that it is still English law that applies to these damages.[59]

Still, s. 12 of the 1995 Act allows a departure from the *lex loci damni* if another law substantially more appropriate for application can be found, namely with regard to the events which constitute the tort or the circumstances or consequences of the events. It does seem that pollution on the high seas giving rise to a claim for damages may be a good example for a justified departure from the given rules, in order to make the general rules of the 1995 Act applicable to these torts as well.[60] Without taking recourse to the displacement of the general rule under s. 12 of the 1995 Act or giving the wording a substantially different interpretation, the cases of drowning or damage of property on the high seas will, at least in the English courts, have very uncertain outcomes for the parties.

The Rome II Regulation, which is in force in the EU member States since early 2009, does not have unambiguous provisions on the law applicable to torts on the high seas. The realm and the consequences of the Rome II Regulation will be given a closer look *infra* under 4. Considering that the Regulation, in its art. 4 (1), also stresses "the law of the country in which the damage occurred irrespective of the country in which the event giving rise to the damage occurred", the problems under the Regulation are essentially the same as under the 1995 English legislation. Hence, the English solution under common law for coping with the problem of damages on the high seas can thus still be held to be good law, even under the Rome II Regulation.[61]

[58] Winter, '1954 "Maritime Torts"', 119 et seq.; M. Hancock, *Torts in the conflict of laws* (Ann Arbor, Chicago: University of Michigan Press, 1942), p. 272, describing the differentiation between the places of the wrong as "somewhat whimsical".

[59] Fawcett, Carruthers and North, *2008 "Private International Law"*, p. 863; Carbone, '2009 "Conflits de lois"', 207, but also setting out possible exceptions.

[60] George, '2007 "Choice of Law"', 148.

[61] Dicey, Morris and Collins, *2012 "Conflict of Laws"*, para. 35-033.

2. Solutions in French Law

French private international law has dealt with the problem of torts on board ships on the high seas but has come to much more pragmatic and, arguably, less elaborate solutions. The law of the ship's flag is applied in a rather wide extent to many legal relationships involving vessels,[62] and there is no obvious authority who would submit torts involving a single ship on the high seas to any other law than the law of the flag.

This emphasis on the law of the flag may partly be due to the fact that nationality—both of individuals and of corporeal objects—are seen as an important starting point of private international law in France, which centred from its very beginning on the "status" of things rather than on judicial relationships.[63]

The law of the flag has thus first and foremost been established as the *lex in rem* over the ship, i.e. the law by which questions such as ownership, mortgages and other rights *in rem* were governed, and which was considered as the "natural" solution, taking into account that ships were constantly moving and that the law of the place where the ship was situated (the *lex rei sitae* normally applicable to corporeal goods) was thus not of avail.[64] In this context, the role of the law of the flag was already well established before the problem of the law applicable to torts on the high seas emerged, and it was then concluded that the vessel's "nationality" allowed a determination of the law applicable to events taking place on the high seas—and to a certain extent even in the waters of a foreign State.[65]

It should be kept in mind that the solution will be different for collisions on the high seas where there are two flag States and hence two possible laws.[66]

It is interesting that French authorities claim that the *lex loci delicti* on the high seas is "replaced" by the law of the ship's flag,[67] they do not consider—at least not any longer[68]—that the ship as such is the place where the tort was committed or that the ship could thus be seen as being part of the territory of its flag State.[69] It is fairly clear that under French law, the law of the ship's flag is applied to torts on the high seas out of the consideration that there is simply no other law called to apply.[70]

---

[62] Tetley, '1993 "The Law of the Flag"', 155 et seq.

[63] See P. Mayer and V. Heuzé, *Droit international privé*, 10th edn. (Paris: Montchrestien, 2010), pp. 43 et seq. for a historic overview.

[64] H. Batiffol, *Droit international privé*, 4th edn. (Paris: Librairie Générale de Droit, 1967), p. 546; M. Rémond-Gouilloud, *Droit maritime*, 2nd edn. (Paris: A. Pedone, 1993), para. 89.

[65] E. du Pontavice, *Le statut des navires* (Paris: Librairies Techniques, 1976), p. 41.

[66] Rémond-Gouilloud, *1993 "Droit maritime"*, para. 100; Tetley, '1993 "The Law of the Flag"', 156; Mayer and Heuzé, *2010 "Droit international privé"*, para. 684.

[67] Ibid., para. 684.

[68] See Rémond-Gouilloud, *1993 "Droit maritime"*, para. 80 on earlier authority.

[69] Ibid., para. 80.

[70] Ibid., para. 98; Mayer and Heuzé, *2010 "Droit international privé"*, para. 684.

The solution is considered to be consistent with the Rome II Regulation.[71]

3. Solutions in German Law

German private international law features an elaborate codification of the applicable provisions in the conflict of laws, clearly determining connecting factors in a number of sets of facts. Still, the rules in the *Einführungsgesetz zum Bürgerlichen Gesetzbuch* (EGBGB) do not address the question of the law applicable to the high seas or to other areas that are not subject to a State's sovereignty, such as outer space or the Antarctic.[72] Almost all authors writing on German law claim that in the cases where the conflicts of laws rules point to a *territorium nullius*, this should be disregarded and the applicable law should be determined according to an alternative connecting factor.[73]

For torts taking place on ships on the high seas, authors are almost unanimous that the alternative connecting factor should be the ship's flag or registry, if the ship has a valid right to fly it.[74] There are some inconsistencies with regard to the role of the law of the home port,[75] but in any case the connecting factor is determined by the administrative registry of the ship.

It does not seem that the question has been given much attention under German private international law at all.[76]

4. Solutions Under the Rome II Regulation

Within the European Union, the conflict-of-law rules applicable to international torts have been harmonised with the introduction of the Rome II Regulation in early 2009. According to art. 4 (1) of the Rome II Regulation, the law applicable to torts

---

[71] Ibid., para. 684.

[72] G. Kegel and K. Schurig, *Internationales Privatrecht*, 9th edn. (München: Beck, 2004), p. 18; J. Kropholler, *Internationales Privatrecht*, 6th edn. (Tübingen: Mohr Siebeck, 2006), p. 537.

[73] Kegel and Schurig, *2004 "Internationales Privatrecht"*, p. 18.

[74] Ibid., p. 18; Kropholler, *2006 "Internationales Privatrecht"*, p. 537.

[75] B. von Hoffmann, K. Thorn and K. Firsching, *Internationales Privatrecht*, 10th edn. (München: Beck, 2012), p. 493, also see Ramming, '2010 "Internationalprivatrechtliche Fragen"', 296 et seq.

[76] However, see the very elaborate study in E. Lorenz, 'Das anwendbare Recht bei Schiffs- und Flugzeugunfällen', in E. von Caemmerer (ed.), *Vorschläge und Gutachten zur Reform des deutschen internationalen Privatrechts der außervertraglichen Schuldverhältnisse* (Tübingen: Mohr, 1983), pp. 450 et seqq., suggesting that the law applicable to torts in areas without sovereignty should be either the law of the ship acting tortiously or the law of the ship or vehicle having suffered damage, depending on whether the wrongdoer could foresee the consequences. Still, it does not seem that the theory has ever been considered in practice or echoed by other authors, see H. Prüssmann and D. Rabe, *Seehandelsrecht: Fünftes Buch des Handelsgesetzbuches*, 4th edn. (München: C. H. Beck, 2000), p. 977.

is the law of the State in which the damage is sustained, irrespective of where the consequences take place. There is an exception for cases in which the alleged wrongdoer and the alleged victim have their habitual residence in the same country and where it would presumably simply be easier to apply the law of that country, and art. 4 (3) sets out a displacement of the general rule for cases in which the tort is "manifestly more closely connected" with a different country. Under certain circumstances, the parties are allowed to submit their case to the law of their choice. In claims for environmental damage, according to art. 7 of the Rome II Regulation, the law applicable will also be determined according to these rules, but the person seeking compensation has the choice to base the claim on the law of the country in which the event giving rise to the damage occurred. According to art. 17 of the Regulation, in assessing the behaviour of the person allegedly liable, regard shall be given to the rules of safety and conduct that are in force at the place and at the time where the event giving rise to the damage took place. The provision can be seen as a corrective, considering that the Rome II Regulation, in stressing the importance of the place where the damage occurred, has a tendency to disregard the law of the place where the wrongful action was committed if the places do not coincide.[77]

For pollution damage on the high seas, both art. 4 (1) and art. 7 of the Regulation point to a "non-jurisdiction". The Regulation does not, in its general rules, provide for an alternative connecting factor, which is not based on territoriality, in connection with maritime torts. In this context, it is interesting to briefly consider how the case of torts involving ships was treated during the drafting of the Rome II Regulation. The earliest proposals which were produced in the written consultation procedure in May 2002 still contained an express conflicts rule, mostly relying on the law of the flag as a connecting factor.[78] Following feedback that it received during the written consultation procedure, the Commission however considered that the possible cases to be covered by the rule were too manifold, that the rule would be inoperative in collision cases where there were two flag States, and that there was a danger that it would enable parties to submit torts to the law of a flag of convenience State.[79] It therefore deleted this rule in its proposal of 2003, and replaced it with a provision merely setting out in the then art. 18 (2) that for the purposes of the Regulation, a ship on the high seas was to be treated as territory of the flag State. It can thus be inferred that the Commission intended maritime torts to be covered by the general rules of the Regulation. The provision survived the following resolution by the European Parliament, but was, without express reason

[77] A. Dickinson, 'Territory in the Rome I and Rome II Regulations' (2013), Lloyd's Maritime and Commercial Law Quarterly, 86–132 at 113.

[78] Dickinson, *2008 "The Rome II Regulation"*, para. 3.303 et seq.; Fawcett, Carruthers and North, *2008 "Private International Law"*, p. 859; R. Plender and M. Wilderspin, *The European Private International Law of Obligations*, 3rd edn. (London: Sweet & Maxwell, 2009), para. 18-041.

[79] Proposal for a Regulation of the EU Parliament and the Council on the law applicable to non-contractual obligations ("Rome II"), COM (2003) 427 final, Explanatory Memorandum, p. 27.

or explanation,[80] deleted in the subsequent Commission proposal and in any further documentation, presumably because the subject matter was perceived to be rather public than private law.[81] The Rome II Regulation thus omits any reference to maritime torts and the law that applies if the general rules point to a *territorium nullius*. Following this omission, it has actually been questioned whether the EU legislator still wanted maritime torts to be covered by the Regulation, rather than by the national laws in question that retained alternative connecting factors.[82] With a view to the purpose of the Regulation, to provide for a uniform regime for conflict-of-law rules for torts throughout Europe, and the consideration of maritime torts in the earlier proposals, this view is hardly to be endorsed.[83] The fact that the connecting factors in the general rules are of no avail in areas beyond national jurisdiction does not mean that there is no possibility to determine an applicable law according to the other rules.

It is true that the wording of art. 4 (3) of the Regulation, providing for the possibility to set aside the initial choice of law in favour of a "more closely connected" law, seems to require that there is an applicable law under art. 4 (1) and that it is simply not as closely connected with the case.[84] In a pragmatic spirit, it is however often argued that the law applicable to torts on the high seas should be determined according to art. 4 (3) of the Regulation.[85] Still, it is then unclear what the "more closely connected law" should be, the problems arising being essentially the same as under the escape clause of the English 1995 Private International Law (Miscellaneous Provisions) Act.[86]

A "more closely connected law" could either be the law of the ship's flag state or, with regard to the problem of flags of convenience, the law of the owner's nationality or domicile, but which may be hard to determine when there is more

[80] George, '2007 "Choice of Law"', 170; Plender and Wilderspin, *2009 "The European Private International Law"*, para. 18-041.

[81] Dickinson, *2008 "The Rome II Regulation"*, para. 3.305.

[82] Fawcett, Carruthers and North, *2008 "Private International Law"*, p. 859. The question is also raised by George, '2007 "Choice of Law"', 170.

[83] Dickinson, *2008 "The Rome II Regulation"*, para. 3.313; Plender and Wilderspin, *2009 "The European Private International Law"*, para. 18-042; Dicey, Morris and Collins, *2012 "Conflict of Laws"*, para. 35-033.

[84] van der Velde, *2006 "De positie van het zeeschip"*, p. 317; Fawcett, Carruthers and North, *2008 "Private International Law"*, p. 860; Tsimplis, '2010 "Law and jurisdiction"', 311; S. Vrellis, 'The Law Applicable to the Environmental Damage: Some Remarks on Rome II Regulation', in J. Forner Delaygua, C. González Beilfuss and R. Viñas Farré (eds.), *Entre Bruselas y La Haya: Estudios sobre la unificación internacional y regional del Derecho internacional privado* (Madrid, Barcelona: Marcial Pons, 2013), p. 871.

[85] Huber P. (ed.), *Rome II Regulation: Pocket commentary* (Munich: Sellier European Law Publishers, 2011), p. 91; Dickinson, '2013 "Territory"', 127.

[86] R. J. Weintraub, 'Rome II and the Tension Between Predictability and Flexibility', in H.-E. Rasmussen-Bonne, R. Freer, W. Lüke and W. Weitnauer (eds.), *Balancing of Interests: Liber Amicorum Peter Hay zum 70. Geburtstag* (Frankfurt am Main: Verlag Recht und Wirtschaft, 2005), p. 459.

than one owner or where ownership changes.[87] Another possibility could be the personal law of the polluter, which would however be extremely difficult to determine given the hierarchical structure of command among ships' multinational crews. A more practical consideration refers to the fact that the community of people on board a ship will most likely have contractual relationships and that any torts taking place in connection with these relationships should also be governed by the law governing the underlying contracts.[88] Whereas this is a valid thought in connection with torts occurring among the crew and where there may be a link to the respective seamen's pay contracts or passenger contracts, there is hardly a contractual relationship in the event of accidental loss of polluting substances on the high seas. It does not seem that there is a law that is at first sight better placed to govern torts of this kind than the law of the flag State.

The arguments that have been put forward under English law to favour the law of the flag as the *lex loci delicti* before the coming into force of the 1995 Act also hold true here: the law of the flag is not susceptible to change without documentation and is easy to determine, it is independent from the ship's geographical position and will normally not change throughout a journey,[89] and not least, it will be the only evident jurisdiction having an interest in the ship while it is situated on the high seas,[90] and, thus, the only obvious connecting point keeping the ship from becoming "subject to no law at all".[91]

For environmental damage, the victim could choose to base the claim on the law of the country in which the event giving rise to the damage occurred (art. 7 of the Rome II Regulation), which also could be considered to point to the polluting ship's flag state.

Still, the relationship between the law applicable according to art. 4 (3) and art. 7 of the Regulation is unclear. Art. 7 only refers to art. 4 (1) and as it has been argued that art. 7 should only serve as a territorial rule replacing another territorial rule[92] and not replace the laws determined by art. 4 (2) or (3) of the Regulation.[93]

---

[87] See the early criticism in Winter, '1954 "Maritime Torts"', 120.

[88] Basedow, *2012 "Die private Haftung"*, p. 1871.

[89] Ibid., p. 1888, but favouring the "law of the ship's nationality" over the "law of the flag" in order to cover cases of flags of convenience and open registers.

[90] Hancock, *1942 "Torts"*, p. 262; Tetley, '1993 "The Law of the Flag"', 161; Dickinson, '2013 "Territory"', 127.

[91] "*R v. Keyn*", 2 Law Reports Exchequer Division 63 (Court for Crown Cases Reserved, Exchequer Division, 13 November 1876), p. 94, by Lindley J.

[92] B. Dutoit, 'Le droit international privé des obligations non contractuelles à l'heure européenne: Le règlement Rome II', in G. Venturini and S. Bariatti (eds.), *Liber Fausto Pocar* (Milano: Giuffrè, 2009), p. 320; M. Bogdan, 'Some Reflections regarding Environmental Damage and the Rome II Regulation', in G. Venturini and S. Bariatti (eds.), *Liber Fausto Pocar* (Milano: Giuffrè, 2009), p. 100.

[93] E. Guinchard and S. Lamont-Black, 'Environmental Law - the Black Sheep in Rome II's Drive for Legal Certainty?: Article 7 of Regulation (EC) No. 864/2007 on the Law Applicable to Non-Contractual Obligations in Context' (2009) 11, Environmental Law Review, 161–72 at 168; Ramming, '2010 "Internationalprivatrechtliche Fragen"', 293.

Still, it seems hardly acceptable to rule out the applicability of art. 7 just because the general rule in art. 4 (1) of the Regulation points to a *territorium nullius*.[94] The Rome II Regulation is, after all, meant to be of universal applicability to torts occurring on the territory of member States[95] and should promote uniformity of law. It does seem from the wording of art. 1 (2) of the Regulation that the list of non-contractual obligations excluded from its scope of applicability is exhaustive and that the regulation applies to all other cases.

It must therefore be concluded that the Rome II Regulation can thus be considered to offer rules on the law applicable to pollution cases on the high seas, even if it does not consider the case explicitly. However, the rules offered by the Rome II Regulation do not unambiguously point to a given jurisdiction, and even if the law of the flag—or the ship's nationality for borderline cases—may have a strong standing in general, this result is by no means obvious, and could be circumvented by circumstances establishing a closer link with a different jurisdiction.

5. Solutions in US Law

The United States are effectively the only State to depart very extensively from the rule of the *lex loci delicti* being applicable to torts, and rather determine the law applicable with regard to the relationship of the parties.[96] This development, which replaces principles of territoriality with principles of personality,[97] began at a rather early point in time for land-based torts.[98] An analysis of the state of the law in the United States is difficult, given the fact that each State has its own common law. Maritime cases are however in most cases governed by federal law,[99] in the form of Acts or by the federal general maritime law. In fact, the general maritime law is the only federal common law. Questions of the interplay of federal and State law are thus left aside here.

In admiralty cases, legal opinion on the state of the law declared in 1934 that the law of the flag was applicable to liability for an alleged tort in the navigation of a vessel on the high seas outside the territorial waters of any state.[100] This assertion was however not repeated in later statements, and the 1986 Third Restatement of

---

[94] Bogdan, *2009 "Some Reflections"*, p. 105, favours the application of the law of the flag under art. 7 and only resorts to a different law determined according to art. 4 (3) of the Regulation as an *ultima ratio*.

[95] Recital (14) of the Rome II Regulation, (EC) No. 864/2007.

[96] See Symeonides, *2005 "Accommodative Unilateralism"* for a critical assessment of the two approaches.

[97] See Symeonides, *2004 "Territoriality and Personality"*.

[98] Ibid., p. 404, detecting the first occurrences in 1953.

[99] E. F. Scoles, P. Hay, P. J. Borchers and S. C. Symeonides, *Conflict of Laws*, 4th edn. (St. Paul: Thomson/West, 2004), p. 890.

[100] *Restatement of the Law of Conflict of Laws* (St. Paul: American Law Institute Publishers, 1934), rule 408.

Foreign Relations Law only alluded to States' competence over the ship as expressed by the law of the sea.[101]

Most precedents dealing with the conflict-of-law rules for torts occurring on ships have been triggered by the applicants' wish to have the case submitted to American law, in most cases because they wanted to avail themselves of the rather generous federal legislation on damages for injured seamen, the so-called Jones Act. The cases that are covered by this legislation almost invariably concern torts that are confined to one ship and are therefore dealt with in the respective point of this study (see *infra* under III.). It should just briefly be mentioned that in these cases of maritime torts, the courts in the United States depart very openly from the *lex loci delicti* rule and replace it with the law of the "most significant relationship", which is to be determined following a consideration of a number of possible connecting factors.

There is very little precedent for the conflicts rule that applies to maritime torts that cause damages outside a ship. If only for the sake of citing a very famous case, the decision of the United States Supreme Court following the sinking of the *Titanic*[102] should not go unmentioned. The *Titanic* had struck an iceberg on the high seas, and the United States Supreme Court was adamant in its assertion that the basis for obtaining compensation for the tragedy was to be found under British law, the *Titanic* having flown the British flag. According to the court, "the act of Congress does not control or profess to control the conduct of a British ship on the high seas".[103] It was only for the matter of the limitation of liability, qualified as a procedural matter by the court, that American law applied. The case was however treated by the court as a tort involving only one ship,[104] and the difficulty of the circumstance that the *Titanic's* passengers had drowned on the high seas rather than on the ship was not evoked. The same result was arrived at in *O'Neill v. Cunard White Star*,[105] concerning the case of a seaman washed overboard a British-flagged ship. The court was here assertive of the fact that if the claim was brought in tort, it would be governed by British law, as the law applicable to "acts done upon the high seas" was the law of the flag.[106]

A more recent and important decision in this matter, in the *Amoco Cadiz* case, has already been dealt with at an earlier point in this study (see *supra*, Chap. 3, A. I.

---

[101] *Restatement of the Law (Third) - The Foreign Relations Law of the United States* (St. Paul: American Law Institute Publishers, 1986), § 402, comment (h).

[102] "*Oceanic Steam Navigation Co. Ltd. v. Mellor* (*The Titanic*)", 1998 American Maritime Cases 2699 (US Supreme Court, 25 May 1914).

[103] Ibid., p. 2701.

[104] The court effectively drew a comparison to a collision between two ships flying the same flag that would be governed by the law of the common flag, and pointed out that "if the foreign law would govern in that case it must govern in this", ibid., p. 2701.

[105] "*O'Neill v. Cunard White Star*", 160 Federal Reporter 2nd Series 446 (US Circuit Court of Appeals, Second Circuit, 05 March 1947).

[106] Ibid., p. 448.

1. a)). In that decision, the court paid "lip-service"[107] to the *lex loci damni* rule by maintaining that French law should be applicable to a claim in tort following a dramatic pollution of the French coastline. It is however not clear to which extent the court considered the geographical position of the *Amoco Cadiz* during the incident. The ship was originally outside French waters when it experienced first difficulties, but quickly drifted into French territorial waters so that French law could have applied following both the *lex loci delicti commissi* and the *lex loci damni* rules. It has in effect been submitted that the court could also have come to the application of French law by applying the law of the closest relationship, even if the wording of the respective passage can be more easily interpreted as a *lex loci damni* approach.[108] The application of Liberian law, as the law of the *Amoco Cadiz's* flag State, was obviously never envisaged.

The decision in the *Prestige* case, in which the court designated the law applicable to the oil spill caused by the tanker following the principle of the closest relationship, is only of limited help in this context. In the *Prestige* case, the tort under scrutiny consisted in an earlier wrongful classification of the vessel while in harbour, not in a nautical fault or a tort that would have taken place on the high seas. The question of the law applicable to torts of this kind is dealt with *infra* under D., but it must however be kept in mind that the case presents one more example of a "closest relationship" approach in a maritime tort case where geographical connecting factors somehow point to arbitrary results.

It is thus rather hard to suggest which law would be applicable to a tort on the high seas under US law. As far as it can be analysed here, it seems that the American courts are more likely to determine the law applicable following an analysis of the facts of the case in order to establish the law that has the closest relationship with the case. This may be the law of the flag in cases where this law effectively governs the operation and maintenance of the ship, such as in the *Titanic* case, but it seems more probable that US court will come to a different result if the vessel flies a flag of convenience, or if the case has stronger links with a different legal system.

6. Conclusion: In Favour of the Law of the Flag State

Under the assumption that a tort causing damage outside the ship on the high seas must be justiciable onshore,[109] there must be a law designated to govern it.[110] If one rules out the connecting factors that can only be established after some research into the facts of the event, such as the polluter's personal law or the shipowner's

[107] Kloss, '1989 "Der Untergang der "Amoco Cadiz""', 185.

[108] Tomljenović, '1999 "Maritime Torts"', 289.

[109] Hancock, *1942 "Torts"*, p. 272; Finlayson, '1986 "Shipboard torts"', 143.

[110] P. Bonassies, 'La loi du pavillon et les conflits de droit maritime' (1969) 128, Recueil des Cours, 511–628 at 516.

domicile, and the rather lopsided solution of the English "general maritime law", there will only be two possible and evident options: the law of the ship's flag or the *lex fori* of the competent court. This conundrum was recognised at an early point in time,[111] and the answer is neither easier nor more satisfying today. Furthermore, statutory instruments on conflict-of-law rules such as the Rome II Regulation or even the 1995 Private International Law (Miscellaneous Provisions) Act in the United Kingdom[112] which are of universal applicability, but do not set out specific rules for maritime torts, do not make it easier to find a consistent solution.

Very generally, courts may be tempted to apply the *lex fori*, perhaps already out of the dogmatic consideration that there is no conflict of laws as there is no other sovereign State involved that may demand the application of its national rules to the given set of facts.[113] On the other hand, and apart from the doubts as to the logic[114] and the legitimacy of this outcome under public international law (see *supra* in Chap. 6), this implies a very unwelcome occasion for the claimants to choose the applicable law along with the competent jurisdiction and thus indirectly influence the outcome. It also completely disregards the fact that the events took place on a ship—possibly even flying a foreign flag—and on the high seas,[115] probably involving many other international factors, and thus bars the application of conflict-of-law rules to a set of facts where it seems most justified.

The application of the law of the ship's flag offers the advantage of being recognised in a considerable number of legal systems and being an obvious connecting factor,[116] which is easily identified and will not change without documentation.[117] Without embarking on the question whether the law of the flag applies by virtue of a territorial link—i.e. if the ship can be equalled to the flag States' land territory for the matter of conflicts rules—or rather by virtue of a personal link, the flag can always be considered as an alternative connecting factor[118] for maritime torts. It can be reconciled—albeit with minor difficulties—to the existing conflict-of-law rules, namely with the Rome II Regulation. Bearing in mind that the persons present on the ship will most likely be in a contractual relationship with the shipowner or carrier, on which the law of the ship's flag may have an influence, and are at least aware of the flag, it seems fair towards them to

[111] Hancock, *1942 "Torts"*, p. 272; Winter, '1954 "Maritime Torts"', 120.

[112] See George, '2007 "Choice of Law"', 148 on the difficulties in reconciling the previous solution to the problem of drowning on the high seas with the provisions of the 1995 Act.

[113] Basedow, *2012 "Die private Haftung"*, p. 1893, referring to the problem of collisions on the high seas.

[114] See Finlayson, '1986 "Shipboard torts"', 143, describing the outcome as "quite strange".

[115] Winter, '1954 "Maritime Torts"', 120.

[116] Bonassies, '1969 "La loi du pavillon"', 547; Ramming, '2010 "Internationalprivatrechtliche Fragen"', 295.

[117] Carbone, '2009 "Conflits de lois"', 201, promoting this at least for the cases where the law is to be determined according to the *lex loci delicti commissi*.

[118] Brækhus, '1979 "Choice of Law Problems"', 276.

also submit any possible claims in tort to the law of the flag.[119] The solution also does away with the problem that a certain behaviour may not be regarded as wrongful according to the laws of the place where it is committed, whereas it may represent a tort according to the laws of the place where the damage is sustained. This consideration has, under the Rome II Regulation, been addressed by art. 17, advocating that in assessing a person's behaviour, "account shall be taken, as a matter of fact and so far as is appropriate" of the rules of safety and conduct of the place where the action was committed. The provision is however not very clear in its scope. In maritime torts, the problem is arguably significantly more important than in land-based torts, because the jurisdiction affected by his deeds will be hard for a wrongdoer on a ship to foresee, and in the absence of international regulation, there is hardly any other legal system to assess his behaviour than the law of the flag. It thus seems desirable to assess maritime torts to a wide degree according to the law of the flag, rather than to introduce the flag State's rules on conduct and safety, for European cases, through the uncertain provision of art. 17 of the Rome II Regulation.

Moreover, the application of the law of the flag also seems justified by a closer consideration of the cases where passengers or crewmembers fall off the ship and drown on the high seas, especially if compared with cases of total loss of the ship such as the *Titanic* tragedy. In cases of actual shipwreck, as opposed to the cases in which persons are lost at sea, the question of whether the passenger died on the ship when the vessel was still afloat, or later at sea, may become very blurred and even impossible to inquire into when the circumstances of death are uncertain. It is however rather well accepted that high seas torts entirely taking place on one ship and which do not bear consequences outside the vessel are submitted to the law of the flag—the limits and exceptions of this principle will be shown *infra* under III. It should hardly make a difference whether a person on board a sinking ship drowned "on" the ship, for instance in a flooded compartment of a vessel that otherwise remained afloat, or whether he drowned on the high seas, when the ship was already lost—both fatalities should be submitted to the same law. Furthermore, it would be very hard to determine the moment when the ship ceased to be a ship during the time of the sinking and thus the moment when all further damages must be considered to have happened on the high seas. It thus seems desirable to submit both cases to the same law, which is another argument for the uniform application of the law of the flag to damages that take place on the high seas, but originate on a ship.

Admittedly, there is the problem of flags of convenience and the lack of an actual relationship between the tort and the place where the ship may be registered.[120]

---

[119] Carbone, '2009 "Conflits de lois"', 207; the thought is also advocated by Tomljenović, '1999 "Maritime Torts"', 258, for torts bearing consequences outside the ship, as the "law of the common maritime enterprise".

[120] George, '2007 "Choice of Law"', 149, giving an illustrative account of how English courts have dealt with this problem and also referring to the cases in which the registration has been deemed to be overruled by the shipowner's nationality.

However, this is only a part of the much greater problem that flags of convenience—and, arguably, the other States' inaptitude to adequately address the problem—have created. As has been shown, the law of the flag is a very valuable connecting factor for conferring a legal status to a ship on the high seas and extremely difficult to replace by another one.[121] If, under certain circumstances, a ship's flag can be considered to be a mere sham and is consciously chosen for unlawful—as opposed to purely economic—reasons, the law of the flag should be disregarded and replaced by a more appropriate legal system. It is submitted that this would presuppose consistent decisions on when this should be considered to be the case and which would be "just" motives for flagging a ship out.

Even if the solution to apply the law of the flag to high seas torts such as oil pollution outside the CLC may be problematic in some respects, it seems that it is the only workable solution so far. Still, and namely with regard to the Rome II Regulation, it does have its dogmatic weaknesses which always bring about the risk of a court applying a different, "more appropriate" law in a given case. This does not seem desirable with regard to problems of predictability and, in the case of the Rome II Regulation, uniform application throughout the EU's member States.

It is submitted that problems with flags of convenience should be dealt with on a different level and by means that present a more global solution to this problem.

III. Damages on or Directly Concerning the Ship

When drawing the distinction between damages on the high seas that bear consequences outside the ship and damages that only concern the ship itself, it may seem at first sight that the cases that are confined to one ship present fewer problems. If the application of the law of the flag is advocated for damages on the high seas, it should be all the more so in the cases where the consequences of the tort are only perceived on the vessel. This is to some extent true, but problems may occur in other respects.

1. The Distinction Between Shipboard and Outside Torts

First of all, it may be very hard to define to what extent the damage is really limited to one ship, especially in the cases in which abstract rights are violated. Industrial action among sailors on board a ship may not be perceived in the immediate surroundings of the ship, but the damage will extend to the employer's headquarters. In the case that a passenger or crewmember is hurt or killed at sea, the damage will be perceived on board the ship, but also after the end of the voyage at his or his

[121] Marsden, Gault, Hazelwood and Tettenborn, *2003 "Collisions at sea"*, para. 17-05.

relatives' domicile.[122] With regard to this, there may also be a need to take the law of the domicile or the place of incorporation of the victim into account, which may blur the clear line of distinction between torts "internal" or "external" to the ship. It does however seem that the distinction can be made according to a different line of thought: in the torts "external" to the ship, the geographical position of the ship does matter—either to the coastal State that will later be faced with the environmental pollution, or to the owner of property, such as fishing nets, brought to the high seas. If the ship had sustained the damage at a different geographical position, the victims would not be concerned by the tort. In the cases of torts "internal" to the ship, the victims always perceive the damage in any case, irrespective of where the vessel is located at that moment[123]—the injured passenger or his relatives will always be faced with the consequences of the tort, irrespective of the ship's position, and the shipping line will sustain the damage following industrial action by the crew in any case.

A similar difficulty arises with the so-called allisions, the cases in which a vessel collides with an obstacle on the high seas, such as the iceberg in the *Titanic* case. The problem of allisions with man-made structures has been dealt with *supra*, under I. In allision cases, the geographical position of the ship and of the damaged structure of course do matter, but the consequences following the damage to the structure will most likely be perceived in the country that benefits from the operation, which may be rather far away from the place where the allision occurred. Furthermore, the actual physical damage done to the offshore installation and the cost of repair will most likely be trifling in comparison to the consequential damage in the breakdown of the power supply or another resource in the beneficiary State. This is apparently one of the reasons why it is often advocated that cases of this kind should not be governed by the law of the ship's flag but rather by the laws of the State in which the most important damage arises (see *supra* under I.). It must however be repeated that the actual damage consists in the broken cable or the damaged installation and that all consequential damage is nothing but indirect damage, notwithstanding its magnitude. In the interest of foreseeability and legal certainty, the place of occurrence of indirect damage should be disregarded; persons on board a ship on the high seas possibly committing errors in navigation can hardly expect that their case will be governed by the laws of a distant State because it happens to be the main beneficiary of an offshore installation. It is with good reason that in general, remote or indirect damages are disregarded in a conflict-of-law consideration of the *lex loci delicti*, which is, within Europe, now also expressly codified in art. 4 (1) of the Rome II Regulation.

[122] George, '2007 "Choice of Law"', 149, considering this as a circumstance that could allow to set aside the law of the flag under an escape clause, if the law of the victim's domicile is "substantially more appropriate" under s. 12 of the 1995 Act, but see the English courts' decision of this very case in "*Roerig v. Valiant Trawlers*" 2002 1 Lloyd's Law Reports 681 (Court of Appeal (England), 28 January 2002).

[123] In this sense Basedow, *2012 "Die private Haftung"*, p. 1887.

2. Solutions Under Domestic Legal Systems

For tort claims arising out of a situation that took place on board a ship and the consequences of which are not influenced by the geographical position of the ship, the applicability of the law of the flag State is generally well established. Historically, at least in maritime nations, it appeared logical to submit cases of this kind to the law of the flag, as the profession of sailor was reserved to nationals,[124] so that cases would be allocated in the home jurisdiction of the parties involved. It should of course not be forgotten that seamen's claims for injuries suffered in the course of their employment may be subject to the provisions of their working contract[125] and thus, rather a matter of contract[126] than of tort law. There is certainly also an important difference between mere contractual provisions and mandatory provisions of labour protection,[127] which are to some extent internationally harmonised by the ILO's 2006 International Maritime Labour Law Convention. The way in which sailors' claims for death and injury are treated under the national law also influences the importance that is accorded to the conflicts rule for shipboard torts. Legal systems in which there is the possibility to successfully sue in tort, such as the United States (see *infra* under e)), will of course have more precedents on the law applicable than other legal systems in which claims of this kind are usually treated as contractual claims, such as France.[128] A thorough analysis of questions of maritime labour law is however outside the scope of this study.

a) Solutions in English Law

Under English law, before the 1995 reforms of the conflict-of-law rules, the double actionability rule applied, replacing the *lex loci delicti commissi* by the law of the flag State.[129] The respective precedents are however not very numerous,[130] but they are unanimous in their assertion, and criminal law precedents confirm this view. The flag of the vessel remained the pertinent connecting factor after the 1995

---

[124] P. Chaumette, 'L'internationalisation du travail maritime' (1994) 46, Droit Maritime Français, 675–94 at 675; Tomljenović, '1999 "Maritime Torts"', 279.

[125] Ibid., p. 278.

[126] See M. Perakis, 'Modern Tendencies towards a Disruption of the Bond between the Ship's Flag and the Applicable Law' (2011), Annuaire de droit maritime et océanique, 341–57 at 354 et seq.

[127] See the discussion in R. Lagoni, 'Arbeitsschutz auf fremdflaggigen Seeschiffen im Hafen' (2010) 48, Archiv des Völkerrechtes, 58–104 at 69 et seq.

[128] A thorough analysis of the characteristics of maritime labour law is outside the scope of this study. See A. Charbonneau, *Marché international du travail maritime - Un cadre juridique en formation* (Aix-en-Provence: Presses Universitaires d'Aix-Marseille, 2009) on the international market for maritime labour law, from a French point of view.

[129] George, '2007 "Choice of Law"', 138 et seq; Fawcett, Carruthers and North, *2008 "Private International Law"*, p. 861.

[130] George, '2007 "Choice of Law"', 139.

reforms,[131] and as will be seen, the solution is reconcilable with the provisions of the Rome II Regulation. There is however one Canadian case,[132] referring to the English state of the law, that has been considered to put this rather simple solution into question. The case is interesting as it shows in how many ways it may be hard to categorise a maritime tort.

The claim in *Gronlund v. Hansen* was brought under a statutory compensation head of damage by the widow of a sailor for the death of her husband. The deceased had been a crew member of a fish trawler, the *Aleutian Queen*, registered in the port of Vancouver. The vessel was involved in a collision, for which the crew of the *Aleutian Queen* was to blame. The vessel was summarily checked and found to have sustained only light damage. When she pursued her course, however, she started to take water and listed, finally sinking in international waters after about 15 min. The claimant's husband died during the evacuation of the vessel.

The decision in question was an appeal from a decision of the Supreme Court of British Columbia against the owners of the *Aleutian Queen*. In the first instance, the court had granted the widow's claim, based on the legislation of the province of British Columbia as the law of the "most significant connexion", referring in this context, *inter alia*, to the vessel's registry in the province.[133] On appeal, the British Columbia Court of Appeal applied the pre-1995 English double actionability rule, according to which the tort must be actionable under the *lex fori*, the law of British Columbia, and the *lex loci delicti*. In determining the *lex loci delicti*, the court referred to the fact that the law applicable to the high seas was "the general maritime law of all civilized nations as it is administered in Canada",[134] i.e., according to English common law, English law. For procedural reasons, however, as the applicability of English law was not pleaded, the court assumed that it was identical with the *lex fori* and decided the case according to the law of British Columbia. Hence the appeal failed.

The court of Appeal's contention that the *lex loci delicti* in the case was the general maritime law rather than the law of British Columbia where the vessel was registered has been criticised as inconsistent with English law.[135] It has been put forward that the case concerned a tort that is internal to one ship, the *Aleutian Queen*, and that thus, the law of the flag should have been designated as the *lex loci delicti*.[136] This argument is to some extent convincing, as the defendants were to blame both for the collision and for having failed to detect the damage sustained by

---

[131] Ibid., p. 146; Fawcett, Carruthers and North, *2008 "Private International Law"*, p. 861; Carbone, '2009 "Conflits de lois"', 204; Dicey, Morris and Collins, *2012 "Conflict of Laws"*, para. 35-033.

[132] "*Gronlund v. Hansen*", 4 Dominion Law Reports, 3rd Series 435 (British Columbia Court of Appeal, 02 April 1969).

[133] "*Gronlund v. Hansen*", 69 Dominion Law Reports, 2nd Series 598 (British Columbia Supreme Court, 11 July 1968), p. 603.

[134] *British Columbia Court of Appeal, 02.04.1969*, "*Gronlund v. Hansen*", p. 443.

[135] George, '2007 "Choice of Law"', 141.

[136] Ibid., p. 141.

the vessel, and thus, all torts occurred on board of one and the same ship.[137] Under these circumstances, the presence of another vessel in the collision was not of importance. On the other hand, the court explicitly found that the sinking of the *Aleutian Queen* was a direct cause of the collision that had occurred beforehand, and that the chain of causality was not broken by the master's failure to detect the real extent of the damages.[138] Thus, the general maritime law could have applied as the law then applicable to collisions on the high seas[139] (see *infra* under V.).

It is thus submitted that the case cannot serve to question the general principle that the law of the flag applies to torts the consequences of which are confined to one vessel, and there may be good reasons to consider the case as a collision case even if all torts occurred on the *Aleutian Queen*. Under maritime law, the owners of the fish trawler may be entitled to limit their liability per incident—the LLMC, in art. VI para. 1, sets out its limits "arising on any distinct occasion". If it was considered that the chain of causality was broken by the master's negligent decision to proceed on his course without checking the damage more carefully, the case may be considered as consisting of two incidents that would need to be separately compensated by two different compensation funds. It is not clear whether this thought influenced the decision of the court, but the qualification of the case as a single-ship tort could definitely later influence the assessment of liability under limitation regimes.

b) Solutions in French Law

Under French law, the applicability of the law of the flag to torts taking place on a ship on the high seas is well established,[140] and, similar to the case of torts taking place outside the ship, not questioned. Under the Rome II Regulation, the territorial connecting factor will be replaced with the law of the flag of the ship. Even if this stretches the wording of art. 4 of the Regulation, it presents a pragmatic solution that coincides with the way these torts have been treated before the coming into force of the Rome II Regulation. The same solution is reached under a statutory rule in Italy, art. 5 of the *Codice della Navigazione* providing for the applicability of the law of the flag or registry to all the acts that take place on board the vessel, even if the vessel sails in the territorial waters of a foreign State.[141]

---

[137] Ibid., p. 141; Dicey, Morris and Collins, *2012 "Conflict of Laws"*, para. 35-033, categorising the case as a tort involving only one ship without considering the inconsistent outcome.

[138] *British Columbia Court of Appeal, 02.04.1969, "Gronlund v. Hansen"*, pp. 437 et seq.

[139] Fawcett, Carruthers and North, *2008 "Private International Law"*, p. 862, footnote 720.

[140] Carbone, '2009 "Conflits de lois"', 206.

[141] See ibid., p. 211; the latter part of the provision however gives rise to debate.

c) Solutions in German Law

German law comes to the same conclusion, both for cases before the coming into force of the Rome II Regulation and after.[142] It is however noteworthy that a number of German authors seem to be rather hostile towards the application of the law of the flag as the "national law of the ship" and often advocate the application of the law of the State in which the shipowner or the charterer or manager has his place of business,[143] or the law of the home port, or a system in which the law of the flag only presents a presumption as to the nationality of the ship, that can be rebutted and replaced by the law of the shipowner's place of business, the law of the home port or any other suitable connecting factor.[144]

d) Solutions Under the Rome II Regulation

Within the jurisdictions implementing the Rome II Regulation, there is thus a clear tendency to either apply art. 4 (1) and to consider that the connecting factor of the "law of the country where the damage occurs" points to the law of the ship's flag, equating the ship with the territory of the State whose flag it flies,[145] or to apply the law of the flag as the "manifestly more closely connected law" under art. 4 (3) of the Regulation. It is submitted that under art. 4 (3), the more closely connected law could also be different from the law of the flag, if the ship is in fact subject to the law of another state, especially in cases of bareboat charter or where the ship flies a flag of convenience without being duly registered. The uncertainty this solution entails[146] is in some way the price for a certain flexibility in the connecting factor, which allows it to cover all possible sets of facts that can occur in maritime torts. Of course, the choice of the law of the flag, if possible escape clauses are disregarded, allows the creation of a continuous regime for all cases on board the ship, on which the passengers and crewmembers can to some extent rely.[147] On a ship on the high seas, it does not seem that there is very often a law better placed to govern an internal tort of this kind, and courts can make use of the escape clause in art. 4 (3) of the Regulation if there is obviously a more appropriate law to be found.

---

[142] Ramming, '2010 "Internationalprivatrechtliche Fragen"', 297; Magnus, *2011 "Seearbeitsverhältnisse"*, p. 457.

[143] Hartenstein, '2008 "Rom I-Entwurf und Rom II-Verordnung"', 153; see the discussion in Ramming, '2010 "Internationalprivatrechtliche Fragen"', 296 et seq.

[144] Basedow, *2012 "Die private Haftung"*, p. 1888.

[145] H. Boonk, 'De betekenis van Rome II voor het zeerecht' (2008), Nederlands Internationaal Privaatrecht, 469–80 at 480, referring to the respective provision in art. 18 of the draft Regulation that was, without obvious reason, not implemented into the final version.

[146] George, '2007 "Choice of Law"', 149, with respect to the corresponding solution under English law.

[147] Carbone, '2009 "Conflits de lois"', 203.

e) Solutions in US Law

As has already been pointed out above, the problem of the law applicable to torts occurring and bearing consequences on one and the same ship is of considerable interest under the law of the United States, both in federal and state law. The relationship between the two is left out of the picture here,[148] since this study focuses on principles influencing the connecting factor in the choice of law. Most cases in which the question of the law applicable to maritime torts arose have concerned the applicability of the so-called Jones Act,[149] a federal piece of legislation setting out rather generous possibilities of compensation for claims arising out of seamen's death or injury in the scope of their employment. Claims under the Jones Act are claims for negligence in tort, the act being one of only a few statutes under which employees can sue their employer in tort and for heads of damages other than mere compensation, such as pain and suffering or mental anguish.[150] This legal basis, which is furthermore not the only footing available,[151] of course widens the field of applicability for conflict-of-law rules under United States law. Furthermore, there is in any case not a clear-cut differentiation between the conflicts rules for tort and contract cases, so that the rules set out *infra* will also take into account a possible contractual relationship between, e.g., a seaman and his employer or a passenger and the carrier[152] without considering whether the claim is based on an contractual basis or has a footing in tort.[153]

Following the rules of applicability of the Jones Act, apart from the fact that it only applies to "seamen" without providing a definition,[154] the Act only applies to US sailors and US ships. It remains to some extent questionable whether it can apply to foreign seamen, possibly injured on foreign vessels in foreign waters.[155] It has actually been argued that the application of the Act to a case that had little or no connection with the United States would not be reconcilable with the principles of

[148] See Simson, *2011 "Rethinking Choice of Law"*, p. 240, for a brief outline on the interplay between federal and State law in the *Lauritzen* case.

[149] Act of June 5, 1920.

[150] I. J. Warshauer and S. C. Dittman, 'The Uniqueness of Maritime Personal Injury and Death Law' (2005) 79, Tulane Law Review, 1163–225 at 1164.

[151] See ibid. on further remedies, such as unseaworthiness and statutory claims for maritime wrongful death.

[152] See "*Hamilton v. S. S. Nassau*" 1955 American Maritime Cases 595 (US District Court for the Southern District of New York, 10 February 1955), p. 599, the Court maintaining *obiter dicta* that a passenger's tort claim for injuries sustained on a ship flying the Panamanian flag would "appear to be" subject to Panamanian law, this being stated without the court having considered the underlying contract of passage.

[153] Bonassies, '1969 "La loi du pavillon"', 547.

[154] See discussion of the problem in Warshauer and Dittman, '2005 "The Uniqueness"', 1173 et seq. and J. T. Gillelan, '"Vessel" Status and Admiralty Tort Jurisdiction' (2005) 3, Benedict's Maritime Bulletin, 138–48.

[155] See Scoles, Hay, Borchers and Symeonides, *2004 "Conflict of Laws"*, p. 891.

public international law,[156] as the United States lacked competence to submit the set of facts to their legislative powers. In cases in which the court decides against the applicability of the Jones Act, it is rather likely that the defence will bring a motion for *forum non conveniens* so that the non-applicability of US law also eventually causes a refusal of jurisdiction as a matter of procedure.[157] This approach, to argue for the applicability or non-applicability of US law rather than the neutral consideration of a case and the application of a conflicts rule after having established jurisdiction,[158] intertwines questions of jurisdiction with questions of choice-of-law (also see *supra* in Chap. 6) to a bigger extent than the European approach to this matter and may account for the differences in dealing with this problem under US law.

The first case in which the applicability of the Jones Act to a maritime setting was considered was *Lauritzen*.[159] It concerned an action brought by a Danish seaman for injuries that he suffered on board a Danish ship while the vessel was in Cuban territorial waters, the sailor working under a Danish pay contract. The court, in considering whether the claimant should be able to invoke the Jones Act as a piece of federal US legislation, proceeded on an analysis of the case; it ascertained the points of contact that there were in the claimant's situation and listed seven pertinent factors, namely the place of the wrongful act, the law of the flag, the claimant's domicile, the defendant's allegiance, the place of the contract, the possibility of bringing the claim in another forum, and the law of the forum. Generally, the place of the wrongful act is given little regard as, in a ship perpetually sailing the waters of different States, it is rather fortuitous,[160] and for cases on the high seas, where there is no legal system in place at the location of the wrongful act, this connecting factor is apparently simply disregarded. Therefore, the conflicts considerations for shipboard torts, at least concerning the applicability of statutory legislation such as the Jones Act, remain the same for territorial waters and areas beyond national sovereignty. In *Lauritzen*, the court finally held that the Jones Act, as American federal legislation, could not apply as the only circumstances that pointed towards the United States were the place where the employment contract was signed and the fact that the claimant had joined the ship in New York.

The court's reasoning considering the seven pertinent factors is revealing for its conflicts approach in that it focuses on the decision whether or not the Jones Act applies, not on a conflicts analysis as it is understood in Europe, where following the application of a conflicts rule a given jurisdiction should be identified as applicable. The seven pertinent factors listed by the court actually also contain

[156] Akehurst, '1972 "Jurisdiction"', 184.

[157] See J. L. Allbritton, 'Choice of Law in a Maritime Personal Injury Setting: The Domestic Jurisprudence' (1983) 43, Louisiana Law Review, 879–911 at 880 et seq.

[158] See Symeonides, *2005 "Accommodative Unilateralism"*.

[159] "*Lauritzen v. Larsen*", 345 United States Reports 571 (US Court of Appeals for the Second Circuit, 25 May 1953).

[160] Allbritton, '1983 "Choice of Law in a Maritime Personal Injury Setting"', 882.

considerations of jurisdiction, such as the possibility of alternative fora,[161] that are much more result-orientated than the *lex loci delicti* approach.[162] The *lex loci delicti* and the law of the flag are only two factors to be taken into account by United States courts.

The reasoning of the court in the *Lauritzen* case was further pursued in *Romero v. International Terminal Operating Co.*,[163] which also concerned a foreign sailor on a foreign-flagged ship, but he had sustained the injury while the ship was in American waters. In considering the factors that pleaded for an applicability of the Jones Act, the court did not attach much importance to the fact that the place of injury was in the territorial waters of the United States and held the Jones Act to be inapplicable. The court also found that the same criteria applied to the question of whether the American general maritime law, as opposed to the more specific legislation of the Jones Act, could govern the case.[164] The third case to flesh out the conflicts rule for shipboard torts was *Hellenic Lines Ltd. v. Rhoditis*.[165] The set of facts was similar to the *Romero* case, the claimants being Greeks who had been injured while the ship, flying the Greek flag, was lying in New Orleans port. *Rhoditis* however added a new circumstance to the factors to be taken into consideration, as the defendant shipowner, even if also of Greek nationality, was a resident of the United States, operated the ship in American waters and thus generated his whole income through his offices in New York and New Orleans. The court recognised the defendant's place of business as an eighth connecting factor and decided that in this case, the American "base of operations" was the most important circumstance triggering the applicability of United States law. This case has, rather soon after the *Rhoditis* case, given rise to further decisions and to considerable debate about the determination and the role of the "American base of operations".[166] The factor continues to be of a great importance today,[167] not least in the context of claims against cruise lines which often have their headquarters in the United States.

---

[161] See S. C. Symeonides, 'Choice of Law in the American Courts in 2002: Sixteenth Annual Survey' (2003) 51, American Journal of Comparative Law, 1–88 at 51, maintaining that this factor should only be used in an analysis of jurisdiction such as a consideration of *forum non conveniens*, not in a conflict-of-law analysis.

[162] See Simson, *2011 "Rethinking Choice of Law"*, pp. 242, 248, footnote 42, considering whether the reasoning in *Lauritzen* was driven by considerations of the efficiency of international commerce.

[163] "*Romero v. International Terminal Co.*", 358 United States Reports 354 (US Court of Appeals for the Second Circuit, 24 February 1959).

[164] Allbritton, '1983 "Choice of Law in a Maritime Personal Injury Setting"', 884.

[165] "*Hellenic Lines v. Rhoditis*", 398 United States Reports 306 (US Court of Appeals for the Fifth Circuit, 08 June 1970), p. 358.

[166] See Allbritton, '1983 "Choice of Law in a Maritime Personal Injury Setting"', 887 et seqq. for a comprehensive overview and a critique of the developments shortly after *Rhoditis*.

[167] Peltz, '2012 "The Athens Convention Revisited"', 503.

It is obvious—and has been explicitly admitted by the court in *Lauritzen*—that the connecting factors can vary in importance and that a given circumstance can carry the case away notwithstanding the fact that all other circumstances point to another law.[168] The test under the law of the United States is thus rather unpredictable. In some cases it may be extremely difficult to identify a law that presents the most significant relationship with the case.[169] On the other hand, it is for instance submitted that as soon as the vessel in question flies the flag of the United States, the applicability of US law is almost certain.[170] Similar results have, with restrictions,[171] been found for an American base of operations.[172] In cases in which the law of the base of operation, the law of the flag and the claimants' and defendants' allegiance do not coincide at all, courts have sometimes come to arguably fortuitous or policy-driven results.[173]

The tragic case of the murder on board the *Achille Lauro* however suggests that in claims that do not fall under the Jones Act, the importance given to respective connecting factors may be very differently weighed. In the proceedings that were started against the shipowner in the United States, for the murder of a US citizen by terrorist hijackers on board a cruise ship on the high seas, the court decided to apply Italian law as the law of the flag.[174] This decision was taken despite the fact that the victim and his dependants were United States citizens and that the contract of carriage under which the passenger travelled had been concluded in the United States. It however contained both a clause to the effect that the law governing the contract was the law of Italy and a jurisdiction clause pointing to the courts of Naples. This may also have been a factor of considerable importance for the court's decision,[175] especially as there had been previous lawsuits brought by other passengers of the *Achille Lauro*,[176] in which the contract's choice of Italian law and jurisdiction had been held up against the claimants' contention that the clause violated American public policy.[177] In this context, since the conflicts rules under US law do not really differentiate between claims in contract and tort, it is difficult to weigh the importance given to the connecting factors. Giving preference to the

---

168 Scoles, Hay, Borchers and Symeonides, *2004 "Conflict of Laws"*, p. 892.

169 Carbone, '2009 "Conflits de lois"', 204.

170 Scoles, Hay, Borchers and Symeonides, *2004 "Conflict of Laws"*, p. 893.

171 See Allbritton, '1983 "Choice of Law in a Maritime Personal Injury Setting"', 887 et seq. on the early, rather inconsistent development with regard to the "base of operation", especially with regard to the mere US ownership.

172 Scoles, Hay, Borchers and Symeonides, *2004 "Conflict of Laws"*, p. 893.

173 In this sense, Allbritton, '1983 "Choice of Law in a Maritime Personal Injury Setting"', 884 et seq.

174 "*Klinghoffer v. Achille Lauro Lines*" 1993 American Maritime Cases 1387 (US District Court, Southern District of New York, 23 July 1992).

175 Peltz, '2012 "The Athens Convention Revisited"', 510.

176 "*Hodes v. S. N. C. Achille Lauro*", 858 Federal Reporter 905 (US Court of Appeals for the Third Circuit, 22 September 1988).

177 See Peltz, '2012 "The Athens Convention Revisited"', 512, 496 et seq.

law of the flag over the law of the contract may amount to overruling the law of the contract and thus the intention of the parties, for which it may be necessary to find an overarching justification such as public policy or consumer protection. It is thus submitted that contractual clauses will, as a rule, be upheld in commercial contracts between two parties of equal bargaining power.[178] However, the law of the flag or the law under which a vessel is effectively operated may again be invoked in order to consider whether the choice of law in the contract was a fair clause. The US Court of Appeals for the Third Circuit, adjudicating on the choice of Italian jurisdiction and law in the tickets that had been issued for the cruise on the *Achille Lauro*, upheld the clauses as they were considered fair and reasonable for disputes arising out of a voyage on an Italian vessel sailing from and to Italy, and carrying passengers from all over the world.[179]

In the *Klinghoffer* case, the court however stressed in its reasoning that it was the connecting factor of the flag, rather than the law of the contract,[180] that was of greatest relevance for the case, and that it was the only solution that would produce consistent results for all events that had taken place on board,[181] thus emphasising one of the big advantages of the law of the flag doctrine.[182]

3. Conclusion

On the whole, and notwithstanding the advantage of flexibility that the solution under United States law entails, the applicability of the law of the flag to torts on board ships on the high seas can hardly be called into question. Even taking into account the difficulties that may arise with flags of convenience and bareboat charter parties, it seems by all means more desirable to set aside these cases through the escape clauses of art. 4 (3) of the Rome II Regulation than to consider other connecting factors that may render the choice of law considerably more arbitrary. It has already been recognised that under the choice of law consideration as applied by the US courts, forum shopping or manipulation of the sets of facts in order to obtain the non-applicability of a given law is quite common.[183] Territoriality is an important issue in the localisation of torts, not least because many torts are only defined by contravention of local rules and provisions. Where there can be no law applicable based on a territorial connecting factor, the law must be designated according to the factor that comes closest, and that is the law of the flag.

---

178 In this sense, albeit with doubts, ibid., pp. 506 et seq.

179 *US Court of Appeals for the Third Circuit, 22.09.1988, "Hodes v. S. N. C. Achille Lauro"*, p. 913.

180 *US District Court, Southern District of New York, 23.07.1992, "Klinghoffer v. Achille Lauro Lines"*, p. 1394.

181 Ibid., p. 1393.

182 Bonassies, '1969 "La loi du pavillon"', 518; Carbone, '2009 "Conflits de lois"', 203.

183 Scoles, Hay, Borchers and Symeonides, *2004 "Conflict of Laws"*, p. 895.

IV. Environmental Damage on the High Seas

Other than damage to property or the causation of death and injury to passengers, incidents on the high seas can of course be the source of environmental damage. Considering that the high seas, according to arts. 87 and 89 UNCLOS, are subject to the freedom of, *inter alia*, navigation and overflight and not subject to the sovereignty of any State, there are of course problems at more than one level to the establishment of civil responsibility for damages caused. Environmental damage is already hard to establish and to quantify in a terrestrial context, and in maritime environmental damage, the additional problem of jurisdiction and standing arises.[184]

Matters of jurisdiction have been the topic of Chap. 6 in this part of the study. It should just be repeated that, generally, the resources of the high seas do not belong to anyone, and that, therefore, it is difficult to establish standing to hold persons or entities liable for the damages caused to the environment.[185] UNCLOS, in art. 235, allows and even encourages contracting States to develop and implement international law in order to protect the marine environment, but it is very questionable to what extent States can act unilaterally to this end, especially in the cases in which the ship flies a foreign flag and can invoke the freedom of navigation. On the other hand, it would be hard to argue that the law of the sea thus completely opposes the protection of the resources of the high seas outside the scope of application of international conventions,[186] and there is at least one case in which a domestic court has held that pollution on the high seas was litigable under certain circumstances. Reviewing the way in which the conflict-of-law problem has been treated in this case, it must be stated that the jurisdictional decision to assume competence over the case seems to have an impact on the choice of law.

It is however rather obvious why domestic cases considering environmental damage on the high seas are not numerous. First, many cases of environmental pollution eventually have perceivable consequences in the territorial waters or the EEZ of coastal States, which generally triggers the applicability of one of the civil liability regimes. Secondly, the problems of jurisdiction and venue of course deter claimants from invoking claims for damages that may be considered as remote or indirect under domestic law. Apart from these legal problems, States may for a practical reason be reluctant to establish civil liability for pollution damage in areas beyond national jurisdiction. As long as there is no international civil liability regime of sufficiently wide applicability that would cover this case, the prosecution of a given type of high seas pollution in one State disadvantages the industries in this State. Unilateral action can therefore be detrimental to a State's economy.[187]

---

[184] Vrellis, *2013 "The Law Applicable"*, p. 875.

[185] M. Rémond-Gouilloud, 'Note TGI Bastia 08.12.1976' (1977), Recueil Dalloz, 429–32 at 430; C. Huglo, 'La réparation du dommage écologique au milieu marin à travers deux expériences judiciaires' (1992), La Gazette du Palais (Doctrine), 582–90 at 584.

[186] Kiss, '1975 "Un cas de pollution internationale"', 223.

[187] Ibid., p. 218.

It is in the so-called "Red Sludge Case" (*l'affaire des boues rouges*) that has already been presented *supra*, Chap. 6, B. II. 1. a) aa), that France however decided on the civil liability of an Italian company for a case of pollution on the high seas. The enterprise ran a factory in Scarlino, south of Livorno in Tuscany, producing titan bioxyde. Toxic wastes of the production process were dumped at sea in considerable quantities, with permission from the Italian authorities. The dumpings were effectuated by two ships specially constructed for this purpose and took place on the high seas of the Mediterranean, about 70 km from the coast of Tuscany but only 40 km from the French island of Corsica.[188]

Notwithstanding the fact that the dumpings as such had been permitted by the Italian authorities, four directors of the company were later criminally persecuted and convicted to suspended prison sentences by the court of Livorno.[189] The decision was however overturned on appeal, following an amendment of the law that, retrospectively, withdrew authorised dumpings from the scope of the criminal law under certain circumstances. The fishermen's corporation (*Prud'hommie des pêcheurs*)[190] of Bastia, Corsica had tried to join the proceedings in Livorno and put forward a claim for damages in an adhesive procedure, but the request was not granted and the fishermen's committee was referred to the civil courts to claim compensation for the damage. Following the acquittal in the appeal proceedings, the *Prud'hommie* seized the court of Bastia for their civil liability claim. In a decision of 8 December 1976, the *Tribunal de Grande Instance* of Bastia gave judgment on the merits of the claim.[191] The court argued that the Italian defendant society was liable for the consequences of the dumpings, so that, generally, the *Prud'hommie* was entitled to the reparation of the damage caused to its interests.

The judgment is on the merits of the claim only. The court did not quantify the damage but referred the task of calculating the actual value of the claim to authorised experts. The court by no means departed from the general idea that there must be an economic dimension to the environmental damage,[192] which had been perceived in Corsica. No mention is made of a compensation uniquely paid for the damage to the environment. However, the description of the damage that was actually caused to the tourism and fishing industries of Corsica remained most abstract. Even in the criminal judgment of the court of Livorno which condemned the four directors in the first instance, it was only established that the introduction of the chemicals into the sea had had adverse consequences to the place of the dumpings and that it was thus probable that Italian islands were affected by it,

---

[188] See ibid., pp. 207 et seq. for a comprehensive outline of the facts.

[189] *Pretura de Livorno, 27.04.1974*, "*Affaire Montecatini Edison*" (French translation), p. 237.

[190] See Huglo, '1992 "La réparation du dommage écologique"', 586, for a short explanation of the status of the *Prud'hommie* in the Mediterranean, conferring it special rights superior to those of a trade union.

[191] *Tribunal de Grande Instance de Bastia, 08.12.1976*, "*La Prud'hommie des pêcheurs de Bastia v. Montedison*".

[192] Rémond-Gouilloud, '1977 "Note TGI Bastia 08.12.1976"', 430 et seq.

with only one control sample proving an impact on the plankton in the territorial waters of the island of Gorgona.[193] A quantifiable damage that could be measured in terms of loss of income could thus hardly be proven, neither for the Italian coastline nor for the island of Corsica. Without explicitly mentioning it,[194] the judgment of the court of Bastia is thus in fact one on environmental damage, concerning the depreciation of the environment, the economic impact of which is merely assumed.

As to the choice of law in the proceedings, the court did not undertake a conflict-of-law analysis, analysing the different foreign elements of the case. Considerations of this kind are completely omitted from the judgment. This has driven certain authors to claim that the court simply applied French law as the law of the forum,[195] or as the *lex loci damni*,[196] or possibly also as an option of choice between the law of the tortious act and the consequences,[197] given the way in which the court stressed the territorial aspect of the damage in order to establish jurisdiction. The *Tribunal de Grande Instance* of Bastia, interestingly, did not specify the law applicable at all; it merely stated that the defendant society's behaviour rendered it liable. It is noteworthy that the court did not give a reasoning on which provision of French law it considered applicable, not even specifying whether the liability was based on negligence or if it was a strict liability triggered by the danger inherent to the toxic substances.[198]

In any case, the applicability of Italian law as the law of the flag of the dumping ships was never envisaged. This is not surprising, as the application of the law of the flag of the polluting ship would amount to a benefit to the tortfeasor in cases of intentional pollution. Even in the cases where the damage is caused accidentally, courts in an affected country will generally be reluctant, not least for policy reasons, to apply the law of the flag.[199] In comparison to the claims for damage that is done to property brought to the high seas or for death or injury, the link between the violation and the country that deems it necessary to adjudicate a case of environmental damage is significantly stronger. It is submitted that following the difficulties of establishing jurisdiction for environmental damage on the high seas, the courts seized will almost invariably be found in a country in which consequences of the damage, even if only remote consequences, are perceived. This justifies the application of the law of the forum, either as the law of the *lex loci damni* or as the

---

[193] Kiss, '1975 "Un cas de pollution internationale"', 215.

[194] Rémond-Gouilloud, '1977 "Note TGI Bastia 08.12.1976"', 429.

[195] Ballarino, '1990 "Questions de droit international privé"', 374, but recognising that French law was also the *lex loci damni*.

[196] In this sense, referring to the fact that French law applied according to the conflict-of-law rules without further explanation Huglo, '1992 "La réparation du dommage écologique"', 585. Also see Audit, '1978 "Note C. Cass. 03.04.1978"', 367, citing this as the solution favoured by most authors.

[197] Ibid., p. 367.

[198] Rémond-Gouilloud, '1977 "Note TGI Bastia 08.12.1976"', 429.

[199] Tomljenović, '1999 "Maritime Torts"', 285.

law of the court that is competent to rule on the matter. In the cases where the environmental damage results in loss of income to the fishery or tourism industry, it could again be argued that the *lex loci damni* is the law of the affected coastal State. This is without prejudice to the fact that, in general, distant indirect damages are disregarded when determining the *lex loci delicti* (see e.g. art. 4 (1) of the Rome II Regulation). The "indirect" damages referred to there presuppose that there is a litigable direct damage in a given jurisdiction. In the rather special case of high seas pollution, this direct damage, which would determine the applicable law in the first place, is inexistent.[200] Still, once the court can determine a consequential economic damage suffered by the environment in its own territory or by legal or natural persons there, it will be easy to consider it subject to the law of that State under a territorial connecting factor.

The following judgment of the *Tribunal de Grande Instance* de Bastia[201] on the compensation payable can actually be interpreted in this way. The court granted compensation to the fishermen's corporation for the computed damage done to their fishing grounds,[202] but restricted the claim of the cities and municipalities of Bastia to compensation done to the trademark of the region. This head of damage, fully compensable[203] and rather popular under French law (see *supra*, Chap. 3, G. on claims under this head of damage outside the CLC regime) is only indirectly one for environmental damage. In its capacity as a compensation remedy for damages done to an existing immaterial right it is economic,[204] and only remotely linked to the deterioration of the high seas environment as such. From the point of view of a conflict-of-law analysis, the *lex loci damni* of the violation of this right is rather to be found on land, in the region covered by and benefitting from the trademark. The judgment on the compensation payable thus derogates from the conclusions found under the judgment on the merits of the claim. The judgment on the merits has been confirmed in its totality by the *Cour de Cassation*.[205]

Notwithstanding the particular language of the judgment on the merits of the claim, boldly asserting the possibility of a claim for environmental damages, it is, and perhaps rightly so, submitted that cases of this kind can only be seen as a property claim of a different kind, namely for business expectations. There is no head of damage under French law for the depreciation of the high seas as a source of well-being,[206] without an economic dimension. In fact, the findings of law as to the compensation of environmental damage have only found a few and isolated echoes

---

[200] See Vrellis, *2013 "The Law Applicable"*, p. 875.

[201] 04.07.1985, unpublished. The contents of the decision is discussed in Huglo, '1992 "La réparation du dommage écologique"'.

[202] See ibid., p. 587 on the calculation, based on the damage done to the marine microorganisms.

[203] de Raulin, '1993 "L'épopée judiciaire"', 68.

[204] Huglo, '1992 "La réparation du dommage écologique"', 583.

[205] "*Montedison v. Département de la Haute Corse et autres*" 1978 Recueil Dalloz 367 (Cour de Cassation (France), 03 April 1978).

[206] Rémond-Gouilloud, '1977 "Note TGI Bastia 08.12.1976"', 432.

in the French courts,[207] and apparently as few in other countries. The set of facts that gave rise to the case would today be subject to the provisions of the London Dumping Convention of 1972,[208] barring the Italian authorities from encouraging sea disposal of this kind. However, the case is illustrative of the fact that for economic damages on the high seas other than mere property claims, it seems right to establish a different conflict-of-law rule, disregarding the law of the flag. At the time the judgment was rendered, there was yet virtually no precedent regarding the conflict-of- law rule applicable to environmental damage, and the first efforts to establish a specific rule date from as late as the 1990s.[209] On a global level, despite the efforts of the Hague Conference on Private International Law, there is still no consensus as to the conflicts considerations for environmental damage.[210]

Under the conflicts rules of the Rome II Regulation, environmental damage is covered, as long as it can be considered a "civil and commercial matter" within the meaning of art. 1 (1) of the Regulation. This means that the claims of States or environmental protection organisations for pure ecological damage without an economic dimension will most likely[211] not fall under the Regulation,[212] notwithstanding the fact that the Regulation itself, in no. 24 of the introductory considerations, defines "environmental damage" as the mere impairment of the normal functioning of the interplay of natural resources. This favours the conclusion that the term of "environmental damage" also encompasses economic damage that is caused by the environmental damage,[213] such as the typical claims of the fishing and tourism industry in a maritime pollution context.

Art. 7 of the Rome II Regulation, setting out a separate conflicts rule for environmental damage, repeats the general rule in art. 4 but additionally gives the victim the choice of the *lex loci delicti commissi*. If the high seas pollution is perceived in the territorial waters of a State, the case can thus be submitted to the domestic law of the affected State—*the lex loci damni* according to the general rule of art. 4 (1) of the Rome II Regulation. This was the solution followed by the French court in the case of the *boues rouges* and which has been followed also in other

[207] Huglo, '1992 "La réparation du dommage écologique"', 589.

[208] Kiss, '1975 "Un cas de pollution internationale"', 228.

[209] See C. von Bar, 'Les dix points d'Osnabrück' (1994) 83, Revue critique du droit international privé, 853–5 on one of the first initiatives to this end.

[210] Opertiti Badán, *2011 "Derecho internacional público"*, p. 66.

[211] See however Vrellis, *2013 "The Law Applicable"*, p. 875, arguing that art. 7 of the Rome II Regulation widens the scope of art. 1 (1) for environmental damage.

[212] Bogdan, *2009 "Some Reflections"*, p. 101; Dicey, Morris and Collins, *2012 "Conflict of Laws"*, para. 35-066.

[213] Bogdan, *2009 "Some Reflections"*, p. 101; Vrellis, *2013 "The Law Applicable"*, p. 873.

cases of environmental disasters,[214] as the somehow logical choice of law.[215] The perception that the law of the flag has here less reason to apply than in cases of property damage is perhaps to be explained by a different approach to the compensation of environmental damage. It seems in effect more important to compensate the damage according to the law of the geographical area in question than to punish a tortious behaviour according to the law of the flag.[216] As the geographical zone that is affected is subject to the territorial sovereignty of the coastal State, it appears perfectly admissible to submit the set of facts to the law that is in force in this area and is in line with art. 4 (1) of the Rome II Regulation.

Under the additional rule of art. 7 of the Regulation allowing the choice of the *lex loci delicti commissi*, it however also seems admissible to identify the law of the place where the tort was committed with the law of the flag of the polluting ship.[217] Considering the way maritime torts are treated under the general rule of art. 4, where it is rather widely accepted to equate the law of the flag with the territorial connecting factor of the law of the country where the events took place, the same solution should be adopted under art. 7. It has also been argued that in cases of cross-border environmental damage, the law of the State in which the wrongful conduct takes place should be given the chance to apply, in order to give States the chance to regulate activities with an environmental impact, even if the consequences take place in another country.[218] Given the fact that in a maritime context the application of the law of the flag is however likely to favour the tortfeasor who has either been accorded the flag as the nationality of his home country or has chosen it consciously, this may in most cases not be an advantageous solution for victims. The choice given to them in art. 7 of the Regulation however allows an analysis of the two judicial systems and a choice of the more favourable one, ensuring at the same time that, in any case, the polluter has to comply with the higher standards.[219]

---

[214] K. Siehr, 'The Rome II Regulation and Specific Maritime Torts: Product Liability, Environmental Damage, Industrial Action' (2010) 74, Rabels Zeitschrift für ausländisches und internationales Privatrecht, 1390-153 at 146, footnote 15.

[215] See A. Szakats, 'Cross Border Pollution - Private International Law Problems in Claiming Compensation' (2001) 32, Victoria University of Wellington Law Review, 609–26 at 616, hinting at that a damage is a necessary prerequisite for a tort.

[216] Carbone, '2009 "Conflits de lois"', 232.

[217] Ramming, '2010 "Internationalprivatrechtliche Fragen"', 293; Vrellis, *2013 "The Law Applicable"*, p. 884.

[218] S. C. Symeonides, 'Rome II and Tort Conflicts: A Missed Opportunity' (2008) 56, American Journal of Comparative Law, 173–222 at 191.

[219] Ibid., p. 210; Vrellis, *2013 "The Law Applicable"*, p. 881.

V. Collisions on the High Seas

Among the possible tort scenarios that can occur on the high seas, collisions between two ships present a special problem. First, a collision creates a situation in which the legal statuses of two vessels come into play on a level playing field. Assuming that, with regard to the provisions of arts. 91, 92 and 97 UNCLOS, ships on the high seas should be governed by the law of their flag, there are, in a collision case, always two legal systems neither of which is in a better or worse position to apply. Furthermore, a collision can entail consequences that might be internal and external to both ships, damage to the vessels themselves or to the cargo, death or injury to crewmembers and passengers, and environmental damage in a wide sense. Following the extremely high number of connecting factors that may exist in a conflict-of-law analysis of a collision, there may be just as many possible countries which could have competence to deal with the case.

Collisions between vessels are by no means a rare occurrence, and were even less so in earlier times when vessels were smaller, more numerous and considerably harder to manoeuvre than today. On the other hand, collisions in modern times tend to be a lot more damaging to the vessels involved, as these are much bigger and more powerful.[220] Even if most collisions do not occur in areas beyond national jurisdiction, but rather closer to land in areas with dense ship traffic, there have always been cases of high seas collisions. The problem of the law applicable to collision cases had thus been a matter of concern for courts at a rather early point in time,[221] and was among one of the earliest issues subject to attempts of harmonisation.[222] The consequences of maritime collisions are addressed in the 1910 Brussels Convention for the Unification of Certain Rules regarding Collisions[223] (Collision Convention), the 1952 Brussels Conventions on Civil Jurisdiction Regarding Collisions[224] and on Penal Jurisdiction Regarding Collisions.[225] The 1972 International Rules for the Prevention of Collisions at Sea (COLREGS),[226] a reenactment of the earlier Rules of the Road with regard to the reality of modern shipping,[227] deal with collisions from a preventive point of view.

The 1910 Convention is however the only convention among these dealing with civil liability from a material point of view, and it does leave a certain amount of questions unanswered.[228] The Collision Convention does not address the problem

---

220 R. Rodière and E. du Pontavice, *Droit maritime*, 12th edn. (Paris: Dalloz, 1997), p. 32.

221 Basedow, *2012 "Die private Haftung"*, p. 1890.

222 Tomljenović, '1999 "Maritime Torts"', 270.

223 UKTS 4 (1913) 6677; RGBl. 1913, p. 49, 89.

224 439 UNTS 233; BGBl. 1972 II, 653, 668.

225 439 UNTS 233; BGBl. 1972 II, 653, 668.

226 1050 UNTS 16, 22-37; BGBl. 1976 II, p. 1017.

227 Prüssmann and Rabe, *2000 "Seehandelsrecht"*, p. 972; J. M. Lennon, 'The Law of Collision and the United States Navy' (2002) 50, Buffalo Law Review, 981–1016 at 990.

228 Ballarino, '1990 "Questions de droit international privé"', 380.

of the law applicable but sets out general principles as to the apportionment of liability in collision cases.[229] The instrument draws a distinction between, firstly, collisions that are accidental, i.e. caused by *force majeure* or that have an uncertain cause; secondly, collisions caused by the fault of one vessel; and thirdly, collisions that are caused by the respective fault of both or all the vessels (arts. II–IV of the Collision Convention). Generally, the liability is borne by the vessel which has committed the fault to the degree that it was at fault.[230] In accidental collisions, or those that are caused by *force majeure* or have an uncertain cause, each vessel has to bear its respective damage. The Convention applies, according to its art. XII, in all cases in which all the vessels concerned belong to contracting States, but there is a codified "national exception" in art XII para. 2 providing for the application of national law where all persons involved belong to the same State.

These simple outlines of liability of course guarantee that the overall outcome regarding the extent of liability of either party will be identical in all contracting States.[231] The Convention however does not contain provisions on the more detailed questions of liability,[232] such as the attribution of the crew's possible negligence to the shipowner, or the limitation of liability or the damages that can be put forward.[233] Problems have even arisen over the understanding of what "fault" is.[234] In general, it will thus be necessary to determine the law applicable to the event that gave rise to the damage, even in cases that are covered by the scope of the Collision Convention.[235] The question concerning the law applicable to the collision arises, of course, all the more in cases in which one of the ships does not fly the flag of a contracting State, so that the Convention does not apply.

Following the wide-spread global tendency to submit torts to the law of the country where the events took place, collisions on the high seas gave rise to considerable problems. If the ships flew the same flag, it was still conceivable to submit the case to the law of the common flag. If the ships however flew different flags, it was hardly possible to submit the whole case and the claims that arose on both ships to the law of the flag of either of them, as the choice would necessarily be

---

[229] See Lennon, '2002 "The Law of Collision"', 993.

[230] See Berlingieri, '1977 "Jurisdiction and the Choice of Law in Collision Cases"', 873 et seq. for more information on apportionment.

[231] Rodière and du Pontavice, *1997 "Droit maritime"*, p. 432.

[232] van der Velde, *2006 "De positie van het zeeschip"*, pp. 199 et seq.

[233] G. H. Roth and K. Plett, 'Schiffszusammenstöße im deutschen Internationalen Privatrecht' (1978) 42, Rabels Zeitschrift für ausländisches und internationales Privatrecht, 662–95 at 667; Prüssmann and Rabe, *2000 "Seehandelsrecht"*, p. 975.

[234] Bonassies, '1969 "La loi du pavillon"', 512; G. Kegel, 'L'abordage en haute mer en droit international privé' (1968), Revue critique du droit international privé, 393–417 at 394. Also see the case of "*El Salazar v. Falkanger*" (Summary in French) 1969 Journal du Droit International (Clunet) 990 (Rechtbank Rotterdam, 18 June 1963) on a collision between a Spanish and a Dutch vessel caused by a technical failure in the Spanish vessel's helm, which represented a "fault" under Dutch but not under Spanish law.

[235] Kegel, '1968 "L'abordage"', 395; Prüssmann and Rabe, *2000 "Seehandelsrecht"*, p. 975.

arbitrary. There is, moreover, no territorial *lex loci delicti* on the high seas that could strike the balance.[236] It has thus sometimes been argued that collision cases should be governed by the law of the flag of the ship that is not at fault, or that is least at fault.[237] Whereas it seems fair to privilege the party that is less to blame, it may in some cases not be possible to determine the respective degrees of fault without first determining the law under which this fault shall be considered.[238] Furthermore, the criterion is of course not of any use in cases of purely accidental collisions, or those with an uncertain cause.

Under English law, it was originally this problem of collisions on the high seas that gave rise to the doctrine of the "general maritime law", i.e. English law, applicable to the high seas[239] (see *supra* under II. 1.). The application of English law, not in its capacity as the law of the forum, but simply as the law applicable to the high seas, allowed the English courts to rather elegantly solve the problem and to find precedents in the legal system of a country that had a particular interest in maritime law.

Even if it has been argued that, essentially, the doctrine amounted to nothing more than a blatant application of the law of the forum by disregarding the foreign elements, the doctrine certainly has the benefit of certainty[240] and foreseeability. It has been echoed in the second instance of the Canadian case of *Gronlund v. Hansen*[241] (also see *supra*, III. 2. a). Under this doctrine, English law applied to collisions on the high seas even in the cases where both ships flew the same flag.[242] The leading case[243] concerned a collision between two ships, both flying the Dutch flag and actually even belonging to the same owners. Brett L. J., when considering the case in tort, quite explicitly referred to the fact that the "injury was committed [...], not in any foreign country, but on the high seas, which are subject to the jurisdiction of all countries, the question of negligence in a collision raised in a suit in this country is to be tried, not indeed by the common law of England, but by the maritime law, which is part of the common law of England as administered in this country".[244] Lindley L. J. stated it even clearer: "I am not aware of any decision in this country to the effect that where two ships come into collision on the high seas the rights and liabilities of their owners have been held to

236 Tomljenović, '1999 "Maritime Torts"', 272.

237 Discussed by George, '2007 "Choice of Law"', 169, as a possible solution under an earlier draft of the Rome II Regulation.

238 Tomljenović, '1999 "Maritime Torts"', 272, footnote 111.

239 Berlingieri, '1977 "Jurisdiction and the Choice of Law in Collision Cases"', 871; George, '2007 "Choice of Law"', 141, 154.

240 Winter, '1954 "Maritime Torts"', 123.

241 *British Columbia Court of Appeal, 02.04.1969*, "*Gronlund v. Hansen*", p. 441.

242 Tetley, '1993 "The Law of the Flag"', 169; Marsden, Gault, Hazelwood and Tettenborn, *2003 "Collisions at sea"*, para. 17-01.

243 "*Chartered Mercantile Bank v. The Netherlands India Steam Navigation*" Law Reports, Queen's Bench Division 521 (High Court (England), Queen's Bench Division, 17 January 1883).

244 Ibid., p. 537.

depend on the laws of the respective flags of the ships. The law applicable in this country to cases of collisions on the high seas is the maritime law as administered in England and not the laws of the flags".[245]

Even if this rule has been criticised for being nationalistic,[246] it clearly remained a part of English law at least until the coming into force of the Rome II Regulation. The general rules in art. 4 of the Regulation, not creating a specific conflicts rule for maritime torts, do not provide guidance on the law applicable to high seas collisions. It is therefore sometimes argued that the "general maritime law" rule could still be good law under the Rome II Regulation,[247] the European legislation in any case not setting out a rule to the contrary. On the other hand, as can be seen by the drafting history of the Regulation that originally contained some provisions of applicability to ships (see *supra*, II. 4.), the legislation was meant to cover maritime torts. It would be hardly in line with the purpose of the Regulation if English courts continued to apply English law as the "law applicable to the high seas". On the other hand, as will be shown *infra*, the applicability of the law of the forum may be an appropriate solution to the problem of collisions on the high seas, even under the provisions of the Rome II Regulation. In this context, the application of English law by English courts could, as a result, at least not be criticised.

In France, the law applicable to collisions on the high seas is the law of the flag if both vessels are flying the same flag.[248] In all other cases, in the absence of any objective connecting factor that could be given regard to,[249] the *lex fori* applies. This has been the uncontested result in the few cases found,[250] without any apparent findings to the contrary. It is admitted by French legal opinion that this solution is somehow the last resort for submitting a collision on the high seas to a valid rule.[251] It is however apparently not envisaged to possibly choose another law under certain circumstances, not even in cases where the set of facts presents stronger links with another legal system.

This is different under German law. In decisions dating from as far back as the early twentieth century, precedents of the *Reichsgericht* were anything but unanimous[252] on this and, respectively, considered applying either the *lex fori*, or the law

---

[245] Ibid., p. 545.

[246] George, '2007 "Choice of Law"', 159.

[247] Dicey, Morris and Collins, *2012 "Conflict of Laws"*, para. 35-033.

[248] Rodière and du Pontavice, *1997 "Droit maritime"*, p. 432.

[249] Berlingieri, '1977 "Jurisdiction and the Choice of Law in Collision Cases"', 871; Rémond-Gouilloud, *1993 "The Future of the Compensation System"*, para. 100; Mayer and Heuzé, *2010 "Droit international privé"*, p. 517.

[250] *"Vigouroux v. Lenten"* Recueil Dalloz 154 (Cour d'Appel de Rennes, 13 March 1961), referring to earlier precedents but nonetheless also applying the *lex fori* for policy reasons; *"Lenten v. Vigouroux"* 1966 Recueil Dalloz 577 (Cour de Cassation (France), 09 March 1966), p. 577.

[251] Bonassies, '1969 "La loi du pavillon"', 549 et seq.; Rodière and du Pontavice, *1997 "Droit maritime"*, p. 432: "as the law of subsidiary application".

[252] G. Beitzke, 'Questions d'abordage en droit international privé allemand', in: *Mélanges offerts à Jacques Maury* (Paris: Dalloz et Sirey, 1960), p. 59; Kegel, '1968 "L'abordage"', 410 et seq.; Roth and Plett, '1978 "Schiffszusammenstöße"', 672, still speaking of a "change of legal practice".

of the ship that was to blame for the collision and subsequently envisaged a choice between the two laws of the flag, to be made in favour of the law that was more favourable to the claimant.[253] This last solution, rather unique from a comparative point of view, can still be explained dogmatically. If it is assumed that the *lex loci delicti* in torts involving a ship is to be equalled with the law of the flag, collision cases present two possible laws.[254] In accord with the former German rule (*Ubiquitätsprinzip*) to offer a claimant the choice of the two laws if the *lex loci delicti commissi* and the *lex loci damni* differ, the rule allows the choice of the more favourable law.[255] Of course, the problems that arise in cases where both parties are possibly to blame[256] and bring claims against each other, and the practical problems of determining the more favourable law[257] render this solution quite impractical. It has been, however, still advocated in more recent times.[258]

The scope of the precedents cited is however not overly clear,[259] as the courts have generally shown a certain tendency to arrive at the application of German law[260] without openly maintaining the application of the *lex fori*. There is evidently a certain reluctance of courts to be observed in applying the law of the forum,[261] as a rather arbitrary connecting factor,[262] and German courts have sacrificed the benefit of certainty to solutions that may be dogmatically convincing but have left the problem without clear-cut solution. There is however almost[263] unanimity in both jurisprudence and legal opinion that the law of the common flag will apply to cases in which the colliding ships fly the same flag.[264]

Concerning the problem of a collision on the high seas, the general rules of the Rome II Regulation are even less useful than in cases involving just one ship on the high seas. The territorial connecting factor of the *lex loci damni* set out in art. 4 (1) of the Regulation can hardly be of any use in a collision in which there are two ships flying different flags that are both likely to have suffered damage. It has been

---

[253] "*Casablanca*", 138 Entscheidungen des Reichsgeriches in Zivilsachen 243 (Reichsgericht (Germany), 12 November 1932), p. 243.

[254] R. Jambu-Merlin, 'Note CA Rennes 13.03.1961' (1962), Recueil Dalloz, 155–7 at 156; Bonassies, '1969 "La loi du pavillon"', 549; Prüssmann and Rabe, *2000 "Seehandelsrecht"*, p. 977.

[255] Beitzke, *1960 "Questions d'abordage"*, p. 61; Kegel, '1968 "L'abordage"', 414 et seq.

[256] Beitzke, *1960 "Questions d'abordage"*, p. 62.

[257] Jambu-Merlin, '1962 "Note CA Rennes 13.03.1961"', 156; Roth and Plett, '1978 "Schiffszusammenstöße"', 688.

[258] Prüssmann and Rabe, *2000 "Seehandelsrecht"*, p. 979.

[259] Ibid., p. 977.

[260] Beitzke, *1960 "Questions d'abordage"*, p. 60; Jambu-Merlin, '1962 "Note CA Rennes 13.03.1961"', 156; Roth and Plett, '1978 "Schiffszusammenstöße"', 673.

[261] See Beitzke, *1960 "Questions d'abordage"*, p. 61, also giving possible reasons.

[262] Roth and Plett, '1978 "Schiffszusammenstöße"', 692.

[263] See Beitzke, *1960 "Questions d'abordage"*, p. 65, on early contrary opinions.

[264] Ibid., p. 61; Roth and Plett, '1978 "Schiffszusammenstöße"', 673; Prüssmann and Rabe, *2000 "Seehandelsrecht"*, p. 979.

submitted that in the cases in which one ship was entirely to blame for the collision, art. 4 (1) of the Rome II Regulation could be taken to point towards the ship that was not at fault.[265] This idea however confuses questions of liability with the question of the place where the damage occurred. Even in a collision in which only one of the ships is to blame for the incident, damage will most likely be caused to both ships.[266] In this context, it has also been submitted that the law applicable according to art. 4 of the Regulation should be the law of the flag of the ship sustaining the damage,[267] which is certainly better in line with the wording of the provision but creates considerable difficulties in practice.[268] In all likelihood, both ships will be damaged in a collision, and, moreover, the Rome II Regulation is rather hostile to the so-called *dépeçage*, i.e. the splitting of a set of facts in two legal relationships to which different laws would apply.[269] Treating the damage that occurred on each ship according to the law of that ship's flag means in practice that there will be two decisions on the circumstances under which the collision arose. It does not seem impossible that the two courts, due to differences in their procedural laws for admission of evidence, or simply when weighing the evidence, may come to different decisions as to the party that is to blame for the accident. In the compensation stage, the liable party would have to set up two limitation funds in different countries. There would consequentially be considerable difficulties in setting off the claims between them. If considerations of contributory negligence are added to this scenario, one of the ships being mostly or entirely to blame for the collision, it appears almost impossible to ensure that all claims are proportionally satisfied.[270]

The rules of the Rome II Regulation can, for collisions on the high seas, only find application if both parties can be considered to have their habitual residence in the same country (art. 4 (2) of the Regulation)[271] or where the circumstances of the case suggest that there is a law that is manifestly very closely connected, pursuant to art. 4 (3) of the Regulation. This solution can however only be followed if it is disregarded at the same time that the provision's wording presupposes that there is a law that can be determined from art. 4 paras. 1 and 2. This could be the case if both ships are flying the same flag,[272] always under the condition that this law is in

---

[265] Plender and Wilderspin, *2009 "The European Private International Law"*, para. 18-048.

[266] Carbone, '2009 "Conflits de lois"', 222.

[267] Dickinson, *2008 "The Rome II Regulation"*, para. 4.56; Ramming, '2010 "Internationalprivatrechtliche Fragen"', 298; Hille, Schröder, Dettmer and Visser, '2010 "Offshore-Windkraftanlagen"', 589.

[268] Dickinson, '2013 "Territory"', 128.

[269] Basedow, *2012 "Die private Haftung"*, p. 1892. See Symeonides, '2008 "Rome II and Tort Conflicts"', 185 et seq. on *dépeçage* under the Rome II Regulation in general.

[270] Carbone, '2009 "Conflits de lois"', 222; Ramming, '2010 "Internationalprivatrechtliche Fragen"', however advocating this solution.

[271] Hartenstein, '2008 "Rom I-Entwurf und Rom II-Verordnung"', 153.

[272] Carbone, '2009 "Conflits de lois"', 221, however claims that a possibly common nationality of both ships "will lose all importance" under the Rome II Regulation.

reality closely connected to the case and not just a coincidence, e.g. in a situation in which both ships are flying the same flag of convenience.

Considering the fact that the provisions of the Rome II Regulation can thus in most cases not even be applied *mutatis mutandis*, it has been submitted that collisions on the high seas are not covered by the connecting factors of the Regulation and therefore need not to be treated according to its provisions. One can of course consider that this uselessness of the rules of the Rome II Regulation withdraws the set of facts from the Regulation's scope of application, as a collision case on the high seas generally does not present a conflict of laws but rather a situation in which no law, as opposed to more than one law, is called to apply. This thought also supports the application of the *lex fori*, in the way that a national court will naturally apply it in cases without any foreign element. In any case, none of the rules of the Regulation requires the application of a given law following a certain connecting factor in a collision case,[273] so that it is difficult to disagree with authors who claim that the Regulation does not oppose a given domestic legal system's solution. It can however also be considered that following its negotiation history that included considerations of maritime torts, the Rome II Regulation is fully applicable,[274] but as the connecting factors of the existing rules point to nothingness, the law applicable should be determined by art. 4 (3), meaning that the *lex fori* applies in the absence of a more closely connected law.

This however causes the problem that the law applicable to the collision will, at least regarding one ship, not be the law of the flag. This means that the conduct of the persons on board may be assessed according to a law that they were not aware of. The difficulties that this can cause are illustrated by the historic case of *The Zollverein*,[275] concerning a collision between two sailing-ships on the high seas. One of them, the *Pet*, flew the English flag, and the other, the *Zollverein*, flew the Prussian flag. The *Zollverein* was essentially to blame for the collision, having sailed too close to another light vessel, which had obscured the view of the *Pet's* lights. Under the generally accepted Rules of the Road at that time, the *Pet* was entitled to maintain her course, and, as she was thus not to blame for the incident, her owners were to receive compensation for the damage sustained. On the other hand, a British Act of Parliament in force at that time provided for a departure from the governing rule if the circumstances of the case made it necessary. In this case, the court established evidence to the fact that the *Pet's* crew had observed the *Zollverein* before the collision and could have avoided the collision by slightly changing course. The *Pet's* crew had thus violated the British Act of Parliament obliging it to avoid a collision even in the case where it was not to blame for its behaviour under the generally accepted Rules of the Road. The case hinged on whether the *Zollverein* parties could plead the *Pet's* violation of a British Act of

[273] Ibid., p. 221; Basedow, *2012 "Die private Haftung"*, p. 1893.

[274] Boonk, '2008 "De betekenis van Rome II"', 471.

[275] "*The Zollverein*", 166 English Reports 1038 (High Court of Admiralty (England), 19 April 1865).

Parliament on the high seas to avoid liability. Even if English law as the "general maritime law" applied to the case, Dr. Lushington, as a judge the High Court of Admiralty struck a clear distinction between the rules that were considered to be the generally applicable maritime law, such as the Rules of the Road, and the Act of Parliament, only applicable to vessels flying the British flag. Dr. Lushington pointed out that under the circumstances that gave rise to the present case, a foreigner could always recover against a defendant that had abided by the generally applicable law since "the legislature of this country has no power to bind foreign vessels in such a condition".[276] If in the reverse case, the British ship could not recover because of the existence of stricter English legislation, this would be "contrary to equity",[277] which may implicitly mean that British ships, due to the strict legislation applicable by Act of Parliament, would be at a disadvantage in comparison to foreign ships. It was therefore held that the *Pet* could recover against the *Zollverein*.

The case dates from the early times of the development of the general maritime law in England and is revealing of the tendency to consider certain English rules, such as the Rules of the Road, as internationally applicable[278] (see *supra*, II. 1.). It is however also illustrative of the difficulties of assessing behaviour on board a ship according to rules that are not the rules, or not the only rules, applicable in the flag State. In this case, at first glimpse, the *Pet* did not comply with the higher British standards of conduct it was bound by as a British ship. The argument that the vessel could either be liable or not liable in the same collision, depending on whether the other ship was British or not, did not go unnoticed,[279] and the case was decided in equity after all, the application of the letter of the law not resulting in a fair outcome.

Under the Rome II Regulation, if the law applicable to the collision is not the law of the flag State, rules and regulations of the flag State could still be given regard to according to art. 17 of the Regulation, as the rules of safety and conduct of the place where the event giving rise to the damage occurred. The way in which the provision is worded ("account shall be taken, as a matter of fact and so far as is appropriate") would also allow for a very measured application of these rules. This creates some legal uncertainty, but if the collision is to be governed by only one law, it is impossible to apply the rules of conduct of both States anyway.

The application of the *lex fori* to collisions between two vessels flying different flags on the high seas, which is rather widely accepted in most European legal systems,[280] could thus be a pragmatic, easy way to determine a uniform law applicable. It would, even disregarding the importance the Rome II Regulation accords to territorial connecting factors, create a uniform solution that appears

---

[276] Ibid., p. 1040.

[277] Ibid., p. 1040.

[278] O'Connell, *1979 "The English Choice of Law Rule"*.

[279] *High Court of Admiralty (England), 19.04.1865, "The Zollverein"*, 166, p. 1040.

[280] Bonassies, '1969 "La loi du pavillon"', 549; Tomljenović, '1999 "Maritime Torts"', 273.

acceptable under the legal systems that have been presented in this part of the study. In cases in which both ships fly the same flag, the law applicable should however be the law of this common flag, either under art. 4 (1) if one considers the law of the flag to be the connecting factor that comes closest to the territorial factor, or under art. 4 (3) of the Regulation if the law of the common flag really appears to have strong links with the case.

This solution—which is essentially the French approach—would interestingly also mirror the conflicts rule as set out by the law of the United States.[281] The law applicable to collisions on the high seas between vessels flying the same flag is the law of this common flag, and in all other cases, in the absence of a better connecting factor, the law applicable is the *lex fori*.[282] The rule has been set out by precedent[283] and then in legal literature by the Restatement[284] explicitly and at a rather early point in time. Even though the scope of the law of the flag has been rephrased in the later Restatements, the findings on the law applicable to collisions on the high seas do not seem to have been seriously questioned.

This of course encourages forum shopping and a certain race of the parties to the courts of their respective countries, in the hope of securing the venue in which their claims will be treated under their own law. This appears however as the price that needs to be paid for a minimal degree of certainty in the treatment of high seas collisions. In these accidents, it is very difficult to see how certainty should otherwise be achieved.[285] It is however needless to say that the application of the *lex fori* only presents a good solution where the jurisdiction of the court that will apply its own law is well established and acceptable under the jurisdiction rules and the principles of the law of the sea. The 1952 Civil Jurisdiction Convention, setting out uniform judicial venues for its contracting States might here be of help. Considering that the *lex fori* in a collision case may be a law that does not coincide with either of the laws of the flag, regard must in any case be given to the circumstances of the case when determining the wrongful behaviour on board the ships, especially when it comes to the provisions on construction, manning and equipment,[286] to the extent that these are not governed by international rules.

[281] Marsden, Gault, Hazelwood and Tettenborn, *2003 "Collisions at sea"*, para. 10-17, footnote 27.

[282] Tetley, '1993 "The Law of the Flag"', 154, 169; Tomljenović, '1999 "Maritime Torts"', 275; George, '2007 "Choice of Law"', 159.

[283] *"The Scotland"*, 105 United States Reports 24 (US Supreme Court), pp. 29 et seq.

[284] *1934 "Restatement of the Law of Conflict of Laws"*, § 410.

[285] Roth and Plett, '1978 "Schiffszusammenstöße"', 682.

[286] Carbone, '2009 "Conflits de lois"', 227.

B. The Exclusive Economic Zone

The Exclusive Economic Zone is a concept that has emerged rather late in the history of the law of the sea, and the invention of which has been triggered by coastal States' eagerness to submit to their control the resources of waters beyond the territorial sea.[287] Having been codified in the UNCLOS, the concept of the EEZ has changed the regime for coastal States' territorial claims over the sea considerably, but the emerging claims over the use of the ocean exclusively concern the living and non-living resources of the waters and of the marine subsoil and other activities for the economic exploration and exploitation of these waters (art. 56 para. 1 (a) UNCLOS). Similarly, the coastal State exercises sovereign rights over the resources of the marine subsoil of the continental shelf in the cases where the continental shelf must be considered as the natural prolongation, in geographic and geological terms, of the coastal States' EEZ (arts. 76 et seq. UNCLOS). The introduction and development of these zones however uniquely concerned the coastal States' right to explore and exploit the resources mentioned and, in the EEZ, the right to establish facilities for the generation of energy from wind and water.

Certainly, with a view to protect the EEZ from pollution or exploitation by foreign entities, coastal States have tried to extend their competences over these waters. These attempts however must, according to UNCLOS (art. 58), be without prejudice to other States' rights to navigation and overflight.

I. Navigation

Since the introduction of the EEZ concept into UNCLOS, coastal States have increasingly made use of the rights conferred by the law of the sea. It is today widely accepted that the coastal State has sovereign rights beyond the territorial waters. The exploitation of marine resources, namely of oil and gas in the marine subsoil, has become an important economic factor in many coastal States and may in some areas be of a greater importance than the wealth created by shipping. This encroachment of States' claims over marine areas that were once part of the high seas has certainly led to a change of attitude towards waters in the vicinity of the coast, and there is widespread recognition of the coastal States' interest in the state and the status of the EEZ. According to art. 220 paras. 5 and 6 UNCLOS, the coastal State has, under certain circumstances, judicial jurisdiction over pollution cases in these waters. The increasingly well accepted competence of coastal States has actually resulted in suggestions that the competence should also encompass judicial

[287] See B. H. Oxman, 'The Territorial Temptation: A Siren Song at Sea' (2006) 100, American Journal of International Law, 830–51, and Lagoni, *2006 "Festlandsockel"*, para. 203 et seqq. for an overview.

jurisdiction for other matters. Considering conflict-of-law rules, it is sometimes advocated that the law applicable to torts in the exclusive economic zone should be the law of the coastal State,[288] given the strong actual links between the State and the waters subject to its functional sovereign rights.

Still, it must be emphasised that the regime applicable to navigation in the EEZ is the regime of the high seas, only subject to the restricted rights that the coastal State enjoys to protect the environment and the resources of the EEZ (art. 56 UNCLOS) and which essentially need to have a footing either in UNCLOS or in other internationally accepted agreements. Outside of this realm, it is the high seas regime on navigation and overflight that applies,[289] and coastal States cannot interfere with foreign-flagged vessels. There is thus nothing in the law of the sea that would suggest that the coastal State has judicial competence over torts taking place on passing ships. For torts that do not have consequences outside the ship, it is certainly the law of the flag that should apply following the *lex loci delicti* consideration.[290] Even if the coastal State enjoys certain sovereign rights over the EEZ, they do not concern navigation, and under the presupposition that judicial competence—both concerning jurisdiction and conflicts of laws—should be reconcilable with the States' legislative and enforcement jurisdiction under the law of the sea, there is nothing to suggest that the coastal State should have competence to adjudicate nor to apply its own laws under the conflicts rule of *lex loci delicti*.

It is however questionable whether the State can invoke a certain competence if there are damages outside the ship, namely, environmental damage caused to the EEZ only and not spreading to territorial waters. Art. 60 UNCLOS provides for the coastal States' competence for the protection and preservation of the marine environment. In order to exercise this competence, the State must have the power to legislate and to enforce, and art. 56 para. 1 (b) (iii) UNCLOS grants "jurisdiction" to this end, the scope and extent of which is described in more detail in arts. 192 to 237 of the Part XII on marine pollution.[291] Within this general part, art. 194 UNCLOS actually obliges States to take all necessary measures in order to combat maritime pollution from all sources, having however regard to the interests of other States and their rights under the convention (art. 194 para. 4). Art. 211 para. 5 UNCLOS, dealing with legislative jurisdiction, consequentially allows States to enact laws on pollution prevention and reduction in their EEZ, but generally only within the internationally accepted standards. The question of what these standards

---

[288] Hartenstein, '2008 "Rom I-Entwurf und Rom II-Verordnung"', 153.

[289] Brækhus, '1979 "Choice of Law Problems"', 275 et seq.; Brownlie, *2003 "Principles of Public International Law"*, p. 223.

[290] Carbone, '2009 "Conflits de lois"', 212.

[291] M. H. Nordquist, S. N. Nandan and S. Rosenne, *United Nations Convention on the Law of the Sea, 1982. A Commentary*, Volume II (Dordrecht, Boston: Martinus Nijhoff Publishers, 1993), para. 56.11(d).

are has been a subject of considerable debate,[292] but there is however unanimity that the protection standards and health and safety regulations established by the widely ratified conventions of MARPOL[293] and SOLAS[294] form part of it. In summary, the coastal State thus has a legislative competence over pollution caused by vessels in its EEZ, but it is restricted to these international standards. As to enforcement jurisdiction for pollution in the EEZ, art. 220 paras. 3 to 6 UNCLOS provide for the coastal State having certain powers, giving it the right to interfere with the vessel. This competence is however carefully structured according to the gravity of the pollution and the degree of suspicion that the coastal State can reasonably have against a given vessel. In the most severe cases, the coastal State is expressly given the power to "institute proceedings", which points to a judicial competence, arguably in both civil and criminal law.

It should however not be forgotten that UNCLOS favours flag State jurisdiction over vessels polluting the EEZ, as is generally witnessed by art. 217, and art. 220 para. 7, requiring coastal States to allow the vessel to proceed in all cases in which financial security for the claims arising has been guaranteed by the flag State. More generally, art. 228 UNCLOS requires all States to stay proceedings where the flag State is effectively starting proceedings for the same infringement, except in very grave cases or where the flag State has already shown itself unable to ensure compliance. Notwithstanding the assertiveness of UNCLOS considering the coastal State's jurisdiction in the matter of the protection of the marine environment, it however seems that the coastal States powers are, in this respect, rather restricted[295] and with regard to legislative competence, subject to international standards.

This however means that a full-scale judicial jurisdiction is subject to these standards, too, as courts could however only subject vessels to laws and regulations that are lawfully enacted. The enforcement powers of art. 220 para. 6 comprise, for the worst cases of maritime pollution clearly attributable to a given ship, the competence of "instituting proceedings" immediately, including the arrest and detention of the ship. The competence to adjudicate in all other cases, where proceedings are not started by a governmental enforcement action but by a private claim, does not appear problematic in this way. It should furthermore be stressed that Part XII of UNCLOS accords considerable importance to the possibility of establishing a compensation claim, as the enforcement powers of art. 220 paras. 3 to 6 are directed at obtaining the necessary information, such as the ship's identity and next port of call. Moreover, art. 229 UNCLOS states that the right to institute civil

[292] See B. H. Oxman, 'The Duty to Respect Generally Accepted International Standards' (1991) 24, New York University Journal of International Law and Politics, 109–59; C. Tomuschat, 'Obligations arising for States without or against their Will' (1993) 241, Recueil des Cours, 195–374 at 348 et seq.

[293] A. E. Boyle, 'Marine Pollution under the Law of the Sea Convention' (1985) 79, American Journal of International Law, 347–72 at 361.

[294] V. Frank, 'Consequences of the "Prestige" Sinking for European and International Law' (2005) 20, International Journal of Marine and Coastal Law, 1–64 at 8.

[295] Boyle, '1985 "Marine Pollution"', 354.

proceedings remains unaffected by the Convention's provisions. The provision however only clarifies that the institution of civil proceedings remains possible[296]; it does not create a jurisdictional venue itself.

Considering that the coastal States' sovereign rights in the EEZ are carefully balanced against the interest in free navigation within the geographical limits of the EEZ, and that legislative competence furthermore needs to be compatible with international standards anyway, it appears right to consider that the geographical connecting factor of the *lex loci delicti* points to the coastal State where there has been pollution in the EEZ. This finding holds true for art. 7 and for art. 4 (1) of the Rome II Regulation, the latter provision being referred to in the former.

With a view to that, it appears legitimate that many States have adopted domestic legislation for the protection of the marine environment in zones for which they can put forward competences to this end.[297] As long as States move within the realm of their legislative and judicial powers under public international law, they can decide to waive the application of a conflicts of laws rule and adopt rules of immediate application.

The judicial competence on civil liability claims for damages caused to the marine environment does however only encompass claims for environmental damages—all other damages caused outside the ship, such as the deterioration of property brought into the EEZ, and in the last consequence also the cases of drowning outside of ships, are not covered by the competences of art. 56 UNCLOS. This may lead to problems where there is a pollution damage which is not, properly speaking, an environmental damage, such as contamination of the stocks of aquaculture.[298] It is very debatable whether aquaculture can be regarded as a use of the EEZ for which a coastal State is given sovereign rights for under art. 56 UNCLOS and whether an impediment of this activity would trigger the application of its law, at least under the Rome II Regulation's *lex loci damni* rule (see *infra* under II.). To the extent that this is considered not to be the case, and to the extent that the pollution damage is effectively not covered by an international civil liability regime, it has been suggested that the application of the coastal State's law should be brought about by an analogy to the pollution conventions which normally apply to pollution damages in a coastal State's EEZ.[299] This is however very doubtful, since an analogy that goes beyond the well-defined scope of an international agreement violates the intention of the parties, who, for instance, negotiated an agreement for oil pollution and not for any other kind of contamination. To the extent that a resource in the EEZ or property brought to these waters cannot be considered as part of an activity covered by a coastal States' functional sovereign rights, the unilateral application of the coastal States' law is contrary to the law of the sea. The conundrum of property damages in the EEZ not covered by the existing

---

[296] Nordquist, Rosenne and Yankov, *1991 "UNCLOS Commentary Vol. IV"*, para. 229.5.

[297] Tomljenović, '1999 "Maritime Torts"', 288.

[298] Basedow, *2012 "Die private Haftung"*, p. 1889.

[299] Ibid., p. 1889.

liability regimes is better left under the high seas regime—with all the negative consequences to the owner—until the international community feels the need to submit it to an international liability regime according to art. 235 para. 3 UNCLOS.

The application of the coastal State's law to environmental damage in the EEZ has the advantage that it is, essentially, the victims' law that applies, considering that the coastal State has a marked interest in the environmental state of the EEZ. Thus, a tortfeasor cannot evade application of that law by choosing a flag State that is particularly lenient in enforcing environmental norms.[300] The arguments which have been set out *supra* under A. IV. in arguing for the application of the *lex fori* to environmental damage on the high seas also hold true, if not to a greater degree, for the application of the coastal State's law to environmental damage in the EEZ. Under the Rome II Regulation, to the extent that environmental damage can be considered a "civil and commercial matter" within the scope of the Regulation, the victims of environmental damage effectively have the choice between the *lex loci damni* under the general rule of art. 4 (1) and the *lex loci delicti commissi* under art. (7), which is a fair outcome both for territorial and maritime environmental torts in the EEZ.[301]

II. Offshore Installations

In contrast to what has been set out *supra* under I. on navigation, the coastal State's exclusive sovereign rights are well established for the exploration and exploitation of the EEZ's resources, both in the waters and in the marine subsoil. According to art. 60 para. 2 UNCLOS, the coastal State enjoys "exclusive jurisdiction" over artificial islands, installations and structures in the exclusive economic zone, "including jurisdiction with regard to customs, fiscal, health, safety and immigration laws and regulations." From the provision's wording, the conferral of jurisdiction is general, the fields of customs, fiscal, health etc. that are set out in the second part of the sentence only presenting examples.[302] Furthermore, art. 208 UNCLOS encourages States to adopt domestic legislation for the protection of the marine environment from seabed activities subject to national jurisdiction. With regard to that, it would be hard to argue that the coastal State's jurisdiction does not encompass judicial competence and the power to apply its law to these installations,[303] including its conflict-of-law rules. With regard to a possible environmental pollution resulting from offshore activities, this solution already appears justified from the point of view that it will most likely be the coastal State that suffers most

[300] Carbone, '2009 "Conflits de lois"', 233.

[301] Ibid., pp. 235 et seq.

[302] In this sense Nordquist, Nandan and Rosenne, *1993 "UNCLOS Commentary Vol. II*", para. 60.15(d), also referring to criminal jurisdiction.

[303] Dickinson, *2008 "The Rome II Regulation*", para. 3.306.

from any adverse consequences of an accident. Problems can however occur where the pollution also has consequences on the waters of another, neighbouring State.[304] This was already the case in the *Ixtoc 1* oil spill in 1979, that was caused by a well blow-out in Mexican waters but significantly affected the coastline of the United States.[305] As exploration and exploitation of marine subsoil resources, namely oil, is today carried out in increasingly bigger distance to the shore, the danger to other States becomes even more important.[306] In effect, there have even be fears that the *Deepwater Horizon* oil spill in the Gulf of Mexico could have an impact on European waters as pollutants could be carried across the Atlantic with the Gulf Stream.[307]

In this respect, it must briefly be mentioned that as of now, there is no international civil liability regime for pollution damages caused by offshore activities, even though there have been considerable attempts[308] in the past to come to an international agreement.[309] In Europe, and namely on the North Sea, there is a voluntary agreement between operators[310] supported by the UK government, the Offshore Pollution Liability Agreement 1974 (OPOL). The OPOL establishes a strict liability regime as against member operators, with a guarantee of the other members if the responsible party fails to honour its obligation.[311] It applies to offshore facilities within the European Union coastal State jurisdiction and Norway,[312] so that there is a certain harmonisation as to liability issues. However, the agreement only covers compensation for the cost of pollution damage and for remedial measures to deal with a pollution that has occurred, not the cost of the technical measures of stopping the blow-out as such.[313] Even for the damages covered, the regime is not binding and does not have a channelling effect, so that victims still have to decide whether they want to sue under the strict liability regime for limited compensation or bring a possibly lengthy and complicated claim in

---

[304] E. Tarelli, 'International Efforts to Establish Rules on Liability for Offshore Activities', in P. Ehlers and R. Lagoni (eds.), *Responsibility and Liability in the Maritime Context* (Münster: LIT Verlag, 2009), p. 188.

[305] See J. E. Fender, 'Trouble Over Oiled Waters: Pollution Litigation or Arbitration: The Ixtoc I Oil Well Blow-out' (1980) 4, Suffolk Transnational Law Journal, 281–301 at 282 et seq.

[306] Allen, '2011 "A Global Oil Stain"'.

[307] As quoted by R. Abeyratne, 'The Deepwater Horizon Disaster - Some Liability Issues' (2010) 35, Tulane Maritime Law Journal, 125–52 at 140.

[308] See M. Kashubsky, 'Marine Pollution from the Offshore Oil and Gas Industry: Review of Major Conventions and Russian Law (Part I)' (2006), Maritime Studies, 1–11 at 5 et seq; Tarelli, *2009 "International Efforts"*, p. 203; Allen, '2011 "A Global Oil Stain"', 91 et seq. for an overview.

[309] See Tarelli, *2009 "International Efforts"*, p. 203.

[310] Allen, '2011 "A Global Oil Stain"', 94.

[311] See Smith, '2011 "The Deepwater Horizon Disaster"', 1491 et seq. for an overview.

[312] See Tarelli, *2009 "International Efforts"*, pp. 197 et seq. for more details.

[313] Smets, *1985 "Indemnisation des dommages"*, p. 320.

tort.[314] The question whether there is a need for more regional cooperation[315] or a truly international agreement[316] that could also apply to areas beyond national jurisdiction[317] is, just like the reasons for the international reluctance to come to terms with this question[318] and the possible rules of a future agreement[319] outside the scope of this study. It however tries to give an analysis of coastal State jurisdiction under the law of the sea as it is currently in force and the conflict-of-law considerations that follow from it.

The debate on international liability for offshore pollution has certainly been kindled by the *Deepwater Horizon* catastrophe in the Gulf of Mexico in 2010, and at the same time, this incident is also revealing of how a national jurisdiction may deal with a marine disaster of this scale.

Apart from these major catastrophes, conflict-of-law problems may arise in connection with other, rather less spectacular incidents that can, nevertheless, be as tragic for the persons involved. Death and injury cases that take place on an offshore installation can be subject to very conflicting views if the installation is operated by an entity from a country other than the coastal State in the waters of which it is situated. Considerations of this kind will also be given regard to in the context of labour disputes and have there already been the subject of precedents in the ECJ.

1. Incidents Outside of or Involving the Installation

It has been pointed out that the coastal State, according to art. 60 para. 1 UNCLOS, enjoys in its EEZ the "exclusive right" to construct, authorise and regulate the construction, operation and use of, on the one hand, artificial islands, and, on the other hand, installations and structures for economic purposes. Art. 60 para. 2 confers "exclusive jurisdiction" over all matters connected with them. Still, the provision must be read in connection with art. 55 UNCLOS, so that the conferral of rights only operates to the extent in which the installations are used for economic purposes.[320] Under the law of the sea, only allowing the exercise of power to the

[314] B. Bollecker-Stern, 'A propos de l'accident Ekofisk - Problèmes posés par la pollution provoquée par les installations de production pétrolière off-shore' (1978) 24, Annuaire français de droit international, 772–91 at 787.

[315] Advocated by D. D. Caron, 'Liability for Transnational Pollution Arising from Offshore Oil Development: A Methodological Approach' (1983) 10, Ecology Law Quarterly, 641–83 at 643, in the wake of the *Ixtoc 1* oil spill in 1979 off the coast of Campeche, Mexico.

[316] Advocated by Allen, '2011 "A Global Oil Stain"', 103 et seq.

[317] Smith, '2011 "The Deepwater Horizon Disaster"', 1511.

[318] See Kashubsky, '2006 "Marine Pollution"', 8, for some suggestions.

[319] See the discussion in Smith, '2011 "The Deepwater Horizon Disaster"', 1507 et seqq.

[320] Churchill and Lowe, *1999 "The Law of the Sea"*, pp. 167 et seq.; Pestke, *2008 "Offshore-Windfarmen"*, p. 79; I. Vella, 'A new advent for renewable offshore resources', in N. A. Martínez Gutiérrez (ed.), *Serving the rule of International Maritime Law: Essays in Honour of Professor David Joseph Attard* (Abingdon: Routledge, 2010), pp. 146 et seq.

extent that the living and non-living resources of the EEZ are made use of, this is certainly correct. In all other respects, namely navigation, the high seas regime applies. This means that the coastal State can only put forward a sovereign right allowing it to submit the activity in question to its jurisdiction and the application of its own laws, including the conflicts rules, if the activity concerns the exploration and exploitation of the EEZ's resources.[321]

This may lead to difficulties in differentiation. First, it can always be questioned whether the case concerned has a connection with the coastal State's sovereign rights or whether there is no real link to these functional rights and obligations. This problem can arise with regard to the coastal State's criminal competence, as a criminal action such as an assault that only fortuitously takes place on an oil rig rather than on land may not be covered by the coastal State's competence over the EEZ. One can however argue that the presence of the persons involved in the crime is generally due to the coastal State's exercise of its functional rights.[322] The same question can effectively be asked with regard to certain activities that may not be as clearly covered by the coastal State's sovereign rights as subsoil mining,[323] such as bunkering.[324] Another example is aquafarming in the EEZ (also see *supra* under I.)—to the extent that this can actually cause damage to third parties. If the use of the EEZ for the purposes of breeding, rather than catching, fish (catch of wildlife stocks being covered by art. 61 UNCLOS) is not seen as an activity the coastal State can assert sovereign rights over, the activity would not be covered by the competences of the coastal State and there could not be a jurisdictional venue here, nor could the coastal State's law be applied as the *lex loci delicti*. If there is no vessel involved in the incident, so that there is no flag State either, damages caused by aquafarming would actually not be justiciable under the law of any State—there will however almost certainly be a judicial venue in the home jurisdiction of the defendant, which may decide to apply the *lex fori* in the absence of any other reasonable solution.

Another important question concerns allisions between a ship and offshore structures—or falling and floating parts of these[325]—in the EEZ, where there could be a possible conflict between the ship's law of the flag and the law of the coastal State which exercises sovereign rights by allowing the exploration or

[321] Dicey, Morris and Collins, *2012 "Conflict of Laws"*, para. 35-033.

[322] Esmaeili, *2001 "The Legal Regime"*, p. 93, also discussing the justiciability of pirate broadcasting from offshore installations.

[323] See D. Anderson, 'Coastal State Jurisdiction and High Seas Freedoms in the EEZ in the Light of the "Saiga" Case', in C. R. Symmons (ed.), *Selected Contemporary Issues in the Law of the Sea*, Publications on ocean development (Leiden, Boston: Martinus Nijhoff Publishers, 2011), on a general overview of activities in the EEZ with regard to which competence is debatable.

[324] See R. Lagoni, 'Offshore Bunkering in the Exclusive Economic Zone', in T. M. Ndiaye and R. Wolfrum (eds.), *Law of the sea, environmental law and settlement of disputes* (Leiden, Boston: M. Nijhoff, 2007); Anderson, *2011 "Coastal State Jurisdiction"*, pp. 110 et seqq.

[325] Scenario considered by Pestke, *2008 "Offshore-Windfarmen"*, p. 52; Hille, Schröder, Dettmer and Visser, '2010 "Offshore-Windkraftanlagen"', 586.

exploitation of the marine subsoil. The situation has been considered *supra* under A. for allisions on the high seas, where the laying of devices such as cables or pipelines is a general freedom of the high seas that can be exercised by any State, so that as a result, the law of the ship's flag must prevail as the more explicit connecting factor. This is of course different with regard to the offshore installations in the EEZ that are discussed here—they have been installed by virtue of the coastal State's functional rights. Even if the allision has not been directly caused by the economic activity in the EEZ, but rather by the mere presence of the installation, the coastal State has the right to put up the structure—or license its installation—and can thus assert a territorial sovereign right that should prevail over the weaker connecting factor of the ship's flag. Under Acts of Parliament in the United Kingdom, it is even considered that the national legislation applies directly,[326] as a provision of immediate application (also see *infra*), without previous consideration of the conflict-of-law problem. The solution to apply the law of the coastal State also presents the advantage of rendering the victim's law applicable in most cases, as it is considerably more likely that the allision, at least in the cases where the structure is immobile, was caused by the ship than by the structure. According to art. 60 paras. 4 and 5 UNCLOS, the coastal State has the right to establish safety zones of up to 500 m around the installation, in which measures can be taken to ensure the safety of the installation on the one hand and the safety of navigation on the other. If such a zone is established, the fact that the allision occurred within a zone in which the coastal State can enact certain rules also points to the fact that the law of the coastal State is called to apply to the case.[327]

Splitting the conflict-of-law regime for the EEZ in activities for which the coastal State can assume sovereign rights, such as seabed mining or fishing, and others such as navigation, coincides with the solution that the ECJ has set out for the conferral of jurisdictional competence following the *forum delicti* rule under art. 5 (3) of the Brussels Regulation (see *supra*, Chap. 6, B. II. 1.) in the *Weber v. Ogden* case. The claimant was held to have had his place of habitual employment in the coastal State because the activity he carried out for the defendant employer over the coastal State's continental shelf was closely linked to the exploration and exploitation of natural resources in the marine subsoil, over which the coastal State exercised sovereign rights.[328] The Advocate General explicitly clarified that the regime for navigation would therefore in all likelihood be different.[329] Even if the decision concerned the allocation of jurisdiction rather than the application of national law, it is here of the utmost desirability that the *forum delicti* and the *lex loci delicti* coincide because, especially in the field of offshore operations, many jurisdictions neglect the applicability of conflicts rules. In fact, most States in which

[326] Marsden, Gault, Hazelwood and Tettenborn, *2003 "Collisions at sea"*, para. 17-09: "statute seems to pre-empt the matter."

[327] Ramming, '2010 "Internationalprivatrechtliche Fragen"', 298.

[328] *European Court of Justice, 27.02.2002, Weber v. Ogden*, p. 2043, pt. 34.

[329] Ibid., p. 2023, pt. 30.

there is an existing offshore industry have elaborate national legislation on licensing, operation and in many cases also on liability for possible damages, which is most often specifically drafted for application in the EEZ.[330] Courts will then not embark on a conflict-of-law analysis when considering a case involving an offshore activity, but directly apply the respective domestic legislation, without considering conflict-of-law rules. In this manner, it has never been questioned that Norwegian law was to govern the Ekofisk accident on the Norwegian continental shelf,[331] that Australian law applied to the Montara Oil Spill off the Australian coast in August 2009, and that the *Deepwater Horizon* catastrophe had to be dealt with under US federal legislation on oil pollution. Following the *Ixtoc I* spill in 1979, which occurred off the Mexican coast but caused damage to the coast of southern US states, it was hinted at and seemingly accepted by US legal opinion that there might not be a legal way of holding the Mexican entities operating the well accountable for the damage under American law, as there were already multiple problems with regard to the jurisdiction of US courts over the companies.[332] Not least, the case hinged on the adjudgment of activities that were undertaken by Mexico in exercise of sovereign rights over the marine resources.[333] It was thus accepted that, consequentially, there would be problems of sovereign immunity possibly ruling out the jurisdiction of the American courts.

The realm and the character of "norms of immediate application" or *lois de police* in this field of law is often not overly clear,[334] and their existence and application can in some cases certainly be criticised for disregarding foreign elements in the case. To the extent to which, in practice, each State tends to apply its own laws to individuals in a direct relationship between the State and the person, such as administrative law, criminal law, social security law etc., their application is certainly justified[335] and the problem is rather one of the differentiation between a direct relationship with the State and a relationship between private entities that is merely influenced by overriding State law.[336] It does seem that the liability for offshore activities, considering that the rights to exploitation are

[330] See Allen, '2011 "A Global Oil Stain"', 95 et seq. on the Australian and United States federal legislation on licensing, prevention and liability; A. W. Moreira, C. Y. Strickland and D. G. Henley, 'Liability for Marine Pollution from Offshore Operations' (2003) 26, Dalhousie Law Journal, 429–67 at 431 et seq. on Canadian legislation.

[331] See Bollecker-Stern, '1978 "A propos de l'accident Ekofisk"', 781.

[332] See J. M. West, 'The Ixtoc I Oil Spill Litigation: Jurisdictional Disputes at the Threshold of Transnational Pollution Responsibility' (1981) 16, Texas International Law Journal, 475–532 at 491 for details.

[333] Fender, '1980 "Trouble Over Oiled Waters"', 300; W. J. Macdonald, 'Ixtoc I: International and Domestic Remedies for Transboundary Pollution Injury' (1980) 49, Fordham Law Review, 404–31 at 406.

[334] de Maekelt, *2011 "Relaciones"*, pp. 48 et seq.

[335] Ibid., p. 48.

[336] Ibid., p. 52.

granted by the coastal State to private or semi-private entities, can be one of these situations.

Unlike the regime for vessel-source pollution in which there are important safeguards to ensure free navigation, the regulation of offshore activities in the EEZ is a matter in which the coastal State has far-reaching legislative competences. Even if these competences can be put forward for the activities over which the coastal State has sovereign rights, so that there is no general extension of the coastal State's laws to the EEZ,[337] the offshore activities discussed here are covered by the competence of the coastal State. Under the law of the sea, the exploration and exploitation of the marine subsoil and the generation of energy is subject to the coastal State's functional sovereign rights, and the State has both the right to legislate in this field[338] and adjudicate on cases that come up in connection with it.[339] It does not seem that a State is obliged to apply conflict-of-law rules to cases involving offshore installations in its EEZ. In the absence of any other foreign factor than the mere fact that the events took place in the EEZ rather than on land territory, it might even be argued that there is not even any other law called to apply. In any case, application of the *lex loci delicti*, following what has been set out above, would result in the application of the domestic law of the coastal State. Under these circumstances, it seems admissible to apply rules of immediate application without a previous conflict-of-law analysis. Even if it can be put forward that EU member States have a duty to apply the rules of the Rome II Regulation, this argument can be countered, first, by the consideration that the case is not even a conflicts case if there is not any other foreign factor and, second, by considering that the outcome would in any case be the same.

In fact, it is the coastal State that assigns the places in which private entities undertake exploration and exploitation activities,[340] and the initial competence to do so is rooted in the coastal State's sovereign rights. It thus only seems fair that the coastal State also assumes full responsibility over these activities[341] and decides which rules will be applied in court. The big difference between offshore activities and shipping is that the exploration and exploitation of the EEZ's resources can hardly be undertaken without the States' consent. A marked exception in this respect is, without doubt, fisheries, where the control of illegal or undesired action is difficult.[342] A thorough analysis of this problem is outside the scope of this study. It will however be sufficient to point out that, even if there are considerable problems in practice, the control of the EEZ and the fishing activities carried out

[337] Hille, Schröder, Dettmer and Visser, '2010 "Offshore-Windkraftanlagen"', 590 et seq.

[338] Moreira, Strickland and Henley, '2003 "Liability for Marine Pollution"', 444, also considering the situation for the continental shelf.

[339] Ramming, '2010 "Internationalprivatrechtliche Fragen"', 297.

[340] Abeyratne, '2010 "The Deepwater Horizon Disaster"', 143.

[341] Pestke, *2008 "Offshore-Windfarmen"*, p. 81.

[342] See Anderson, *2011 "Coastal State Jurisdiction"*, p. 113 on possible ways to control whether fishing vessels are effectively fishing or only transporting catches.

there are within the competence of the coastal State, which, even if conferred to a regional organisation and then subject to harmonised decision-making processes, generally justifies the application of the coastal State's domestic law.

It is however needless to say that this solution has a marked disadvantage with regard to damages in the waters or the territory of another State. It allows coastal States embarking on offshore activities to set standards unilaterally and without regard to the interests of other States, a problem that was recognised as early as 1979 following the *Ixtoc I* accident.[343]

However, this is the state of the law as of present. Transboundary pollution to the detriment of individuals in another State may be a problem that is litigated under public international law, as between States, or dealt with in diplomacy. A State is barred from unilaterally assuming jurisdiction or the application of its domestic law to a case that concerns the exercise of sovereign rights by another State. Until the time where there is an international agreement by which States submit civil liability claims for pollution caused by offshore activities to a uniform regime, these activities fall under the sole competence of the coastal State, which should be reflected by the application of the coastal State's laws, either in the shape of conflict-of-law rules or of norms of immediate application.

2. Incidents on the Installation

As has been shown, maritime incidents involving an offshore installation that have an impact on the environment or on any other asset in the EEZ are clearly covered by the coastal State's domestic law. This could be different, from a practical point of view, for incidents whose consequences are restricted to the installation itself. Examples of this are accidents on the installation resulting in death or injury to the persons present, or other incidents in which the coastal State may not have a vital interest.

In practice, many offshore installations especially in developing countries may be licensed by the coastal State but are effectively run by an entity from another country, and the licensees are likely to subject employment contracts and other legal relationships on board the platform to a law other than that of the coastal State.[344] An illustration of this problem is the rather old English precedent of *Sayers v. International Drilling Co. N.V.*[345] The claimant, an English oil rig worker, had entered employment with a Dutch company that subsequently sent him to work on an oil rig off the Nigerian coast. In the course of his employment, the claimant was

---

[343] A. T. Leonhard, 'Ixtoc I: A Test for the Emerging Concept of the Patrimonial Sea' (1980) 17, San Diego Law Review, 617–27 at 625.

[344] M. White, 'Offshore Craft and Structures: A Proposed International Convention' (1999) 18, Australian Mining & Petroleum Law Journal, 21–7 at 24.

[345] "*Sayers v. International Drilling Co.*" 1971 Weekly Law Reports 1176 (Court of Appeal (England), 10 May 1971).

seriously hurt in a work accident caused by the negligence of his fellow employees. The claimant sued the Dutch company in England, based on an English statute and essentially brought the claim in tort. The defendant company however raised defences under the claimant's work contract, which contained provisions on compensation for work accidents but did not contain a choice-of-law clause. In determining the law that was to govern the relationship between the parties, the court ruled out Nigerian law very quickly. Lord Denning M. R. argued that the defendant company was Dutch and "the Nigerians had nothing to do with the rig",[346] Salmon L. J. pointed out that the contract stated that the defendant company had the right to change the claimant's place of work at any moment[347] and that the location where the accident occurred was thus rather fortuitous. Even if it is true that the court thus disregarded the fact that Nigeria exercised sovereign rights over the oil rig, even if run by a Dutch company, it is not certain whether the case can really serve as a precedent for the applicability of the licensee company's law. The case hinged on the question which law applied to the work contract and whether the provisions on compensation were valid, and it was thus decided in contract after all.

To the extent that labour law disputes are, in most countries, contract disputes most likely submitted to a given legal system by a choice-of-law clause, statutory conflict-of-law rules may in fact be of a reduced importance here. In any case, offshore installations generally do not fly a flag[348] and the attribution of nationality by registry, e.g. as "ships",[349] or by other means that are effectuated in certain States[350] is not internationally consistent.[351] In the cases where the claim is really brought in tort and governed by the *lex loci delicti*, the territorial link to the coastal State is therefore not overly likely to be called into question, even if the installation is registered in another country.

Within Europe, it does not seem that the applicability of the coastal State's laws to all matters concerning the offshore installation, civil and criminal[352] and including labour law and social security matters,[353] is seriously questioned. The applicability of the laws of the operating company's place of business is thus ruled out.[354] Authors often refer to the *Weber v. Odgen* case, which, even if it was decided as a matter of jurisdiction under the Brussels I Regulation, is often taken to point out a solution for the conflict-of-law problem, too. In any case, aligning the *forum delicti* and the applicable law under the Rome I Regulation—and arguably also the Rome

[346] Ibid., p. 1181.

[347] Ibid., p. 1183.

[348] Bonassies, '1969 "La loi du pavillon"', 518.

[349] See Esmaeili, *2001 "The Legal Regime*", pp. 21 et seq. on this problem.

[350] Kaye, '2007 "International Measures"', 388.

[351] White, '1999 "Offshore Craft and Structures"', 23.

[352] U. Jenisch, 'Offshore-Windenergieanlagen im Seerecht' (1997), Natur und Recht, 373–81 at 375.

[353] Pestke, *2008 "Offshore-Windfarmen*", p. 80.

[354] Jenisch, '1997 "Offshore-Windenergieanlagen"', 375.

II Regulation—is a sound solution in terms of legitimacy under public international law, especially for tort cases. Under maritime labour law, problems would then rather arise with regard to the question whether a clause in the labour contract is reconcilable with the law that the conflicts rule points to.

This is, again, different under the law of the United States. Under the doctrine of "the proper law of the tort", and especially in cases where actions in both contract and tort possible, conflicts considerations will also be taken into account for maritime personal injury claims (see *supra*, A. III. 2. e)). The Jones Act can under certain circumstances also apply to workers on offshore installations, the so-called "brownwater" as opposed to "bluewater" seamen, namely where the installation can be considered as a "ship" for the purposes of the Act. Still, at a rather early point in time, it was held the pertinent connecting factors test set out by the *Lauritzen-Rhoditis* precedents does not apply to "brownwater" seamen to the same extent.[355] This was mainly because offshore installations such as oil rigs were stable at a given geographical point so that the place of the wrong was foreseeable, and because offshore workers, at least at that time, were generally all hired under the same circumstances, unlike "bluewater" seamen hired by different manning agencies in different countries.[356] The Jones Act was subsequently amended in order to exclude claims from foreign "brownwater" seamen for injuries sustained outside US waters, except for the cases where there is not any other legal remedy for them under the law of the foreign coastal State or under the laws of their country of residence or citizenship.[357] There are other pieces of US legislation dealing with death and injury claims, some of them explicitly applying to activities of exploring and exploiting the marine subsoil,[358] but they are, in general, exclusively applicable to activities taking place in waters over which the United States claim functional sovereign rights. It is interesting to note that the US legislation in this way also emphasises the importance of the laws of the coastal State when dealing with torts that occurred on the offshore installations in the waters of that State. In the *Deepwater Horizon* disaster, the applicability of domestic law to the fatalities and injury cases on the platform[359] was never seriously questioned, regardless of the platform's registry in the Marshall Islands,[360] since the accident took place during drilling operations in waters over which the United States claim sovereign rights. In the same spirit, it is unanimously held that US law applies to helicopter accidents during landing on offshore platforms, even if the victims die at sea rather than on

---

[355] Allbritton, '1983 "Choice of Law in a Maritime Personal Injury Setting"', 895 et seq.

[356] Ibid., p. 897.

[357] Scoles, Hay, Borchers and Symeonides, *2004 "Conflict of Laws"*, p. 891, footnote 6.

[358] See Warshauer and Dittman, '2005 "The Uniqueness"', 1219 et seq.; M. Davies, 'Liability Issues raised by the Deepwater Horizon Blowout' (2011) 25, Australian and New Zealand Maritime Law Journal, 35–48 at 46 et seq.

[359] See R. Perry, 'The Deepwater Horizon Oil Spill and the Limits of Civil Liability' (2011) 86, Washington Law Review, 1–68 at 8 et seq. for more information.

[360] Davies, '2011 "Liability Issues"', 45.

the platform. The question of the applicable statute as such is however fervently debated.[361] The question whether the death occurred as a consequence of the maritime activity of transport—if the helicopter pilot is concerned—or as a consequence of the exploitation activities if the victims are oil rig workers does matter for the applicability of the US statute in question,[362] but not for the question whether US law or the law of any other country should apply.

In conclusion, it must be stated that the exploration and exploitation of the EEZ's resources is a matter uniquely submitted to the sovereign rights of the coastal State, and that the competence to deal, as a matter of legislation, enforcement and judicial jurisdiction, with the legal relationships that derive from it is rather far-reaching and, depending on the view taken by the forum, may encompass the matter of limitation of liability.[363] In this context, it must be stressed that the LLMC only applies to ships and, by virtue of its art. XV para. 5 (b), explicitly excludes its application to floating platforms serving the purpose of exploring or exploiting seabed or subsoil resources. Therefore, even the limitation of liability is left to the unilateral legislative initiative of the coastal State. It may decide on a rather strict regulation of these activities, or a very lenient one, in order to incentivise investors to finance the exploitation of resources.[364] The existence of different attitudes in this matter may be regretted, but is however reality and may be one of the reasons that has it made difficult for the international community to come to terms in this matter.

C. Territorial Waters

The coastal State's territorial waters, extending 12 nm from its baselines, are, by virtue of the law of the sea, subject to the State's territorial sovereignty. In contrast to the regime applicable to the EEZ, the coastal State's territorial sovereignty, according to art. 2 UNCLOS, extends to these waters, giving the coastal State, at first sight, rather wide powers, albeit with some exceptions.

The most important[365] exception is without doubt foreign-flagged ships' right to innocent passage,[366] which has already been discussed *supra*, Chap. 6, B. II. 1. c). In general, the coastal State is not entitled to deny or impair the passage of a ship

[361] See R. Hackney, 'If a Helicopter Hits an Offshore Platform and Crashes at Sea, Where Do You Bury the Survivors?' (2011) 42, Journal of Maritime Law and Commerce, 231–54.

[362] Ibid., pp. 236 et seqq.

[363] C. A. Fleischer, 'Liability for Oil Pollution Damage Resulting from Offshore Operations' (1976) 20, Scandinavian Studies in Law, 107–43 at 134 et seq.

[364] Ibid., p. 136.

[365] Ngantcha, *1990 "The Right of Innocent Passage"*, p. 38.

[366] See K. Hakapää and E. J. Molenaar, 'Innocent passage - past and present' (1999) 23, Marine Policy, 131–45 for an overview.

through its territorial waters, unless the ship's transit has adverse impacts on the coastal State's peace, good order and security, which is however a quite substantial threshold to cross. The exemplary list of activities in art. 19 para. 2 UNCLOS that rule out innocent passage is also illustrative in that respect, as in general, the ship needs to embark on activities that clearly give proof of a disregard of the coastal State's sovereignty. With regard to possible marine pollution, it must be emphasised that art. 19 para. 2 (h) UNCLOS only enunciates "wilful and serious pollution", so that, following the wording of the provision, a ship polluting foreign territorial waters accidentally or below the threshold of a "serious" pollution is still in innocent passage.[367]

UNCLOS does not contain detailed provisions on judicial competence for torts happening on ships in innocent passage through the territorial sea, let alone for the conflict-of-law rules that should apply. In view of the close interplay between legislative, enforcement and judicial jurisdiction, it seems however logical to consider the impact that the right of innocent passage has on legislative and enforcement jurisdiction and draw conclusions for judicial competence and the conflict-of-law rules, too.

Concerning the legislative competence, art. 21 UNCLOS enables the coastal State to adopt laws and regulations relating to innocent passage through the territorial sea. There are however limits to the matters the coastal State can regulate—domestic laws and regulations on design, construction, equipment and manning must conform with internationally accepted standards, such as those set out in SOLAS and MARPOL. This is in the interest of protecting free navigation, as design, construction, equipment and manning are matters that the vessel cannot change *en route* when crossing different coastal State's territorial waters.[368] It can however be inferred from art. 21 para. 1 UNCLOS that for all other matters relating to innocent passage, especially concerning the behaviour of ships' crews and operational matters, the coastal State is free to legislate. With regard to indirect legislation, not immediately regulating ships, art. 24 UNCLOS is generally considered to allow the coastal State to enact indirect legislation,[369] not directly addressing ships but which will also be applicable to them,[370] as long as the rules adopted do not have the practical consequence of impairing or denying innocent passage and are not discriminatory. If this is true, the coastal State can submit a ship in its territorial waters to its own laws on civil jurisdiction, including the conflict-of-law rules on the law applicable.

---

[367] Boyle, '1985 "Marine Pollution"', 359; A. K.-J. Tan, *Vessel-Source Marine Pollution* (Cambridge, New York: Cambridge University Press, 2006), p. 207.

[368] Boyle, '1985 "Marine Pollution"', 360; E. J. Molenaar, 'Port State Jurisdiction: Toward Comprehensive, Mandatory and Global Coverage' (2007) 38, Ocean Development and International Law, 225–57 at 230.

[369] Nordquist, Nandan and Rosenne, *1993 "UNCLOS Commentary Vol. II"*, para. 24.1.

[370] See D. P. O'Connell and I. A. Shearer, *The International Law of the Sea* (*Vol. II*), (Oxford: Clarendon Press, 1984), p. 741, presenting the problem from the point of view of English law.

This far-reaching legislative competence of the coastal State also has an impact on applicable jurisdiction that actually can characterise a certain behaviour as a tort. Whereas navigation on the high seas and in the EEZ is not subject to the domestic legislation of any State and wrongful behaviour must be defined according to internationally accepted standards, such as, for instance, the COLREGS, the coastal State is free to define behavioural standards in its territorial waters[371]—but not technical ones for construction, design, manning and equipment of ships which must be internationally accepted in order not to restrict foreign-flagged ships' right to innocent passage. The coastal State can thus, for instance, enact domestic collision avoidance rules.[372]

Concerning the enforcement jurisdiction of the coastal State, arts. 27 and 28 UNCLOS recognise a rather wide competence of the coastal State to enforce criminal and civil jurisdiction over persons but, at the same time, encourage the coastal State to voluntary refrain from exercising it unless the case affects its own interests. Art. 27 para. 1 UNCLOS states that the coastal State "should not" exercise its enforcement jurisdiction unless the consequences of the crime extend to the coastal State, the crime is of a kind to disturb the peace of the country or the good order of the territorial sea, or if the ship's master or a representative of the flag State has requested the coastal State's public authorities' assistance, or, fourthly, for drug offenses where there arguably is a universal goal of cooperation between States. In all other cases, UNCLOS calls upon coastal States to refrain from the exercise of enforcement jurisdiction, without however rendering this an international obligation. The situation is very similar for civil jurisdiction, where UNCLOS provides in art. 28 para. 1 that the coastal State "should not" interfere with a ship's innocent passage to enforce a civil claim against a person on board. A differentiation between acts internal to the ship, not affecting the coastal State, and acts external to the ship is again made in art. 28 para. 2 for the civil jurisdiction as against ships. The coastal State "may not" arrest a ship in innocent passage, the interdiction here being absolute,[373] unless the civil claim was incurred by the ship during the passage through the territorial waters or for the purpose of the passage—i.e., if it concerns an interest of the coastal State. This differentiation for enforcement jurisdiction must be considered to have an impact on judicial jurisdiction, too, as has been shown *supra*, Chap. 6, B. II. 1. c), and arguably also on the conflict-of-law rules. It is in fact desirable that in the majority of cases, the *forum delicti* and the *lex loci delicti* should coincide, so that the court that is competent by virtue of its

[371] Jambu-Merlin, '1962 "Note CA Rennes 13.03.1961"', 156; Carbone, '2009 "Conflits de lois"', 229, with regard to domestic collision prevention rules.

[372] See the English precedent of "*The Eland*" *and* "*The Monte Urquiola*" 2 Lloyd's Law Reports 328 (Admiralty Division, 11 July 1969), a collision case between a Spanish and a Liberian vessel in Argentine waters, where the court found the *Eland* to be guilty based on Argentine laws on maximum speed and collision avoidance rules, the English court actually interpreting the Argentine provisions.

[373] Nordquist, Nandan and Rosenne, *1993* "*UNCLOS Commentary Vol. II*", para. 28.4(b); Churchill and Lowe, *1999* "*The Law of the Sea*", p. 97.

geographical competence for the place of the tort, should also apply its own law to the case.[374]

In any case, events that are bearing consequences outside the ship and thus have an impact on the waters under the territorial sovereignty of the coastal State, are under the jurisdiction of that State, and it is rather likely that the coastal State will choose to apply its own laws under the respective conflicts rules. This is also the solution under arts. 4 (1) and 7 of the Rome II Regulation.

In contrast to that, events occurring exclusively on board a ship deserve a separate discussion. To the extent that a ship in innocent passage should not be submitted to the coastal State's enforcement mechanisms, it is still arguable that incidents that take place among the crewmembers or passengers on board should not be submitted to the law of the coastal State either, since the ship's position is in these cases a rather fortuitous circumstance. It is remarkable that, as widely as it is often asserted that the coastal State's laws apply to all events taking place in territorial waters, this result is often mildly put into question for certain cases and under certain circumstances,[375] most often in considering the conflicting realm of the flag State's laws.[376] Again, collisions present a rather special problem and will be discussed separately.

I. Applicability of the Coastal State's Laws in Territorial Waters

When considering the coastal State's territorial sovereignty with regard to navigation on the one hand and with regard to fixed platforms and other offshore installations on the other hand, it must first be stressed that the coastal State enjoys full sovereignty over these installations in its territorial sea and that, hence, they are fully submitted to its jurisdiction.[377] Correspondingly, the *lex loci delicti* rule in the territorial sea points to the laws of the coastal State.[378] As UNCLOS does not oblige States to maintain a register for offshore installations and recognises that the attribution of nationality may be effectuated in a different way (see *supra* under A. I.), coastal States are thus completely free to submit offshore installations in their territorial waters to their powers and to apply their domestic rules to them or to refrain from this. There is no conflicting legal regime the role of which would be recognised by the law of the sea, such as the flag for vessels. The fact that it is hardly possible to question the application of the coastal State's laws to offshore

[374] Wagner, '2006 "Internationales Deliktsrecht"', 375.

[375] Bonassies, '1969 "La loi du pavillon"', 550: "one should not conclude this quickly".

[376] See Brækhus, '1979 "Choice of Law Problems"', 282, considering that the law of the flag gives a certain "protection" to the ship.

[377] Esmaeili, *2001 "The Legal Regime"*, pp. 71, 100.

[378] Hille, Schröder, Dettmer and Visser, '2010 "Offshore-Windkraftanlagen"', 589.

installations, whereas there are some doubts with regard to ships, is revealing of the fact that it is after all the special status of vessels under the law of the sea that could result in a deviation from the full and uncontested application of the coastal State's laws to all cases involving ships.

However, and very generally, the coastal State's interest in the state of its territorial waters is considerable,[379] and this from historical times.[380] It is certainly not lightly that one could question the applicability of the coastal State's laws following a *lex loci delicti* analysis.

Under English law and the original double actionability rule established by *Philips v. Eyre* (see *supra* under A. II. 1.), a tort, in order to be justiciable under English law, needed to be actionable both under English law as the *lex fori* and under the law of the place where the tort occurred. For maritime torts taking place in the territorial waters of a foreign State, this *lex loci delicti* was taken to be the law of the coastal State,[381] so that the law of the vessel's flag would be entirely disregarded. This was the result found in a number of collision cases[382] (see *infra* under II.) and for cases in which there was damage done to port installations in foreign countries,[383] and it was suggested that torts occurring on one ship only should be solved in the same way.[384]

Precedents to support this result are however scarce and sometimes not overly clear in their scope of applicability. In the 1910 case of *Carr v. Fracis Times & Co.*,[385] an action was brought for events which had taken place in the territorial waters of Muscat, then under the rule of the Sultan of Muscat, "an absolute and despotic ruler". An officer of the British Navy had seized ammunition belonging to the claimants, upon orders from the British government, which were in turn authorised by the orders of the Sultan. When the seizure was disputed, the defendant put forward that according to the law of Muscat—the *lex loci delicti*, by virtue of the Sultan's orders—there was a valid defence, so that under the double actionability rule the claim must fail. The law of the flag of the ship on which the goods had been transported was not taken into account.

---

[379] Carbone, '2009 "Conflits de lois"', 211.

[380] Tetley, '1993 "The Law of the Flag"', 145.

[381] Finlayson, '1986 "Shipboard torts"', 120; George, '2007 "Choice of Law"', 160.

[382] "*The Halley*", 2 Law Reports Privy Council Appeals 193 (Privy Council, 17 June 1868); "*The Arum*" Law Reports, Probate Division 12 (High Court (England), Probate, Divorce and Admiralty Division, 15 October 1920); "*The Waziristan*" Weekly Law Reports 1446 (High Court (England), Probate, Divorce and Admiralty Division, 19 October 1953).

[383] "*The Mary Moxham*", 1 Law Reports, Probate Division 107 (High Court (England), Probate, Divorce and Admiralty Division, 03 May 1876).

[384] George, '2007 "Choice of Law"', 164. See however the *dicta* in *High Court (England), Probate, Divorce and Admiralty Division, 03.05.1876, "The Mary Moxham"*, 1, p. 112 by Mellish L. J.: "I agree that as to acts done on board a ship itself, no doubt the English ship carries English law with it".

[385] "*Carr v. Fracis Times & Co.*" Law Reports, Appeal Cases 176 (House of Lords, 08 July 1901).

The applicability of the laws of the coastal State to torts solely occurring on the ship was also arrived at in a Scottish precedent[386] of 1954. The case concerned a crewmember of a ship flying the British flag who had been seriously injured in a shipboard accident while the vessel was lying at anchor in the territorial waters of the Dominican Republic. In application of the double actionability rule, the case hinged on the question whether the claimant needed to show the actionability of his claim under the laws of the Dominican Republic, or whether the case should be considered as uniquely domestic, as the ship was registered in a Scottish port and the proceedings were brought before the Scottish court of Sessions. The judges, in considering the case, showed a considerably sympathy for the claimant's argument that the laws of the Dominican Republic should not be taken into account as the tort had exclusively taken place on board the ship and the geographical position of the vessel was purely fortuitous. Carmont L. J. admitted that it was in fact strange that the case should depend on the laws of the Dominican Republic under these circumstances, and this only due to the fact that the ship was within the territorial sea of that country, whereas the outcome would be completely different if the ship had been four nautical miles from the shore,[387] the territorial sea then only extending three miles from the baselines. Sorn L. J. agreed that the application of a fortuitous *loci delicti* "produced results which had an element of absurdity".[388] The judges were however unanimous in holding that the applicability of the coastal State's laws was the state of the law and that they could not hold judgment to the contrary.

The Scottish court's findings have further been quoted as common law precedent in a New South Wales case of 2002[389] in which the set of facts was essentially the same, a seaman on a ship registered in New Zealand having been injured during unloading operations in a New South Wales port. The court, after very detailed analysis of the problem, was adamant that the applicability of the laws of the coastal State was a principle from which it could not depart. In this context, the court also considered the precedents under US law, namely *Lauritzen* (see *supra*, A. III. 2. e)), in which the outcome had been held to be different following the analysis of the seven or eight "pertinent factors". The Australian court however rejected this argument in favour of the applicability of New South Wales Law. In comments at a later point in time, there have however sometimes been attempts to interpret the outcome of the case more narrowly, by suggesting that the principle only applied where the ship was in fact in port and unloading its cargo and had thus a physical link with the coastal State, so that the outcome might be different if the vessel was merely passing through the foreign territorial waters.[390]

---

[386] "*MacKinnon v. The Iberia Shipping Co.*" 1955 Court of Session Cases 20 (Court of Session (Scotland), 26 October 1954).

[387] Ibid., p. 28.

[388] Ibid., p. 36.

[389] "*Union Shipping New Zealand Ltd. v. Morgan*", 54 New South Wales Law Reports 690 (Court of Appeal of the Supreme Court of New South Wales, 14 June 2002).

[390] P. R. Beaumont, 'The Contribution of Alexander (Sandy) Anton to the Development of Private International Law' (2006), Juridical Review, 1–28 at 26, footnote 89; George, '2007 "Choice of Law"', 163 et seq.

Even if it was thus to some extent recognised that the application of the *lex loci delicti* to torts within one vessel in territorial waters created doubtful results,[391] there were no attempts to change this undesirable result under English law. This attitude was continued by the Law Commission when evaluating the tort law before the 1995 reforms in which the double actionability rule was abolished. The Commission declared that application of the coastal State's laws to torts in territorial waters was "unattractive",[392] but there were finally no steps taken to alter this outcome under English law.

This result is thus presumably unchanged by the 1995 reforms to the English conflict-of-law rules, only subject to the exception provisions under which a different law can apply[393] if it is "manifestly better placed to govern the case".

Very similarly, authors discussing the problem under German law before the coming into force of the Rome II Regulation argued that a distinction should be made between cases in which the geographical position of the ship was of the essence and others where this was not the case.[394] It was argued that in all cases in which the incident occurred while the ship was merely in transit and thus did not have a closer link to the State whose territorial waters it was crossing, the applicable law should be the law of the flag rather than the law of the coastal State.

The validity of this pragmatic approach can however be questioned with the coming into force of the Rome II Regulation. It must be stated that art. 4 (1) clearly points to a geographical connecting factor and that in territorial waters, in contrast to the high seas, this factor is meaningful. The coastal State does enjoy territorial sovereignty over its territorial waters, hence, the *lex loci damni* rule can be considered to point to a "country".[395] Still, some doubts can be expressed as to whether the applicability of the coastal State's laws is justified in cases where the ship merely passes through or fortuitously anchors in the territorial waters[396] as opposed to actually calling in a port, thus creating a certain link with the port State that may justify the application of the coastal State's law.[397] This is a concern that has also been addressed in connection with the New South Wales case cited above,[398] but it was in this context merely pointed out that the coastal State's law can possibly be ruled out under the escape clauses of the Rome II Regulation.[399]

---

[391] Also see Finlayson, '1986 "Shipboard torts"', 123.

[392] *1984 "Working Paper No. 87"*, para. 2.112.

[393] George, '2007 "Choice of Law"', 161.

[394] Kegel and Schurig, *2004 "Internationales Privatrecht"*, p. 19; Kropholler, *2006 "Internationales Privatrecht"*, p. 537.

[395] George, '2007 "Choice of Law"', 169; Fawcett, Carruthers and North, *2008 "Private International Law"*, p. 859; Dickinson, *2008 "The Rome II Regulation"*, para. 3.306; Dicey, Morris and Collins, *2012 "Conflict of Laws"*, para. 35-033; Ramming, '2010 "Internationalprivatrechtliche Fragen"', 294.

[396] Fawcett, Carruthers and North, *2008 "Private International Law"*, p. 860.

[397] Carbone, '2009 "Conflits de lois"', 210 et seq.

[398] Beaumont, '2006 "The Contribution"', 26, footnote 89.

[399] Fawcett, Carruthers and North, *2008 "Private International Law"*, p. 860.

The derogation of the *lex loci rule* for torts occurring in territorial waters was not advocated.

The jurisdiction of the ECJ in the *DFDS v. Sjöfolk* decision discussed *supra* in Chap. 6, B. II. 1. c) could however be taken to point to the fact that for torts purely internal to the ship, the law of the flag rather than the ship's fortuitous geographical position may be the pertinent connecting factor.[400] It must however be kept in mind that the judgment concerned the establishment of the *forum delicti* under the Brussels Regulation, not the choice of law under the Rome II Regulation. Conclusions can only been drawn from the case if one considers that the rules on the *forum delicti* and the *lex loci delicti* should necessarily point to the same State. This is however strongly suggested,[401] considering that the Rome II Regulation was held to be another important step in the harmonisation of jurisdiction and conflict-of-law rules after the Brussels Regulation and the Rome I Regulation, the three instruments being meant to create a uniform regime throughout Europe.[402]

The same underlying uncertainty as to whether the territorial sovereignty of the coastal State should really encompass the application of domestic law to events purely internal to the ship can be witnessed in France. There is no widely recognised precedent as to the applicability of the coastal State's laws to torts purely internal to the ship that could be considered to answer the question. The reason for this may be found in the fact that French law tends to consider cases of shipboard death and injury more often as a matter of contractual law, so that the law applicable will rather be found in the contractual terms and the circumstances under which the contract was concluded than under a *lex loci delicti* analysis. Based on the solutions that have been found for high seas torts, it has been suggested that the same differentiation should be made in territorial waters, external torts being governed by the law of the coastal State whereas torts entirely confined to the ship should remain subject to the law of the flag.[403] In any case, the fact that the law of the flag may apply to matters which do not regard the interests of the coastal State was recognised at an early point in time.[404] Following the coming into force of the Rome II Regulation, there is apparently even more uncertainty as to the law that should apply to maritime torts in territorial waters of a State. Notwithstanding the clear codification of the *lex loci damni* rule in the Rome II Regulation, it has been suggested that in the cases in which a ship merely sails through the territorial waters of a coastal State, it should "perhaps" still be the law of the flag that applies to the events on board, in the same way that this is the rule in international criminal law.[405]

---

[400] Dickinson, *2008 "The Rome II Regulation"*, para. 4.55; Basedow, *2012 "Die private Haftung"*, p. 1887.

[401] Wagner, '2006 "Internationales Deliktsrecht"', 375; Dickinson, '2013 "Territory"', 97.

[402] Plender and Wilderspin, *2009 "The European Private International Law"*, para. 1-028.

[403] Tetley, '1993 "The Law of the Flag"', 156.

[404] G. Ripert, *Droit maritime (Tome I)*, 4th edn., (Paris: Éditions Rousseau, 1950), p. 324.

[405] Mayer and Heuzé, *2010 "Droit international privé"*, para. 685.

As there is, under US conflicts law, no clear-cut rule as to the one factor determining the law applicable to maritime torts, the solutions found under the *Lauritzen-Rhoditis* principle of multiple pertinent factors will also hold true for torts in the territorial waters of a State, the geographical position of the ship being just one factor amongst others.

Considering the solutions that have been identified under European legislation bound by the Rome II Regulation, it is noteworthy that namely in English law and other common law legal systems, the applicability of the law of the coastal State has been asserted but that many authors express their discontent with it and try to qualify or narrow down the scope of the *lex loci damni* rule. The disadvantages of this outcome have already been pointed out: it seems difficult to apply different laws to one and the same set of facts, depending on whether it takes place just inside or just outside territorial waters,[406] even if due to development in technology, it is today generally possible to always determine the ship's exact position. One can however witness a certain discontent with the literal application of the *lex loci damni* rule to all torts in territorial waters, even to those that are completely confined to the ship and where the coastal State does not have any interest to adjudicate, in the same way that it does not have interest to legislate or "should not" exercise its enforcement jurisdiction under the provisions of UNCLOS. It is most probably by this thought that the ever re-emerging differentiation between torts on a ship actually only passing and a ship lying in port or carrying out loading and unloading operations[407] is to be explained: under the law of the sea, the coastal State assumes a greater territorial sovereignty over its internal waters and ports, as there is no right of innocent passage through these waters.

The importance of the right of innocent passage and its implications for coastal State sovereignty should not be overlooked. Events that only take place on the ship, without the coastal State's interest being affected, are, following the idea of the right of innocent passage, only under the competence of the flag State. The coastal State does not have an interest in the adjudication of these incidents, and almost the same arguments that hold true in questioning jurisdictional competence under the Brussels I Regulation also apply under the Rome II Regulation. The purpose of applying the *lex loci delicti* to torts is the fact that the event happened in a given geographical area, the rules and laws of which were familiar to the parties, who did not know that they would have to deal with each other. In a ship, passing through different maritime zones and possibly through the territorial waters of more than one State, this link between the place where the tort took place and the expectations of the parties does not exist. To the contrary, the persons present on board will

---

[406] George, '2007 "Choice of Law"', 137; also pointed out by Carmont L. J. in *Court of Session (Scotland), 26.10.1954*, "*MacKinnon v. The Iberia Shipping Co.*", p. 29.

[407] See Finlayson, '1986 "Shipboard torts"', 126; Beaumont, '2006 "The Contribution"', 26, footnote 89; George, '2007 "Choice of Law"', 164.

rather look to the law of the flag as a stable legal system governing their relationships with each other. Even under a flag of convenience, the legal system is at least certain, whereas the position of the ship is, for everything that remains inside the vessel's hull, entirely fortuitous. With the advent of the problem of flags of convenience, ships may not any longer be subjected to very strict control by their flag States, and the coastal State may have an interest in intervening in internal health and security questions on board,[408] but only if they have implications for the flag State and are thus not strictly "internal" to the ship any longer.

This is also consistent with the regime for coastal State enforcement jurisdiction as set out by arts. 27 and 28 UNCLOS,[409] seeking that the coastal State refrain from enforcement where there is no coastal State interest to do so. These considerations should also influence the choice of law—if the coastal State "should not" enforce criminal law and civil law against foreign-flagged ships in territorial waters, it "should not" deem its own law applicable either. The Rome II Regulation should accordingly be interpreted with regard to the underlying law of the sea, in the way that this is necessarily done in the EEZ and on the high seas, too. It can either be considered that art. 4 (1) of the Rome II Regulation does not apply in its literal wording, because the territorial waters cannot be deemed a "country" in the sense of the provision with regard to the right of innocent passage, or it can be maintained that the law of the flag should apply as the law "manifestly better placed" to apply under art. 4 (3) of the Regulation.[410] In any case, the strict and literal application of art. 4 (1) of the Regulation, even if there is territorial sovereignty over territorial waters, does not seem an appropriate solution for torts purely internal to the ship.

II. Collisions in Territorial Waters

In comparison to events in which there is only one ship involved and where the unique conflict is between the coastal State's laws and the laws of the flag State, the situation is more complicated in collision cases, namely if the ships involved fly different flags. The conflict here is between the laws of the coastal State on the one hand and the ships' flag States' laws on the other hand, with the additional problem that, of course, at first sight neither of the flag laws can prevail over the other.

Again, the 1910 Collision Convention, generally providing for an apportionment of liability between the ships to the extent that either of them was at fault, does not contain rules as to the law that should be applicable to all questions left open by the Convention (see *supra*, A. V. on the corresponding problem for collisions on the high seas).

[408] Brækhus, '1979 "Choice of Law Problems"', 284.

[409] Mayer and Heuzé, *2010 "Droit international privé"*, para. 685.

[410] Envisaged by Fawcett, Carruthers and North, *2008 "Private International Law"*, p. 860.

For collisions in territorial waters, the application of the laws of the coastal State may be the easiest way to sort out the conflict and is in fact a well-established rule in many legal systems.[411] It is in line with the considerations on the impact of the right to innocent passage as described *supra* under I. if one considers that a collision is "external" to the ships and touches the interests of the coastal State.

The applicability of the coastal State's laws is in fact the solution under English law,[412] without any apparent authority to the contrary. A very early precedent for it, still decided under the double actionability rule, is the case of *The Halley*.[413] In this case, the court applied Belgian law to a collision between a Norwegian and a British ship that occurred in Belgian waters. The vessel owned by the defendants had been under mandatory pilotage at the moment of the collision. Since it was the pilot who was to blame for the accident, the defendants were exonerated from liability under Belgian law, where compulsory pilotage offered a valid defence, but not under English law, where the shipowners would have remained liable notwithstanding the pilotage. The applicability of the laws of the coastal State to collisions occurring in territorial waters was also confirmed in later decisions.[414]

As much certainty as there is under English law, there is however one Scottish precedent that has sometimes been considered to cast doubts on the clearness of the rule under common law. In *The Reresby v. The Cobetas*,[415] the Scottish Court of Session held that the law applicable to a collision between a British and a Spanish ship in French territorial waters was the *lex fori*, i.e. Scottish law, and not French law, which the defendants had pleaded. Blackburn L. J. in fact considered the case of *The Halley*, and distinguished it on the basis of that its set of facts took place in waters *intra fauces terrae*,[416]i.e. in internal waters. In the same spirit, he considered the case of *Carr v. Fracis Times*, on the seizure of goods in the territorial waters of Muscat, to which the authorisation of the Sultan of Muscat had been held to be a good defence, and distinguished it on the fact that this authorisation had been an act of State,[417] whereas the question here was on the applicability of domestic legislation to a relationship between two ships in passage.

The courts' findings in *The Reresby* are somehow difficult to reconcile with the earlier precedents on conflicts of laws in territorial waters. The distinction of *The Halley* is questionable in that the court in that case never explicitly stated the

[411] Bonassies, '1969 "La loi du pavillon"', 550; Berlingieri, '1977 "Jurisdiction and the Choice of Law in Collision Cases"', 870, maintaining this at least with regard to proceedings taking place in the coastal State; Tomljenović, '1999 "Maritime Torts"', 271.

[412] Marsden, Gault, Hazelwood and Tettenborn, *2003 "Collisions at sea"*, para. 17-17.

[413] *Privy Council, 17.6.1868, "The Halley"*, p. 202.

[414] *High Court (England), Probate, Divorce and Admiralty Division, 15.10.1920, "The Arum"; High Court (England), Probate, Divorce and Admiralty Division, 19.10.1953, "The Waziristan".*

[415] *"The Reresby v. The Cobetas"* 1923 Scots Law Times 719 (Court of Session (Scotland), Outer House, 16 October 1923).

[416] Ibid., p. 722.

[417] Ibid., p. 722.

collision occurred in internal waters but only decided on the basis that they considered the place of the collision to be in a foreign country.[418] This can in fact also cover territorial waters. On the other hand, the case of *The Halley* hinged on the question whether the compulsory pilotage ordered by the Belgian statute constituted a valid defence, and as it was by virtue of Belgian regulations that the *Halley* was under pilotage, it seems to be the logical consequence that Belgian law should also decide on whether this exonerated the shipowner from liability. It is true that the case of *The Halley* thus had a closer connection to the Belgian sovereignty than a mere collision case between two vessels in territorial waters. The findings of Blackburn L. J. in *The Reresby* are, to a certain extent, in line with the authors who suggest that there should be a difference in the conflicts rule for shipboard torts on ships that are effectively in port[419] or engaged in loading or unloading operations, having thus a real relationship to the port State as opposed to ships merely passing through territorial waters (see *supra* under I.). This argument is to some extent convincing, especially if, for instance, the incident happens on the very limits of the territorial waters and the damages to the ships themselves are insignificant in comparison to the damages that are caused to persons and cargo on the ships.[420] Considering the distinction of *Carr v. Fracis Times*, this case of course concerns a very specific set of facts. It is not entirely inconceivable that there should be a difference between acts by which a State exercises power over subjects that happen to be in the realm of its sovereignty, thus comparable to rules of immediate application, and mere civil law applicable to the interrelationship between two private parties in this area.

Even if *The Reresby* is widely criticised and often considered as wrongly decided,[421] the concerns it raises are by no means far-fetched, and it could just as easily be quoted as authority for the theory that the coastal State's laws only apply in its territorial waters to the extent that the coastal State is really affected by the events, which does not seem to have been the case in *The Reresby*. In modern times, a collision between two ships in territorial waters would affect the coastal State more likely than not—because of fears of pollution by oil or other substances, and because the coastal State generally assumes the duty to provide search and rescue services for persons in distress in territorial waters. It must nevertheless be kept in mind that with regard to jurisdiction, the 1952 Civil Jurisdiction Convention does not provide for a judicial venue in the courts of the coastal State if the collision occurred in territorial waters (see *supra*, Chap. 6, B. II. 1. c)). This may lead to the result that the case is adjudicated outside the coastal State where the Convention is applicable and that jurisdiction and choice of law do not coincide.

---

[418] *Privy Council, 17.6.1868*, "*The Halley*", pp. 202 et seq.

[419] Marsden, Gault, Hazelwood and Tettenborn, *2003* "*Collisions at sea*", para. 17-17, footnote 56.

[420] Carbone, '2009 "Conflits de lois"', 225.

[421] *Court of Session (Scotland), 26.10.1954*, "*MacKinnon v. The Iberia Shipping Co.*", p. 27 by Migdale L; ibid., p. 30 by Carmont L.

In the United States, collisions in territorial waters are generally considered as being governed by the law of the coastal State,[422] but there are, here again, some hints that courts and authors are reluctant to assert this result in each any every case. In the case of *The Scotland*, which was a case of collision on the high seas, the court also considered, as *dicta*, the law that would be applicable to a collision in territorial waters and found that "if a collision should occur in British waters, at least between British ships [...] we would administer justice according to the British law",[423] ostensibly avoiding having to clarify the question whether British law would be applicable regardless of the ship's flags. There is however no apparent authority explicitly ruling out the applicability of the coastal State's law to collisions in territorial waters.

The same result is also advocated under Belgian law.[424] In France, even if it is generally asserted that the law of the coastal State applies to collisions taking place in the territorial waters, there is at least one source also pointing to a possible differentiation between collisions taking place in internal waters on the one hand and territorial waters on the other hand, recalling that the territorial link appears "quite fragile".[425]

Under German law, the law applicable to collisions in territorial waters is generally considered to be the law of the coastal State,[426] unless both ships are German, in which case a German court will generally apply German law even if the collision occurred in the territorial waters of a foreign State.[427] There are however many old precedents in which these principles have been set aside, usually in favour of the application of German law.[428] Therefore, it is not clear whether it would generally be held that a flag State common to both ships would trigger the applicability of the law of the flag, prevailing over the *lex loci delicti*. It is however submitted that the taking into account of flag States' laws is difficult to reconcile with the provisions of the Rome II Regulation, and that the likelihood of German courts deciding in this way is probably reduced after the Regulation's coming into force.

In allisions in which a vessel collides with a bridge or another fixed structure in the territorial sea, the application of a law other than that of the coastal State is hardly conceivable.[429] Even under the pertinent factors test in United States conflicts laws it has been recognised that for allisions in territorial waters, the law of the coastal State should be more thoroughly taken into account, as the fixed

[422] *1934 "Restatement of the Law of Conflict of Laws"*, pp. § 410; N. J. Healy and J. C. Sweeney, 'Basic Principles of the Law of Collision' (1991) 22, Journal of Maritime Law and Commerce, 359–404 at 390; Carbone, '2009 "Conflits de lois"', 224.

[423] *US Supreme Court*, *"The Scotland"*, p. 29.

[424] *"Axeldijk"* 1963 Pasicrisie 374 (Cour de Cassation (Belgium), 23 November 1962); also see Bonassies, '1969 "La loi du pavillon"', 550; Carbone, '2009 "Conflits de lois"', 224.

[425] Rodière and du Pontavice, *1997 "Droit maritime"*, p. 432.

[426] Roth and Plett, '1978 "Schiffszusammenstöße"', 670.

[427] Beitzke, *1960 "Questions d'abordage"*, p. 63; Prüssmann and Rabe, *2000 "Seehandelsrecht"*, p. 976.

[428] See Beitzke, *1960 "Questions d'abordage"*, pp. 63 et seq.

[429] Brækhus, '1979 "Choice of Law Problems"', 275.

structure in the territorial sea is geographically more closely linked to the coastal State[430] than a ship could ever be. In any case, the result advocated for damages in connection with fixed structures in the EEZ must be even more valid for offshore facilities in territorial waters given the extreme nearness to the coast.

III. Conclusion: Conflict of Laws in Territorial Waters

If doctrine and precedents on conflict-of-law rules in territorial waters are given a closer look, it is obvious that there is some reluctance to admit that torts in territorial waters are always and under all circumstances submitted to the law of the coastal State. It seems that this uncertainty as to the scope of the territorial competence mainly stems from the importance of the right to innocent passage that foreign-flagged ships enjoy in territorial waters. If the coastal State cannot interfere with a ship's passage except under the circumstances set out by art. 19 para. 2 UNCLOS, it does not seem justified to submit the ship to the coastal State's laws, at least not in cases where the coastal State is not affected by the events.

This underlying thought probably triggers the consideration to differentiate between acts in ports and other internal waters and acts in the territorial sea, a distinction that has been made obvious in the court's findings in the case of *The Reresby*. As there is no innocent passage in internal waters, it seems, under the law of the sea, more justified to submit the ship to the coastal State's laws in these waters than farther out at sea. The distinction is sometimes also drawn according to the fact of whether the ship actually loads or unloads and thus has "a real link" with the coastal State[431] that would justify the State's exercise of power.

The consideration of the right to innocent passage most probably also triggers the wish to differentiate between events taking place entirely on board the ship and events outside the vessel. This mirrors the corresponding differentiation for events taking place on the high seas, most often advocated under English law where the "general maritime law doctrine" (see *supra* under A. II. 1.) rendered it necessary to distinguish between acts taking place on the high seas—submitted to the "general maritime law"—and events on the ship governed by the law of the flag. Considering that in contrast to the high seas, territorial waters are submitted to the sovereignty of a coastal State, it may seem that the differentiation between "internal" and "external" events is here even more important and that these events should be submitted to different laws. This is evidently the line of thoughts behind the language found in the ECJ's *DFDS v. Sjöfolk* decision—however only *obiter dicta* and with regard to jurisdiction—that can be interpreted as meaning that it is the flag state that has jurisdiction for events on board a ship even if the ship is in port. With all the caveats

[430] Tetley, '1993 "The Law of the Flag"', 167.

[431] Marsden, Gault, Hazelwood and Tettenborn, *2003 "Collisions at sea"*, para. 17-17, footnote 56.

linked to the judgment's interpretation, the scope of this argument is however unclear.

It must be pointed out that it can often be very difficult to distinguish between events internal and external to the ship, especially as, given the geographical nearness of territorial waters to the shore, events on board a ship, even if physically confined to the ship, are more likely to affect the coastal State than the same acts taking place on a ship in the EEZ. Examples for this can be the admittedly outdated phenomenon of illegal broadcasting or even the mere presence of a vessel in internal waters. Events happening on the ship are thus very likely to affect the coastal State in any case, which can justify the application of the coastal State's laws.

The rule as to the differentiation of "internal" and "external" acts can also apply, with the same considerations as to legitimacy, to collisions in territorial waters. In general, in the event of a collision between two vessels, endangering the shipping routes in the territorial waters and in all likelihood entailing a danger of environmental pollution, the coastal State has a valid interest in seeing its laws applied. With regard to this, it seems also desirable to let the coastal State's laws prevail over the law of the flag, even where the colliding ships fly a common flag[432] and reinforce the territorial connection of the case. This would do away with a possible problem if both ships, coincidentally, fly the same flag of convenience. In other cases, where the flags coincide with the owners' nationalities or principal places of business, the laws of the flag would, at least under the escape clause of the Rome II Regulation, art. 4 (2), be applied as the law of the common habitual residence.

Splitting the conflict-of-law rule for events "internal" and "external" to the ship could at first glimpse bring about a certain risk of a *dépeçage*, the parallel application of two laws to one and the same set of facts, a phenomenon that the Rome II Regulation is rather hostile to.[433] It must in this context be stressed that the differentiation is of course to be made for the case as a whole, not for certain claims within one event. Once it is decided that the event is "external" to the ship, triggering the application of the coastal State's laws, all tort claims connected with these events must be submitted to the law, even the heads of damage that only concern damages on the ship. This will render the application of the laws of the coastal State rather the rule, the law of the flag State really only being applied to events that take place within the vessel's hull and which do not have an impact on the way in which the transit passage through the territorial sea is carried out. In this way, the interests of the coastal State and the flag State with regard to the right of innocent passage are given regard to.

The practical difficulties discussed thus do not oppose the splitting of the conflict-of-law rules regarding acts outside and inside the ship. The consideration of interests under the law of the sea and practical implications rather call for the differentiation, one that should take effect even under the Rome II Regulation.

---

[432] Beitzke, *1960 "Questions d'abordage"*, p. 69.

[433] Basedow, *2012 "Die private Haftung"*, p. 1892.

D. Torts with Regard to Shipbuilding and Classification

As a last point in the analysis of conflict-of-law problems for maritime torts, the questions that arise in the context of third party claims against shipyards, classification societies and other entities shall not go unmentioned, given the importance of these claims following a catastrophe at sea. As the cases of the *Amoco Cadiz*, the *Erika* and the *Prestige* have shown, shipyards and classification societies will, in catastrophic accidents, very likely be confronted with claims against them. To the extent that they cannot invoke channelling provisions of civil liability regimes in their favour—and it must be kept in mind that the BOPC, for instance, does not contain channelling provisions at all (*supra*, Chap. 4, A. III.)—they will generally be faced with tort claims under domestic law.

The legal relationships created by classification[434] are multiple, and there are different attitudes in different legal systems concerning a classification society's public function[435] and its impact on obligations to third parties, pollution victims or passengers that may suffer from an accident.[436] Furthermore, there is another major problem with the question whether classification societies fulfil duties of control that are essentially of a public character and should thus enjoy the immunity from civil claims that is accorded to foreign governments.[437] Considerations of substantive law—the justiciability of claims—are however not the topic of this study, and the possible conflict with civil liability regimes has been dealt with *supra*, Chap. 3, E. I. 4.). The purpose of this part is an overview of the problems that can arise in the conflicts of law, especially with regard to the *lex loci delicti* rule under tort law, and of the solutions that have been retained so far. Questions of contract law, based on the shipbuilding or classification agreement, are left out of the picture, as these will most likely be subject to contractual clauses regarding jurisdiction and choice of law. It is however true that under certain circumstances, a court may decide that the law of the contract also has an impact on third parties, even if they are not party to the agreement.[438]

When considering the legal relationship between a classification society and the victims of an accident, it must be stressed that establishing causation between a

---

[434] See J. Harrison, 'The Accountability of Classification Societies: The Role of Classification and Market-oriented and Policy Issues (The "Erika")' (2000) 7, Journal of International Maritime Law, 299–302 at 299 et seq.; A. M. Tettenborn, 'The liabilities of classification societies - more awkward than it looks?', in D. R. Thomas (ed.), *Liability regimes in contemporary maritime law* (London: Informa, 2007), para. 7.2 et seq.

[435] See J. L. Pulido Begines, 'The EU Law on Classification Societies: Scope and Liability Issues' (2005) 36, Journal of Maritime Law and Commerce, 487–543 at 487 et seqq.

[436] For a comprehensive overview on selected jurisdictions, see Lagoni, *2007 "The liability of classification societies"*, Part 2.

[437] See M. Alba Fernández, 'Establishing Causation in Tort Liability of Classification Societies: What Can a Civil Law mind Learn from American and English Cases?' (2006), Il Diritto Marittimo, 339–81 at 348.

[438] Davies, *2007 "Classification society liability"*, para. 8.4 et seqq.

wrongful classification and an incident at sea may be difficult.[439] The classification as such will hardly be the only cause of an accident,[440] and it is obvious that a faulty construction or a neglect of maintenance will generally be recognisable as the more direct cause. In many cases, it may thus be considered that the link between the allegedly wrongful behaviour and the accident is too remote, especially in the cases where time has passed before the accident occurred.[441] Considering that negligence in construction or maintenance happen over a longer period of time, it will be difficult to localise them, even without wanting to reconcile that *locus delicti* found with the place of the wrongful classification.

With regard to the law applicable to an allegedly wrongful classification as such, especially after a devastating accident, it seems that solutions found so far are anything but unanimous. This is certainly partly due to the fact that there are multiple connecting factors, the classification being a process consisting of, possibly, one or more inspections of the ship, in all likelihood in different ports, and then the issuing of the certificate in the headquarters or a branch of the classification society, to be sent to the shipowner, who will most likely have his place of business in another country. The accident itself will in all likelihood have occurred in yet another jurisdiction. This was in fact the situation in the *Prestige* proceedings before the US courts, where the Kingdom of Spain brought, *inter alia*, a claim against the *Prestige's* US-based classification society, ABS. The defendant company argued that the law applicable to this tort claim should either be the law of the Bahamas, as that was the law of the vessel's flag, or the law of either the United Arab Emirates or China, which were the places where the *Prestige's* last two surveys had taken place. Since the case concerned a maritime tort, the US District Court for the Southern District of New York considered the *Lauritzen/Rhoditis* factors in its conflicts of laws considerations,[442] even though the allegedly wrongful behaviour, the inspections and the subsequent issue of the class certificate had taken place in port and on land after all. The court obviously considered that the tortious act consisted in the fact that ABS issued class documents that did not reflect the vessel's actual condition rather than ABS's failure to survey the *Prestige* more carefully. The fact that the wrong documents were issued through ABS's headquarters in the US was thus important, whereas the court rather disregarded the location

[439] Alba Fernández, '2006 "Establishing Causation"', 349 et seqq.

[440] See H. G. Payer, 'Insurer and Class and Marine Accidents', in M. H. Nordquist and J. N. Moore (eds.), *Current Marine Environmental Issues and the International Tribunal for the Law of the Sea* (The Hague, London: Martinus Nijhoff Publishers, 2001), p. 296.

[441] See West, '1981 "The Ixtoc I Oil Spill Litigation"', 505 on the claim against the Texan producer of equipment that was suspected to have caused the or contributed to the accident on the *Ixtoc I* platform.

[442] *US District Court, Southern District of New York, 03.08.2010*, "*Reino de España v. American Bureau of Shipping* (*The Prestige*)" (*after remand*), pp. 1881 et seqq.

where the accident took place and the consequences to the Spanish coastline.[443] The Court of Appeal upheld this reasoning.[444]

Bearing in mind that the pertinent factors according to *Lauritzen/Rhoditis* also encompass considerations as to the underlying contract, the law applicable to a case is likely to be the same, no matter whether the case is brought in tort or in contract. This means that some conclusions can be drawn from cases in which claims were brought by persons directly or indirectly benefitting from the classification, such as a subsequent buyer of a recently surveyed vessel. There is heated debate as to whether the conflicts test of *Lauritzen/Rhoditis* and the analysis of the seven or eight pertinent factors should really be applied in claims against classification societies; it is often argued that the test is not very suitable for these proceedings, especially where the claim is not for death or injury, but for rather commercial damage.[445] The test has effectively been modified for application to classification societies.[446] The law of the contract in the cases where there is no true contractual relationship between the parties, the accessibility of foreign fora and the law of the forum are often disregarded, the last two factors presumably being shunned as they do not weigh as heavily in a commercial suit as in the cases dealing with the application of the Jones Act to injured seamen which had originally given rise to the test.

In the 1996 case of *Carbotrade v. Bureau Veritas*[447] in which the charterer brought a claim in tort against the French classification society that had allegedly failed to detect the vessel's bad condition, this problem was discussed in great length. The ship was owned by a corporation that had its headquarters in Gibraltar, and the ship was also registered in Gibraltar. Consequentially, the court in the first instance had held that the law of Gibraltar applied to the claim against Bureau Veritas. On appeal, however, the court of Appeals for the Second Circuit considered the registry in Gibraltar as a flag of convenience and decided that Greek law should be applicable on the basis that the allegedly wrongful classification certificates had been issued in Greece, through the classification societies' branch offices in the port of Piraeus, and that the shipowner, even if formally registered in Gibraltar, was controlled by Greek interests. These findings were however heavily criticised by a dissenting judge,[448] for giving too much weight to the place of the wrongful act, for too readily piercing the veil of the owner company, which was duly registered in Gibraltar, and for disregarding the importance of the connecting factor of the ship's

---

[443] Ibid., pp. 1886 et seq.

[444] *US Court of Appeals for the Second Circuit, 29.08.2012*, "*Reino de España v. ABS (The Prestige)*", pp. 2121 et seqq.

[445] Scoles, Hay, Borchers and Symeonides, *2004* "*Conflict of Laws*", pp. 895 et seq.

[446] Symeonides, '2003 "Choice of Law"', 53; also see the extensive discussion in Daniel, '2006 "Potential Liability"', 198 et seqq.

[447] "*Carbotrade v. Bureau Veritas*" 1997 American Maritime Cases 98 (US Court of Appeals for the Second Circuit, 31 October 1996).

[448] Ibid., pp. 109 et seq.

flag, since Gibraltar as a British Overseas Territory could not exactly be called a typical flag of convenience.

Similarly, in the 2002 case of *Sealord Marine v. ABS*,[449] the classification society was sued by a subsequent buyer of the vessel for having failed to detect the real condition of the ship before the sale. The court did not attach much importance to the fact that the vessel had also been inspected in Spain and in Texas after the first inspection in Piraeus and favoured the application of Greek law because both the claimant and the original seller were incorporated in Greece and the certificates had thus been issued there, through the defendant company's Piraeus branch.

The flexibility of the factors makes it very difficult to predict the law that will be applicable, and, arguably, the final choice may be influenced by political considerations.[450] Thus, in the *Prestige* case, the court finally decided to apply US law, holding that the most pertinent circumstance was the fact that the class document had been issued in the classification society's headquarters in the US. In the same way, the issuing of the certificates in the port of Piraeus by the Greek offices of the otherwise French classification society were considered decisive in *Carbotrade v. Bureau Veritas* and was of similar influence in *Sealord Marine v. ABS*, even if there are also cases in which other factors were deemed to be more important.[451] While the place of issuing of the certificates is a valid consideration in itself, it would render the legal system of the classification society's main place of business generally applicable if it was given the same weight in all cases, and the classification society could influence the choice of law by setting up independent offices in different jurisdictions. This can, with regard to the global dimension of the classification societies' work—possibly causing damages all over the world—hardly be a desirable outcome.

Under the stricter *lex loci delicti* rule, it is however equally hard to advocate a given connecting factor, as classification is a process almost always taking place in different jurisdictions, and as it is difficult to judge the importance of the various steps. In France, precedents are not especially unanimous as to the law applicable to a wrongful classification. In a case that was decided before the coming into force of the Rome II Regulation and which concerned the foundering of a drilling rig in Scottish waters following a negligent classification by an English society,[452] the French *Cour de Cassation* considered a number of possible legal systems that could have applied by the virtue of the *lex loci* rule, as different parts of the rig had been constructed and inspected in various countries. The court, without clearly stating

---

[449] "*Sealord Marine Co. v. American Bureau of Shipping*", 220 Federal Supplement 2nd Series 260 (US District Court for the Southern District of New York, 13 September 2002).

[450] See Han and Yu, '2006 "New developments"', 247, 249, submitting that the reluctance to hold classification societies liable under the law of the US is due to the fact that liability for these very considerable damages would be unlimited.

[451] See Daniel, '2006 "Potential Liability"', 211 et seq. for examples.

[452] "*Mobil North Sea v. Compagnie française d'entreprises métalliques*", (2000) 89 Revue critique du droit international privé 199 (Cour de Cassation (France), 11 May 1999).

that one of the events was most important,[453] then proceeded with the question as to which of these laws had the closest connection with the case and held it to be Scottish law as Scottish territorial waters were the place where the accident finally occurred. The approach can be criticised as not applying a clear-cut rule, but rather determining the "proper law" on a consideration of the facts of the case, in which the place of the actual damage was only coincidentally the one that was considered most important.[454] It seems however that this was done by virtue of the circumstances as the tort was simply too complex, having connections with too many countries, so that the court was forced to exercise a choice between the potentially applicable laws.[455]

Different results have been reached by the *Cour de Cassation* when applying the same method in a different case.[456] In a 2007 case in which the French classification society Bureau Veritas was the defendant,[457] the place where the accident had taken place, the waters off Madagascar, were disregarded as "purely fortuitous". In the presence of multiple laws that could possibly apply to the wrongful classification resulting in an accident, this time the closest link was held to exist with France, as the place where the guidelines for classification had been established and the wrongful documents had been issued,[458] triggering the application of Bureau Veritas's own legal system. The decision may[459] however have had a political dimension[460] in the same way that this was suspected in the US *Prestige* case, albeit with the opposite outcome, Bureau Veritas being held liable under French law.

It has furthermore, and rightly, been pointed out that this result is rather difficult or impossible to reconcile with the *lex loci damni* approach of the Rome II Regulation[461] that would generally call for the application of the law of the place

---

[453] See J.-M. Bischoff, 'Note C.Cass. 11.05.1999' (2000) 89, Revue critique du droit international privé, 201–6 at 202 et seqq. for a discussion; A. Junker and D. Schramm, 'Die gekenterte Bohrplattform - Vertragliche und deliktische Haftung bei Distanzdelikten in der französischen Rechtsprechung' (2001), Praxis des Internationalen Privat- und Verfahrensrechts, 482–5 at 484.

[454] G. Légier, 'Note C.Cass. 11.05.1999' (1999), Journal du Droit International (Clunet), 1050–7 at 1054; B. Audit, 'Observations C. Cass. 11.05.1999' (1999), Recueil Dalloz (Sommaires Commentés), 295–6 at 296; Bischoff, '2000 "Note C.Cass. 11.05.1999"', 203.

[455] Légier, '1999 "Note C.Cass. 11.05.1999"', 1055.

[456] D. Bureau, 'Note C. Cass. 27.03.2007' (2007) 96, Revue critique du droit international privé, 405–12 at 955.

[457] "*Wellborn*", (2007) 59 Droit Maritime Français 750 (Cour de Cassation (France), 27 March 2007).

[458] P. Bonassies, 'Observations C. Cass. 27.03.2007' (2007) 59, Droit Maritime Français, 753–6 at 754; G. Légier, 'Note C. Cass. 27.03.2007' (2007), Journal du Droit International (Clunet), 949–56 at 954.

[459] See however Bureau, '2007 "Note C. Cass. 27.03.2007"', 411.

[460] See 'Note C. Cass. 27.05.2010' (2011) 63, Droit Maritime Français, 39 at 39: "the Bureau Veritas being affected".

[461] Bureau, '2007 "Note C. Cass. 27.03.2007"', 412; Bonassies, '2007 "Observations C. Cass. 27.03.2007"', 754; Légier, '2007 "Note C. Cass. 27.03.2007"', 951; Carbone and Di Schiano Pepe, '2009 "Uniform Law and Conflicts"', 79.

where the damage is sustained, except for the compensation of environmental damage under art. 7 of the Regulation. Even under the escape clause of art. 4 (3) of the Regulation, allowing for the application of a law better placed to govern the case, the mere fact that the classification society's headquarters were in France—and that the classification decision was most probably[462] taken there—hardly justifies to rule out the *lex loci damni*.[463]

On the other hand, in a decision of 2010,[464] the *Cour de Cassation* had to decide in a case in which a ship had been constructed, sold and subsequently adapted for another use in a Malaysian shipyard. After the classification, in which the ship was declared ready for use, inherent defects in the replaced parts were discovered, and the shipowner sued the classification society for compensation for the necessary repairs. The case is however exceptional in that the parties agreed to have the case adjudicated in tort whereas there was of course an underlying contract to the classification.[465] In this case the court held that the tort consisted in the negligent inspections, which had taken place in Malaysia. It stressed the fact that the lawsuit was not about issuing the documents—which had taken place in France—nor about the extent or the guidelines of the classification, but uniquely consisted in the failure to detect the weaknesses, an act that should have taken place in Malaysia.[466] The court did not base its decision on the fact that the shortcomings in the shipyard's work had been detected in the United Arab Emirates, and that the losses had finally been suffered by a party in Qatar.[467] Even if this decision goes beyond the simple applicability of French law as the law of the place where the class had been issued, it does not present a satisfying solution to the general problem of weighing the importance of connecting factors in classification processes either. It is furthermore of limited applicability, since the court very explicitly held that the tort here consisted in an objective shortcoming in the inspections according to the defendant company's own classification guidelines,[468]and not in the question whether these guidelines were appropriate for really guaranteeing the ship's safety. This remark could easily be interpreted to mean that in the opposite case, the tort would have been held to have been committed in France, triggering the application of French law. As refined as the reasoning may be in distinguishing between the tort of having poor rules of classification and the tort of following good rules badly, it should not be forgotten that there was, initially, a contract between the parties obliging the

[462] See Bureau, '2007 "Note C. Cass. 27.03.2007"', 411, rightly pointing out that the court apparently only assumed that this was the case.

[463] Carbone, '2009 "Conflits de lois"', 216.

[464] "*Meranti 38*", 62 Droit Maritime Français 701 (Cour de Cassation (France), 27 May 2010).

[465] L. d'Avout, 'Observations C. Cass. 27.05.2010' (2010) 62, Droit Maritime Français, 704–12 at 704.

[466] Ibid., p. 705.

[467] D. Bureau, 'Note C. Cass. 27.05.2010' (2010) 99, Revue critique du droit international privé, 704–6 at 706.

[468] *Cour de Cassation (France), 27.05.2010, "Meranti 38"*, p. 704.

classification society to follow these rules, which was most probably governed by a choice-of-law clause pointing to French law.[469] By agreeing to bring the claim in tort, the parties opted out of this contractual agreement, which in any event shows a considerable disregard for the classification contract, including the rules.[470] Due to this particularity of the case, it is not clear to which extent it can really serve as a precedent.

The set of facts of this case also reveals that there may be a considerable problem in localising the law of the place where the damage occurred if the damage does not consist in an actual accident, but in additional repair work. This can be carried out in different ports, at the discretion of the shipowner or any other interested party, and it would seem strange if the place where the repairs were carried out would determine the *locus damni*—but equally strange if the place where the defects were discovered would be the pertinent factor. The situation becomes even more complicated where there is a subsequent sale of the ship and the damage consists, for the buyer, in the acquisition of the defective ship.[471] In these situations it effectively seems that the *locus damni* should be found in the place where the insufficient condition of the ship was brought about, so that the *Cour de Cassation's* decision in this case can hardly be criticised.

It can thus be seen that a complex delictual situation of this kind, even under the Rome II Regulation's *lex loci damni* rule, will most likely always give rise to discussions about which place will be pertinent in localising the damage.[472] In the cases in which the damage merely consists in the ship's loss in value, regard should also be given to the underlying contract of classification and the law by which it is governed, following art. 11 of the Rome II Regulation.[473] In the cases where there effectively is an accident following poor repairs or a wrongful classification, the damage should be localised in the place where the accident occurred, giving regard to the law of the flag State in accordance with what has been set out *supra* under A.–C., in the interest of a homogeneous treatment of all heads of damage such as environmental claims, passenger and cargo compensation.

It can certainly be argued that the place of the accident in these cases is purely fortuitous[474] and the strict application of the *lex loci damni* rule may lead to harsh results where the accident takes place in the waters of a State that is particularly hostile to generous compensation, whereas the defendants would be faced with more extensive claims under the laws of their home jurisdiction. Uncertainties of this kind are however due to the special character of tort law,[475] dealing with cases that the parties can never anticipate, neither the location where they take place nor

---

[469] d'Avout, '2010 "Observations C. Cass. 27.05.2010"', 706 et seq.

[470] Ibid., p. 707.

[471] Ibid., p. 707.

[472] See ibid., pp. 709 et seq.

[473] Ibid., p. 711.

[474] *Cour de Cassation (France), 27.03.2007*, "*Wellborn*", p. 407.

[475] Bureau, '2007 "Note C. Cass. 27.03.2007"', 410 et seq.

the law that might apply. Holding that the tort consists in the issue of class documents, triggering the application of the law of the classification society's home jurisdiction, can hardly be advocated as a general solution, considering that this would leave victims with the difficult situation of always having to sue in the alleged tortfeasor's home jurisdiction. Concededly this is less pressing in cases where the claimant is a shipowner, charterer or other maritime party, perhaps even having its headquarters in the same jurisdiction as the classification society's branch issuing the documents. It however appears as blatantly unfair where the claimants are individuals that have suffered from an accident at sea and are entirely outside the classification contract.

From a pragmatic point of view, it could be argued that most large classification societies are based in developed countries with a rather generous compensation legislation for environmental pollution and death and personal injury, but this, firstly, does not hold true for all classification societies,[476] and secondly, this can turn out to be severe where classification societies are held liable for political reasons. Whereas French courts have been rather strict with the French *Bureau Veritas* in the two decisions cited, the New York courts absolutely ruled out the American classification society's liability in the *Prestige* case.

With regard to shipyards, the number of possible laws which may apply is however slightly smaller, and matters of causation may not be as pressing. The tortious behaviour will almost always consist in a faulty repair or another shortcoming in the shipyard's performance, and the link of causation with a possible accident or damage will in all likelihood be more direct. It however depends on the legal system in question whether third parties, not involved in the shipbuilding contract, can put forward claims against the shipyard.[477]

Under the Rome II Regulation, a claim against a shipyard may possibly be governed by art. 5 of the Regulation,[478] dealing with product liability.[479] The law applicable to the claim would then be determined following a cascade of possible connecting factors and hence possible applicable laws if any of them coincides with the country in which the product was marketed. The provision aims at allowing the party aggrieved by the faulty product to sue under a familiar jurisdiction while maintaining a certain foreseeability for the producer, who can only be sued under the law of countries in which the product was actually marketed. The provision

---

476 I. Corbier, 'Les sociétés de classification et la sécurité maritime' (2005), Jurisprudence du Port d'Anvers, 3–12 at 11.

477 See M. H. Bagot, 'The Ebb and Flow of "East River": Consideration of the Supreme Court's Decision on Products Liability in Shipbuilding Contracts' (2006) 30, Tulane Maritime Law Journal, 137–62 for an overview of the problem under US Law.

478 Envisaged by K. Siehr, 'The Rome II Regulation and Specific Maritime Torts: Product Liability, Environmental Damage, Industrial Action' (2010) 74, Rabels Zeitschrift für ausländisches und internationales Privatrecht, 139–53 at 141 et seq.

479 See J. M. Alcántara, 'Shipbuilding and Classification of Ships. Liability to Third Parties' (2008), Zbornik pravnog Fakulteta u Zagrebu, 135–45 at 137, considering shipbuilders' responsibilities under product liability legislation in a wider sense.

aims at the protection of consumers who may face difficulties when wanting to bring a claim against a producer with significantly superior economic power, having its place of business in another country. It is however very debatable if goods like ships, which are produced under commercial contracts for a given purpose and most often according to the buyer's wishes, and which are sold under commercial contracts, are covered by art. 5 of the Rome II Regulation. In any case, as services are doubtlessly outside the scope of art. 5,[480] mere repair works in a shipyard will not trigger the applicability of the provision. With regard to the condition of the "product" being marketed in a given country, any applicability at all of the Rome II Regulation's product liability provisions at all appears all the more doubtful, as a new-built ship is normally built for a given buyer, and can thus hardly be considered as "marketed".

It thus seems that, in general, tort claims against shipbuilders will be governed by the general *lex loci damni* rule of art. 4 (1) of the Rome II Regulation. It is likely that in the advent of an actual accident, it will be the place of that incident at sea which is considered as the *locus damni*, and for events that occur on the high seas or under the navigation regime of the high seas, the decisive factor will be the flag of the ship. This, again, allows the submission of all damages caused by the accident to one legal regime. For damages that do not materialise in physical harm, but rather in repair costs or other additional expenses, the question is as difficult as concerning classification societies, and here again, it could be advocated that the *locus damni* should then be localised at the place where the ship's defective condition was caused, which will in most cases be the home jurisdiction of the shipyard. It is however submitted that in these cases, where the ship's condition does not result in actual physical harm but rather in additional expenses to one of the parties involved, the contracts between the parties will be of a marked importance, probably also having an impact on the relationship with the aggrieved parties.

The questions surrounding conflicts in a classical tort relationship with an otherwise uninvolved party that are triggered by an actual accident can easily be solved following the *lex loci damni* rule of the Rome II Regulation, establishing considerable legal certainty. Even if the cases of wrongful classification and liability for shipbuilding and repairs largely take place on land, it does seem right to submit them to the general rule of the Rome II Regulation and let them be governed by the law applicable in the respective maritime zone where they have caused or contributed to an accident at sea. This also allows having a homogeneous regime of liability in the rather probable cases where the accident is caused by more than one act or omission, submitting the respective negligent acts and any circumstance of contributory negligence to one and the same law.

[480] Plender and Wilderspin, *2009 "The European Private International Law"*, para. 19-031.

E. Conclusion

As has been shown, the application of conflict-of-law rules in a maritime context creates a significant number of problems and has intrigued authors and puzzled courts for centuries. However, bearing in mind the special context of the overarching law of the sea and with some slight adaptations, it appears possible to submit maritime torts to the general conflict-of-law rules, namely the Rome II Regulation. Notwithstanding the Regulation's silence as to shipboard torts and other events at sea, its application to maritime torts has been envisaged in the Regulation's drafting history, and historical solutions that have been found can still be of pertinence today when applying the Regulation.

In very general terms, and mirroring the findings that have been set out *supra* in Chap. 6 with regard to jurisdiction, it seems important that torts should be governed by the *lex loci delicti*, the law of the sea dictating the realm of a coastal State's legislative competence in a given maritime zone. Admittedly, the zonal approach can create problems where incidents take place very near territorial boundaries or in uncertain places,[481] but this is a problem that the conflicts rules for torts with their territorial connecting factors have always struggled with. Given that tort law is mostly about judgment of actual behaviour, the judgment of an action according to the laws of the place where it was carried out is the only valid solution, having regard to State's territorial sovereignty that generally allows them to prohibit or encourage almost every behaviour in the realm of their territory.

Consequently, the conflicts of laws at sea, i.e. the application of a given State's laws to an event, can only operate in accordance with the underlying public international law—ultimately, the interests of other States—if territorial sovereignty and flag State sovereignty are given regard to. It goes without saying that there are cases in which the rules do not present the best solution and where the application of another law under a so-called "escape clause" will be justified, but here again, the scope and the importance of escape clauses have been previously defined, and these guidelines as to their application should not be departed from without good cause.

The impression that conflicts rules for maritime torts are a field of rather arbitrary and case-by-case decisions should not be added to. To the contrary, it should be highlighted that at least the European jurisdictions that have been presented in this study have identified and solved the problems in very similar ways, even before the coming into force of the Rome II Regulation.

481 Brækhus, '1979 "Choice of Law Problems"', 276.

Chapter 8
Conclusions

Notwithstanding developments in technology and an increasing awareness for matters of safety and environmental protection, accidents at sea continue to occur and are unlikely to ever be a thing of the past. Numerous international and domestic regulations are concerned with accidents, treating them as undesirable events that have certain adverse consequences. Rules and international agreements have been enacted for certain purposes, such as the protection of life at sea, the protection of the marine environment or the protection of economic interests. In this context, it can be submitted that the underlying aim of regulation is not the avoidance of accidents as such, but the avoidance of their undesired consequences or the compensation of damage that they entail. Even if there is, today, a rather comprehensive framework governing human activities at sea, it comprises important lacunae where adverse consequences of accidents at sea have not received sufficient attention, either because there is no entity wanting to assume responsibility for them or because the interests are too conflicting for the concerned parties to come to terms. Examples of the first phenomenon are certainly the protection of the marine environment on the high seas or the protection of life at sea where there is no State willing to assume responsibility for the persons in danger. With regard to the second phenomenon, conflicting interests between coastal States and flag States have arguably caused some uncertainties in the navigational regime in territorial waters; and the perception that offshore exploitation of non-living resources in waters under national jurisdiction is purely a matter of coastal state sovereignty has impeded international solutions even where damage is caused to the waters of more than one State. The law of the sea does not present an exception when it comes to one of the general difficulties of public international law: States will cooperate where this is in their common interest, and where there are directly conflicting interests, negotiations will never be undertaken or will soon come to a deadlock. Some States may revert to unilateral action, but in any case, lacunae in the international system cannot always be filled by domestic law, at least not always in a way that would be reconcilable with the principles of international legitimacy.

S.F. Gahlen, *Civil Liability for Accidents at Sea*, Hamburg Studies on Maritime Affairs 32, DOI 10.1007/978-3-662-45555-5_8

International and domestic regulation have addressed undesired events at sea both from a preventive point of view and with regard to the questions of liability. It is however obvious that there cannot always be a clear-cut differentiation between the two. Rules on civil liability in private law are very often not only considered as a means of compensation to victims, but also as a mechanism of determent, discouraging potential tortfeasors from behaviour that may harm third parties' interests and eventually result in a costly compensation claim. In the maritime domain, this aim can be harder to achieve, since maritime law and namely the limitation of liability still serve to encourage the maritime enterprise and to render the corresponding risks calculable and insurable. The civil responsibility regimes for the compensation of environmental damage, but also the regime on passenger claims that have been discussed in this study, first and foremost aim at providing victims with quick and easy mechanisms of compensation. On the whole, they do not contain important deterrence mechanisms that would incentivise shipowners and other maritime players to avoid accidents in the best possible way. The existence of liability regimes in which civil liability is limited and mandatorily insured is often perceived as an unjustified subsidy to the shipping interests, at the cost of coastal States and communities.

Still, the maritime industry is in this respect very different from business on land. The fact that ships and shipping businesses operate globally and can rather easily avoid being effectively held liable renders international civil liability regimes such as the CLC the only way to reliably provide for compensation funds at all. Without these liability regimes, shipowners could in many cases only be sued in domestic tort proceedings, with all the difficulties of establishing a judicial venue and determining the law applicable that this entails. Furthermore, the practical difficulties of effectively holding a foreign entity liable in a given court and in enforcing the judgment, especially in view of the general possibility of global limitation for maritime claims, should not be overlooked. Considering everything, it is obvious that the existing civil liability regimes are of a considerable value.

From a point of view of international law, the conventions discussed in Part II of this study are of a very peculiar character, considering that they represent international law, as agreed between States, but address the matter of civil liability for environmental pollution or passenger claims through the mechanisms of private law. These liability frameworks provide for compensation to be paid by private entities, and they make use of domestic courts to deal with the claims that arise. This renders domestic law a very important factor in the application of the conventions' provisions. Uniform application of the international rules in all contracting States is a goal that is not, and arguably can never be, perfectly achieved.

The international civil liability regimes for environmental pollution and passenger claims are not meant to be more than civil liability regimes after all. Their general purpose is to provide victims of maritime accidents with financial compensation from a limitation fund, and they do not regulate beyond this. Even where they hint to the possibility of proceedings outside their realm, e.g. where the limitation of liability is lost or where claims are pursued against persons clearly not benefitting from channelling mechanisms, these claims need to be brought in national law and on a legal basis outside the convention.

If domestic legal systems have stepped in to create footings for claims outside the liability regime or to interpret certain provisions overly liberally, this can certainly be criticised as a departure from the international agreement. It can however also be considered as a legitimate way to unilaterally foster deterrent mechanisms and to punish irresponsible behaviour, a purpose entirely left out of the scope of the international maritime liability regimes.

It should never be forgotten that these conventions are the result of negotiations between States with well-defined economic and political interests, which submit their industries to considerable costs when ratifying the agreement and which, on the other hand, restrict the claims of victims under their jurisdiction to the liability limits. States will only adhere to a convention of this kind if it is perceived that the benefits of the regulation outweigh its costs. The need to incentivise States to ratify these agreements has also caused certain conventions to be rather reduced in their importance or effectiveness. In this respect, the CLC and FUND, which have been perceived as advantageous to their interests by the oil industry, are remarkable achievements.

This does not mean that the system as it is, even without the problem of competing or parallel claims in domestic law, would be without flaws. The fragmentation of the system according to different pollutants is often roundly condemned, but has apparently not yet caused important difficulties in practice. It is certainly a consequence of historic development, the international community having tackled liability for environmental pollution substance by substance. Today, the character of the system has become a matter of path dependency. The tendency for every new convention to rely on existing mechanisms and identical wordings has also caused the system to have certain gaps and uncertainties that are not well adapted to the modern realities of shipping.

It seems that the existence of civil liability regimes through which the vast majority of cases can be handled has however obstructed the view of the very important question of state responsibility for maritime accidents. International technical standards for ships serve first and foremost the protection of life at sea and of the marine environment, and by ratifying the respective conventions, flag States have committed themselves to effectively controlling their vessels' compliance with these norms. It is by no means a new thought that, in theory, every accident caused by poor compliance with technical standards or any other violation of internationally applicable norms could engage the international responsibility of a State. Again, shipping is in this respect very different from industrial activities on land, where violation of safety standards will first and foremost affect the State exercising territorial sovereignty over the industry, and which therefore has a marked interest in controlling compliance. The question of state responsibility for shipping accidents is neglected to a very wide extent, even if it is debated in theory in many cases. The problem of State responsibility for omissions in flag State control is often discussed as an aspect of the problem of classification societies' liability, a claim that is in most legal systems already extremely difficult to establish. The question of state responsibility for the conduct of port and coastal authorities when assisting ships in distress has received some attention, but liability is in most cases ruled out as the authorities' omissions were not the initial cause of the accident. In this context, it does not seem very beneficial that civil liability regimes aim to create exclusive judicial venues in domestic civil courts for proceedings against private entities. These are hardly the right forum to effectively address questions of State responsibility, and as long as victims have the chance to obtain compensation for their claims in these proceedings, it is unlikely that questions of State responsibility for these accidents will be discussed in more appropriate fora.

Questions of international legitimacy, jurisdiction and choice of law are of a particular interest for maritime incidents at sea. Allocation of jurisdiction at sea can on the one hand lead to overlapping claims and on the other hand to a jurisdictional vacuum. The assumption of judicial jurisdiction for civil claims and the choice of law can be extremely difficult, considering that the UNCLOS does not contain explicit rules on this. The law of the sea however sets out very specific competences for States in the respective maritime zones, and with a view to the interests of States and the balance established between them, judicial jurisdiction must follow the rules of the law of the sea. To the extent that these rules are either derived from UNCLOS or part of customary international law, the principles guiding the allocation of jurisdiction should, after all, be internationally uniform. In contrast to that, rules on conflicts of law are domestic law, and States can enact them in any possible way, at least where they are not bound by international or regional law. It is only with a view to international comity and fairness towards foreign entities and individuals that the applicable law is very often determined according to the same factors that also guide the allocation of jurisdiction.

There is thus a difficult interplay between States' rights and obligations as pointed out by UNCLOS, the allocation of judicial jurisdiction that can be derived from it, and the enactment of conflicts rules. It seems that it is first and foremost this complex interaction that has caused questions regarding jurisdiction and conflict of laws for incidents at sea to be an important source of debate and international disparity. Whereas supranational legislation such as the EU rules on jurisdiction and conflict of laws are normally very valuable in overcoming discrepancies of this kind, the EU Regulations' focus on territorial factors renders it very difficult to apply them to incidents at sea. There is a marked need to interpret them in a way that is reconcilable with the allocation of States' rights and obligations under the law of the sea, but in meeting this challenge it then becomes possible to point out rather clear solutions to the complex questions of jurisdiction and conflicts of law.

It is the fascination of both maritime law and the law of the sea that the cases they are concerned with take place at sea, outside the clearly defined realm of territorial sovereignty. This is where the world community meets: States in their capacity as coastal States, flag States or the national States of private entities or individuals, and vessels, persons and companies from all over the world. Whereas many legal relationships are governed by contracts and agreements, accidents will invariably give rise to legal relationships between persons who never envisaged being affected by each other's conduct. It is remarkable that the principles of international maritime law and the law of the sea allow answering questions of responsibility and liability even in these cases, and even outside the realm of States' sovereignty. The sea still presents important freedoms, but it is not without law.

Bibliography

Books, Articles, Contributions

Droit de la mer. Études dédiées au doyen Claude-Albert Colliard (1992). Paris: Pedone.

Note C. Cass. 27.05.2010 (2011). In *Droit Maritime Français* 63, p. 39.

Abecassis, David W.; Jarashow, Richard L. (1985): Oil Pollution from Ships. International, United Kingdom and United States law and practice. 2nd ed. London: Stevens.

Abeyratne, Ruwantissa (2010): The Deepwater Horizon Disaster - Some Liability Issues. In *Tulane Maritime Law Journal* 35, pp. 125–152.

Akehurst, Michael (1972): Jurisdiction in International Law. In *British Yearbook of International Law* 46, pp. 145–257.

Alba Fernández, Manuel (2006): Establishing Causation in Tort Liability of Classification Societies: What Can a Civil Law Mind Learn from American and English Cases? In *Il Diritto Marittimo*, pp. 339–381.

Alcántara, José M. (2008): Shipbuilding and Classification of Ships. Liability to Third Parties. In *Zbornik pravnog Fakulteta u Zagrebu*, pp. 135–145.

Allbritton, Jack L. (1983): Choice of Law in a Maritime Personal Injury Setting: The Domestic Jurisprudence. In *Louisiana Law Review* 43, pp. 879–911.

Allen, Craig H. (1998): The ISM Code and Shipowner Records: Shared Safety Goals vs. Industry's Privacy Needs. In *University of San Francisco Maritime Law Journal* 11, pp. 1–46.

Allen, Jacqueline (2011): A Global Oil Stain - Cleaning Up International Conventions for Liability and Compensation for Oil Exploration/Production. In *Australian and New Zealand Maritime Law Journal* 25, pp. 90–107.

Altfuldisch, Rainer (2007): Haftung und Entschädigung nach Tankerunfällen auf See. Bestandsaufnahme, Rechtsvergleich und Überlegungen de lege ferenda. Berlin, New York: Springer.

American Law Institute (1934): Restatement of the Law of Conflict of Laws. St. Paul: American Law Institute Publishers.

American Law Institute (1986): Restatement of the Law (Third) - The Foreign Relations Law of the United States. St. Paul: American Law Institute Publishers.

Anderson, Charles B.; Rue, Colin de la (2001): Liability of Charterers and Cargo Owners for Pollution from Ships. In *Tulane Maritime Law Journal* 26, pp. 1–60.

Anderson, David (2011): Coastal State Jurisdiction and High Seas Freedoms in the EEZ in the Light of the "Saiga" Case. In Clive Ralph Symmons (Ed.): Selected Contemporary Issues in the Law of the Sea. Leiden, Boston: Martinus Nijhoff Publishers, pp. 105–115.

S.F. Gahlen, *Civil Liability for Accidents at Sea*, Hamburg Studies on Maritime Affairs 32, DOI 10.1007/978-3-662-45555-5

Anderson, Michael (2002): Transnational Corporations and Environmental Damage: Is Tort Law the Answer? In *Washburn Law Journal* 41, pp. 399–425.

Antapassis, Anthony M. (2007): Liability of Classification Societies. In Katharina Boele-Woelki, Sjef van Erp (Eds.): General reports of the XVIIth congress of the International Academy of Comparative Law. Bruxelles, Utrecht: Bruylant Eleven, pp. 631–684.

Arroyo, Ignacio (1977): The Application of the International Convention on Civil Liability for Oil Pollution Damage to the "Urquiola" Case. In *Lloyd's Maritime and Commercial Law Quarterly*, pp. 337–343.

Arroyo Martínez, Ignacio (1986): International Conventions and Domestic Law. In *Yearbook of Maritime Law* 2, pp. 5–50.

Arroyo Martínez, Ignacio (1996): The Contract of Maritime Passage - A Commentary on the Athens Convention Relating to the Carriage of Passengers and their Luggage by Sea, 1974. In Peter Wetterstein, Anders Beijer (Eds.): Essays in Honour of Hugo Tiberg. Stockholm: Juristförlaget, pp. 15–54.

Arroyo Martínez, Ignacio (2009): El caso "Prestige". Consideraciones sobre el auto de 18 de Marzo de 2009 del Juzgado de Concurbión, A Coruña (España). In *Anuario de Derecho Maritimo* 26, pp. 205–216.

Arroyo Martínez, Ignacio (2009): Para esta viaje no se necesitan alforjas: El caso "Prestige". Comentario de la sentencia de 2 de enero de 2008 de la corte del distrito sur de Nueva York. In *Anuario de Derecho Maritimo* 26, pp. 81–100.

Arsanjani, Mahnoush H.; Cogan, Jacob Katz; Sloane, Robert D.; Wiessner, Siegfried (Eds.) (2011): Looking to the Future. Essays on international law in honor of W. Michael Reisman. Leiden, Boston: Martinus Nijhoff Publishers.

Ashford, Megan (2006): A Jet Ski: Vessel, Boat or Ship? R. v. Goodwin [2006] 1 Lloyd's Rep 432. In *Australian and New Zealand Maritime Law Journal* 20, pp. 64–68.

Audit, Bernard (1978): Note C. Cass. 03.04.1978. In *Recueil Dalloz (Informations Rapides)*, pp. 367–368.

Audit, Bernard (1999): Observations C. Cass. 11.05.1999. In *Recueil Dalloz (Sommaires Commentés)*, pp. 295–296.

Avout, Louis d' (2010): Observations C. Cass. 27.05.2010. In *Droit Maritime Français* 62, pp. 704–712.

Baatz, Yvonne (Ed.) (2008): Southampton on shipping law. London: Informa.

Baatz, Yvonne (2011): Maritime Law. 2nd ed. London: Sweet & Maxwell.

Bagot, Michael H. (2006): The Ebb and Flow of "East River": Consideration of the Supreme Court's Decision on Products Liability in Shipbuilding Contracts. In *Tulane Maritime Law Journal* 30, pp. 137–162.

Balkin, Rosalie P. (2003): Some Future Developments in Liability and Compensation for Environmental Damage at Sea. In Myron H. Nordquist, John Norton Moore, Said Mahmoudi (Eds.): The Stockholm Declaration and Law of the Marine Environment. The Hague, London: Martinus Nijhoff Publishers, pp. 437–454.

Balkin, Rosalie (2007): The Rise and Rise of IMO's Liability and Compensation Regimes. In Tafsir Malick Ndiaye, Rüdiger Wolfrum (Eds.): Law of the sea, environmental law and settlement of disputes. Leiden, Boston: M. Nijhoff, pp. 687–712.

Ballarino, Tito (1990): Questions de droit international privé et dommages catastrophiques. In *Recueil des Cours* 220 (I), pp. 299–387.

Bar, Christian von (1994): Les dix points d'Osnabrück. In *Revue critique du droit international privé* 83, pp. 853–855.

Barnes, Richard (2004): Refugee Law at Sea. In *International and Comparative Law Quarterly* 53, pp. 47–77.

Bartlett, James W. (1985): In re Oil Spill by the "Amoco Cadiz" - Choice of Law and a Pierced Corporate Veil defeat the 1969 Civil Liability Convention. In *Maritime Lawyer* 10, pp. 1–23.

Basedow, Jürgen (2012): Die private Haftung aus Seedelikten zwischen Völkerrecht und internationalem Privatrecht. In Holger Hestermeyer, Doris König, Nele Matz-Lück, Volker

Röben, Anja Seibert-Fohr, Peter-Tobias Stoll, Silja Vöner (Eds.): Coexistence, Cooperation and Solidarity. Liber Amicorum Rüdiger Wolfrum. Leiden, Boston: Martinus Nijhoff Publishers, pp. 1869–1896.

Basedow, Jürgen; Magnus, Ulrich (Eds.) (2007): Pollution of the Sea. Prevention and compensation. Berlin, New York: Springer.

Basedow, Jürgen; Magnus, Ulrich; Wolfrum, Rüdiger (Eds.) (2010): The Hamburg Lectures on Maritime Affairs 2007 & 2008. Berlin, Heidelberg: Springer.

Basedow, Jürgen; Wurmnest, Wolfgang (2005): Third-party liability of classification societies. Berlin, New York: Springer.

Bates, John H.; Benson, Charles (1993): Marine environment law. London: Lloyd's of London Press.

Batiffol, Henri (1967): Droit international privé. 4th ed. Paris: Librairie Générale de Droit.

Beaumont, Paul R. (2006): The Contribution of Alexander (Sandy) Anton to the Development of Private International Law. In *Juridical Review*, pp. 1–28.

Bederman, David J. (1986): Dead in the Water: International Law, Diplomacy, and Compensation for Chemical Pollution at Sea. In *Virginia Journal of International Law* 26 (2), pp. 485–514.

Béguin, Jacques; Bourel, Pierre (Eds.) (1994): L'internationalisation du droit. Mélanges en l'honneur de Yvon Loussouarn. Paris: Dalloz.

Beitzke, Günther (1960): Questions d'abordage en droit international privé allemand. In: Mélanges offerts à Jacques Maury. Paris: Dalloz et Sirey, pp. 59–73.

Berlingieri, Francesco (1977): Jurisdiction and the Choice of Law in Collision Cases and an Overview of the Concept of Fault and its Apportionment. In *Tulane Law Review* 51, pp. 866–879.

Berlingieri, Francesco (Ed.) (2000): The Travaux Préparatoires of the LLMC Convention, 1976 and of the Protocol of 1996. CMI. Antwerpen: CMI.

Berlingieri, Francesco (2011): Arrest of ships. A commentary on the 1952 and 1999 Arrest Convention. 5th ed. London: Informa.

Berlingieri, Francesco (2011): The Athens Convention on the Carriage of Passengers and Their Luggage by Sea and the European Parliament Regulation (EC) No. 392/2009. In *Il Diritto Marittimo* 113, pp. 1125–1140.

Berlingieri, Francesco (2012): Le contrat de croisière en droit italien et la responsabilité du transporteur en cas d'accident. In *Droit Maritime Français* 64, pp. 301–305.

Berlingieri, Francesco (2012): Les sociétés de classification peuvent-elles bénéficier de la canalisation prévue à l'article III.2 de la CLC 1992? In *Droit Maritime Français* 64, pp. 1015–1019.

Berlingieri, Francesco (2012): The 1952 and 1999 Arrest Conventions. A Comparison. In *Il Diritto Marittimo*, pp. 367–396.

Berlingieri, Giorgio; Boglione, Angelo; Carbone, Sergio M.; Siccardi, Francesco (Eds.) (2010): Scritti in onore di Francesco Berlingieri. Volume I. Genova: Il Diritto Marittimo.

Berlingieri, Giorgio; Boglione, Angelo; Carbone, Sergio M.; Siccardi, Francesco (Eds.) (2010): Scritti in onore di Francesco Berlingieri. Volume II. Genova: Il Diritto Marittimo.

Beurier, Jean-Pierre (2003): Le transport maritime, le droit et le désordre économique international. In Yves Daudet, Pierre Michel Eisemann (Eds.): La mer et son droit. Paris: A. Pedone, pp. 87–100.

Bièvre, Aline F. M. de (1986): Liability and Compensation for Damage in Connection With the Carriage of Hazardous and Noxious Substances by Sea. In *Journal of Maritime Law and Commerce* 17, pp. 61–88.

Birnie, Patricia W.; Boyle, Alan E.; Redgwell, Catherine (2009): International law and the environment. 3rd ed. Oxford, New York: Oxford University Press.

Bischoff, Jean-Marc (2000): Note C. Cass. 11.05.1999. In *Revue critique du droit international privé* 89, pp. 201–206.

Bishop, Archie (2007): The liability of salvors for pollution. In D. Rhidian Thomas (Ed.): Liability regimes in contemporary maritime law. London: Informa, pp. 99–105.

Bishop, Archie (2010): Places of refuge - Environmental salvage. In Norman A. Martínez Gutiérrez (Ed.): Serving the rule of International Maritime Law. Essays in Honour of Professor David Joseph Attard. Abingdon: Routledge, pp. 352–362.

Bissell, William Kenneth (1976): Intervention on the High Seas: an American Approach Employing Community Standards. In *Journal of Maritime Law and Commerce* 7, pp. 718–735.

Blackburn, Elizabeth (1988): Lis alibi pendens and forum non conveniens in collision actions after the Civil Jurisdiction and Judgments Act 1982. In *Lloyd's Maritime and Commercial Law Quarterly*, pp. 91–102.

Boele-Woelki, Katharina; Erp, Sjef van (Eds.) (2007): General reports of the XVIIth congress of the International Academy of Comparative Law. Bruxelles, Utrecht: Bruylant Eleven.

Boer, Ted M. de (2010): Living Apart Together: The Relationship between Public and Private International Law. In *Netherlands International Law Review* 62, pp. 183–207.

Bogdan, Michael (2009): Some Reflections regarding Environmental Damage and the Rome II Regulation. In Gabriella Venturini, Stefania Bariatti (Eds.): Liber Fausto Pocar. Milano: Giuffrè, pp. 95–105.

Boisson, Philippe (1994): Classification Societies and Safety at Sea. In *Marine Policy* 18 (5), pp. 363–377.

Boisson, Philippe (1996): La Convention SNDP de 1996 et l'indemnisation des dommages causés par le transport maritime de marchandises dangereuses. In *Droit Maritime Français* 48, pp. 979–985.

Boisson, Philippe (2001): L'OMI adopte une nouvelle convention pour indemniser les dommages dus à la pollution par les soutes. In *Droit Maritime Français* 53, pp. 659–667.

Boisson, Philippe (2003): Navigabilité du navire et responsabilité des États. In Anne Knops (Ed.): Liber Amicorum Roger Roland. Gent: Larcier, pp. 57–73.

Boisson, Philippe (2008): La société de classification bénéficie-t-elle de l'exclusion prévue par l'article III (4) de la Convention CLC? In *Droit Maritime Français* 60, pp. 696–703.

Boisson, Philippe (2010): La responsabilité de la société de classification dans l'arrêt "Erika". In *Droit Maritime Français* 62, pp. 897–902.

Boisson, Philippe (2012): Du "Titanic" au "Concordia": 100 ans de droit de la sécurité des navires de croisière. In *Droit Maritime Français* 64, pp. 328–337.

Bollecker-Stern, Brigitte (1978): A propos de l'accident Ekofisk - Problèmes posés par la pollution provoquée par les installations de production pétrolière off-shore. In *Annuaire français de droit international* 24, pp. 772–791.

Bonassies, Pierre (1969): La loi du pavillon et les conflits de droit maritime. In *Recueil des Cours* 128, pp. 511–628.

Bonassies, Pierre (1984): La décision "Amoco Cadiz". In *Droit Maritime Français* 36, pp. 688–698.

Bonassies, Pierre (1993): Problèmes et avenir de la limitation de responsabilité. In *Droit Maritime Français* 45, pp. 95–107.

Bonassies, Pierre (2000): Après l'Erika: Les quatre niveaux de réparation des dommages résultant d'une pollution maritime par hydrocarbures. In *Revue de Droit Commercial, Maritime, Aérien et des Transports*, pp. 140–145.

Bonassies, Pierre (2001): Observations CA Caen 02.10.2001. In *Droit Maritime Français* 53, pp. 989–993.

Bonassies, Pierre (2002): Aspects nouveaux de la responsabilité du capitaine. In *Droit Maritime Français* 54, pp. 3–13.

Bonassies, Pierre (2002): Note Tribunal de Commerce de Saint-Nazaire, 16.12.2000. In *Droit Maritime Français (Hors série n° 6)* 54, pp. 43–45.

Bonassies, Pierre (2002): Observations CA Aix-en-Provence, 10.10.2001. In *Droit Maritime Français* 54, pp. 156–159.

Bonassies, Pierre (2003): Conventions internationales. Protocole à la Convention d'Athènes de 1974 sur le transport de passagers. In *Droit Maritime Français (Hors série n° 7)* 55, pp. 7–9.

Bonassies, Pierre (2007): Observations C. Cass. 27.03.2007. In *Droit Maritime Français* 59, pp. 753–756.
Bonassies, Pierre (2008): Note TGI Paris 16.01.2008. In *Droit Maritime Français (Hors série n° 12)* 60, pp. 19–29.
Bonassies, Pierre (2008): Réflections d'un juriste français sur le jugement "Erika". In *Il Diritto Marittimo* 110, pp. 249–257.
Bonassies, Pierre (2008): Sociétés de classification et Convention de 1969/1992 sur la responsabilité pour les dommages dus à la pollution par les hydrocarbures. In *Droit Maritime Français* 60, pp. 691–696.
Bonassies, Pierre (2009): Observations CJE 24.08.2008. In *Droit Maritime Français* 61, pp. 17–21.
Bonassies, Pierre (2010): La Cour de Cassation française et la Convention de 1952 sur la sasie conservatoire des navires de mer. In Giorgio Berlingieri, Angelo Boglione, Sergio M. Carbone, Francesco Siccardi (Eds.): Scritti in onore di Francesco Berlingieri. Volume I. Genova: Il Diritto Marittimo, pp. 216–223.
Bonassies, Pierre (2012): L'arrêt "Erika" et la compétence des juridictions françaises. In *Droit Maritime Français* 64, pp. 995–999.
Bonassies, Pierre (2013): Après la décision "Erika": Observations sur la compétence des juridictions pénales françaises après abordage en haute mer. In *Droit Maritime Français* 65, pp. 195–200.
Bonassies, Pierre (2013): Observations CA Aix-en-Provence 04.10.2012. In *Droit Maritime Français* 65, pp. 312–314.
Bonassies, Pierre (2013): Observations CA Aix-en-Provence 11.10.2012. In *Droit Maritime Français* 65, pp. 224–227.
Bonassies, Pierre; Scapel, Christian (2010): Droit maritime. 2nd ed. Paris: Librairie Générale de Droit.
Boonk, Herman (2008): De betekenis van Rome II voor het zeerecht. In *Nederlands Internationaal Privaatrecht*, pp. 469–480.
Borić, Tomislav; Lurger, Brigitta; Schwarzenegger, Peter; Terlitza, Ulfried (Eds.) (2011): Öffnung und Wandel - Die internationale Dimension des Rechts I. Festschrift für Willibald Posch zum 65. Geburtstag. Wien: LexisNexis.
Bowett, D. W. (1983): Jurisdiction: Changing Patterns of Authority over Activities and Resources. In *British Yearbook of International Law* 53, pp. 1–26.
Boyle, Alan E. (1985): Marine Pollution under the Law of the Sea Convention. In *American Journal of International Law* 79, pp. 347–372.
Brækhus, Sjur (1979): Choice of Law Problems in International Shipping (Recent Developments). In *Recueil des Cours* 164, pp. 259–330.
Briggs, Adrian (1995): The Brussels Convention Tames the Arrest Convention. In *Lloyd's Maritime and Commercial Law Quarterly*, pp. 161–166.
Briggs, Adrian (2008): The conflict of laws. 2nd ed. Oxford, New York: Oxford University Press.
Briggs, Adrian (2009): Civil Jurisdiction and Judgments. 5th ed. London: Informa (Lloyd's commercial law library).
Brown, E. D. (1968): The Lessons of the "Torrey Canyon". In *Current Legal Problems* 21, pp. 113–136.
Brownlie, Ian (2003): Principles of Public International Law. 6th ed. Oxford, New York: Oxford University Press.
Bureau, Dominique (2007): Note C. Cass. 27.03.2007. In *Revue critique du droit international privé* 96, pp. 405–412.
Bureau, Dominique (2010): Note C. Cass. 27.05.2010. In *Revue critique du droit international privé* 99, pp. 704–706.
Cachard, Olivier (2008): Observations CJCE 24.06.2008. In *Droit Maritime Français* 60, pp. 712–719.

Cachard, Olivier (2013): Observations CA Bordeaux 14.01.2013. In *Droit Maritime Français* 65, pp. 212–220.

Caddell, Richard (2009): Expanding the Ambit of Liability for Oil Pollution Damage from Tankers: The Charterer's Position under EU Law. In *Journal of International Maritime Law* 15, pp. 219–223.

Caemmerer, Ernst von (Ed.) (1983): Vorschläge und Gutachten zur Reform des deutschen internationalen Privatrechts der außervertraglichen Schuldverhältnisse. Tübingen: Mohr.

Cane, Peter F. (1994): The liability of classification societies. In *Lloyd's Maritime and Commercial Law Quarterly*, pp. 363–376.

Carbone, Sergio M. (2009): Conflits de lois en droit maritime. In *Recueil des Cours* 340, pp. 73–269.

Carbone, Sergio M.; Di Schiano Pepe, Lorenzo (2009): Uniform Law and Conflicts in Private Enforcement of Environmental Law: The Maritime Sector and Beyond. In *Il Diritto Marittimo* 111, pp. 50–80.

Caron, David D. (1983): Liability for Transnational Pollution Arising from Offshore Oil Development: A Methodological Approach. In *Ecology Law Quarterly* 10, pp. 641–683.

Casado Raigón, Rafael; Cataldi, Giuseppe (Eds.) (2009): Mélanges de droit de la mer offerts à Daniel Vignes. Brussels: Bruylant.

Castillo Daudí, Mireya (1998): La responsabilidad civil por daños de contaminación por hidrocarburos y el siniestro del buque "Aegean Sea". In *Revista Española de Derecho Internacional* 50 (2), pp. 69–92.

Castillo Daudí, Mireya (2001): El siniestro del buque "Aegean Sea" y la indemnización de los daños. In *Anuario de Derecho Maritimo* 18, pp. 261–301.

Charbonneau, Alexandre (2009): Marché international du travail maritime - Un cadre juridique en formation. Aix-en-Provence: Presses Universitaires d'Aix-Marseille.

Chaumette, Patrick (1994): L'internationalisation du travail maritime. In *Droit Maritime Français* 46, pp. 675–694.

Chen, Qi (2010): On the Application of CLC 92 in China. In Michael Faure, Lixin Han, Hongjun Shan (Eds.): Maritime Pollution Liability and Policy. China, Europe and the US. Alphen aan den Rijn: Kluwer Law International, pp. 347–358.

Churchill, R. R.; Lowe, A. V. (1999): The Law of the Sea. 3rd ed. Yonkers (NY), Manchester: Manchester University Press.

Churchill, Robin R. (2005): The Impact of State Practice on the Jurisdictional Framework Contained in the LOS Convention. In Alex G. Oude Elferink (Ed.): Stability and change in the law of the sea. The role of the LOS convention. Leiden, Boston: Martinus Nijhoff, pp. 91–143.

Clarke, Malcolm (Ed.) (2013): Maritime Law Evolving. Thirty Years at Southampton. Oxford, Portland: Hart Publishing.

Clemens-Jones, Timothy (1993): "Heidberg": malfaiteur ou victime d'une injustice? In *Droit Maritime Français* 45, pp. 712–717.

Corbier, Isabelle (2002): La faute inexcusable de l'armateur ou du droit de l'armateur à limiter sa responsabilité. In *Droit Maritime Français* 54, pp. 403–417.

Corbier, Isabelle (2005): Les sociétés de classification et la sécurité maritime. In *Jurisprudence du Port d'Anvers*, pp. 3–12.

Crespo Hernández, Ana (1996): Nota a la Sentencia del Juzgado de lo Penal núm. 2 de La Coruña de 30 abril de 1996. In *Revista Española de Derecho Internacional* 47, pp. 372–375.

Cudennec, Annie (Ed.) (2006): Le droit penal et la mer. Rennes: Presses universitaires de Rennes (Collection l'univers des normes).

Cudennec, Annie (2009): La politique communautaire de préservation du milieu marin: Quelle place pour le droit international? In Rafael Casado Raigón, Giuseppe Cataldi (Eds.): Mélanges de droit de la mer offerts à Daniel Vignes. Brussels: Bruylant, pp. 199–224.

Czerwenka, Beate (2010): Neue Haftungs- und Entschädigungsregeln in der Personenschiffahrt - Harmonisierung durch Europarecht. In *Transportrecht*, pp. 165–174.

Damar, Duygu (2011): Wilful misconduct in international transport law. Berlin, Heidelberg: Springer-Verlag.

Daniel, B. D. (2006): Potential Liability of Marine Classification Societies to Non-Contracting Parties. In *University of San Francisco Maritime Law Journal* 19, pp. 185–296.

Daudet, Yves; Eisemann, Pierre Michel (Eds.) (2003): La mer et son droit. Paris: A. Pedone.

Davies, Martin (2003): Obligations and implications for ships encountering persons in need of assistance at sea. In *Pacific Rim Law & Policy Journal* 12 (1), pp. 109–141.

Davies, Martin (2007): Classification society liability in the United States. In D. Rhidian Thomas (Ed.): Liability regimes in contemporary maritime law. London: Informa, pp. 129–147.

Davies, Martin (2011): Liability Issues raised by the Deepwater Horizon Blowout. In *Australian and New Zealand Maritime Law Journal* 25, pp. 35–48.

Delebecque, Philippe (2000): Responsabilité et indemnisation des dommages dus à la pollution par des hydrocarbures. À propos de la catastrophe de l'"Erika". In *La Semaine Juridique Édition Générale*, pp. 125–127.

Delebecque, Philippe (2008): L'évolution du transport de passagers. In Jacques Muriel Fabre-Magnan, Jean-Sébastien Borghetti, Olivier Deshayes, Cécile Péres (Eds.): Études offertes à Geneviève Viney. Paris: Librairie Générale de Droit, pp. 307–318.

Delebecque, Philippe (2010): Aspects civils de l'arrêt Erika. In *Droit Maritime Français* 62, pp. 878–885.

Delebecque, Philippe (2012): Aspects civils de la décision "Erika". In *Droit Maritime Français* 64, pp. 1000–1006.

Dicey, Albert Venn; Berriedale Keith, A. (1922): The Conflict of Laws. 3rd ed. London: Stevens.

Dicey, Albert Venn; Morris, John Humphrey Carlile; Collins, Lawrence (2012): Dicey, Morris and Collins on the Conflict of Laws. 15th ed. London: Sweet & Maxwell.

Dickinson, Andrew (2008): The Rome II Regulation: The Law Applicable To Non-Contractual Obligations. Oxford: Oxford University Press.

Dickinson, Andrew (2013): Territory in the Rome I and Rome II Regulations. In *Lloyd's Maritime and Commercial Law Quarterly*, pp. 86–132.

Dörfelt, Felix (2009): Gerichtsstand sowie Anerkennung und Vollstreckung nach dem Bunkeröl-Übereinkommen. In *Praxis des Internationalen Privat- und Verfahrensrechts*, pp. 470–474.

Dorsey, William R. (2003): The 2002 Protocol to the Athens Convention Relating to the Carriage of Passengers and Their Luggage by Sea. In *Benedict's Maritime Bulletin* 1, pp. 12–16.

Doud, Alden Lowell (1973): Compensation for Oil Pollution Damage: Further Comment on the Civil Liability and Compensation Fund Conventions. In *Journal of Maritime Law and Commerce* 4, pp. 525–542.

Droz, Georges A. L. (1991): Regards sur le droit international privé comparé. In *Recueil des Cours* 229 (IV), pp. 25–421.

Dupuy, Pierre-Marie; Fassbender, Bardo; Shaw, Malcolm N.; Sommermann, Karl-Peter (Eds.) (2006): Völkerrecht als Wertordnung. Festschrift für Christian Tomuschat. Kehl: N.P. Engel.

Dupuy, René-Jean (Ed.) (1985): L'avenir du droit international de l'environnement. Académie de Droit International de La Haye. Dordrecht: Nijhoff.

Dutoit, Bernard (2009): Le droit international privé des obligations non contractuelles à l'heure européenne: Le règlement Rome II. In Gabriella Venturini, Stefania Bariatti (Eds.): Liber Fausto Pocar. Milano: Giuffrè, pp. 309–329.

Dutson, Stuart (1997): The Conflict of Laws and Statutes: The International Operation of Legislation Dealing With Matters of Civil Law in the United Kingdom and Australia. In *Modern Law Review* 60, pp. 668–689.

Egler, Philipp (2011): Seeprivatrechtliche Streitigkeiten unter der EuGVVO. Berlin: Springer-Verlag.

Ehlers, Peter (2013): Meeresfreiheit und aquitoriale Ordnung. Zur Entwicklung des Seerechts. In *Verwaltungsarchiv*, pp. 406–428.

Ehlers, Peter; Lagoni, Rainer (Eds.) (2009): Responsibility and Liability in the Maritime Context. Münster: LIT Verlag.

Einhorn, Talia; Siehr, Kurt (Eds.) (2004): International Cooperation Through Private International Law. The Hague: T. M. C. Asser Press.
Eskenazi, Nancy J. (1993): Forum Non Conveniens and Choice of Law In Re: The "Amoco Cadiz" Oil Spill. In *Journal of Maritime Law and Commerce* 24, pp. 371–398.
Esmaeili, Hossein (2001): The Legal Regime of Offshore Oil Rigs in International Law. Aldershot, Burlington: Ashgate/Dartmouth.
Fajardo del Castillo, Teresa (2009): La catastrophe du "Prestige": Quelques leçons pour la Communauté Européenne et l'OMI. In Rafael Casado Raigón, Giuseppe Cataldi (Eds.): Mélanges de droit de la mer offerts à Daniel Vignes. Brussels: Bruylant, pp. 455–471.
Falkanger, Aage Thor (2011): Maritime Casualties and Intervention - coastal state measures when casualties pose the threat of pollution. Bergen: Fagbokforlaget.
Faure, Michael; Han, Lixin; Shan, Hongjun (Eds.) (2010): Maritime Pollution Liability and Policy. China, Europe and the US. Alphen aan den Rijn: Kluwer Law International.
Faure, Michael; Walt, André van der (Eds.) (2010): Globalization and private law. The way forward. Cheltenham: Edward Elgar.
Faure, Michael; Ying, Song (Eds.) (2008): China and International Environmental Liability. Legal Remedies for Transboundary Pollution. Cheltenham: Edward Elgar.
Fawcett, J. J.; Carruthers, Janeen M.; North, Peter M. (2008): Cheshire, North & Fawcett - Private International Law. 14th ed. Oxford, New York: Oxford University Press.
Fender, James E. (1980): Trouble Over Oiled Waters: Pollution Litigation or Arbitration. The Ixtoc I Oil Well Blow-out. In *Suffolk Transnational Law Journal* 4, pp. 281–301.
Fernández Arroyo, Diego P. (2004): Exorbitant and Exclusive Grounds of Jurisdiction in European Private International Law: Will They Ever Survive? In Heinz-Peter Mansel, Thomas Pfeiffer, Herbert Kronke, Christian Kohler, Rainer Hausmann (Eds.): Festschrift für Erik Jayme. München: Sellier, pp. 169–186.
Fernández Arroyo, Diego P.; Marques, Claudia Lima (Eds.) (2011): Derecho internacional privado y derecho internacional público. Asunción: CEDEP.
Fife, Rolf Einar (2003): The Duty to Render Assistance at Sea: Some Reflections after "Tampa". In Jarna Petman, Jan Klabbers (Eds.): Nordic cosmopolitanism. Essays in international law for Martti Koskenniemi. Leiden, Boston: Martinus Nijhoff Publishers, pp. 469–484.
Finlayson, C. F. (1986): Shipboard torts and the conflict of laws. In *Victoria University of Wellington Law Review* 16, pp. 119–145.
Fleischer, Carl August (1976): Liability for Oil Pollution Damage Resulting from Offshore Operations. In *Scandinavian Studies in Law* 20, pp. 107–143.
Fontaine, Emmanuel (1993): The French Experience: "Tanio" and "Amoco Cadiz" incidents compared. In Colin M. de la Rue (Ed.): Liability for damage to the marine environment. London, New York: Lloyd's of London Press, pp. 101–108.
Forner Delaygua, Joaquim; González Beilfuss, Cristina; Viñas Farré, Ramón (Eds.) (2013): Entre Bruselas y La Haya. Estudios sobre la unificación internacional y regional del Derecho internacional privado. Madrid, Barcelona: Marcial Pons.
Foyer, Jacques; Meunier, Jacques; Jault-Seseke, Fabienne; Bergé, Jean Sylvestre; Dauriac, Isabelle (Eds.) (2012): Le droit entre tradition et modernité. Mélanges à la mémoire de Patrick Courbe. Paris: Dalloz-Sirey.
Frank, Veronica (2005): Consequences of the "Prestige" Sinking for European and International Law. In *International Journal of Marine and Coastal Law* 20, pp. 1–64.
French, Duncan (2006): Treaty Interpretation and the Incorporation of Extraneous Legal Rules. In *International and Comparative Law Quarterly* 55, pp. 281–314.
Gabaldón García, José Luis; Ruiz Soroa, José María (2006): Manual de Derecho de la Navegación Marítima. 3rd ed. Madrid, Barcelona: Marcial Pons.
Gadow-Stephani, Inken von (2006): Der Zugang zu Nothäfen und sonstigen Notliegeplätzen für Schiffe in Seenot. Berlin, New York: Springer.
Ganten, Reinhard H. (2008): Developments in Oil Pollution Liability. In *Environmental Policy and Law* 38, pp. 312–315.

Garber, Thomas (2012): Zum Begriff des gewöhnlichen Arbeitsortes i. s. d. Art. 19 Abs. 2 lit. a EuGVVO insbes. bei der Verrichtung der arbeitsvertraglichen Tätigkeit an Bord eines Schiffes. In Reinhold Geimer, Rolf A. Schütze (Eds.): Recht ohne Grenzen. Festschrift für Athanassios Kaissis zum 65. Geburtstag. München: Sellier European Law Publishers, pp. 221–236.

García Pérez, Marta; Sanz Larruga, Francisco Javier (Eds.) (2006): Seguridad marítima y medio ambiente. Oleiros: Instituto Universitario de Estudios Marítimos.

García-Pita y Lastres, José Luis; Quintáns Eiras, Rocío; La Díaz Rosa, Angélica de (Eds.) (2012): Estudios de Derecho Marítimo. Cizur Menor: Thomson Aranzadi.

Gaskell, Nicholas (1986): Lessons of the Mont Louis. Part Two: Compensation for Hybrid Accidents. In *International Journal of Estuarine and Coastal Law* 1, pp. 269–307.

Gaskell, Nicholas (1987): The Zeebrugge Disaster: Application of the Athens Convention 1974. In *New Law Journal*, pp. 285–288.

Gaskell, Nicholas (1996): The Draft Convention on Liability and Compensation for Damage resulting from the Carriage of Hazardous and Noxious Substances. In Peter Wetterstein, Anders Beijer (Eds.): Essays in Honour of Hugo Tiberg. Stockholm: Juristförlaget, pp. 225–296.

Gaskell, Nicholas (1998): New limits for passengers and others in the United Kingdom. In *Lloyd's Maritime and Commercial Law Quarterly*, pp. 312–336.

Gaskell, Nicholas (2000): Pollution, Limitation and Carriage in "The Aegean Sea". In Francis D. Rose (Ed.): Lex Mercatoria. Essays on International Commercial Law in Honour of Francis Reynolds. London: LLP, pp. 71–103.

Gaskell, Nicholas (2009): The Bunker Pollution Convention 2001 and limitation of liability. In *Journal of International Maritime Law* 15, pp. 477–494.

Gaskell, Nicholas (2013): Compensation for Offshore Pollution: Ships and Platforms. In Malcolm Clarke (Ed.): Maritime Law Evolving. Thirty Years at Southampton. Oxford, Portland: Hart Publishing, pp. 63–93.

Gaskell, Nicholas; Forrest, Craig (2008): Marine Pollution Damage in Australia: Implementing the Bunker Oil Convention 2001 and the Supplementary Fund Protocol 2003. In *University of Queensland Law Journal* 27, pp. 103–165.

Gaskell, Nicholas; Shaw, Richard (1999): The Arrest Convention 1999. In *Lloyd's Maritime and Commercial Law Quarterly*, pp. 470–485.

Gauci, Gotthard (1995): Limitation of liability in maritime law: an anachronism? In *Marine Policy* 19, pp. 65–74.

Gauci, Gotthard M. (1999): Protection of the Marine Environment through the International Ship-Source Oil Pollution Regimes. In *Review of European Community and International Environmental Law* 8, pp. 29–36.

Geimer, Reinhold; Schütze, Rolf A. (Eds.) (2012): Recht ohne Grenzen. Festschrift für Athanassios Kaissis zum 65. Geburtstag. München: Sellier European Law Publishers.

George, Martin P. (2007): Choice of Law in Maritime Torts. In *Journal of Private International Law* 3, pp. 137–171.

Gillelan, Joshua T. (2005): "Vessel" Status and Admiralty Tort Jurisdiction. In *Benedict's Maritime Bulletin* 3, pp. 138–148.

Girvin, Stephen; Stephens, Toby (2007): Liabilities of salvors. In D. Rhidian Thomas (Ed.): Liability regimes in contemporary maritime law. London: Informa, pp. 81–98.

Godin, Ph.; Vialard, Antoine (1998): Cinquième journée Ripert, deuxième intervention. Vers une révision de la Convention d'Athènes de 1974 sur le transport de passagers par mer. In *Droit Maritime Français* 50, pp. 733–734.

Gold, Edgar (2002): Gard Handbook on P & I Insurance. 5th ed. Norway: GARD.

Goñi, José Luis (2012): L'arrêt de la Cour d'appel des États-Unis du 29 août 2012 dans l'affaire du "Prestige" et la responsabilité de la société de classification. In *Droit Maritime Français* 64, pp. 1023–1026.

Göransson, Magnus (1997): The HNS Convention. In *Uniform Law Review* 2, pp. 249–270.

Gorton, Lars; Ramberg, Jan; Sandström, Jan (Eds.) (1991): Festskrift till Kurt Grönfors. Stockholm: Norstedts.

Gottwald, Peter; Nagel, Heinrich (2007): Internationales Zivilprozessrecht. 6th ed. Köln: O. Schmidt.

Gouritin, Armelle (2011): The International Regime for the Compensation of Oil-pollution Damage: A Good Candidate to Have a Human Rights Law Approach? In *Review of European Community and International Environmental Law* 20 (2), pp. 194–207.

Grammenos, Costas Th (Ed.) (2010): The Handbook of Maritime Economics and Business. 2nd ed. London: Lloyd's List.

Griggs, Patrick (1996): Extending the frontiers of liability - the proposed Hazardous Noxious Substances Convention and its effect on ship, cargo and insurance interests. In *Lloyd's Maritime and Commercial Law Quarterly*, pp. 145–155.

Griggs, Patrick (1996): The HNS Convention 1996. In *International Journal of Shipping Law*, pp. 254–260.

Griggs, Patrick (2001): International Convention on Civil Liability for Bunker Oil Pollution Damage, 2001. British Maritime Law Association. Available online at http://www.bmla.org.uk/.

Griggs, Patrick (2002): Le protocole d'Athènes. In *Droit Maritime Français* 54, pp. 291–305.

Griggs, Patrick (2003): Obstacles to Uniformity of Maritime Law. The Nicholas J. Healy Lecture. In *Journal of Maritime Law and Commerce* 34, pp. 191–208.

Griggs, Patrick (2010): Making Maritime Law - Do Conventions Work? In Giorgio Berlingieri, Angelo Boglione, Sergio M. Carbone, Francesco Siccardi (Eds.): Scritti in onore di Francesco Berlingieri. Volume I. Genova: Il Diritto Marittimo, pp. 540–546.

Griggs, Patrick; Williams, Richard; Farr, Jeremy (2005): Limitation of Liability for Maritime Claims. 4th ed. London: LLP.

Grunvald, Sylvie; Saas, Claire (2009): Regards croisés sur les catastrophes pétrolières: De l'Amoco Cadiz à l'Erika, le regard du pénaliste. In *Annuaire de droit maritime et océanique*, pp. 379–396.

Guinchard, Emmanuel; Lamont-Black, Simone (2009): Environmental Law - the Black Sheep in Rome II's Drive for Legal Certainty? Article 7 of Regulation (EC) No. 864/2007 on the Law Applicable to Non-Contractual Obligations in Context. In *Environmental Law Review* 11, pp. 161–172.

Gunasekera, Dan Malika (2010): Civil liability for bunker oil pollution damage. Frankfurt am Main: Peter Lang.

Hackney, Ryan (2011): If a Helicopter Hits an Offshore Platform and Crashes at Sea, Where Do You Bury the Survivors? In *Journal of Maritime Law and Commerce* 42 (2), pp. 231–254.

Haddon-Cave, Charles (2000): Limitation against passenger claims: Medieval, unbreakable, unconscionable. In *CMI Yearbook*, pp. 234–243.

Hakapää, Kari (1981): Marine pollution in international law. Material obligations and jurisdiction. Helsinki: Suomalainen Tiedeakatemia.

Hakapää, Kari; Molenaar, Eric Jaap (1999): Innocent passage - past and present. In *Marine Policy* 23, pp. 131–145.

Halpern, Joseph (1983): "Exorbitant Jurisdiction" and the Brussels Convention: Toward a Theory of Restraint. In *Yale Journal of World Public Order* 9, pp. 369–387.

Han, Lixin; Yu, Ping (2006): New developments regarding the liability of classification societies. In *Journal of International Maritime Law* 12, pp. 243–250.

Hancock, Moffatt (1942): Torts in the conflict of laws. Ann Arbor, Chicago: University of Michigan Press.

Hansen, Robin F. (2008): Multinational Enterprise Pursuit of Minimized Liability: Law, International Business Theory and the "Prestige" Oil Spill. In *Berkeley Journal of International Law* 26, pp. 410–451.

Harrison, James (2008): Conflicting Interpretations - The "Slops" Incident and the Application of the International Oil Pollution Liability and Compensation Regime to Offshore Storage and Transfer Operations. In *Journal of Environmental Law* 20, pp. 455–464.

Harrison, Jim (2000): The Accountability of Classification Societies: The Role of Classification and Market-oriented and Policy Issues (The "Erika"). In *Journal of International Maritime Law* 7, pp. 299–302.

Hartenstein, Olaf (2008): Rom I-Entwurf und Rom II-Verordnung: Zur Bedeutung zukünftiger Änderungen im Internationalen Privatrecht für das Seerecht. In *Transportrecht* (4), pp. 143–161.

Healy, Nicholas J.; Sweeney, Joseph C. (1991): Basic Principles of the Law of Collision. In *Journal of Maritime Law and Commerce* 22, pp. 359–404.

Heiser, Anthony (2002): Border Protection; UNCLOS and the M. V. Tampa Incident 2001. In *Australian and New Zealand Maritime Law Journal* 16, pp. 84–121.

Hestermeyer, Holger; König, Doris; Matz-Lück, Nele; Röben, Volker; Seibert-Fohr, Anja; Stoll, Peter-Tobias; Vöner, Silja (Eds.) (2012): Coexistence, Cooperation and Solidarity. Liber Amicorum Rüdiger Wolfrum. Leiden, Boston: Martinus Nijhoff Publishers.

Hetherington, Stuart (2003): "Prestige" - Can the Law Assist? In *CMI Yearbook*, pp. 361–374.

Hille, Alexander; Schröder, Christoph; Dettmer, Michael; Visser, Marco (2010): Offshore-Windkraftanlagen - Haftung und Haftpflichtversicherung. In *Versicherungsrecht*, pp. 585–598.

Hoffmann, Bernd von; Thorn, Karsten; Firsching, Karl (2012): Internationales Privatrecht. Einschließlich der Grundzüge des Internationalen Zivilverfahrensrechts. 10th ed. München: Beck.

Hofstetter, Karl A. (1995): Sachgerechte Haftungsregeln für multinationale Konzerne. Tübingen: Mohr Siebeck.

Honka, Hannu (1994): The Classification System and its Problems with Special Reference to the Liability of Classification Societies. In *Tulane Maritime Law Journal* 19, pp. 1–36.

Honka, Hannu (1998): Nordic Reactions to Maritime Disasters: The Estonia Case. In Harald Koch, Armin Willingmann (Eds.): Großschäden - Complex Damages. Rechtliche und alternative Regulierungsstrategien im In- und Ausland. 1st ed. Baden-Baden: Nomos, pp. 73–90.

Honka, Hannu (2008): Modern Comprehensive Approach to Safety at Sea. In Ulrich Karpen, Djoko Lukic, Sonja Heinze (Eds.): Maritime Security - Current Problems in the Baltic Sea. Baden-Baden: Nomos, pp. 107–115.

Hovanesian, Archie (1970): Post "Torrey Canyon": Toward a New Solution to the Problem of Traumatic Oil Spillage. In *Connecticut Law Review* 2 (2), pp. 632–647.

Huber, Peter (Ed.) (2011): Rome II Regulation. Pocket commentary. Munich: Sellier European Law Publishers.

Huglo, Christian (1992): La réparation du dommage écologique au milieu marin à travers deux expériences judiciaires. In *La Gazette du Palais (Doctrine)* (2), pp. 582–590.

Hume, Lauren E. (2011): Are We Sailing in Occupied Waters?: Rethinking the Availability of Punitive Damages under the Oil Pollution Act 1990. In *New York University Law Review* 86, pp. 1444–1481.

IMCO (1973): Official records of the international Legal Conference on Marine Pollution Damage, 1969. London.

IMCO (1978): Official records of the Conference on the Establishment of an International Compensation Fund for Oil Pollution Damage, 1971. London.

Ipsen, Hans Peter; Necker, Karl-Hartmann (Eds.) (1979): Recht über See. Festschrift Rolf Stödter zum 70. Geburtstag. Hamburg, Heidelberg: R. v. Decker's Verlag.

Jacobsen, Douglas A.; Yellen, James D. (1984): Oil Pollution: The 1984 London Protocols and the "Amoco Cadiz". In *Journal of Maritime Law and Commerce* 15 (4), pp. 467–488.

Jacobsson, Måns (2007): L'expérience française du FIPOL. In *Droit Maritime Français* 59, pp. 968–984.

Jacobsson, Måns (2007): Ship source pollution - Oil and hazardous and noxious substances. In D. Rhidian Thomas (Ed.): Liability regimes in contemporary maritime law. London: Informa, pp. 3–18.

Jacobsson, Måns (2007): The International Liability and Compensation Regime for Oil Pollution from Ships - International Solutions for a Global Problem. In *Tulane Maritime Law Journal* 32, pp. 1–33.

Jacobsson, Måns (2007): Uniform Application of the International Regime on Liability and Compensation for Oil Pollution Damage. In Tafsir Malick Ndiaye, Rüdiger Wolfrum (Eds.): Law of the sea, environmental law and settlement of disputes. Leiden, Boston: M. Nijhoff, pp. 421–436.

Jacobsson, Måns (2009): Bunkers Convention in force. In *Journal of International Maritime Law* 15, pp. 21–36.

Jacobsson, Måns (2009): The HNS Convention - Prospects for its entry into force. In *CMI Yearbook*, pp. 418–430.

Jacobsson, Måns (2010): Dispute resolution in oil pollution cases. Are there alternatives to litigation in national courts? In Norman A. Martínez Gutiérrez (Ed.): Serving the rule of International Maritime Law. Essays in Honour of Professor David Joseph Attard. Abingdon: Routledge, pp. 328–341.

Jacobsson, Måns (2010): How clean is clean? The concept of "reasonableness" in the response to tanker oil spills. In Giorgio Berlingieri, Angelo Boglione, Sergio M. Carbone, Francesco Siccardi (Eds.): Scritti in onore di Francesco Berlingieri. Volume I. Genova: Il Diritto Marittimo, pp. 565–585.

Jacobsson, Måns (2010): L'indemnisation des dommages résultant des atteintes à l'environnement dans le cadre du régime international CLC/FIPOL. In *Droit Maritime Français* 62, pp. 463–480.

Jacobsson, Måns (2011): An ideal international scheme for compensation for marine pollution damage. In *Journal of International Maritime Law* 17, pp. 263–273.

Jacobsson, Måns (2012): Perspective of the Global Compensation Regimes; The Relationship between EU Legislation and Maritime Liability Conventions. In *European Journal of Commercial Contract Law*, pp. 63–76.

Jambu-Merlin, R. (1962): Note CA Rennes 13.03.1961. In *Recueil Dalloz*, pp. 155–157.

Jenisch, Uwe (1997): Offshore-Windenergieanlagen im Seerecht. In *Natur und Recht*, pp. 373–381.

Jessen, Henning (2013): Neues zur Dritthaftung privater Klassifikationsgesellschaften nach "Erika" und "Prestige"? In *Recht der Transportwirtschaft*, pp. 387–393.

Jones, John Paul (2009): The Sky Has Not Fallen Yet on Punitive Damages in Admiralty Cases. In *Tulane Law Review* 83, pp. 1289–1302.

Juenger, Friedrich K. (1989): Mass Disasters and the Conflict of Laws. In *University of Illinois Law Review*, pp. 105–128.

Junker, Abbo; Schramm, Dorothee (2001): Die gekenterte Bohrplattform - Vertragliche und deliktische Haftung bei Distanzdelikten in der französischen Rechtsprechung. In *Praxis des Internationalen Privat- und Verfahrensrechts*, pp. 482–485.

Juste-Ruíz, José (2010): Compensation for pollution damage caused by oil tanker accidents: from "Erika" to "Prestige". In *Aegean Review of the law of the sea and maritime law* 1, pp. 37–60.

Kappet, Liliane C. (2006): Tankerunfälle und der Ersatz ökologischer Schäden. Hamburg, Münster: Lit.

Karpen, Ulrich; Lukic, Djoko; Heinze, Sonja (Eds.) (2008): Maritime Security - Current Problems in the Baltic Sea. 1st ed. Baden-Baden: Nomos.

Karsten, Jens (2009): Im Fahrwasser der Athener Verordnung zu Seereisenden: Neuere Entwicklungen des europäischen Passagierrechts. In *Verbraucher und Recht*, pp. 213–225.

Kashubsky, Mikhail (2006): Marine Pollution from the Offshore Oil and Gas Industry: Review of Major Conventions and Russian Law (Part I). In *Maritime Studies*, pp. 1–11.

Kaufmann, Sven G. (2013): Reformbedarf auf internationaler Ebene für die rechtliche Aufarbeitung von Tankerkatastrophen: Das Erika-Urteil des französischen Kassationsgerichtes vom 25. September 2012. In *Zeitschrift für ausländisches öffentliches Recht und Völkerrecht*, pp. 509–541.

Kaye, Stuart (2007): International Measures To Protect Oil Platforms, Pipelines, and Submarine Cables from Attack. In *Tulane Maritime Law Journal* 31, pp. 377–423.

Kbaier, Rouchdy; Sebek, Viktor (1985): New trends in compensation for oil pollution damage. "Amoco Cadiz" legal proceedings and the 1984 diplomatic conference on liability and compensation. In *Marine Policy*, pp. 269–279.

Kegel, Gerhard (1968): L'abordage en haute mer en droit international privé. In *Revue critique du droit international privé*, pp. 393–417.

Kegel, Gerhard; Schurig, Klaus (2004): Internationales Privatrecht. 9th ed. München: Beck.

Kenney, Frederick J.; Tasikas, Vasilios (2003): The "Tampa" incident: IMO perspectives and responses on the treatment of persons rescued at sea. In *Pacific Rim Law & Policy Journal* 12 (1), pp. 143–177.

Kiran, R. Bhanu Krishna (2010): Liability and Compensation for Oil Pollution Damage: An Examination of IMO Conventions. In *National University of Juridical Sciences Law Journal* 3, pp. 399–421.

Kiss, Alexandre (1975): Un cas de pollution internationale: l'affaire des boues rouges. In *Journal du Droit International (Clunet)* 102, pp. 207–248.

Kiss, Alexandre (1985): L'affaire de l'"Amoco Cadiz": responsabilité pour une catastrophe écologique. In *Journal du Droit International (Clunet)*, pp. 575–589.

Kloss, Claudia (1989): Der Untergang der "Amoco Cadiz" vor amerikanischen Gerichten. In *Praxis des Internationalen Privat- und Verfahrensrechts* (3), pp. 184–188.

Knops, Anne (Ed.) (2003): Liber Amicorum Roger Roland. Gent: Larcier.

Koch, Harald; Willingmann, Armin (Eds.) (1998): Großschäden - Complex Damages. Rechtliche und alternative Regulierungsstrategien im In- und Ausland. Baden-Baden: Nomos.

Kopela, Sophia (2011): Civil and Criminal Liability as Mechanisms for the Prevention of Oil Marine Pollution: The "Erika" Case. In *Review of European Community and International Environmental Law* 20, pp. 313–324.

Kraska, James (2011): Maritime power and the law of the sea. Expeditionary operations in world politics. Oxford, New York: Oxford University Press.

Kröger, Bernd (2001): Passengers Carried by Sea - Should they be Granted the Same Rights as Airline Passengers? In *CMI Yearbook*, pp. 244–252.

Kropholler, Jan (2006): Internationales Privatrecht. 6th ed. Tübingen: Mohr Siebeck.

Krüger, Kai (1991): Fault Liability for Classification Societies towards Third parties? In Lars Gorton, Jan Ramberg, Jan Sandström (Eds.): Festskrift till Kurt Grönfors. Stockholm: Norstedts, pp. 271–296.

Labrot, Véronique (2006): Compétences des États et Droit de la mer: regard sur les compétences pénales. In Annie Cudennec (Ed.): Le droit penal et la mer. Rennes: Presses universitaires de Rennes (Collection l'univers des normes), pp. 19–41.

Lagarde, Paul (1991): Le nouveau droit international privé des contrats après l'entrée en vigueur de la Convention de Rome du 19 juin 1980. In *Revue critique du droit international privé*, pp. 287–340.

Lagoni, Nicolai (2007): The liability of classification societies. Berlin, New York: Springer.

Lagoni, Rainer (2007): Offshore Bunkering in the Exclusive Economic Zone. In Tafsir Malick Ndiaye, Rüdiger Wolfrum (Eds.): Law of the sea, environmental law and settlement of disputes. Leiden, Boston: M. Nijhoff, pp. 613–627.

Lagoni, Rainer (2008): Das Wrackbeseitigungs-Übereinkommen und das Völkerrecht. In *Transportrecht*, pp. 352–357.

Lagoni, Rainer (2010): Arbeitsschutz auf fremdflaggigen Seeschiffen im Hafen. In *Archiv des Völkerrechtes* 48, pp. 58–104.

Lagoni, Rainer (2012): Cable and Pipeline Surveys at Sea. In Holger Hestermeyer, Doris König, Nele Matz-Lück, Volker Röben, Anja Seibert-Fohr, Peter-Tobias Stoll, Silja Vöner (Eds.):

Coexistence, Cooperation and Solidarity. Liber Amicorum Rüdiger Wolfrum. Leiden, Boston: Martinus Nijhoff Publishers, pp. 933–955.

Lagoni, Rainer; Paschke, Marian (Eds.) (1999): Seehandelsrecht und Seerecht. Festschrift für Rolf Herber zum 70. Geburtstag. Hamburg: Lit.

Lagoni, Rainer; Proelß, Alexander (2006): Festlandsockel und Ausschließliche Wirtschaftszone. In Wolfgang Graf Vizthum (Ed.): Handbuch des Seerechts. München: Beck, pp. 161–286.

Laing, Lionel H. (1932): American Fishing Vessels in Canadian Waters of the Pacific. In *American Journal of International Law* 26, pp. 374–379.

Lang, John (2001): The Investigation of Marine Accidents: A U. K. Accident Investigator's View. In *Tulane Law Review* 75, pp. 1673–1685.

Law Commission; Scottish Law Commission (1984): Working Paper No. 87. London: Her Majesty's Stationery Office.

Law Commission; Scottish Law Commission (1990): Private International Law. Choice of Law in Tort and Delict. No. 193. London: Her Majesty's Stationery Office.

Layton, Alexander; Mercer, Hugh (2004): European Civil Practice. 2 volumes. London: Thomson/ Sweet & Maxwell (1).

Le Bihan-Guénolé, Martine (2007): Une cargaison d'hydrocarbure accidentellement déversée au cours de son transport maritime peut-elle être qualifiée de déchet? In *Droit Maritime Français* 59, pp. 740–749.

Le Couviour, Karine (1999): La Convention SNPD (1): quelques réflexions sur la dernière pièce du dispositif. In *Droit Maritime Français* 51, pp. 1075–1094.

Le Couviour, Karine (2002): Responsabilités pour pollutions majeures résultant du transport maritime d'hydrocarbures. In *La Semaine Juridique Édition Générale*, pp. 2269–2277.

Le Couviour, Karine (2008): Catastrophe de l'Erika: premiers commentaires d'un jugement déjà exemplaire. In *La Semaine Juridique Édition Générale*, pp. 3–5.

Le Couviour, Karine (2010): Erika: décryptage d'un arrêt peu conventionnel. In *La Semaine Juridique Édition Générale* (II), pp. 804–807.

Le Couviour, Karine (2012): Erika: l'arrêt salvateur de la Cour de Cassation. In *La Semaine Juridique Édition Générale*, pp. 2095–2099.

Legendre, Claire (1976): La Convention d'Athènes relative au transport par mer de passagers et de leurs bagages. In *Droit Maritime Français* 28, pp. 451–461.

Légier, Gérard (1999): Note C. Cass. 11.05.1999. In *Journal du Droit International (Clunet)*, pp. 1050–1057.

Légier, Gérard (2007): Note C. Cass. 27.03.2007. In *Journal du Droit International (Clunet)*, pp. 949–956.

Legros, Cécile (2012): L'intégration des conventions internationales dans le droit dérivé de l'Union Européenne: L'exemple du droit des transports. In Jacques Foyer, Jacques Meunier, Fabienne Jault-Seseke, Jean Sylvestre Bergé, Isabelle Dauriac (Eds.): Le droit entre tradition et modernité. Mélanges à la mémoire de Patrick Courbe. Paris: Dalloz-Sirey, pp. 367–387.

Lennon, J. Michael (2002): The Law of Collision and the United States Navy. In *Buffalo Law Review* 50, pp. 981–1016.

Leonhard, Alan T. (1980): Ixtoc I: A Test for the Emerging Concept of the Patrimonial Sea. In *San Diego Law Review* 17, pp. 617–627.

Lewins, Kate (2004): The Cruise Ship Industry - Liabilities to Passengers for Breach of s52 and s74 "Trade Practices Act 1974" (Cth). In *Australian and New Zealand Maritime Law Journal* 18, pp. 30–54.

Lewins, Kate (2010): Australian cruise passengers travel in legal equivalent of steerage - considering the merits of a passenger liability regime for Australia. In *Australian Business Law Review* 38, pp. 127–142.

Li, K. X.; Ng, Jim Mi (2002): International Maritime Conventions: Seafarers' Safety and Human Rights. In *Journal of Maritime Law and Commerce* 33, pp. 381–404.

Little, Gavin (1998): The Hazardous and Noxious Substances Convention: a new horizon in the regulation of marine pollution. In *Lloyd's Maritime and Commercial Law Quarterly*, pp. 554–567.

Looks, Volker; Sinkus, Dirk (1994): Die Kosten für Ölbekämpfungsmaßnahmen und die Haftungsbeschränkung unter der Londoner Konvention von 1976. In *Transportrecht*, pp. 263–266.

Lopusky, Jan (1993): El Convenio sobre sustancias peligrosas y nocivas (H.N.S.): ¿Estamos en el buen camino? In *Anuario de Derecho Maritimo* 10, pp. 235–256.

Lorenz, Egon (1983): Das anwendbare Recht bei Schiffs- und Flugzeugunfällen. In Ernst von Caemmerer (Ed.): Vorschläge und Gutachten zur Reform des deutschen internationalen Privatrechts der außervertraglichen Schuldverhältnisse. Tübingen: Mohr, pp. 440–463.

Lubach, Carl (2013): Zur Havarie der "Costa Concordia": Erste Gerichtsentscheidung in Spanien. In *Transportrecht*, pp. 405–407.

Lucchini, Laurent (1970): La pollution des mers par les hydrocarbures: Les Conventions de Bruxelles de novembre 1969 ou les fissures du droit international classique. In *Journal du Droit International Privé* 97 (4), pp. 795–827.

Lucchini, Laurent (1978): A propos de l'"Amoco Cadiz". La lutte contre la pollution des mers: Evolution ou Révolution du droit international. In *Annuaire français de droit international* 24, pp. 721–754.

Lucchini, Laurent (1985): Le procès de l'"Amoco Cadiz": Présent et voies du futur. In *Annuaire français de droit international* 31, pp. 762–782.

Macdonald, William J. (1980): Ixtoc I: International and Domestic Remedies for Transboundary Pollution Injury. In *Fordham Law Review* 49, pp. 404–431.

Maekelt, Tatiana B. de (2011): Relaciones entre el derecho internacional privado y el derecho internacional público. In Diego P. Fernández Arroyo, Claudia Lima Marques (Eds.): Derecho internacional privado y derecho internacional público. Asunción: CEDEP, pp. 27–57.

Magnus, Ulrich (2011): Seearbeitsverhältnisse und die Rom I- und II-Verordnungen. In Tomislav Borić, Brigitta Lurger, Peter Schwarzenegger, Ulfried Terlitza (Eds.): Öffnung und Wandel - Die internationale Dimension des Rechts I. Festschrift für Willibald Posch zum 65. Geburtstag. With assistance of Willibald Posch. Wien: LexisNexis, pp. 443–459.

Maier, Harold G.; McCoy, Thomas R. (1991): A Unifying Theory for Judicial Jurisdiction and Choice of Law. In *American Journal of Comparative Law* 39, pp. 249–292.

Manchuk, Guy (2007): The Law of the Flag and Maritime Criminal Jurisdiction: A New Rule To Replace an Outdated, Inconvenient Doctrine. In *Tulane Maritime Law Journal* 32, pp. 221–248.

Mandaraka-Sheppard, Aleka (2009): Modern Maritime Law and Risk Management. London: Informa.

Mankowski, Peter (2003): Europäisches Internationales Arbeitsprozessrecht - Weiteres zum gewöhnlichen Arbeitsort. In *Praxis des Internationalen Privat- und Verfahrensrechts*, pp. 21–28.

Mankowski, Peter (2005): Das Internationale Arbeitsrecht und die See - Die Fortsetzung folgt. In *Praxis des Internationalen Privat- und Verfahrensrechts*, pp. 58–61.

Mann, Francis A. (1964): The Doctrine of International Jurisdiction Revisited after Twenty Years. In *Recueil des Cours* 111 (I), pp. 19–115.

Mansel, Heinz-Peter; Pfeiffer, Thomas; Kronke, Herbert; Kohler, Christian; Hausmann, Rainer (Eds.) (2004): Festschrift für Erik Jayme. München: Sellier.

Marsden, Reginald G.; Gault, Simon; Hazelwood, Steven J.; Tettenborn, A. M. (2003): Marsden on collisions at sea. 13th ed. London: Sweet & Maxwell.

Marten, Bevan (2006): Limitation of Liability for Personal Injury in New Zealand: ACC Meets the Sea. In *Australian and New Zealand Maritime Law Journal* 16, pp. 16–24.

Marten, Bevan (2011): The "Rena" and liability. In *New Zealand Law Journal*, pp. 341–344.

Martínez Gutiérrez, Norman A. (Ed.) (2010): Serving the rule of International Maritime Law. Essays in Honour of Professor David Joseph Attard. Abingdon: Routledge.

Martínez Gutiérrez, Norman A. (2011): Limitation of Liability in International Maritime Conventions. London, New York: Routledge.

Martínez Gutiérrez, Norman A. (2012): New European rules on the liability of carriers of passengers by sea in the event of accidents. In *Journal of International Maritime Law* 18, pp. 293–305.

Martínez Gutiérrez, Norman A. (2012): The Bunkers Convention and the Shipowner's Right to Limit Liability. In *Journal of Maritime Law and Commerce* 43 (2), pp. 235–254.

Maseda Rodríguez, Javier (2007): Responsabilidad civil por vertido de hidrocarburos en el mar: competencia judicial internacional y actividad procesal del naviero y víctimas. In José Manuel Sobrino Heredia (Ed.): Mares y Océanos en un mundo de cambio. Tendencias jurídicas, actores y factores. Valencia: Tirant lo Blanch, pp. 525–571.

Mason, Michael (2003): Civil liability for oil pollution damage: examining the evolving scope for environmental compensation in the international regime. In *Marine Policy* 27, pp. 1–12.

Mayer, Pierre; Heuzé, Vincent (2010): Droit international privé. 10th ed. Paris: Montchrestien.

Mehren, Arthur Taylor von (1983): Adjudicatory Jurisdiction: General Theories Compared and Evaluated. In *Boston University Law Review* 63 (2), pp. 279–340.

Meilán Gil, José Luis (2005): Problemas Jurídico-Administrativos Planteados por el "Prestige". Cizur Menor: Thomson Aranzadi.

Mensah, Thomas A. (1999): International Regime on Liability and Compensation for Marine Environmental Damage from Ships: Genesis and Achievements. In Rainer Lagoni, Marian Paschke (Eds.): Seehandelsrecht und Seerecht. Festschrift für Rolf Herber zum 70. Geburtstag. Hamburg: Lit, pp. 401–461.

Mensah, Thomas A. (2007): Prevention of Marine Pollution: The Contribution of IMO. In Jürgen Basedow, Ulrich Magnus (Eds.): Pollution of the Sea. Prevention and compensation. Berlin, New York: Springer, pp. 41–61.

Mensah, Thomas A. (2010): Can the "Slops" be considered as a ship for the purposes of the 1992 Civil Liability Convention and the 1992 Fund Convention? In *Aegean Review of the law of the sea and maritime law* 1, pp. 145–155.

Michinel, Miguel (2007): The "Prestige" in the courts. In *Australian and New Zealand Maritime Law Journal* 21, pp. 171–177.

Miller, Michael D. (2005): Marine War Risks. 3rd ed. London: LLP.

Mills, Alex (2009): The Confluence of Public and Private International Law. Justice, Pluralism and Subsidiarity in the International Constitutional Ordering of Private Law. Cambridge, New York: Cambridge University Press.

Miribel, Stéphane (2011): L'affaire "Rosa Delmas": Limitation et faute inexcusable, une nouvelle approche? In *Droit Maritime Français* 63, pp. 863–869.

Molenaar, Eric Jaap (2007): Port State Jurisdiction: Toward Comprehensive, Mandatory and Global Coverage. In *Ocean Development and International Law* 38, pp. 225–257.

Moreira, A. William; Strickland, Cecily Y.; Henley, David G. (2003): Liability for Marine Pollution from Offshore Operations. In *Dalhousie Law Journal* 26, pp. 429–467.

Morin, Michel (1997): Les rapports entre droit international public et droits internes: l'exemple du FIPOL. In *Droit Maritime Français* 49, pp. 325–335.

Morse, C. G. J. (Robin) (1996): Torts in Private International Law: A New Statutory Framework. In *International and Comparative Law Quarterly* 45, pp. 888–902.

Mortensen, Reid (2006): Homing Devices in Choice of Tort Law: Australian, British and Canadian Approaches. In *International and Comparative Law Quarterly* 55, pp. 839–878.

Mukherjee, Proshanto K. (2007): The Penal Law of Ship-Source Marine Pollution: Selected Issues in Perspective. In Tafsir Malick Ndiaye, Rüdiger Wolfrum (Eds.): Law of the sea, environmental law and settlement of disputes. Leiden, Boston: M. Nijhoff, pp. 463–496.

Müller, Walter (1999): Faut-il réviser la Convention d'Athènes de 1974 sur le transport de passagers par mer? In *Droit Maritime Français* 51, pp. 5–8.

Müller, Walter (2000): Passengers Carried by Sea. Should the Athens Convention 1974 be modified and adapted to the liability regime in air-law? In *CMI Yearbook*, pp. 665–668.

Muriel Fabre-Magnan, Jacques; Borghetti, Jean-Sébastien; Deshayes, Olivier; Péres, Cécile (Eds.) (2008): Études offertes à Geneviève Viney. Paris: Librairie Générale de Droit.

Naeemullah, Imran (2013): A Decade Later, $1 Billion Saved. The Second Circuit Relieves a Maritime Classification Society of Unprecedented Liability for Environmental and Economic Damages in "Reino de España v. American Bureau of Shipping, Inc.". In *Tulane Maritime Law Journal* 37, pp. 639–654.

Nanda, Ved P. (1967): The "Torrey Canyon" Disaster: Some Legal Aspects. In *Denver Law Journal* 44, pp. 400–425.

Ndendé, Martin (2007): L'accident de l'Erika. Procédures d'indemnisation des victimes et enjeux judiciaires autour d'une catastrophe pétrolière. In *Revue de Droit des Transports* (Février), pp. 11–17.

Ndendé, Martin (2008): La politique d'indemnisation des victimes des marées noires devant le FIPOL. In *Revue de Droit des Transports*, pp. 27–35.

Ndendé, Martin (2008): Les enseignements du jugement pénal de Paris sur les responsabilités des acteurs mis en cause dans la catastrophe de l'"Erika". In *Annuaire de droit maritime et océanique* 26, pp. 253–272.

Ndendé, Martin (2008): Responsabilités pour pollutions marines par hydrocarbures. In *Revue de Droit des Transports* (Mars), pp. 32–34.

Ndiaye, Tafsir Malick; Wolfrum, Rüdiger (Eds.) (2007): Law of the sea, environmental law and settlement of disputes. Leiden, Boston: M. Nijhoff.

Neyret, Laurent (2008): Naufrage de l'"Erika": vers un droit commun de la réparation des atteintes à l'environnement. In *Recueil Dalloz Chronique*, pp. 2681–2689.

Neyret, Laurent (2010): L'affaire "Erika": moteur d'évolution des responsabilités civile et pénale. In *Recueil Dalloz*, pp. 2238–2246.

Ngantcha, Francis (1990): The Right of Innocent Passage and the Evolution of the International Law of the Sea. London, New York: Pinter.

Nordisk institutt for sjørett (2000): Legislative approaches in maritime law. European Colloquium on Maritime Law. Oslo, 7.-8.12.2000. Oslo: Sjørettsfondet (MarIus, 283). Available online at www.jus.uio.no/nifs.

Nordquist, Myron H.; Moore, John Norton (Eds.) (2001): Current Marine Environmental Issues and the International Tribunal for the Law of the Sea. Center for Oceans Law and Policy. The Hague, London: Martinus Nijhoff Publishers.

Nordquist, Myron H.; Moore, John Norton; Mahmoudi, Said (Eds.) (2003): The Stockholm Declaration and Law of the Marine Environment. The Hague, London: Martinus Nijhoff Publishers.

Nordquist, Myron H.; Nandan, Satya N.; Rosenne, Shabtai (1993): United Nations Convention on the Law of the Sea, 1982. A Commentary. Volume II. Dordrecht, Boston: Martinus Nijhoff Publishers.

Nordquist, Myron H.; Nandan, Satya N.; Rosenne, Shabtai (2002): United Nations Convention on the Law of the Sea, 1982. A Commentary. Volume III. Dordrecht, Boston: Martinus Nijhoff Publishers.

Nordquist, Myron H.; Rosenne, Shabtai; Yankov, Alexander (1991): United Nations Convention on the Law of the Sea, 1982. A Commentary. Volume IV. Dordrecht, Boston: Martinus Nijhoff Publishers.

O'Connell, Daniel P. (1979): The English Choice of Law Rule in Collision Suits and Shipboard Torts. In Hans Peter Ipsen, Karl-Hartmann Necker (Eds.): Recht über See. Festschrift Rolf Stödter zum 70. Geburtstag. Hamburg, Heidelberg: R. v. Decker's Verlag, pp. 101–107.

O'Connell, Daniel Patrick; Shearer, Ivan Anthony (1984): The International Law of the Sea (Vol. II). Oxford: Clarendon Press.

O'Connell, Dennis M. (1970): Reflections on Brussels: IMCO and the 1969 Pollution Conventions. In *Cornell International Law Journal* 3 (2), pp. 161–188.

Odier, Françoise (1992): L' assistance maritime - une institution en quête de reconnaissance. In: Droit de la mer. Études dédiées au doyen Claude-Albert Colliard. Paris: Pedone, pp. 65–71.

Oosterveen, Willem (2004): Some recent developments regarding liability for damage resulting from oil pollution - from the perspective of an EU Member State. In *Environmental Law Review* 6, pp. 223–239.

Opertiti Badán, Didier (2011): Derecho internacional público y derecho internacional privado. Hacia un diálogo renovado. In Diego P. Fernández Arroyo, Claudia Lima Marques (Eds.): Derecho internacional privado y derecho internacional público. Asunción: CEDEP, pp. 59–80.

Oude Elferink, Alex G. (Ed.) (2005): Stability and change in the law of the sea. The role of the LOS convention. Leiden, Boston: Martinus Nijhoff.

Oxman, Bernard H. (1991): The Duty to Respect Generally Accepted International Standards. In *New York University Journal of International Law and Politics* 24, pp. 109–159.

Oxman, Bernard H. (1998): Human Rights and the United Nations Convention on the Law of the Sea. In *Columbia Journal of Transnational Law* 36, pp. 399–429.

Oxman, Bernard H. (2006): The Territorial Temptation: A Siren Song at Sea. In *American Journal of International Law* 100, pp. 830–851.

Özçayir, Z. Oya (1998): Liability for oil pollution and collisions. London: LLP.

Palandt, Otto (2014): Bürgerliches Gesetzbuch. 73rd ed. München: Beck.

Pallis, Mark (2002): Obligations of States towards Asylum Seekers at Sea: Interactions and Conflicts Between Legal Regimes. In *International Journal of Refugee Law* 14, pp. 329–364.

Papadopoulou, Danai (2009): The Role of French Environmental Associations in Civil Liability for Environmental Harm: Courtesy of "Erika". In *Journal of Environmental Law* 21 (1), pp. 87–112.

Papanicolopulu, Irini (2014): A Missing Part of the Law of the Sea Convention: Addressing Issues of State Jurisdiction over Persons at Sea. In Clive H. Schofield (Ed.): The Limits of Maritime Jurisdiction. Leiden: Martinus Nijhoff, pp. 387–404.

Parance, Béatrice (2008): Note TGI Paris, 16.01.2008. In *La Semaine Juridique Édition Générale*, pp. 36–39.

Parry, Clive (1958): The Theory of Restrictive Practices. In *Transactions of the Grotius Society* 44, pp. 109–123.

Payer, Hans G. (2001): Insurer and Class and Marine Accidents. In Myron H. Nordquist, John Norton Moore (Eds.): Current Marine Environmental Issues and the International Tribunal for the Law of the Sea. The Hague, London: Martinus Nijhoff Publishers, pp. 291–299.

Peet, Gerard (1992): The MARPOL Convention: Implementation and Effectiveness. In *International Journal of Estuarine and Coastal Law* 7, pp. 277–295.

Peltz, Robert D. (2012): The Athens Convention Revisited. In *Journal of Maritime Law and Commerce* 43, pp. 491–520.

Peltz, Robert D.; Warger, Vincent J. (2001): Amendments to Athens Convention Threaten US Maritime Law. In *International Travel Law Journal*, pp. 170–191.

Peplowska, Zuzanna (2010): What is a ship? The Policy of the International fund for Compensation for Oil Pollution Damage: the effect of the Greek Supreme Court judgment in the "Slops" case. In *Aegean Review of the law of the sea and maritime law* 1, pp. 157–164.

Perakis, Manolis (2011): Modern Tendencies towards a Disruption of the Bond between the Ship's Flag and the Applicable Law. In *Annuaire de droit maritime et océanique*, pp. 341–357.

Perry, Ronen (2011): The Deepwater Horizon Oil Spill and the Limits of Civil Liability. In *Washington Law Review* 86, pp. 1–68.

Pestke, Silvia (2008): Offshore-Windfarmen in der Ausschliesslichen Wirtschaftszone. Im Zielkonflikt zwischen Klima- und Umweltschutz. Baden-Baden: Nomos.

Petman, Jarna; Klabbers, Jan (Eds.) (2003): Nordic cosmopolitanism. Essays in international law for Martti Koskenniemi. Leiden, Boston: Martinus Nijhoff Publishers.

Pierronnet, François-Xavier; Luquiau, Antoine (2000): L'indemnisation des dommages subis par les passagers transportés par mer: de nouvelles voies possibles? In *Annuaire de droit maritime et océanique* 18, pp. 265–296.

Pineus, Kaj (1983): L'échouement du "Tsesis". In *Droit Maritime Français* 35, pp. 248–249.
Piras, Massimiliano (2012): International Recent Developments: European Union - Maritime Passenger Transport. In *Tulane Maritime Law Journal* 36, pp. 627–639.
Platzöder, Renate (1983): Third United Nations Conference on the Law of the Sea: Documents, Volume IV. Dobbs Ferry (NY): Oceana Publications.
Platzöder, Renate (1986): Third United Nations Conference on the Law of the Sea: Documents, Volume X. Dobbs Ferry (NY): Oceana Publications.
Platzöder, Renate (1987): Third United Nations Conference on the Law of the Sea: Documents, Volume XI. Dobbs Ferry (NY): Oceana Publications.
Plender, Richard; Wilderspin, Michael (2009): The European Private International Law of Obligations. 3rd ed. London: Sweet & Maxwell.
Pontavice, Emmanuel du (1976): Le statut des navires. Paris: Librairies Techniques.
Pontavice, Emmanuel du (1980): Conventions Internationales et Droit Interne. In *Annuaire de droit maritime et aérien* 5, pp. 17–81.
Pontavice, Emmanuel du (1985): Affaire "Droit de l'environnement versus Droit maritime" ou la décision rendue le 18 avril 1984 concernant "l'Amoco Cadiz". In *Annuaire de droit maritime et aérien*, pp. 9–60.
Pozdnakova, Alla (2013): Criminal jurisdiction over perpetrators of ship-source pollution. Leiden, Boston: Martinus Nijhoff.
Prüssmann, Heinz; Rabe, Dieter (2000): Seehandelsrecht. Fünftes Buch des Handelsgesetzbuches. 4th ed. München: C. H. Beck.
Pugash, James Z. (1977): The Dilemma of the Sea Refugee: Rescue Without Refuge. In *Harvard International Law Journal* 18 (3), pp. 577–604.
Pugh, Michael (2004): Drowning not Waving: Boat People and Humanitarianism at Sea. In *Journal of Refugee Studies* 17 (1), pp. 50–69.
Pulido, Juan L. (2007): Compensation by Coastal States - The "Prestige" Disaster. In Jürgen Basedow, Ulrich Magnus (Eds.): Pollution of the Sea. Prevention and compensation. Berlin, New York: Springer, pp. 151–169.
Pulido Begines, Juan L. (2005): The EU Law on Classification Societies: Scope and Liability Issues. In *Journal of Maritime Law and Commerce* 36 (4), pp. 487–543.
Radovic, Jorge; Brandani, Aldo (2013): Protecting the aquatic environment: Are all costs and benefits accounted for? In *Droit Maritime Français* 65, pp. 839–861.
Rah, Sicco (2009): Asylsuchende und Migranten auf See. Berlin: Springer.
Rak, Henrik (2008): Liability for Pollution from Shipwrecks. In Henrik Rak, Peter Wetterstein (Eds.): Shipwrecks in International and National Law - Focus on Wreck Removal and Pollution Prevention. Papers from a seminar 13–17 June 2007, Kasnäs, Finland. Turku, pp. 53–72.
Rak, Henrik; Wetterstein, Peter (Eds.) (2008): Shipwrecks in International and National Law - Focus on Wreck Removal and Pollution Prevention. Papers from a seminar 13–17 June 2007, Kasnäs, Finland. Åbo Akademi University. Turku.
Rak, Henrik; Wetterstein, Peter (Eds.) (2011): Environmental Liabilities in Ports and Coastal Areas - Focus on Public Authorities and Other Actors. Papers from a seminar 11–14 August 2010, Korpoström, Finland. Åbo/Turku: Åbo Akademi University.
Ramming, Klaus (2007): Das Bunkeröl-Übereinkommen. In *Versicherungsrecht*, pp. 306–322.
Ramming, Klaus (2007): Zur Zuständigkeit deutscher Gerichte für Ansprüche wegen Ölverschmutzungsschäden. In *Transportrecht*, pp. 13–18.
Ramming, Klaus (2010): Internationalprivatrechtliche Fragen der Haftung des Reeders. In *Transportrecht*, pp. 284–302.
Rares, Steven (2011): An international Convention on offshore hydrocarbon leaks? In *Lloyd's Maritime and Commercial Law Quarterly*, pp. 361–371.
Rasmussen-Bonne, Hans-Eric; Freer, Richard; Lüke, Wolfgang; Weitnauer, Wolfgang (Eds.) (2005): Balancing of Interests. Liber Amicorum Peter Hay zum 70. Geburtstag. Frankfurt am Main: Verlag Recht und Wirtschaft.

Raulin, Arnaud de (1993): L'épopée judiciaire de l'Amoco Cadiz. In *Journal du Droit International Privé*, pp. 63–96.
Rémond-Gouilloud, M. (1993): The Future of the Compensation System as established by International Convention. In Colin M. de la Rue (Ed.): Liability for damage to the marine environment. London, New York: Lloyd's of London Press, pp. 83–99.
Rémond-Gouilloud, Martine (1977): Note TGI Bastia 08.12.1976. In *Recueil Dalloz*, pp. 429–432.
Rémond-Gouilloud, Martine (1979): Leçons d'un naufrage. In *Recueil Dalloz Chronique*, pp. 133–138.
Rémond-Gouilloud, Martine (1986): Les surprises du "fonds". In *Droit Maritime Français* 38, pp. 89–98.
Rémond-Gouilloud, Martine (1993): Droit maritime. 2nd ed. Paris: A. Pedone.
Rémond-Gouilloud, Martine (2010): Pollution accidentelle et responsabilités. In *Droit Maritime Français* 62, pp. 663–668.
Rémond-Gouilloud, Martine (2012): Sur le préjudice écologique. In *Droit Maritime Français* 64, pp. 1020–1022.
Renger, Reinhard (1993): Haftung und Entschädigung für Ölverschmutzungsschäden auf See. In *Transportrecht*, pp. 132–135.
Renger, Reinhard (1998): Recht und Praxis der Haftung und Entschädigung für Ölverschmutzungsschäden auf See. In Harald Koch, Armin Willingmann (Eds.): Großschäden - Complex Damages. Rechtliche und alternative Regulierungsstrategien im In- und Ausland. Baden-Baden: Nomos, pp. 151–176.
Rengifo, Antonio (1997): The International Convention on Liability and Compensation for Damage in Connection with the Carriage of Hazardous and Noxious Substances by Sea, 1996. In *Review of European Community and International Environmental Law* 6, pp. 191–197.
Requejo Isidro, Marta (2007): Contaminación por hidrocarburos y responsabilidad civil: quién, cuánto, por qué. In José Manuel Sobrino Heredia (Ed.): Mares y Océanos en un mundo de cambio. Tendencias jurídicas, actores y factores. Valencia: Tirant lo Blanch, pp. 473–495.
Ringbom, Henrik (2001): The "Erika" Accident and Its Effects on EU Maritime Regulation. In Myron H. Nordquist, John Norton Moore (Eds.): Current Marine Environmental Issues and the International Tribunal for the Law of the Sea. The Hague, London: Martinus Nijhoff Publishers, pp. 265–290.
Ringbom, Henrik (2004): EU Regulation 44/2001 and its Implications for the International Maritime Liability Conventions. In *Journal of Maritime Law and Commerce* 35, pp. 1–33.
Ringbom, Henrik (2008): The EU Maritime Safety Policy and International Law. Leiden, Boston: Martinus Nijhoff Publishers.
Ringbom, Henrik (2008): Wrecks in International Law. In Henrik Rak, Peter Wetterstein (Eds.): Shipwrecks in International and National Law - Focus on Wreck Removal and Pollution Prevention. Papers from a seminar 13–17 June 2007, Kasnäs, Finland. Turku, pp. 9–34.
Ripert, Georges (1950): Droit maritime (Tome I). 4th ed. 3 volumes. Paris: Éditions Rousseau (1).
Robert, Sabrina (2003): L'"Erika": Responsabilités pour un désastre écologique. Paris: Pedone.
Rodière, René; Pontavice, Emmanuel du (1997): Droit maritime. 12th ed. Paris: Dalloz.
Rodríguez Docampo, José (2012): Alcance de la canalización de la responsabilidad y de las daños ambientales en el convenio internacional sobre responsabilidad civil por contaminación de hidrocarburos: CLC/92. In José Luis García-Pita y Lastres, Rocío Quintáns Eiras, Angélica de La Díaz Rosa (Eds.): Estudios de Derecho Marítimo. Cizur Menor: Thomson Aranzadi, pp. 911–931.
Rodríguez Gayan, Eloy Miguel (2003): Claves de derecho privado en el asunto "Prestige". In *Revista Española de Derecho Internacional*, pp. 117–147.
Rohart, Jean-Serge (2000): Observations CA Paris 19.05.1999. In *Droit Maritime Français* 52, pp. 29–31.

Rohart, Jean-Serge (2006): Les critères du FIPOL sur l'admissibilité des demandes d'indemnisation. In *Droit Maritime Français* 58, pp. 970–976.

Røsæg, Erik (2000): The impact of insurance practices on liability conventions. In: Legislative approaches in maritime law. European Colloquium on Maritime Law. Oslo, 7.-8.12.2000. Nordisk institutt for sjørett. Oslo: Sjørettsfondet (MarIus, 283).

Røsæg, Erik (2004): News under the Athens Sun - New Principles and Lost Opportunities of the Athens Convention 2002. In *Scandinavian Studies in Law* 46, pp. 153–173.

Røsæg, Erik (2007): Passenger liabilities and insurance: Terrorism and war risks. In D. Rhidian Thomas (Ed.): Liability regimes in contemporary maritime law. London: Informa, pp. 207–230.

Røsæg, Erik (2010): The Athens Convention on Passenger Liability and the EU. In Jürgen Basedow, Ulrich Magnus, Rüdiger Wolfrum (Eds.): The Hamburg Lectures on Maritime Affairs 2007 & 2008. Berlin, Heidelberg: Springer, pp. 55–75.

Røsæg, Erik (2010): The Rebirth of the HNS Convention. In Giorgio Berlingieri, Angelo Boglione, Sergio M. Carbone, Francesco Siccardi (Eds.): Scritti in onore di Francesco Berlingieri. Volume II. Genova: Il Diritto Marittimo, pp. 852–860.

Rosas, Allan (2011): EU Law and International Agreements Concluded by EU Member States, with Particular Emphasis on Maritime Law. In Henrik Rak, Peter Wetterstein (Eds.): Environmental Liabilities in Ports and Coastal Areas - Focus on Public Authorities and Other Actors. Papers from a seminar 11–14 August 2010, Korpoström, Finland. Åbo/Turku: Åbo Akademi University, pp. 9–47.

Rose, Francis D. (Ed.) (2000): Lex Mercatoria. Essays on International Commercial Law in Honour of Francis Reynolds. London: LLP.

Roth, Günter H.; Plett, Konstanze (1978): Schiffszusammenstöße im deutschen Internationalen Privatrecht. In *Rabels Zeitschrift für ausländisches und internationales Privatrecht* 42, pp. 662–695.

Rothwell, D. R. (2002): The Law of the Sea and the MV Tampa Incident: Reconciling Maritime Principles with Coastal State Sovereignty. In *Public Law Review* 13, pp. 118–127.

Rue, Colin M. de la (Ed.) (1993): Liability for damage to the marine environment. London, New York: Lloyd's of London Press.

Rue, Colin M. de la; Anderson, Charles B. (2009): Shipping and the environment. Law and practice. 2nd ed. London: Informa Law.

Ruiz Abou-Nigm, Verónica (2011): The Arrest of Ships in Private International Law. Oxford: Oxford University Press.

Sadeleer, Nicolas de (2009): Liability for Oil Pollution Damage versus Liability for Waste Management: The Polluter Pays Principle at the Rescue of the Victims. In *Journal of Environmental Law* 21 (2), pp. 299–307.

Saggerson, Alan (2002): That Sinking Feeling - Psychiatric Injury, Lost Valuables and the Demise of the Cruise Ship "Sun Vista". In *International Travel Law Journal*, pp. 198–202.

Saggerson, Alan (2004): Case Notes: A Tale of Two Conventions. In *International Travel Law Journal*, pp. 14–17.

Scapel, Christian (2012): Le droit positif de la croisière maritime en France. In *Droit Maritime Français* 64, pp. 306–313.

Schaar, Jill A. (2000): The Shipowners' Limitation of Liability Act: Still Afloat or Sinking Fast? In *Tulane Maritime Law Journal* 24, pp. 659–704.

Schelin, Johan (Ed.) (2009): General Trends in Maritime and Transport Law 1929–2009. XI Hässelby Colloquium 2009. Institute of Maritime and Transport Law, University of Stockholm.

Schelin, Johan (2011): The HNS Protocol - The Resurrection of a Convention. In Henrik Rak, Peter Wetterstein (Eds.): Environmental Liabilities in Ports and Coastal Areas - Focus on Public Authorities and Other Actors. Papers from a seminar 11–14 August 2010, Korpoström, Finland. Åbo/Turku: Åbo Akademi University, pp. 287–290.

Schofield, Clive (2012): Iceberg(s) Dead Ahead? In *Internationale Zeitschrift* 8, pp. 8–15.

Schofield, Clive H. (Ed.) (2014): The Limits of Maritime Jurisdiction. Leiden: Martinus Nijhoff.

Schweitzer, Stéphanie (2012): Observations CA Nancy 20.01.2011. In *Droit Maritime Français* 64, pp. 609–612.

Scoles, Eugene F.; Hay, Peter; Borchers, Patrick J.; Symeonides, Symeon C. (2004): Conflict of laws. 4th ed. St. Paul: Thomson/West.

Scott, Shirley V. (2005): The LOS Convention as a Constitutional Regime for the Oceans. In Alex G. Oude Elferink (Ed.): Stability and change in the law of the sea. The role of the LOS convention. Leiden, Boston: Martinus Nijhoff, pp. 9–38.

Severance, Arthur Alan (2006): The Duty to Render Assistance in the Satellite Age. In *California Western International Law Journal* 36, pp. 377–400.

Shaw, Richard (2005): What is a ship in maritime law? In *Journal of International Maritime Law* 11, pp. 247–249.

Shaw, Richard (2008): Carriage of Passengers. In Yvonne Baatz (Ed.): Southampton on shipping law. London: Informa, pp. 147–153.

Shaw, Richard (2010): The 1996 HNS Convention - An Impossible Dream? In Giorgio Berlingieri, Angelo Boglione, Sergio M. Carbone, Francesco Siccardi (Eds.): Scritti in onore di Francesco Berlingieri. Volume II. Genova: Il Diritto Marittimo, pp. 906–912.

Shaw, Richard (2013): Pollution of the Sea by Hazardous and Noxious Substances - Is a Workable International Convention on Compensation an Impossible Dream? In Malcolm Clarke (Ed.): Maritime Law Evolving. Thirty Years at Southampton. Oxford, Portland: Hart Publishing, pp. 43–61.

Shearer, Ivan Anthony (1986): Problems of Jurisdiction and Law Enforcement Against Delinquent Vessels. In *International and Comparative Law Quarterly* 35, pp. 320–343.

Siccardi, Francesco (2005): Pollution Liability and Classification Societies: Is the System a Fair One? In *Il Diritto Marittimo* 107, pp. 691–710.

Siehr, Kurt (2010): The Rome II Regulation and Specific Maritime Torts: Product Liability, Environmental Damage, Industrial Action. In *Rabels Zeitschrift für ausländisches und internationales Privatrecht* 74, pp. 139–153.

Simson, Gary J. (2011): Rethinking Choice of Law: What Role for the Needs of the Interstate and International System? In Mahnoush H. Arsanjani, Jacob Katz Cogan, Robert D. Sloane, Siegfried Wiessner (Eds.): Looking to the Future. Essays on international law in honor of W. Michael Reisman. Leiden, Boston: Martinus Nijhoff Publishers, pp. 235–259.

Smeele, Frank (2010): International Civil Litigation and the Pollution of the Marine Environment. In Jürgen Basedow, Ulrich Magnus, Rüdiger Wolfrum (Eds.): The Hamburg Lectures on Maritime Affairs 2007 & 2008. Berlin, Heidelberg: Springer, pp. 77–118.

Smets, Henri (1985): Indemnisation des dommages exceptionnels à l'environnement causés par les activités industrielles. In René-Jean Dupuy (Ed.): L'avenir du droit international de l'environnement. Dordrecht: Nijhoff, pp. 275–356.

Smith, Marissa (2011): The Deepwater Horizon Disaster: An Examination of the Spill's Impact on the Gap in International Regulation of Oil Pollution from Fixed Platforms. In *Emory International Law review* 25, pp. 1477–1516.

Smith, Timothy (2001): Fighting on the Ocean Blue: New Zealand's Extra-Territorial Jurisdiction and Maritime Protest. In *Victoria University of Wellington Law Review* 32, pp. 499–520.

Smith, Wilbur Holmes (1971): The duty to render assistance at sea: Is it effective or adrift? In *California Western International Law Journal* 2 (2), pp. 146–163.

Sobrino Heredia, José Manuel (Ed.) (2007): Mares y Océanos en un mundo de cambio. Tendencias jurídicas, actores y factores. Valencia: Tirant lo Blanch.

Somers, Eduard; Gonsaeles, Gwendoline (2010): The Consequences of the Sinking of the M/S "Erika" in European Waters: Towards a Total Loss for International Shipping Law? In *Journal of Maritime Law and Commerce* 41, pp. 57–83.

Sotiropoulos, Panayotis K. (1998): Carriage of Passengers and their Luggage by Sea - The Athens Convention 1974 and its Protocols. Synopsis of the Replies to the Questionnaire. In *CMI Yearbook*, pp. 129–144.

Soyer, Bariş (2000): 1996 Protocol to the 1976 Limitation Convention: A More Satisfactory Global Limitation Regime for the Next Millennium? In *Journal of Business Law*, pp. 153–172.

Soyer, Bariş (2002): Sundry Considerations on the Draft Protocol to the Athens Convention Relating to the Carriage of Passengers and Their Luggage by Sea 1974. In *Journal of Maritime Law and Commerce* 33, pp. 519–541.

Soyer, Bariş (2007): Boundaries of the Athens Convention: What you see is not always what you get! In D. Rhidian Thomas (Ed.): Liability regimes in contemporary maritime law. London: Informa, pp. 183–206.

Strauss, Andrew L. (1995): Beyond National Law: The Neglected Role of the International Law of Personal Jurisdiction in Domestic Courts. In *Harvard International Law Journal* 36, pp. 373–424.

Stroink, Frits (2010): The relation between private law and administrative law in view of globalization. In Michael Faure, André van der Walt (Eds.): Globalization and private law. The way forward. Cheltenham: Edward Elgar, pp. 139–143.

Symeonides, Symeon C. (2003): Choice of Law in the American Courts in 2002: Sixteenth Annual Survey. In *American Journal of Comparative Law* 51, pp. 1–88.

Symeonides, Symeon (2004): Territoriality and Personality in Tort Conflicts. In Talia Einhorn, Kurt Siehr (Eds.): International Cooperation Through Private International Law. The Hague: T. M. C. Asser Press, pp. 401–433.

Symeonides, Symeon C. (2005): Accommodative Unilateralism as a Starting Premise in Choice of Law. In Hans-Eric Rasmussen-Bonne, Richard Freer, Wolfgang Lüke, Wolfgang Weitnauer (Eds.): Balancing of Interests. Liber Amicorum Peter Hay zum 70. Geburtstag. Frankfurt am Main: Verlag Recht und Wirtschaft, pp. 417–434.

Symeonides, Symeon (2008): American private international law. Alphen aan den Rijn: Kluwer Law International.

Symeonides, Symeon C. (2008): Rome II and Tort Conflicts: A Missed Opportunity. In *American Journal of Comparative Law* 56, pp. 173–222.

Symmons, Clive Ralph (Ed.) (2011): Selected Contemporary Issues in the Law of the Sea. Leiden, Boston: Martinus Nijhoff Publishers.

Szakats, Alexander (2001): Cross Border Pollution - Private International Law Problems in Claiming Compensation. In *Victoria University of Wellington Law Review* 32, pp. 609–626.

Takahashi, Koji (2003): External Competence Implications of the EC Regulation of Jurisdiction and Judgments. In *International and Comparative Law Quarterly* 52, pp. 529–534.

Talley, Wayne K. (2010): Vessel Safety and Accident Analysis. In Costas Th Grammenos (Ed.): The Handbook of Maritime Economics and Business. 2nd ed. London: Lloyd's List, pp. 519–535.

Tan, Alan Khee-Jin (2006): Vessel-Source Marine Pollution. Cambridge, New York: Cambridge University Press.

Tang, Zheng (2008): The Interrelationship of European Jurisdiction and Choice of Law in Contract. In *Journal of Private International Law* 4, pp. 35–59.

Tarelli, Elis (2009): International Efforts to Establish Rules on Liability for Offshore Activities. In Peter Ehlers, Rainer Lagoni (Eds.): Responsibility and Liability in the Maritime Context. Münster: LIT Verlag.

Tassel, Yves (1997): Le droit maritime - un anachronisme? In *Annuaire de droit maritime et océanique*, pp. 143–157.

Tassel, Yves (2001): Observations Tribunal de Commerce de Saint-Nazaire 16.12.2000. In *Droit Maritime Français* 53, pp. 281–284.

Tassel, Yves (2005): Observations CA Rouen 09.11.2004. In *Droit Maritime Français* 57, pp. 734–737.

Tauman, Jessica E. (2002): Rescued at sea, but nowhere to go: The cloudy legal waters of the "Tampa" crisis. In *Pacific Rim Law & Policy Journal* 11 (2), pp. 461–496.

Tebbens, Harry Duintjer (2008): The European Union and the Athens Convention on Maritime Carriers' Liability for Passengers in Case of Accidents: An Incorporation Adventure. In *Revue Héllenique de Droit International* 61, pp. 653–667.

Tetley, William (1993): The Law of the Flag, "Flag Shopping", and Choice of Law. In *Tulane Maritime Law Journal* 17, pp. 139–184.

Tettenborn, Andrew M. (2007): The liabilities of classification societies - more awkward than it looks? In D. Rhidian Thomas (Ed.): Liability regimes in contemporary maritime law. London: Informa, pp. 109–127.

Thomas, D. Rhidian (Ed.) (2007): Liability regimes in contemporary maritime law. London: Informa.

Thomas, D. Rhidian (2008): Bunker oil pollution - a new liability convention. In *Journal of International Maritime Law* 14, pp. 375–376.

Thomas-Ciora, Geneviève (2006): Le transport des passagers par voie maritime. In *Droit Maritime Français* 58, pp. 977–984.

Tiberg, Hugo (1984): Oil pollution of the sea and the Swedish "Tsesis" decision. In *Lloyd's Maritime and Commercial Law Quarterly*, pp. 218–226.

Tiberg, Hugo (2006): Fraud and Sinking Ships in Marine Insurance. In Hugo Tiberg, Malcolm Clarke, Anders Beskow, Mia Carlsson, Rose-Marie Lundström, Thomas Utterström, Simon Arvmyren (Eds.): Essays on Tort, Insurance Law and Society in Honour of Bill W. Dufwa. Volume II. Stockholm: Jure, pp. 1167–1183.

Tiberg, Hugo; Clarke, Malcolm; Beskow, Anders; Carlsson, Mia; Lundström, Rose-Marie; Utterström, Thomas; Arvmyren, Simon (Eds.) (2006): Essays on Tort, Insurance Law and Society in Honour of Bill W. Dufwa. Volume II. Stockholm: Jure.

Tomljenović, Vesna (1999): Maritime Torts. In *Yearbook of Private International Law* (I), pp. 249–297.

Tomuschat, Christian (1993): Obligations arising for States without or against their Will. In *Recueil des Cours* 241, pp. 195–374.

Tsimplis, Michael N. (2005): The Bunker Pollution Convention 2001: completing and harmonizing the liability regime for oil pollution from ships? In *Lloyd's Maritime and Commercial Law Quarterly*, pp. 83–100.

Tsimplis, Michael N. (2008): Marine pollution for shipping activities. In *Journal of International Maritime Law* 14, pp. 101–152.

Tsimplis, Michael N. (2009): Liability in respect of passenger claims and its limitation. In *Journal of International Maritime Law* 15, pp. 125–150.

Tsimplis, Michael N. (2010): Law and jurisdiction for English limitation of liability proceedings. In *Journal of International Maritime Law* 16, pp. 289–316.

Tsimplis, Michael N. (2011): The effect of European regulations on the jurisdiction and applicable law for limitation proceedings. In *Lloyd's Maritime and Commercial Law Quarterly*, pp. 307–321.

Tu, Guangjian (2009): Finding A Proper Nexus for Constructing Specific (Special) Jurisdiction Regarding Commercial Contract and Tort Cases. In *Journal of Private International Law* 5, pp. 243–279.

UNCTAD: Review of Maritime Transport 2011. Available online at http://www.unctad.org/en/docs/rmt2011ch1_en.pdf.

Vanheule, Bernard (2003): Oil Pollution: The International Liability and Compensation Regime. In *European Transport Law* 38, pp. 547–576.

Veaux, Daniel; Veaux-Fournerie, Paulette (1994): La pénétration en droit français de la thérorie de la faute inexcusable en matière aérienne et maritime sous l'influence des conventions internationales. In Jacques Béguin, Pierre Bourel (Eds.): L'internationalisation du droit. Mélanges en l'honneur de Yvon Loussouarn. Paris: Dalloz, pp. 391–403.

Velde, Welmoed van der (2006): De positie van het zeeschip in het Internationaal Privaatrecht. Groningen.

Vella, Ivan (2010): A new advent for renewable offshore resources. In Norman A. Martínez Gutiérrez (Ed.): Serving the rule of International Maritime Law. Essays in Honour of Professor David Joseph Attard. Abingdon: Routledge, pp. 136–155.

Venturini, Gabriella; Bariatti, Stefania (Eds.) (2009): Liber Fausto Pocar. Milano: Giuffrè.

Vialard, Antoine (1993): L'affaire "Heidberg": gros temps sur la Convention de Londres 1976 sur la limitation de responsabilité en matière de créances maritimes. In *Droit Maritime Français* 45, pp. 706–711.

Vialard, Antoine (2000): "L'apparence" de faute inexcusable comme cause de déchéance "provisoire" du droit à la limitation de responsabilité. In *Droit Maritime Français* 52, pp. 787–793.

Vialard, Antoine (2002): L'évolution de la notion de faute inexcusable et la limitation. In *Droit Maritime Français* 54, pp. 579–585.

Vialard, Antoine (2003): Faut-il réformer le régime d'indemnisation des dommages de pollution par hydrocarbures? In *Droit Maritime Français* 55, pp. 435–449.

Vialard, Antoine (2003): Responsabilité limitée et indemnisation illimitée. In Anne Knops (Ed.): Liber Amicorum Roger Roland. Gent: Larcier, pp. 571–577.

Vialard, Antoine (2005): Observations CA Bordeaux 31.05.2005. In *Droit Maritime Français* 57, pp. 846–850.

Vialard, Antoine (2008): La faute de l'assuré dans le droit français des assurances maritimes. In *Annuaire de droit maritime et océanique*, pp. 605–619.

Vialard, Antoine (2009): La limitation de responsabilité, clé de doute pour le droit maritime du 21e siècle. In *Droit Maritime Français* 61, pp. 21–28.

Vizthum, Wolfgang Graf (2006): Aquitoriale Souveränität. In Pierre-Marie Dupuy, Bardo Fassbender, Malcolm N. Shaw, Karl-Peter Sommermann (Eds.): Völkerrecht als Wertordnung. Festschrift für Christian Tomuschat. Kehl: N.P. Engel, pp. 1067–1086.

Vizthum, Wolfgang Graf (Ed.) (2006): Handbuch des Seerechts. München: Beck.

Volner, Igor (2006): Forum Selection Clauses: Different Regulations from the Perspective of Cruise Ship Passengers. In *European Journal of Law Reform* 8, pp. 439–476.

Vrellis, Spyridon (2013): The Law Applicable to the Environmental Damage. Some Remarks on Rome II Regulation. In Joaquim Forner Delaygua, Cristina González Beilfuss, Ramón Viñas Farré (Eds.): Entre Bruselas y La Haya. Estudios sobre la unificación internacional y regional del Derecho internacional privado. Madrid, Barcelona: Marcial Pons, pp. 869–887.

Wagner, Gerhard (2006): Internationales Deliktsrecht, die Arbeiten an der Rom II-Verordnung und der Europäische Deliktsgerichtsstand. In *Praxis des Internationalen Privat- und Verfahrensrechts*, pp. 372–391.

Wang, Hui (2011): Civil liability for marine oil pollution damage. A comparative and economic study of the international, US and the Chinese compensation regime. Alphen aan den Rijn: Kluwer Law International.

Warshauer, Irving J.; Dittman, Stevan C. (2005): The Uniqueness of Maritime Personal Injury and Death Law. In *Tulane Law Review* 79, pp. 1163–1225.

Weintraub, Russell J. (1997): Admiralty Choice-of-Law Rules for Damages. In *Journal of Maritime Law and Commerce* 28 (2), pp. 237–256.

Weintraub, Russell J. (2005): Rome II and the Tension Between Predictability and Flexibility. In Hans-Eric Rasmussen-Bonne, Richard Freer, Wolfgang Lüke, Wolfgang Weitnauer (Eds.): Balancing of Interests. Liber Amicorum Peter Hay zum 70. Geburtstag. Frankfurt am Main: Verlag Recht und Wirtschaft, pp. 451–461.

Weiss, Jeffrey A. (2001): Maritime Disasters Through the Ages. In *Journal of Maritime Law and Commerce* 32, pp. 215–240.

Wendel, Philipp (2007): State responsibility for interferences with the freedom of navigation in public international law. Berlin, New York: Springer.

West, James M. (1981): The Ixtoc I Oil Spill Litigation: Jurisdictional Disputes at the Threshold of Transnational Pollution Responsibility. In *Texas International Law Journal* 16, pp. 475–532.

Wetterstein, Peter (1997): Carriage of Hazardous Cargoes by Sea - The HNS Convention. In *Georgia Journal of International and Comparative Law* 26, pp. 595–614.

Wetterstein, Peter (2006): Shipowners' Environmental Impairment Liability - Developments and Challenges. In Hugo Tiberg, Malcolm Clarke, Anders Beskow, Mia Carlsson, Rose-Marie Lundström, Thomas Utterström, Simon Arvmyren (Eds.): Essays on Tort, Insurance Law and Society in Honour of Bill W. Dufwa. Volume II. Stockholm: Jure, pp. 1237–1263.

Wetterstein, Peter (2009): Complete Freedom of the Seas or Polluter Pays for Everything? In Johan Schelin (Ed.): General Trends in Maritime and Transport Law 1929–2009. XI Hässelby Colloquium 2009. Institute of Maritime and Transport Law, University of Stockholm, pp. 109–143.

Wetterstein, Peter; Beijer, Anders (Eds.) (1996): Essays in Honour of Hugo Tiberg. Stockholm: Juristförlaget.

White, Michael (1999): Offshore Craft and Structures: A Proposed International Convention. In *Australian Mining & Petroleum Law Journal* 18, pp. 21–27.

Wilhelmsen, Trine-Lise (2003): Misconduct of the Assured and Identification. In *CMI Yearbook*, pp. 540–572.

Winter, D. (1954): Maritime Torts: The Choice-of-Law Principles. In *International and Comparative Law Quarterly* 3 (1), pp. 115–125.

Wiswall, Frank L. (2003): 2002 Protocol to the Athens Convention. In *Benedict's Maritime Bulletin* 1, pp. 17–19.

Woonings, Kerryn (2010): 'The Plot of the Pilot': Pilotage and Limitation of Liability in Maritime Law. In *Australian and New Zealand Maritime Law Journal* 24, pp. 126–137.

Wu, Chao (1996): Pollution from the carriage of oil by sea. Liability and compensation. London, Boston: Kluwer Law International.

Wu, Chao (2002): Liability and Compensation for Bunker Pollution. In *Journal of Maritime Law and Commerce* 33, pp. 553–567.

Wurmnest, Wolfgang (2006): Third Party Liability of Classification Societies in the Context of Shipping Accidents. In Marta García Pérez, Francisco Javier Sanz Larruga (Eds.): Seguridad marítima y medio ambiente. Oleiros: Instituto Universitario de Estudios Marítimos, pp. 111–119.

Yang, Haijiang (2006): Jurisdiction of the Coastal State over Foreign Merchant Ships in Internal Waters and the Territorial Sea. Berlin, New York: Springer.

Yessiou-Faltsi, Pelayia (2011): The 1952 Brussels International Convention on the arrest of ships for maritime claims and on jurisdiction. In *Aegean Review of the law of the sea and maritime law*, pp. 175–180.

Zhang, Liying (2010): Compensation for the Domestic Oil Pollution in China's Coast: Which Law Shall Apply? In Michael Faure, Lixin Han, Hongjun Shan (Eds.): Maritime Pollution Liability and Policy. China, Europe and the US. Alphen aan den Rijn: Kluwer Law International, pp. 359–370.

Zhu, Ling (2006): Compulsory insurance and compensation for bunker oil pollution damage. Berlin, New York: Springer.

Zhu, Ling (2009): Can the Bunkers Convention Ensure Adequate Compensation for Pollution Victims? In *Journal of Maritime Law and Commerce* 40, pp. 203–219.

Zhu, Ling; Ng, Adolf K. Y.; Zhang, Mingzhao (2012): Some thoughts on civil liability for maritime piracy-caused oil pollution. In *Journal of International Maritime Law* 18, pp. 306–315.

Zimmermann, Jaclyn A. (2000): Inadequacies of the Oil Pollution Act of 1990: Why the United States should adopt the Convention on Civil Liability. In *Fordham International Law Journal* 23, pp. 1499–1539.

Cases

Admiralty Division, 11.07.1969, "The Eland" and "The Monte Urquiola". In *1969 2 Lloyd's Law Reports*, pp. 328–340.
Areios Pagos, 06.07.2006, "Marine Environmental Services MC and Environmental Protection Technical SA v. IOPC". In *2006 Epitheorisi Emporikou Dikaiou*, pp. 1797–1801.
British Columbia Court of Appeal, 02.04.1969, "Gronlund v. Hansen" (appeal). In *Dominion Law Reports, 3rd Series* 4, pp. 435–444.
British Columbia Supreme Court, 11.07.1968, "Gronlund v. Hansen". In *Dominion Law Reports, 2nd Series* 69, pp. 598–605.
Central London County Court (Business List), 26.04.2002, "Lee v. Airtours". In *2004 1 Lloyd's Law Reports*, pp. 683–692.
Circuit Court of Appeals, Second Circuit, 05.03.1947, "O'Neill v. Cunard White Star". In *Federal Reporter 2nd Series* 160, pp. 446–449.
Corte di Cassazione (Italy), 17.10.2002, "IOPC v. RINA". In *2003 Rivista di diritto internazionale*, pp. 238–252.
Cour d'Appel d'Aix-en-Provence, 05.11.1998, "Zulu Sea". In *2002 Droit Maritime Français* 54, pp. 125–131.
Cour d'Appel d'Aix-en-Provence, 08.06.2000, "Moldavia". In *2002 Droit Maritime Français* 54, pp. 132–138.
Cour d'Appel d'Aix-en-Provence, 10.01.2001, "Multitank Arcadia". In *2002 Droit Maritime Français* 54, pp. 150–156.
Cour d'Appel d'Aix-en-Provence, 04.10.2012, "Triton II". In *2013 Droit Maritime Français* 65, pp. 310–312.
Cour d'Appel d'Aix-en-Provence, 11.10.2012, "Creux Saint Georges III". In *2012 Droit Maritime Français* 65, pp. 221–224.
Cour d'Appel de Bordeaux, 31.05.2005, "Heidberg". In *2005 Droit Maritime Français* 57, pp. 839–846.
Cour d'Appel de Bordeaux, 14.01.2013, "Heidberg" (after remand). In *2013 Droit Maritime Français* 65, pp. 201–211.
Cour d'Appel de Caen, 02.10.2001, "Johanna Hendrika". In *2001 Droit Maritime Français* 53, pp. 981–989.
Cour d'Appel de Montpellier, 07.12.1999, "Stellaprima". In *2000 Droit Maritime Français* 52, pp. 813–818.
Cour d'Appel de Montpellier, 18.11.2003, "L'Inglais" and "François Delphine". In *2005 Droit Maritime Français* 57, pp. 708–712.
Cour d'Appel de Montpellier, 04.11.2004, "Brescou". In *2005 Droit Maritime Français* 57, pp. 713–716.
Cour d'Appel de Nancy, 20.01.2011, "Sea Diamond". In *2012 Droit Maritime Français* 64, pp. 604–609.
Cour d'Appel de Paris, 19.05.1999, "Prince Laurent". In *2000 Droit Maritime Français* 52, pp. 26–29.
Cour d'Appel de Paris, 30.03.2010, "Erika". In *2010 Droit Maritime Français* 62, pp. 857–877.
Cour d'Appel de Rennes, 13.03.1961, "Vigouroux v. Lenten". In *1962 Recueil Dalloz*, pp. 154–155.
Cour d'Appel de Rennes, 03.10.1985, "Géniteau et Gouvernement Britannique v. Amoco". In *1986 Droit Maritime Français* 38, pp. 99–106.
Cour d'Appel de Rennes, 25.05.2004, "Lebaupain v. FIPOL". In *2004 Droit Maritime Français* 57, pp. 529–535.
Cour d'Appel de Rennes, 24.05.2005, "FIPOL v. Gouzer". In *2006 Droit Maritime Français* 58, pp. 1014–1018.
Cour d'Appel de Rennes, 24.05.2005, "FIPOL v. Hôtel Celtique". In *2006 Droit Maritime Français* 58, pp. 1021–1018.

Cour d'Appel de Rennes, 24.05.2005, "FIPOL v. Simenoff". In *2006 Droit Maritime Français* 58, pp. 1021–1024.
Cour d'Appel de Rennes, 27.06.2006, "Boissons Morbihan v. FIPOL". In *2006 Droit Maritime Français* 58, pp. 1030–1033.
Cour d'Appel de Rouen, 09.11.2004, "Moheli". In *2004 Droit Maritime Français* 56, pp. 727–737.
Cour d'Appel de Rouen, 14.02.2013, "Antares" and "Audrey Marie". In *2013 Droit Maritime Français* 65, pp. 875–884.
Cour d'Appel de Versailles, 09.05.2012, "Costa Concordia". In *2012 Droit Maritime Français* 64, pp. 598–601.
Cour de Cassation (Belgium), 23.11.1962, "Axeldijk". In *1963 Pasicrisie*, pp. 374–381.
Cour de Cassation (France), 09.03.1966, "Lenten v. Vigouroux". In *1966 Recueil Dalloz*, pp. 577–578.
Cour de Cassation (France), 03.04.1978, "Montedison v. Département de la Haute Corse". In *1978 Recueil Dalloz*, p. 367.
Cour de Cassation (France), 20.05.1997, "Johanna Hendrika". In *1997 Droit Maritime Français* 49, pp. 976–987.
Cour de Cassation (France), 11.05.1999, "Mobil North Sea v. Compagnie française d'entreprises metalliques". In *2000 Revue critique du droit international privé* 89, pp. 199–201.
Cour de Cassation (France), 03.04.2002, "Stellaprima". In *2002 Droit Maritime Français* 54, pp. 460–462.
Cour de Cassation (France), 27.03.2007, "Wellborn". In *2007 Droit Maritime Français* 59, pp. 750–753.
Cour de Cassation (France), 27.05.2010, "Meranti 38". In *2010 Droit Maritime Français* 62, pp. 701–703.
Cour de Cassation (France), 25.09.2012, "Erika". In *2012 Droit Maritime Français* 64, pp. 985–994.
Cour de Cassation (France), 11.12.2012, "Dizzy". In *2013 Droit Maritime Français* 65, pp. 142–158.
Court for Crown Cases Reserved, Exchequer Division, 13.11.1876, "R v. Keyn". In *Law Reports Exchequer Division* 2, pp. 63–239.
Court of Appeal (England), 10.05.1971, "Sayers v. International Drilling Co.". In *1971 Weekly Law Reports* (I), pp. 1176–1187.
Court of Appeal (England),27.06.2001, "The Leerort". In *2001 2 Lloyd's Law Reports*, pp. 291–303.
Court of Appeal (England), 28.01.2002, "Roerig v. Valiant Trawlers". In *2002 1 Lloyd's Law Reports*, pp. 681–692.
Court of Appeal (England), 07.02.2003, "Algrete Shipping v. IOPC". In *2003 1 Lloyd's Law Reports*, pp. 327–340.
Court of Appeal (England), 12.02.2004, "The CMA Djakarta". In *2004 1 Lloyd's Law Reports*, pp. 460–470.
Court of Appeal of the Supreme Court of New South Wales, 14.06.2002, "Union Shipping New Zealand Ltd. v. Morgan". In *2002 New South Wales Law Reports* 54, pp. 690–698.
Court of Common Pleas (England), 20.01.1864, "The Sud-Marine Telegraph Company v. Dickson". In *1864 English Reports* 143, pp. 983–991.
Court of Exchequer Chamber (England), 27.11.1865, "Lloyd v. Guibert". In *Law Reports, Queen's Bench Division* 1, pp. 115–130.
Court of Exchequer Chamber (England), 23.07.1870, "Phillips v. Eyre". In *Law Reports, Queen's Bench Division* 6, pp. 1–31.
Court of Session (Scotland), 26.10.1954, "MacKinnon v. The Iberia Shipping Co". In *1955 Court of Session Cases*, pp. 20–38.
Court of Session (Scotland), 19.05.1999, "Landcatch v. IOPC". In *1999 Scots Law Times*, pp. 1208–1224.

Court of Session (Scotland), Outer House, 16.10.1923, "The Reresby v. The Cobetas". In *1923 Scots Law Times*, pp. 719–722.
Court of Session (Scotland), Outer House, 30.07.1998, "Black v. The Braer Corporation". In *1999 Scots Law Times*, pp. 1401–1405.
Court of Session (Scotland), Outer House, 30.07.1998, "Eunson v. The Braer Corporation". In *1999 Scots Law Times*, pp. 1405–1409.
Court of Session (Scotland), Outer House, 29.12.1998, "Gray v. The Braer Corporation". In *1999 Scots Law Times*, pp. 1410–1414.
European Court of Justice, 30.11.1976, "Bier v. Mines de Potasse d'Alsace", In *1976 Reports of Cases (ECJ)*, pp. 1735–1749.
European Court of Justice, 06.12.1994, "The Tatry". In *1994 Reports of Cases (ECJ)* (I), pp. 5442–5482.
European Court of Justice, 27.02.2002, "Weber v. Ogden", In *2002 Reports of Cases (ECJ)* (I), pp. 2016–2054.
European Court of Justice, 05.02.2004, "DFDS Torline v. SEKO Sjöfolk". In *2004 Reports of Cases (ECJ)* (I), pp. 1420–1459.
European Court of Justice, 01.03.2005, "Owusu v. Jackson". In *2005 Reports of Cases (ECJ)* (I), pp. 1445–1464.
European Court of Justice, 24.06.2008, "Commune de Mesquer". In *2008 Droit Maritime Français* 60, pp. 704–712.
European Court of Justice, 04.05.2010, "TNT Express Netherlands v. AXA". In *2010 Reports of Cases (ECJ)* (I), pp. 4137–4164.
Federal Court (Canada), 06.05.2011, "Société Telus v. Peracomo Inc.". In *2011 Federal Court (CanLII)*, p. 494.
Federal Court of Appeal (Canada), 29.06.2012, "Peracomo Inc. v. Société Telus". In *2012 Federal Court of Appeal (CanLII)*, p. 199.
High Court (England), Common Pleas Division, 20.04.1880, "Scaramanga v. Stamp". In *Law Reports, Common Pleas Division* 5, pp. 295–308.
High Court (England), Probate, Divorce and Admiralty Division, 03.05.1876, "The Mary Moxham". In *Law Reports, Probate Division* 1, pp. 107–116.
High Court (England), Probate, Divorce and Admiralty Division, 25.11.1879, "The City of Meccah". In *Law Reports, Probate Division* 5, pp. 28–34.
High Court (England), Probate, Divorce and Admiralty Division, 15.10.1920, "The Arum". In *1921 Law Reports, Probate Division*, pp. 12–22.
High Court (England), Probate, Divorce and Admiralty Division, 19.10.1953, "The Waziristan". In *1953 Weekly Law Reports*, pp. 1446–1457.
High Court (England), Queen's Bench Division, 17.01.1883, "Chartered Mercantile Bank v. The Netherlands India Steam Navigation". In *Law Reports, Queen's Bench Division*, pp. 521–546.
High Court (England), Queen's Bench Division, 28.02.1891, "Phelps, James & Co. v. Hill". In *Law Reports, Queen's Bench Division*, pp. 605–618.
High Court (England), Queen's Bench Division, 20.03.1990, "The Lion". In *1990 2 Lloyd's Law Reports*, pp. 144–153.
High Court (England), Queen's Bench Division, 14.11.1991, "The Breydon Merchant". In 1992 *1 Lloyd's Law Reports*, pp. 373–376.
High Court (England), Queen's Bench Division, 07.04.1998, "The Aegean Sea". In *1998 2 Lloyd's Law Reports*, pp. 39–41.
High Court (England), Queen's Bench Division, 14.07.2000, "MSC Mediterranean Shipping v. Delumar BV (The MSC Rosa M)". In *2000 2 Lloyd's Law Reports*, pp. 399–405.
High Court (England), Queen's Bench Division, 11.11.2002, "Margolle v. Delta Maritime Co. Ltd." In *2003 1 Lloyd's Law Reports*, pp. 203–211.
High Court of Admiralty (England), 19.04.1865, "The Zollverein". In *English Reports* 166, pp. 1038–1040.

House of Lords, 08.07.1901, "Carr v. Fracis Times & Co.". In *1902 Law Reports, Appeal Cases*, pp. 176–186.
International Arbitral Tribunal, 02.04.1929, "Kate A. Hoff v. United Mexican States". In *UN Reports of International Arbitral Awards* 4, pp. 444–449.
Judicial Panel on Multidistrict Litigation, of 4/06/1979. In *American Maritime Cases*, pp. 1811–1841.
Maritime Court of Shanghai (China), 27.10.1987 (French summary). In Journal du Droit International (Clunet), p. 178.
Piraeus Court of Appeal, 16.02.2004, "Marine Environmental Services MC and Environmental Protection Technical SA v. IOPC". In *2004 Peiraki Nomologia*, pp. 188–195.
Plymouth County Court, 21.08.2003, "Norfolk v. MyTravel Group PLC". In *2004 1 Lloyd's Law Reports*, pp. 106–110.
Pretura de Livorno, 27.04.1974, "Affaire Montecatini Edison" (French translation). In *1975 Journal du Droit International (Clunet)* 102, pp. 237–248.
Privy Council, 17.06.1868, "The Halley". In *Law Reports Privy Council Appeals* 2, pp. 193–204.
Rechtbank Rotterdam, 18.06. 1963, "El Salazar v. Falkanger" (French summary). In *1969 Journal du Droit International (Clunet)*, pp. 990–992.
Rechtbank Rotterdam, 08.04.1969, "Torrey Canyon". In *1969 Nederlandse Jurisprudentie*, pp. 1333–1336.
Reichsgericht (Germany), 12.11.1932, "Casablanca". In *Entscheidungen des Reichsgeriches in Zivilsachen* 138, pp. 243–247.
Svea Court of Appeal, 06.06.1981, "Tsesis". In *1983 Nordiske domme i sjøfartsanliggender*, pp. 1–41.
Svea Court of Appeal, 26.11.1986, "José Martí". In *1987 Nordiske domme i sjøfartsanliggender*, pp. 64–81.
Tribunal Correctionnel de Paris, 16.01.2008, "Erika". In *2008 Annuaire de droit maritime et océanique* 26, pp. 701–743.
Tribunal de Commerce de Bordeaux, 23.09.1993, "Heidberg". In *1993 Droit Maritime Français* 45, pp. 731–746.
Tribunal de Commerce de Saint-Nazaire, 16.12.2000, "Erika". In *2001 Droit Maritime Français* 53, pp. 259–284.
Tribunal de Grande Instance de Bastia, 08.12.1976, "La Prud'hommie des pêcheurs de Bastia v. Montedison". In *1977 Recueil Dalloz*, pp. 427–429.
Tribunal de Grande Instance de Nanterre (Référés), 13.02.2012, "Costa Concordia". In *2012 Droit Maritime Français* 64, pp. 750–755.
US Court of Appeals for the Second Circuit, 24.02.1959, "Romero v. International Terminal Co.". In *United States Reports* 358, pp. 354–414.
US Court of Appeals for the Second Circuit, 31.10.1996, "Carbotrade v. Bureau Veritas". In *1997 American Maritime Cases*, pp. 98–114.
US Court of Appeals for the Second Circuit, 12.06.2009, "Reino de España v. ABS (The Prestige)". In *2009 American Maritime Cases*, pp. 2402–2407.
US Court of Appeals for the Second Circuit, 29.08.2012, "Reino de España v. ABS (The Prestige)". In *2012 American Maritime Cases*, pp. 2113–2136.
US Court of Appeals for the Seventh Circuit, 24.01.1992, "In Re Oil Spill Amoco Cadiz". In *1992 American Maritime Cases*, pp. 913–989.
US Court of Appeals for the Third Circuit, 22.09.1988, "Hodes v. S. N. C. Achille Lauro". In *Federal Reporter* 858, pp. 905–916.
US District Court for the Southern District of New York, 10.02.1955, "Hamilton v. S. S. Nassau". In *1955 American Maritime Cases*, pp. 595–599.
US District Court for the Southern District of New York, 13.09.2002, "Sealord Marine Co. v. ABS". In *Federal Supplement 2nd Series* 220, pp. 260–274.
US Supreme Court, October 1881, "The Scotland". In *United States Reports* 105, pp. 24–36.
US Supreme Court, 25.05.1914, "Oceanic Steam Navigatio Co. Ltd. v. Mellor (The Titanic)". In *1998 American Maritime Cases* 1998, pp. 2699–2703.

US Court of Appeals for the Fifth Circuit, 08.06.1970, "Hellenic Lines v. Rhoditis". In *United States Reports* 398, pp. 306–318.
US Court of Appeals for the Second Circuit, 25.05.1953, "Lauritzen v. Larsen". In *United States Reports* 345, pp. 571–593.
US District Court, Northern District of Illinois (Eastern Division), 17.10.1979, "In Re Oil Spill by Amoco Cadiz". In *1979 American Maritime Cases*, pp. 2617–2628.
US District Court, Northern District of Illinois (Eastern Division), 18.04.1984, "In Re Oil Spill by Amoco Cadiz". In *1984 American Maritime Cases*, pp. 2123–2199.
US District Court, Northern District of Illinois (Eastern Division), 11.01.1988, "In Re Oil Spill by Amoco Cadiz" (compensation amounts). In *1989 Il Diritto Marittimo*, pp. 876–894.
US District Court, Southern District of New York, 23.07.1992, "Klinghoffer v. Achille Lauro Lines (The Achille Lauro)". In *1993 American Maritime Cases*, pp. 1387–1394.
US District Court, Southern District of New York, 02.01.2008, "Reino de España v. ABS". In *2008 American Maritime Cases*, pp. 83–91.
US District Court, Southern District of New York, 03.08.2010, "Reino de España v. ABS" (after remand). In *2010 American Maritime Cases*, pp. 1877–1894.

About the International Max Planck Research School for Maritime Affairs at the University of Hamburg

The International Max Planck Research School for Maritime Affairs at the University of Hamburg was established by the Max Planck Society for the Advancement of Science, in co-operation with the Max Planck Institute for Foreign Private Law and Private International Law (Hamburg), the Max Planck Institute for Comparative Foreign Public Law and International Law (Heidelberg), the Max Planck Institute for Meteorology (Hamburg) and the University of Hamburg. The School's research is focused on the legal, economic, and geophysical aspects of the use, protection, and organization of the oceans. Its researchers work in the fields of law, economics, and natural sciences. The School provides extensive research capacities as well as its own teaching curriculum. Currently, the School has 22 Directors who determine the general work of the School, act as supervisors for dissertations, elect applicants for the School's PhD-grants, and are the editors of this book series:

Prof. Dr. Dr. h.c. mult. Jürgen Basedow is Director of the Max Planck Institute for Foreign Private Law and Private International Law; *President and Professor Monika Breuch-Moritz* is the President of the German Federal Maritime and Hydrographic Agency; *Prof. Dr. Dr. h.c. Peter Ehlers* is the Director ret. of the German Federal Maritime and Hydrographic Agency; *Prof. Dr. Dr. h.c. Hartmut Graßl* is Director emeritus of the Max Planck Institute for Meteorology; *Dr. Tatiana Ilyina* is the Leader of the Research Group "Ocean Biogeochemistry" at the Max Planck Institute for Meteorology in Hamburg; *Prof. Dr. Florian Jeßberger* is Head of the International and Comparative Criminal Law Division at the University of Hamburg; *Prof. Dr. Lars Kaleschke* is Junior Professor at the Institute of Oceanography of the University of Hamburg; *Prof. Dr. Hans-Joachim Koch* is Director emeritus of the Seminar of Environmental Law at the University of Hamburg; *Prof. Dr. Robert Koch* is Director of the Institute of Insurance Law at the University of Hamburg; *Prof. Dr. Doris König* is the President of the Bucerius Law School; *Prof. Dr. Rainer Lagoni* is Director emeritus of the Institute of Maritime Law and the Law of the Sea at the University of Hamburg; *Prof. Dr. Gerhard Lammel* is Senior Scientist and Lecturer at the Max Planck Institute for Chemistry, Mainz; *Prof. Dr. Ulrich Magnus* is Managing Director of the

S.F. Gahlen, *Civil Liability for Accidents at Sea*, Hamburg Studies on Maritime Affairs 32, DOI 10.1007/978-3-662-45555-5

Seminar of Foreign Law and Private International Law at the University of Hamburg; *Prof. Dr. Peter Mankowski* is Director of the Seminar of Foreign and Private International Law at the University of Hamburg; *Prof. Stefan Oeter* is Managing Director of the Institute for International Affairs at the University of Hamburg; *Prof. Dr. Marian Paschke* is Managing Director of the Institute of Maritime Law and the Law of the Sea at the University of Hamburg; *PD Dr. Thomas Pohlmann* is Senior Scientist at the Centre for Marine and Climate Research and Member of the Institute of Oceanography at the University of Hamburg; *Dr. Uwe A. Schneider* is Assistant Professor at the Research Unit Sustainability and Global Change of the University of Hamburg; *Prof. Dr. Detlef Stammer* is Professor in Physical Oceanography and Remote Sensing at the Institute of Oceanography of the University of Hamburg; *Prof. Dr. Jürgen Sündermann* is Director emeritus of the Centre for Marine and Climate Research at the University of Hamburg; *Prof. Dr. Rüdiger Wolfrum* is Director emeritus at the Max Planck Institute for Comparative Foreign Public Law and International Law and a judge at the International Tribunal for the Law of the Sea; *Prof. Dr. Wilfried Zahel* is Professor emeritus at the Centre for Marine and Climate Research of the University of Hamburg.

At present, *Prof. Dr. Dr. h.c. Jürgen Basedow* and *Prof. Dr. Ulrich Magnus* serve as speakers of the Research School.

MIX
Papier aus verantwortungsvollen Quellen
Paper from responsible sources
FSC® C105338

If you have any concerns about our products,
you can contact us on
ProductSafety@springernature.com

In case Publisher is established outside the EU,
the EU authorized representative is:
Springer Nature Customer Service Center GmbH
Europaplatz 3, 69115 Heidelberg, Germany

Printed by Libri Plureos GmbH
in Hamburg, Germany